KT-498-498

4th edition

information systems development

methodologies, techniques & tools

David Avison & Guy Fitzgerald

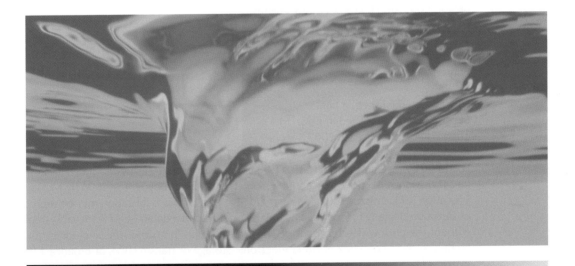

The **McGraw·Hill** Companies

London   Boston   Burr Ridge, IL   Dubuque, IA   Madison, WI   New York   San Francisco
St. Louis   Bangkok   Bogotá   Caracas   Kuala Lumpur   Lisbon   Madrid   Mexico City   Milan
Montreal   New Delhi   Santiago   Seoul   Singapore   Sydney   Taipei   Toronto

Information Systems Development, 4th Edition
David Avison and Guy Fitzgerald
ISBN-13 978-0-07-711417-6
ISBN-10 0-07-7114175

Mixed Sources
Product group from well-managed
forests and other controlled sources
www.fsc.org Cert no. TT-COC-002769
© 1996 Forest Stewardship Council

Published by McGraw-Hill Education
Shoppenhangers Road
Maidenhead
Berkshire
SL6 2QL
Telephone: 44 (0) 1628 502 500
Fax: 44 (0) 1628 770 224
Website: www.mcgraw-hill.co.uk

British Library Cataloguing in Publication Data
A catalogue record for this book is available from the British Library

Library of Congress Cataloguing in Publication Data
The Library of Congress data for this book has been applied for from the Library of Congress

Head of Development: Caroline Prodger
Senior Marketing Manager: Alice Duijser
Senior Production Editor: Beverley Shields

Text Design by Hard Lines
Cover design by Ego Creative
Printed and bound in UK by Bell & Bain Ltd.

First Edition published in 1988
Second Edition published in 1995
Third Edition published in 2002
Reprinted 2008

ISBN-13 978-0-07-711417-6
ISBN-10 0-07-7114175

4th edition

information systems development

methodologies, techniques & tools

Dedicated to
Leone, Marie-Anne and Thomas
and
Lin, Anna and Jane
With love

Brief table of contents

Contents vii

Preface xiv

Guided tour xvi

Technology xviii

Acknowledgements xx

The authors xxi

Part I: Introduction 1

    1.   Context 3

    2.   Information systems development 21

Part II: The life cycle approach 29

    3.   Information systems development life cycle 31

Part III: Themes in information systems development 45

    4.   Organizational themes 51

    5.   People themes 79

    6.   Modelling themes 109

    7.   Rapid and evolutionary development 119

    8.   Engineering themes 151

    9.   External development 175

Part IV: Techniques 195

    10.   Holistic techniques 199

    11.   Data techniques 217

    12.   Process techniques 243

    13.   Object-oriented techniques 273

    14.   Project management techniques 289

    15.   Organizational techniques 295

    16.   People techniques 307

    17.   Techniques in context 315

Part V: **Tools and toolsets**. **327**

         18.   Tools 331

         19.   Toolsets 361

Part VI: **Methodologies** **389**

         20.   Process-oriented methodologies 395

         21.   Blended methodologies 419

         22.   Object-oriented methodologies 451

         23.   Rapid development methodologies 469

         24.   People-oriented methodologies 487

         25.   Organizational-oriented methodologies 507

         26.   Frameworks 537

Part VII: **Methodology issues and comparisons** **565**

         27.   Issues 567

         28.   Methodology comparisons 591

         Bibliography **615**

         Index **631**

Detailed table of contents

Contents   vii
Preface   xiv
Guided tour   xvi
Technology   xviii
Acknowledgements   xx
The authors   xxi

Part I: Introduction   1

      1. Context   3
         1.1   Information systems   3
         1.2   Example information systems   4
         1.3   Environment and context   7
         1.4   Global economy   7
         1.5   Digital economy   8
         1.6   Electronic commerce   9
         1.7   Non-commercial impacts   9
         1.8   Change   10
         1.9   Human dimension   11
         1.10   Organizational aspects   14
         1.11   Professional aspects   14

      2. Information systems development   21
         2.1   Key concepts   21
         2.2   Need for a methodology   23
         2.3   Information systems development methodology 24

Part II: The life cycle approach   29

      3. Information systems development life cycle   31
         3.1   Information systems development life cycle (SDLC) 31
         3.2   Methodology   35
         3.3   Techniques   36
         3.4   Tools   37
         3.5   Potential strengths of SDLC   38
         3.6   Potential weaknesses of SDLC   38
         3.7   Conclusion   43

| Part III: | Themes in information systems development | 45 |

4. Organizational themes — 51

4.1	Systems theory	51
4.2	Information systems strategy	53
4.3	Business process re-engineering (BPR)	62
4.4	Information systems planning	65
4.5	Stages of growth	68
4.6	Flexibility	71
4.7	Project management	74

5. People themes — 79

5.1	Participation	79
5.2	End-user computing	84
5.3	Expert systems	87
5.4	Knowledge management	93
5.5	Customer orientation	96
5.6	Requirements	97

6. Modelling themes — 109

6.1	Modelling	109
6.2	Process modelling	109
6.3	Data modelling	111
6.4	Object modelling	113

7. Rapid and evolutionary development — 119

7.1	Evolutionary development	119
7.2	Prototyping	123
7.3	Rapid application development (RAD)	128
7.4	Agile development	134
7.5	Web-based development	145

8. Engineering themes — 151

8.1	Legacy systems	151
8.2	Software engineering	152
8.3	Automated tools	156

8.4	Method engineering (ME)	158
8.5	Component development	161
8.6	Security issues	163
8.7	Database management	165
8.8	Data warehouse and data mining	170

9. External development — 175

9.1	Application packages	175
9.2	Open source software (OSS)	178
9.3	Enterprise resource planning (ERP)	183
9.4	Outsourcing and offshoring	186

Part IV: Techniques — 195

10. Holistic technique — 199

10.1	Rich pictures	199
10.2	Root definitions	204
10.3	Conceptual models	208
10.4	Cognitive mapping	213

11. Data techniques — 217

11.1	Entity modelling	217
11.2	Normalization	229

12. Process techniques — 243

12.1	Data flow diagramming	243
12.2	Decision trees	251
12.3	Decision tables	253
12.4	Structured English	255
12.5	Structure diagrams	258
12.6	Structured walkthroughs	260
12.7	Matrices	263
12.8	Action diagrams	265
12.9	Entity life cycle	267

13. Object-oriented techniques — 273

13.1	Object orientation	273
13.2	Unified Modelling Language (UML)	279

14. Project management techniques 289

14.1	Estimation techniques	289
14.2	PERT charts	290
14.3	Gantt charts	292

15. Organizational techniques 295

15.1	Lateral thinking	295
15.2	Critical success factors (CSFs)	296
15.3	Scenario planning	299
15.4	Future analysis	300
15.5	Strengths, Weaknesses, Opportunities and Threats analysis (SWOT)	301
15.6	Case-based reasoning	303
15.7	Risk analysis	304

16. People techniques 307

| 16.1 | Stakeholder analysis | 307 |
| 16.2 | Joint application development (JAD) | 310 |

17. Techniques in context 315

17.1	Introduction	315
17.2	Techniques – potential benefits of their use and characteristics	316
17.3	Techniques impact on problem understanding: Potential blocks to problem cognition	317
17.4	Techniques impact on problem understanding: Visual and linguistic influences on problem cognition	318
17.5	Applying lessons from cognitive psychology: A macro analysis of techniques	321
17.6	A two-dimensional classification: Visual/ language and paradigm/process influences	322
17.7	Conclusion	324

Part V: Tools and toolsets 327

18. Tools 331

| 18.1 | Groupware: GroupSystems | 331 |
| 18.2 | Website development: Dreamweaver | 335 |

18.3	Drawing: Microsoft Visio	339
18.4	Project management: Microsoft Project	342
18.5	Database management: Access	348

19. Toolsets — 361

19.1	Introduction	361
19.2	Information Engineering Facility	363
19.3	Oracle	368
19.4	Select Enterprise	372
19.5	Discussion	375
19.6	Framing influences	385

Part VI: Methodologies — 389

20. Process-oriented methodologies — 395

20.1	Structured analysis, design and implementation of information systems (STRADIS)	395
20.2	Yourdon Systems Method (YSM)	402
20.3	Jackson Systems Development (JSD)	407

21. Blended methodologies — 419

21.1	Structured Systems Analysis and Design Method (SSADM)	419
21.2	Merise	426
21.3	Information Engineering (IE)	434
21.4	Welti ERP development	446

22. Object-oriented methodologies — 451

22.1	Object-oriented analysis (OOA)	451
22.2	Rational Unified Process (RUP)	460

23. Rapid development methodologies — 469

23.1	James Martin's RAD	469
23.2	Dynamic Systems Development Method (DSDM)	472
23.3	Extreme Programming (XP)	479
23.4	Web Information Systems Development Methodology (WISDM)	481

24. People-oriented methodologies   487

24.1   Effective technical and human implementation of
       computer-based systems (ETHICS)   487
24.2   KADS   496
24.3   CommonKADS   501

25. Organizational-oriented methodologies 507

25.1   Soft Systems Methodology (SSM)   507
25.2   Information systems work and analysis of
       change (ISAC)   516
25.3   Process Innovation (PI)   526
25.4   Projects in controlled environments (PRINCE)   530
25.5   Renaissance   533

26. Frameworks   537

26.1   Multiview   537
26.2   Strategic Options Development and Analysis
       (SODA)   549
26.3   Capability Maturity Model (CMM)   551
26.4   Euromethod   558

Part VII: Methodology issues and comparisons   565

27. Issues   567

27.1   What is a methodology?   567
27.2   Rationale for a methodology   570
27.3   Adopting a methodology in practice   572
27.4   Evolution and development of methodologies   576

28. Methodology comparisons   591

28.1   Comparison issues   591
28.2   Framework for comparing methodologies   597
28.3   Comparison   604

Bibliography   615

Index   631

Preface

Information systems development is at the core of the IS field and David Avison and Guy Fitzgerald's key text on the subject has been used by lecturers and students worldwide for over 18 years. Information systems development keeps evolving and changing and thus a fourth edition of Avison and Fitzgerald's classic text is now needed. Probably the biggest change in the last few years has been the increasing use of the rapid approaches to developing information systems, so much so that we now have a new theme, 'Rapid and evolutionary development' (Chapter 7). This encompasses a new section on agile development, as well as the revised sections on evolutionary development, prototyping, rapid application development and web-based development. This change is reflected also in revised sections on the methodologies Dynamic Systems Development Method (DSDM) and Extreme Programming (XP) in Chapter 23.

Other changes are related to issues that have been growing in importance, thus security issues, and data warehousing and data mining are two new sections that have been added in Chapter 8 to the Engineering theme. Similarly, the growing importance of the offshore outsourcing of systems development, sometimes referred to as offshoring, has led to the outsourcing section in Chapter 9 being expanded to reflect this important development. Some sections have expanded and this includes sections on component development, Oracle, PRINCE2 and SODA.

Changes for the new edition have been partly driven by existing readers and adopters and we are very grateful to them. Two (very different) topics in information systems development, that of professional aspects and requirements were identified as warranting specific sections rather than mere passing reference, and we have addressed this omission in Sections 1.11 and 5.6 respectively. Sections on component development (in Chapter 8) and open source (in Chapter 9) have also been greatly changed to reflect changes in practice.

In the last two chapters we compare methodologies and discuss issues concerned with their adoption in practice and problems associated with their adoption. At the end of Part V, we discuss the costs and benefits associated with the adoption of tools and toolsets. In previous editions we have not discussed the use of techniques in this way. They are typically seen as benign, very often as simple aids to help carry out a task and are used in many methodologies. But techniques may restrict understanding by framing the ways of thinking about the problem situation and some techniques may limit rather than enhance our understanding. We discuss this issue in a new chapter 17 and again in an expanded section on tools later in Chapter 18.

Of course we have also taken the opportunity to bring all sections up to date and to correct some errors kindly drawn to our attention by readers.

Despite all these changes, we have kept the basic structure of *themes, techniques, tools* and *methodologies,* but this structure does enable readers to use the book in a number of ways.

Because of its broad and effective content base together with its excellent structure, the text provides a sound basis for courses in information systems at all levels, from introductory through to specialist, and is relevant for courses with both an information technology and management perspective. It is of course particularly relevant for specialist courses in information systems development at both undergraduate and postgraduate level. It is both a theoretical and practical text with web-based support material available for both lecturers and students. The Online Learning Centre can be found at: http://www.mcgraw-hill.co.uk/ textbooks/avison-fitzgerald – to learn more about what is available, look at page xviii detailing the technology to enhance teaching and learning.

Guided tour

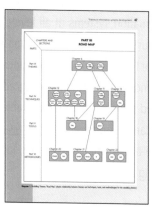

Part introductions

Each part opens with an introduction to the key themes of each section of the book, looking at the topics to be covered and placing them within a roadmap which helps you to navigate through the book.

Key terms

New terms and key words are highlighted in bold throughout the chapter as a useful reference for learning new terminology.

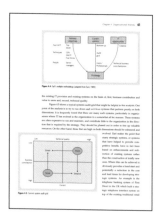

Figures

Each chapter provides numerous figures to help you to visualize the various development models discussed in the book, and to illustrate and summarize important Information Systems concepts.

Summary

- Systems theory has had widespread influence in information systems work. It suggests a holistic approach to viewing organizations rather than a scientific approach.
- Information systems strategy is about the way in which the organization sees the role of information systems in the company and the general attempt to identify better ways of doing things, leading to increased revenues, greater functionality, better products and services, improved presentation or image, improvement to the organization's competitive positioning, etc. to gain competitive advantage. The overall aim is to emphasize effectiveness rather than merely efficiency.
- Business process re-engineering is the fundamental rethinking and radical redesign of business processes to achieve dramatic improvements in critical, contemporary measures of performance, such as cost, quality, service, and speed. In its more recent guise, it has itself been re-engineered and it is less radical.
- Information systems planning involves top manage-

Chapter summary

This briefly reviews and reinforces the main topics you will have covered in each chapter to ensure you have acquired a solid understanding of key themes and issues.

Questions

1. In what ways are the themes of this chapter 'organizational'? What links these themes and what separates them?
2. What is 'strategic' about strategic information systems?
3. Discuss why business process re-engineering has been softened or toned down. Do you think this change has reduced its potential?
4. For an organization of your choice, identify the 'stages of growth' that it passed through and discuss whether these are similar to any SoG model discussed in the text.
5. Discuss the difficulties related to making information systems flexible so that implementing future change is easier.
6. How are large projects controlled in your organization? Address the question in relation to the role of people, techniques and software.

Review questions

These questions focus on important ideas that have emerged in the chapter and encourage you to review and apply the knowledge you have acquired.

Further reading

Cadle, J. and Yeates, D. (2001) *Project Management for Information Systems*, Prentice Hall, Harlow.

Checkland, P. and Scholes, J. (1990) *Soft Systems Methodology in Action*. John Wiley, Chichester.

Currie, W.L. and Galliers, R. (1999) *Rethinking Management Information Systems*, Oxford University Press, Oxford.

Earl, M.J. (ed.) (1996) *Information Management: The Organisational Dimension*, Oxford University Press, Oxford.

Galliers, R.D. and Sutherland, A.R. (1991) Information systems management and strategy formulation: the 'stages of growth model' revisited, *Journal of Information Systems*, Vol. 1, No. 2.

Hammer, M. and Champy, J. (1993) *Re-engineering the Corporation: A Manifesto for Business Revolution*, Harper Business, New York.

Melao, N. and Pidd, M. (2000) A conceptual framework for understanding business processes and business process modelling, *Information Systems Journal*, Vol. 10, No. 2, 105–129.

Further reading

At the end of the chapter, the authors provide a list of further reading, pointing you to key journals, books and other sources where you can research topics in greater depth.

Bibliography

A full bibliography at the end of the book provides a comprehensive scholarly reference list surveying the literature of Information Systems, an invaluable starting point for further research.

Technology to enhance learning and teaching

Visit **www.mcgraw-hill.co.uk/textbooks/avison-fitzgerald** today

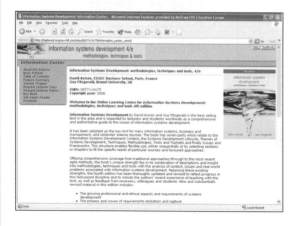

A range of supporting teaching resources is available to accompany this textbook and to aid lecturers in delivering their Information Systems Development modules.

Resources for lecturers include:

- PowerPoint Presentations prepared by the authors to help academics to deliver week by week lectures using artwork from the textbook

- Extra essay questions provide lecturers with a range of pre-prepared questions for tests or assessments

Check the website regularly for updates and new resources.

Lecturers: Customize content for your courses using the McGraw-Hill Primis Content Centre

Now it's incredibly easy to create a flexible, customized solution for your course, using content from both US and European McGraw-Hill Education textbooks, content from our Professional list including Harvard Business Press titles, as well as a selection of over 9,000

cases from Harvard, Insead and Darden. In addition, we can incorporate your own material and course notes.

For more information, please contact your local rep who will discuss the right delivery options for your custom publication – including printed readers, e-Books and CD-ROMs. To see what McGraw-Hill content you can choose from, visit www.primisonline.com.

Study skills

Open University Press publishes guides to study, research and exam skills to help undergraduate and postgraduate students through their university studies.

Visit www.openup.co.uk/ss to see the full selection of study skills titles, and get a **£2 discount** by entering the promotional code **study** when buying online!

Computing skills

If you'd like to brush up on your Computing skills, we have a range of titles covering MS Office applications such as Word, Excel, PowerPoint, Access and more.

Get a £2 discount off these titles by entering the promotional code **app** when ordering online at www.mcgraw-hill.co.uk/app

Acknowledgements

Our thanks go to the following reviewers who commented on the third edition and on our new material for the fourth edition:

Antony Bryant, Leeds Metropolitan University

Stephen Burbidge, University of Middlesex

Umberto Fiaccadori, Lund University, Sweden

Debra Howcroft, Manchester Business School

Sherry Jeary, Bournemouth University

Bruce Knowles, University of Abertay, Dundee

Russell Pearson, University of Middlesex

Christine Urquhart, University of Wales, Aberystwyth

Robert Vieira, University of East Anglia

Julie Williamson, Coventry University

The authors

David Avison is Distinguished Professor of Information Systems at ESSEC Business School, near Paris, France after being Professor at the School of Management at Southampton University for nine years. He is also visiting professor at Brunel University in England. He is joint editor (with Guy Fitzgerald) of Blackwell Science's *Information Systems Journal* now in its 16th volume. He has authored over twenty books as well as a large number of papers in learned journals, edited texts and conference papers. He is Vice Chair of the International Federation of Information Processing (IFIP) Technical Committee 8, previously Chair of its working group 8.2 on the impact of IS/IT on organizations and was past President of the UK Academy for Information Systems. He is joint program chair of International Conference in Information Systems (ICIS) 2005 in Las Vegas and has chaired several other international conferences.

Guy Fitzgerald is Professor of Information Systems at Brunel University and Deputy Head (Research) of the School of Information Systems, Computing and Maths. Before this he was Cable & Wireless Professor of Business Information Systems at Birkbeck College, University of London, and prior to that he was at Templeton College, Oxford University. He has also worked in the computer industry with companies such as British Telecom, Mitsubishi and CACI Inc, International. His research interests are concerned with the effective management and development of information systems and he has published widely in these areas. In addition he has undertaken a number of cases studies in organizations that have used information systems to enable significant organizational transformation. He has also undertaken research in relation to strategy, executive information systems, outsourcing, and flexibility. He is founder and co-editor, with David Avison, of the *Information Systems Journal (ISJ)* from Blackwell Science.

Part I: **Introduction**

Contents

Chapter 1 – Context   p 3

Chapter 2 – Information systems development p 21

I n Part I we introduce the context and main topics of this book. The domain of information systems is an exciting world. In Chapter 1 we discuss the context in which the developments that are the subject of this book take place. For example, the following issues are discussed: the increased rate of technological change, increased globalization of systems, growing importance of the knowledge-based economy, Web and Internet technologies, role of stakeholders and drivers for change. We stress the organizational and human aspects at least as much as the technology, and we also discuss professional issues. We provide early definitions of the major themes of the book. In Chapter 2 we define information systems and give examples of the different types of information system in organizations. Reasons are given for the need to use a methodology to develop information systems and suggestions made regarding the requirements of such a methodology.

1 Context

1.1 Information systems

This text is about information systems and specifically about ways of developing information systems. In this chapter we introduce the nature of information systems and illustrate this with examples. We develop this further and introduce other related concepts in Chapter 2. The main thrust of Chapter 1, however, is the context in which information systems are developed. We tend to stress the human and organization aspects to counteract the stress placed on the technology, which is a feature of many texts on this topic.

An **information system** in an organization provides processes and information useful to its members and clients. These should help it operate more effectively. The information might concern its customers, suppliers, products, equipment, procedures, operations, and so on. Information systems in a bank, for example, might concern the payment of its employees, the operation of its customer accounts, or the efficient running of its branches.

All organizations have information systems. The organization might be a commercial business, a government organization, or a community organization, for example, a bank, trade union, church, hospital, university, library, charity, or co-operative. This book concerns itself with **formalized information systems**. By formalized, we do not mean 'mathematical'. We use the term to distinguish information systems discussed here from less formalized information systems such as the 'grapevine', which consists of rumour, gossip, ideas, and preferences. These informal information systems may also be valid information systems and tend to be intuitive or qualitative in nature. Business discussions at the golf club or over lunch are also valid for disseminating information. However, organizations also need to develop formalized systems that will provide information on a regular basis and in a predefined manner, and these are the concern of this book.

We are mainly concerned with **computer-based information systems**, for the computer can process data (the basic facts) speedily and accurately, and provide information when and where required, which is complete and at the correct level of detail, so that it is useful for some purpose. By comparison, manual systems may be slower (information must be timely to be of value) and unable to deal with large volumes (of customers, suppliers, or transactions). Further, manual systems may be less accurate because checking procedures can be tedious, and not failure-proof (inaccurate 'information' can lead to poor decision making). In some circumstances, however, manual systems can be perfectly adequate or indeed superior to computerized information systems, such as in a small business, or where particular flexibility or human judgement is required.

In modern information systems, the basic data to be processed can include pictures, graphics, video, sound, and text as well as traditional alphanumeric data, such as a customer record. The computer system might be used to store data or convert the data to useful information by producing reports, pictures, graphics, or handling management enquiries. This does not mean that a computerized information system is 'purely' a computer system – there may well be manual (or clerical) aspects, such as keying-in (input) the data or some cross-checking – it simply means that part of the system is likely to be computerized. Nor does it mean that the computer technology itself is the most important aspect of the information system (in the same way that a word processor is not the most important aspect of the system whereby an author produces a novel). Thus, in the main, technology itself is not the major emphasis of this book, and nor should it be. It is mainly about information systems (IS) not information technology (IT).

The information systems of an organization will be required to help it analyse the business, along with its environment, and formulate and check that it achieves its goals. These goals might be related to profitability, long-term survival, service provision, expansion, greater market share, and employee and customer satisfaction. The information system may help the organization to achieve improved efficiency of its operations and effectiveness through better managerial decisions. Information systems are sometimes regarded as providing competitive advantage. Another way of looking at this would be to say that, without good information systems, a business would be at a considerable competitive disadvantage. They are therefore an important organizational resource.

1.2 Example information systems

Although many information systems are unique to particular types of organization – such as a system for placing and paying out bets in a betting office, a system to register voters for a local authority, a ticket reservation system for an airline, or a system for recording lottery choices – many information systems are common to a wide range of organizations. These include payroll, invoicing, project planning, decision-support, and so on. Even systems designed for a particular type of organization can be of a more general type. Therefore, reservation systems, for example, can be found in travel agents, cinemas, theatres, and libraries as well as in airlines. The generic nature of such systems is not always recognized in systems development as many systems have been developed as a one-off for a particular company or industry sector. Application packages, such as word processing or project management, are an exception and they have been developed for generic use in different types of organization.

For the moment we need to think about some examples of information systems before considering how they might be developed. We will outline two example information systems:

- A payroll system is an information system. This was one of the first applications to be computerized. Why? Because producing the payroll manually became very time-consuming and labour-intensive as the number of employees grew in an organization. If the payroll was late or contained errors there would be trouble, possibly even strikes.

Organizations needed to get this right so they turned to computerization. It is difficult to imagine today that such processes could be undertaken manually. All but the smallest organization (or maybe those wishing to avoid taxation!) utilize a computerized payroll package to pay wages and salaries. The raw data of a payroll system might include the number of hours worked, rates of pay, and deductions such as tax, pensions, union subscriptions, national insurance, and savings schemes. The information system typically produces payslips, which contain information about gross wage, net wage, and details of the deductions made, reports for management, historical records, and the provision of information for the tax office, other government departments, banks, etc. The management information reports produced might help management make decisions relating to increasing or decreasing staff levels, incentive schemes, and appraising the performance of employees. Payroll systems are not simple; there exists a wealth of legislation that the system has to cope and comply with, for example, relating to company cars, pensions, employee savings schemes, and share options, above and beyond the basic production of the payroll. A payroll system must also be flexible and maintainable, for almost every few days there are changes, some required by staff (e.g. they get married), some by the company (e.g. they change salaries and rates of pay), or by the government (e.g. they change tax rates). So, payroll is a classic computerized information system; it is used (it may be internal or outsourced) in virtually all organizations; it is clearly of great importance (everybody wants to be paid accurately); but it does not cause many problems or issues (except occasionally after a Budget). It is what is termed a mature application. It is not very exciting, it uses traditional technology, yet over time it has evolved to become a sophisticated system that works well and efficiently. It does not provide any competitive advantage to an organization, it is just a necessary system for an organization to have to enable it to be in business. In terms of development, it was originally developed in-house but today is typically provided and procured as an application package. There are many such mature systems in organizations, e.g. invoicing systems, billing systems, order processing systems, inventory systems and personnel systems. They may not all work as effectively and efficiently as payroll, but they are in application terms mature and traditional.

- Our next example system is very different. It is an electronic auction house (such as eBay). It is relatively new (eBay started in 1995) and exciting, and uses the World Wide Web as its user interface. Yet, essentially, it is also just an information system. It matches buyers with sellers utilizing an auction concept. This is not new, there have been auctions for hundreds of years, but the electronic auction enables buyers and sellers to be geographically distributed across the world, whereas the traditional auction required the buyers and sellers (well, strictly, just the potential buyers), and the product being auctioned, to be together in a particular place at a particular time. The electronic auction breaks the tradition in terms of both time and space. The electronic auction house (known as an intermediary) provides all the necessary support services to enable someone to put an item up for auction and for potential buyers to find the item, examine information about it, and bid for it. The auction house then 'conducts' the auction based on bids received over a period of time, defined by the seller. Actually, the information system usually just records the bids as they are received and makes them available online, so everyone can see the current state of play. If a potential buyer has had their bid beaten by someone else they can see this and bid again. At the end of the auction period, the item goes to the highest bidder, provided the reserve price is met, and the system puts the successful bidder and the seller in touch. They can then arrange payment and delivery. The auction house usually provides payment, insurance, delivery, security, and other services to its clients if required and in general provides an environment that enables people to participate easily and securely in an auction from the comfort of their home or workplace, or indeed anywhere. They also provide information on the seller, and whether he or she can be trusted, gathered from the responses of people who have dealt with him/her previously.

Figure 1.1 shows an example from an auction on eBay.co.uk. Someone has put a Les Paul guitar up for auction, and the seller has provided details of the guitar plus some pictures. It has attracted 35 bids and the current highest bid is £790. The auction has only 1 hour 31 minutes to run, but it may well yet attract further bids. The seller has set a reserve price but we are told that the reserve has been reached. The seller appears to have a good history on eBay and has received many positive responses from people he has previously traded with.

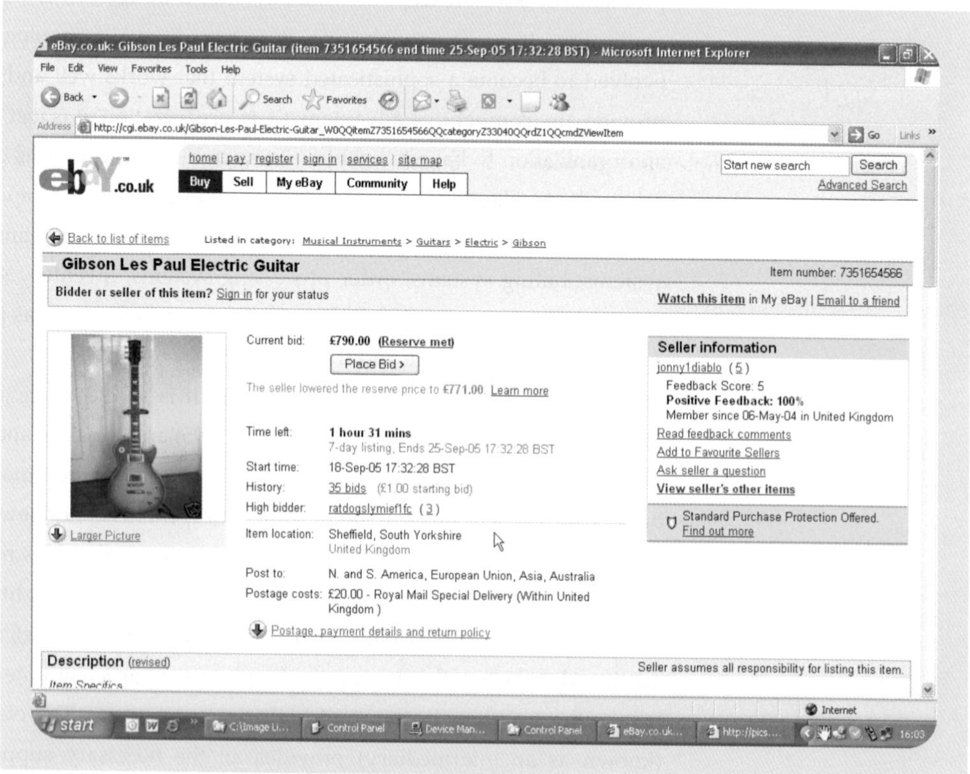

Figure 1.1 An auction on eBay (UK) for a Les Paul guitar

The electronic auction is an information system, comprising people, rules, procedures, technology, software, communications and allied services, such as delivery companies and third-party insurance companies. Such auctions have proved very popular for selling almost anything. eBay began by providing a marketplace for toys and stamps, but quickly embraced the sale of almost anything, goods as well as services, including cars, jewellery, musical instruments, cameras, computers, furniture, sporting goods, boats, and event tickets. It is now the leading online auction place in the world with a turnover (value of goods sold) of over $10 billion per annum. It claims that, on any given day, 'there are millions of items listed for auction across thousands of categories'. It also claims to have over 50 million registered users, although this probably includes people who registered for the site but 'are just looking' and maybe only looked once. Nevertheless, there is clearly a large community of people who like auctions and appear to be having fun, and such electronic auction houses have created a new kind of marketplace that has attracted significant numbers of people and trade. eBay is now making a profit of over $250 million, through its commission on sales. eBay now also provides a live auction facility enabling people to conduct an auction in the more traditional way but with online real-time bidding from home.

We have provided two examples of information systems, at somewhat different ends of the spectrum which we hope will help form the background when we address different ways of developing such information systems. Another important factor is the environment and context in which such information systems are developed.

1.3 Environment and context

Information systems do not exist in a vacuum. Clearly, they are developed and operate within an environmental context that has a significant effect on them. This environment is increasingly complex and dynamic. Businesses today are said to be facing increased competition, global challenges, and market shifts, together with continuing, rapid, technological developments. We will look briefly at some of these factors driving the environment and context in which information systems exist and are developed.

1.4 Global economy

No longer do companies face competition only from other local companies. Competition can come from anywhere in the world. There has been a reduction in barriers to trade and competition. This is a result of an opening up of a number of countries to a more market orientation, for example China and Russia, and specific legislation within, for example, the European Union and as a result of the North American Free Trade Agreement. Continuing deregulation, removal of trading restrictions, and the breaking up of monopolies are a feature of many economies throughout the world and are helping to drive a more global economy. Other barriers are also being removed. For example, the adoption of the Euro is the latest in a series of initiatives to make international trade easier in the European Union and beyond. Another factor is improved physical transport capabilities, for example, developments in motorways, airlines, railways, and packaging. To give one example, UK supermarkets display perishable and exotic produce that has originated in Africa, South America, or indeed from almost anywhere around the world. Together with improved packaging and storage techniques, such produce can now be shipped in timescales that mean they are almost as good in the UK as in the town next to their area of origin. Local seasons for fruit and vegetables have almost disappeared; strawberries are now available at almost any time of the year, although usually at a premium price. Information systems support such transactions.

A further impact of the global economy is that, although markets may be more open and accessible, costs can still differ significantly from country to country, particularly labour costs. So, economies with higher labour costs have to compete directly with those that have lower labour costs. No longer can companies rely on such competition being prevented by tariff barriers or dissipated by large transportation costs. The UK success story of Dyson, an innovative company which designed and developed a revolutionary new cyclonic vacuum cleaner product, announced that manufacturing would cease in the UK and be transferred to a country with cheaper labour costs. The global economy therefore took its toll on what was perceived as an important contribution to UK manufacturing, but Dyson felt that in order to survive it had to react in this way. Information systems support these changes. But the world of information systems has been itself impacted by such things, for example, outsourcing IS development to countries with cheaper labour costs.

1.5 Digital economy

Another element in the dynamic context of business is the digital economy, resulting from the convergence of computing and telecommunications technologies. This has had a significant effect on businesses and society in general and is epitomized by the impact of the Internet and the World Wide Web. Organizations have found that their operations, products, services, information, markets, competition, and economic environment are all potentially affected. Companies are seeking to create more digital content in their products and services and deliver them over digital networks. This has been characterized by the term 'from place to space'. So, for example, banks are attempting to shift their communication and service delivery channels with their customers from personal interactions at branches to call centres and finally to the Internet. The product is mainly digital information and transactions can be undertaken without a high street presence, although this is still seen as some customers are wedded to visiting their branch. However, the impact of this change is enormous for the banking sector. We are currently seeing competition (in the UK) with the existing banks offering a range of Internet-based products and services, but we are also seeing the advent of new entrants who have found they can enter this new digital banking world relatively easily. Egg, for example, is a bank established by the Prudential, traditionally an insurance company. This indicates that the world is under significant change. Further, if banking products and services are essentially digital and can be delivered via the Internet, then they can potentially be delivered in the UK by a non-UK provider, for example a European or American bank, or indeed in America by a British bank. We have not yet seen much evidence of this because of legislation and regulation barriers, but it is surely only a matter of time. The global and digital economies are not universally welcomed and there are some who see this as exploitation, particularly of poorer people. The term 'digital divide' is sometimes used to express this view, although the term is also used to express the division between those economies that are technology-sophisticated and those that are not.

1.6 Electronic commerce

The digital economy includes electronic commerce or e-commerce which is simply the conducting of commercial transactions electronically, typically via the Internet, between geographically separated parties. It may involve some or all parts of the transaction process relating to pre-sale, execution, settlement and after-sale activities. Clearly, e-commerce has been the subject of much debate over recent years, and it has been suggested that it has revolutionized business for the following reasons:

- eases access to global markets (known as reach);
- extends business hours to 24 hours, 7 days a week;
- reduces the costs of transacting business;
- reduces the costs of marketing;
- facilitates customized one-to-one communication with customers;
- shortens the transaction cycle;
- provides a more perfect market; that is, it is easy for customers to find and compare prices (sometimes termed electronic marketplaces).

Whether this is yet a true revolution is debatable, and whether all these benefits necessarily or automatically accrue is open to question. For example, the cost of marketing is thought by some to be higher because of the broader, more diverse markets that need be covered. Further, it is argued that the need for, and thus the cost of, establishing good brands and brand images might be greater than for traditional commerce. However, after a relatively slow start, e-commerce has grown dramatically with an estimated US$2 trillion worth of goods and services currently traded online. There is also a view that e-commerce is really very little different from traditional commerce. The business processes are essentially the same, it is just the way that they are implemented that is different. In a few years nobody will bother to make a distinction in the way that transactions are carried out. Nevertheless, e-commerce is certainly very important and here to stay and it is likely to continue to be influential even if all the early hype is not quite justified.

1.7 Non-commercial impacts

Although e-commerce takes most of the headlines, there are non-commercial impacts of the digital economy that are equally important. In many countries there is a large public sector that is responsible for many activities. In the UK, for example, it covers transportation, education, health, taxation, regulation, employment, customs and excise, the law and the environment, together with 388 English local government authorities. Together, these government departments and associated agencies are responsible for a large section of the economy and a large number of employees.

The UK government has recognized the importance of actively promoting the public sector and government agencies on the Web. They regard this as a key opportunity to provide higher quality services directly to citizens at lower cost and more efficiently 24 hours a day. It is thought that government departments should be able to achieve significant improvements in the provision of information to the public, especially allied with the push to more 'open government' and increasing freedom of information.

But this vision is not just confined to the provision of information. Citizens should be able to conduct business with the government online. According to a report from the National Audit Office it 'requires a fundamental transformation of many central departments' and agencies' business processes' (National Audit Office, 1999). To this end the UK government committed itself to having all public services online and available for citizens and companies by the end of 2005. Progress has been made and government predictions suggest that a significant percentage, although probably not all, services will be online by this date. It seems that all local government services and around 80% of other public services will achieve this target. The government has established something called the Government Gateway which is the hub of its 'joined-up government' initiative, and this has over 6 million citizens registered and enabled to carry out a range of electronic services including, for example, online tax returns, claiming Child Tax Credit, paying parking fines, checking pension entitlement, and booking a blood donor session. These e-government initiatives are potentially important elements of the digital economy alongside the usual more commercial elements.

1.8 Change

The environmental context of this book is also described as increasingly dynamic and turbulent (Wielemaker et al., 2000) with change being endemic and the norm, rather than the exception (Prahalad and Hamel, 1994). Successful organizations are often thought to be those that are capable of dealing with such change and the opportunities it presents. Ciborra (1996) suggests that 'within a firm . . . what seems to matter is the flexibility and adaptive capability in the face of environmental discontinuities'. Modern organizations seek to be responsive, adaptive, and flexible in their operations and strategy.

However, even those organizations that embrace the challenge of change have found that it is not easy to achieve, not least because their information systems (IS) and information technology (IT) are anything but flexible and adaptive. As we have suggested, technology has enabled many new and innovative information systems. Ironically, some managers have found that in practice IT can be a barrier to change as well as an enabler of change. The problems of 'legacy systems' (Section 8.1) (i.e. information systems that have been running for many years) are well known. It is sometimes difficult to be innovative in organizations because the existing systems are so ingrained, represent such large investments, and are difficult to change. Further, business cannot be suspended for a period while these significant changes are implemented. In addition, organizations have found that the cycle of business change is often shorter than the cycle for IT/IS change, which produces bottlenecks. As early as 1990, *The Economist* magazine suggested that 'businessmen have discovered a . . . disconcerting problem: markets change, but computer systems do not', and the challenge for businesses is to break these rules of the past, and structure IS to meet the variety of changing information requirements that businesses are now facing.

So, the environment and context is dynamic. Wielemaker et al. (2000) argue that 'business is moving at 150 miles an hour' and 'the business models are morphing at an incredible speed . . . a year in this business is really like five, six, seven years in traditional business'. Further, decisions are made much faster, often accepting higher risks because less data are available, due to the timescale, to make the decision. The knock-on effect of this on information systems and information systems development is profound. More is demanded from them, they have to be developed quickly, they have to be easily updatable, and they must utilize resources effectively. The pressures are enormous and the effects on the people involved are significant. We look at this human dimension next.

1.9 Human dimension

In this book we stress the human and organizational aspects of information systems development at least as much as the technological aspects. Today, the pressures are such that the human elements are the key to success. Indeed, in many of the information systems failures that have occurred, the conclusion has been placed squarely on human and organizational factors rather than technical ones (see, for example, reports into the failure and later success of the London Ambulance Service: Fitzgerald, 2000). A lack of planning may lead to unexpected costs and project abandonment. Poor training may result in people not co-operating with or supporting the information system. The same result may be caused by the system not making use of the users' business knowledge. Failure may also be due to poor methods, techniques, and tools. We consider the need for an information systems development methodology in Section 2.2 and also distinguish between techniques and tools.

Here we look at the people who will probably be involved in the in-house development of a computer information system. We identify a number of groups or **stakeholders** (see also Section 16.1). First, we identify those on the system development side:

- *Programmers* code and develop a system in a computer programming language.
- *Systems analysts* specify the requirements for a system and the outline designs and solutions that will meet the requirements. Typically, they are the interface or liaison between the business users/analysts and the programmers.
- *Business analysts* understand the complexities of the business and its needs and liaise with the systems analysts. They are typically from the business side of the organization but adopt this role in the context of a particular development project for a specific period.
- *Project managers* manage the project with particular emphasis on schedules and resources.
- *Senior IT management* are responsible for IT and managing it overall within the organization.
- *Chief information officer (CIO)* is responsible for IT, IS, and information strategy and

aligning them to the needs of the business as a whole. Although usually a member of the IT department, it is essential that the CIO is part of the organization's top management team and understands the implications of the global and digital economy (Ross and Feeny, 2000). Sometimes, this position is known as the chief knowledge officer, recognizing the importance of knowledge as well as information (see Section 2.1). CIO (2005) is an excellent website relating to the concerns of the CIO.

The above groups may not exist as distinct groups in all organizations. The boundaries between them have undoubtedly become blurred over the years. In some circumstances one person may undertake a number of roles, or a group may flexibly undertake all roles as needed. The situation is even more confused by the tendency of the IT industry to have a wide and varied range of overlapping job titles for these roles. Further, many organizations no longer have rigid separations between the IT systems development side and the business. Often multi-skilled development teams, capable in both business and IT, are formed for a particular development project, often managed and led by the business units themselves.

Next, we identify those in the business or organization for whom the system is required. This group is often generically known as 'users', but this is misleading as they are not homogeneous and there is a range of different types of user. Indeed, 'users' can also be 'developers'. We break this category down as follows:

- *End-users* use the system in an operational sense. They may be intermediaries between the system and the business users.
- *Business users* are people in the particular business function that have a need for the system. They might or might not physically interact with the system itself. They are interested in its functions and output, as support for achieving their business objectives.
- *Business management* have responsibility for the business function that the system addresses and may have been responsible for commissioning the system and financing it from their budget. They are responsible for the strategic use of IT in their business unit.
- *Business strategy management* are responsible for the overall strategy of the organization and the way that information systems can both support and enable the strategy.

Again, we are describing roles for people here. They may be combined or separated. Sometimes, different categories of user are identified, for example, *regular user* and *occasional* or *casual user*. This categorization is important for determining what type of user interface may be required in a system, or what type of training is needed. Clearly, these will be different for regular users and occasional users. A regular user may not require a lot of help and explanation and just the minimum of interactions, whereas an occasional user will require detailed help and guidance when using the system.

An example is provided by an airline reservation system. If occasional users decide to book a flight on the Internet, they are carefully guided through the process and have to enter exact details, such as the dates, the originating and destination airport, the class, and so on.

Regular and experienced users, such as travel agents, often have a different interface, which enables them to enter shorthand details using airline codes. Once learned, this is much quicker than the occasional user interface, and indeed if the travel agents were forced to interact in the latter way, they would become very frustrated. Travel agents also often have the ability to do wider searches, for ranges of dates, multiple destinations, fare types, etc. Sometimes, information systems must be able to cater for a range of different users including both regular and occasional.

Our next category is external users. These are stakeholders outside the boundaries of the company in which the system exists:

- *Customers or potential customers* use the system to buy products and services, or search for information relating to products. They are generally not employees of the company and thus have a different relationship to the earlier categories of user. Too often, customers are ignored when systems are being designed and developed, even though they are obviously important stakeholders with specific requirements. In some Internet applications this is particularly important. We look at this issue in Section 7.5.

- *Information users* are people external to the organization who may use the system but are not customers, in that they do not buy anything. Users of a government website may just be looking for information on building regulations, for example. This category of user is also often ignored when the system is designed.

- *Trusted external users* have a particular relationship with the organization and may be given special privileges in the system. Suppliers are examples of such users. There are likely to be specific design requirements and security implications for this category of user.

- *Shareholders, other owners or sponsors* are people who have invested in the organization and have a financial interest. They may be only peripherally concerned with the information systems in the organization but they will want to ensure that they are contributing to the financial development and success of the organization.

- *Society* includes those people who may be affected by the system without necessarily being traditional customers or users in any way. This is a broad category and relates to people, or society as a whole, who may be potentially affected by the system in some way. People may be put out of work by a system or it may disseminate inaccurate or personal information. It might be that a system refuses credit to someone based on old or inaccurate information, or it may be that the credit rating system discriminates against particular groups of people in some way. Society is an important stakeholder in systems development and societal impacts also need to be considered.

In general, we believe that it is desirable that all stakeholders of a system are involved in the development process. They all have some kind of stake in the success of the information system.

In the information systems development process, some users might be part of a group, such as the information systems strategy group, the steering committee, and the development team. We will examine this in more detail when looking at organizational aspects. Although we have said that it is desirable that all users are involved in the development process, systems development has in the past been dominated by professional users, in particular computer programmers and computer systems analysts. Some of the approaches to information systems development discussed in this book attempt to redress the balance. Certainly, it helps if these professional users see their task as supporting the other users of the system.

1.10 Organizational aspects

Although methodologies, techniques, and tools are all a necessary part of the infrastructure to develop information systems, so are management aspects. Information systems development as a whole and each individual information systems project need to be managed. Organizations differ, but a common arrangement is to have an information systems strategy group, a steering committee, and a systems development team.

The aim of the **information systems strategy group** is to develop a plan for information systems development in the organization and ensure that the plan is carried out and tuned as circumstances change. It is a high-level group, usually meeting each month, representing top management, heads of the various divisions, and head of the information systems function. We discuss information systems strategy in Section 4.2.

The **steering committee** will oversee each project within the overall plan, ensuring that the wishes of the information systems strategy group are met, and setting its own standards for the project including performance requirements, approving the personnel working on the project, and approving the final system. Project management and control (see Section 4.7), such a major element affecting the success or failure of the information systems project, will largely rest on the steering committee. The head of the information systems function will frequently be a member of both groups. The steering committee may also specify the information systems development approach to be used in a particular project, although this may be a standard for all information systems development in the organization. The steering committee is very likely to include the head of the department affected by the information system being developed and may also include the finance director, personnel manager, and, possibly, outside consultants, as well as the systems development project team leader.

The **systems development team** will concern itself with the day-to-day development of the information system and includes the analysts, programmers, and users working on the project. The composition of the team will differ as the system progresses through the stages of the systems development life cycle (see Chapter 3), although there will normally be one **project team leader** ensuring continuity throughout.

1.11 Professional aspects

One link between the human dimension and organizational aspects concerns the issue of professionalism. We will consider this topic first by looking at an example of failure of practice and then outline codes of practice and conduct that aim to prevent this happening. We will look at the British Computer Society as exemplar, though many countries have good, and in many cases, similar alternatives. The case study we will look at relates to the catastrophic failure of the Australian telecommunications company One.Tel.

Avison and Wilson (2002) argue that the failure of One.Tel in Australia was partially caused by the failure of their information systems, and the invoicing system in particular. Systems development at One.Tel exemplified the 'Initial' level of maturity described by the Capability Maturity Model (see Section 26.3). The characteristics of this level are 'chaotic, ad hoc, heroic; unorganised, uncoordinated; high variance, unpredictable, crisis management' (Paulk et al., 1993a).

The teams of young and highly paid technicians at One.Tel thrived in this environment. Systems were delivered in quick time, but the lack of standard development methods and other standards made these systems difficult to maintain. This lack of discipline was understandable and not unusual at this stage in the growth of the firm (see also Section 4.5), but it was problematical, particularly in the case of the billing system. Companies depend on the unfailing timeliness and accuracy of this system for their cash flow, and One.Tel was no exception. In the long term, some serious flaws in the billing system at One.Tel revealed themselves:

- The first major flaw was a long-term dependence on an inadequate design. The original system was designed and developed by developers, including programmers, under conditions of great stress and urgency. It should have been viewed as only a short-term solution. The system lacked flexibility, and was supported by inadequately designed database tables. It became impossible to accommodate, within the database, the complex sales plans, which were an important part of One.Tel's marketing strategy.

- The second major flaw was a lack of checks and balances. The system failed to provide the most basic financial integrity checks. It was impossible to reconcile the value of bills produced in a billing run, either backwards to the calls loaded from the carriers, or forwards to the value finally posted to the General Ledger. There were no checks at each stage of value loaded, value billed, or value posted.

- The third major flaw was a lack of prioritization and forward planning. Proper priority was not given to major enhancements required by the billing system. Two conspicuous examples of this were the implementation of the Goods and Services Tax (GST) and the introduction of the NextGen mobile service, both in 2000. In the case of GST, not only were these changes implemented one month late, but they were so poorly executed that it caused billing run times to increase by about 50 per cent. The changes to accommodate NextGen mobile were implemented three months behind schedule, which caused the first users of the new phones to wait three months for their first bill. It would appear that sufficient resources were not allocated in time to meet critical deadlines. On each occasion the billing system suffered from these failures to plan, and the result was large numbers of seriously delayed bills.

Perhaps the most damning effect of the failure of the billing system was that it brought the company into serious disrepute as the invoices were often inaccurate as well as delayed.

We may well ask whether IT and IS staff following a professional code of conduct and practice might have improved the situation of One.Tel. Information technology practitioners are required to make calls of judgement related to ethical, technical and professional issues. Professional codes of conduct and codes of ethics are intended to provide guidance in making such calls of judgement. The relevant computer society, the Australian Computer Society (ACS, 2005), has and had such codes at the time, and similar codes are provided by the US Association of Computing Machinery (ACM, 2005) and the British Computer Society (BCS, 2005a) and so on. But in all cases, compliance with the IT professional codes is voluntary, as IT practitioners are normally not required to be a member of a professional society in order to practise.

This Code of the British Computer Society (BCS, 2005b) sets out the professional standards required by the Society as a condition of membership. It applies to members of all grades,

---

THE PUBLIC INTEREST

1. You shall carry out work or study with due care and diligence in accordance with the relevant authority's requirements, and the interests of system users. If your professional judgement is overruled, you shall indicate the likely risks and consequences.

– The crux of the issue here, familiar to all professionals in whatever field, is the potential conflict between full and committed compliance with the relevant authority's wishes, and the independent and considered exercise of your judgement.

– If your judgement is overruled, you are encouraged to seek advice and guidance from a peer or colleague on how best to respond.

2. In your professional role you shall have regard for the public health, safety and environment.

– This is a general responsibility, which may be governed by legislation, convention or protocol.

– If in doubt over the appropriate course of action to take in particular circumstances you should seek the counsel of a peer or colleague.

3. You shall have regard to the legitimate rights of third parties.

– The term 'third Party' includes professional colleagues, or possibly competitors, or members of 'the public' who might be affected by an IS project without their being directly aware of its existence.

4. You shall ensure that within your professional field/s you have knowledge and understanding of relevant legislation, regulations and standards, and that you comply with such requirements.

– As examples, relevant legislation could, in the UK, include The UK Public Disclosure Act, Data Protection or Privacy legislation, Computer Misuse law, legislation concerned with the export or import of technology, possibly for national security reasons, or law relating to intellectual property. This list is not exhaustive, and you should ensure that you are aware of any legislation relevant to your professional responsibilities.

– In the international context, you should be aware of, and understand, the requirements of law specific to the jurisdiction within which you are working, and, where relevant, to supranational legislation such as EU law and regulation. You should seek specialist advice when necessary.

5. You shall conduct your professional activities without discrimination against clients or colleagues.

– Grounds of discrimination include race, colour, ethnic origin, sexual orientation.

– All colleagues have a right to be treated with dignity and respect.

– You should adhere to relevant law within the jurisdiction where you are working and, if appropriate, the European Convention on Human Rights.

– You are encouraged to promote equal access to the benefits of IS by all groups in society, and to avoid and reduce 'social exclusion' from IS wherever opportunities arise.

6. You shall reject any offer of bribery or inducement.

DUTY TO RELEVANT AUTHORITY

7. You shall avoid any situation that may give rise to a conflict of interest between you and your relevant authority. You shall make full and immediate disclosure to them if any conflict is likely to occur or be seen by a third party as likely to occur.

Figure 1.2: BCS Code of Conduct (BCS 2005b)

8. You shall not disclose or authorise to be disclosed, or use for personal gain or to benefit a third party, confidential information except with the permission of your relevant authority, or at the direction of a court of law.

9. You shall not misrepresent or withhold information on the performance of products, systems or services, or take advantage of the lack of relevant knowledge or inexperience of others.

DUTY TO THE PROFESSION

10. You shall uphold the reputation and good standing of the BCS in particular, and the profession in general, and shall seek to improve professional standards through participation in their development, use and enforcement.

– As a Member of the BCS you also have a wider responsibility to promote public understanding of IS – its benefits and pitfalls – and, whenever practical, to counter misinformation that brings or could bring the profession into disrepute.

– You should encourage and support fellow members in their professional development and, where possible, provide opportunities for the professional development of new members, particularly student members. Enlightened mutual assistance between IS professionals furthers the reputation of the profession, and assists individual members.

– You shall not make any public statement in your professional capacity unless you are properly qualified and, where appropriate, authorised to do so. You shall not purport to represent the BCS unless authorised to do so.

– The offering of an opinion in public, holding oneself out to be an expert in the subject in question, is a major personal responsibility and should not be undertaken lightly.

– To give an opinion that subsequently proves ill founded is a disservice to the profession, and to the BCS.

13. You shall notify the Society if convicted of a criminal offence or upon becoming bankrupt or disqualified as Company Director.

PROFESSIONAL COMPETENCE AND INTEGRITY

14. You shall seek to upgrade your professional knowledge and skill, and shall maintain awareness of technological developments, procedures and standards which are relevant to your field, and encourage your subordinates to do likewise.

15. You shall not claim any level of competence that you do not possess. You shall only offer to do work or provide a service that is within your professional competence.

– You can self-assess your professional competence for undertaking a particular job or role by asking, for example,

i. am I familiar with the technology involved, or have I worked with similar technology before?

ii. have I successfully completed similar assignments or roles in the past?

iii. can I demonstrate adequate knowledge of the specific business application and requirements successfully to undertake the work?

16. You shall observe the relevant BCS Codes of Practice and all other standards which, in your judgement, are relevant, and you shall encourage your colleagues to do likewise.

17. You shall accept professional responsibility for your work and for the work of colleagues who are defined in a given context as working under your supervision.

Figure 1.2: Continued

including students, and affiliates, and also non-members who offer their expertise as part of the Society's Professional Advice Register. The BCS code of conduct is shown in Figure 1.2.

It would be interesting to debate whether IT professionals followed this code of conduct (the Australian one is similar) or, if they did not, whether following it would have made any difference (looking at other IT/IS failures would stimulate similar debates).

The 36-page BCS code of practice can be found at BCS (2005c). The BCS also provides a consultancy code of practice for IT consultants contained in its Professionals Advice Register. The Code aims to encourage greater awareness and understanding of the issues relating to consultancy in the field of information systems. The Eight-point Code is as follows:

1. Professional responsibility – Exercise reasonable professional skill and care

2. Law – Know about and comply with the law
3. Conduct – Act in accordance with the Society's Code of Conduct
4. Approach – Maintain a balanced, disciplined and comprehensive approach
5. Management – Contribute effectively to best practice consultancy within your organization
6. Independence and statement of interests – Disclose any circumstances which may compromise your professional objectivity or independence
7. Professional development – Keep up to date by seeking continuing education and training
8. Public awareness – Encourage public understanding of Information Technology issues.

We might also ask whether IT and IS staff following one or more of the methodologies described in this book may have delivered a better invoicing system. The BCS code of good practice (BCS 2005d) requires that professionals keep up their technical competence and later suggests the following amongst its advice entitled *Adhere to Regulations*:

- Follow the standards relevant to the organization's business, technology and development methods; encouraging new standards, where appropriate standards do not exist.
- Use standards in an intelligent and effective manner to achieve well-engineered results.

Later, it suggests under the title *Use Appropriate Methods and Tools*:

- Keep up to date with new methods and the tools to support these methods.
- Promote the effective use of methods and tools within the organization.
- Recommend the adoption of new methods only when they have been demonstrated to be effective for the organization and are supported by suitable tools.
- Explain to non-IT staff the purpose of any methods that have impact on their duties, so that they can understand the outputs and appreciate the benefits.
- Recognize the scope and applicability of methods and resist any pressure to use inappropriate methods.

Under *Manage Your Workload Efficiently*, it suggests:

- Report any overruns to budget or timescales as they become apparent; do not assume that you will be able to recover them later.

There are large sections on project management, relationship management, security, change management, and standards, which all feature in this text and which are regarded by the BCS as *Key IT Practices*. Again, it later discusses codes of practice relating to Requirements Analysis and Specification, Software Development, Documentation, System Installation, Support and Maintenance, all discussed in this book.

The BCS provides educational qualifications that give an appropriate foundation for those who wish to follow a career in computing or information systems. One of their courses is entitled *Analysis, Design, Development of Information Processing Systems*. The BCS also accredit courses at Universities and other Higher Education Institutions and these courses can gain exemption from the BCS Professional Examination.

Following the BCS advice, greater thought to the topics discussed in this book could potentially have produced an invoicing system at One.Tel that did not manifest the problems that were only too clear in the implementation. Decisions about choice of techniques, tools and methods all have ethical as well as professional aspects. We suggest that knowledge of methodologies, tools and techniques should be part of the armoury of any IS and IT professional and student and, at least to some extent, general manager and user, and this is an important contribution of this book.

Summary

- An information system in an organization provides processes and information useful to its members and clients. These should help the organization operate more effectively. The information might concern its customers, suppliers, products, equipment, procedures, operations, and so on.

- Information systems are developed and operate within an environmental context, which has a significant effect on them. This environment is increasingly complex and dynamic. We have discussed the global economy, the digital economy, e-commerce, and change as part of this context.

- Stakeholders closely involved in driving an IS development project include programmers, systems analysts, business analysts, project managers, senior IT management, and the chief information officer.

- Professional codes of ethics and conduct guide programmers, analysts and other IT and IS specialists in their work choices.

- User stakeholders include end-users, business users, business management, and business strategy management.

- Other stakeholders include customers, information users, trusted external users, shareholders, and society as a whole.

Questions

1. We have discussed some issues that we consider as providing environmental 'context' to information systems development. Discuss other issues that form part of this context.

2. Discuss the role of the various stakeholders in an IS development project.

3. Read about an IS failure or success. What part of the story can be attributed to technological, human, and organization aspects? Was 'context' important to success or failure? Did the 'professionals' adhere to any codes of conduct? Were they members of any professional association? What were the ethical issues associated with the various stakeholders?

4. What aspects of local and national government are available on the Internet in your country? For instance, can you see the minutes of a local government meeting, complete a tax form or see your health record?

Further reading

Boafo, K. (ed.) (2003) *Status of Research on the Information Society*, UNESCO Publications for the World Summit on the Information Society (WSIS), United Nations Educational, Scientific and Cultural Organization, Paris, France. Available for download at http://portal.unesco.org

Braa, K., Sorensen, C. and Dahlbom, B. (eds) (2000) *Planet Internet*, Studentliteratur, Lund.

Everard, J. (1999) *Virtual States: Globalisation, Inequality and the Internet*, Routledge, London.

Kogut, B. (ed.) (2004) *The Global Internet Economy*, MIT Press, Boston.

McKnight, L.W. (2002) *Creative Destruction: Business Survival Strategies in the Global Internet Economy*, MIT Press, Boston.

2 Information systems development

2.1 Key concepts

In Chapter 1 we discussed what information systems are and provided some context in which they are developed. In this chapter we discuss the need for an information systems development methodology and the requirements of such a methodology. This leads on to Chapter 3 where we discuss the information systems development life cycle, which has had an enormous influence as a general approach to develop information systems.

However, we start by providing some definitions of the main concepts discussed in the book. Of course, we will provide more 'meat' to these early definitions as we explore them in later chapters. We provide these early definitions to help the reader understand what follows rather than provide 'ideal-type' definitions which everyone can accept.

One problem in this area is that many of the terms used in this section are used differently and inconsistently elsewhere. We will attempt to be consistent in our usage or explain where usage differs. This is a fairly new discipline, and differences of opinion are to be expected, but it does not make our task easy.

Data represent unstructured facts about events, objects, or people. When three 'strings' of data '250796', '78700199', and '19873' are associated, they could be used to give the information that a person whose identity number is 19873 possesses a driving licence (number 78700199), even though that person is under the minimum legal age for driving motor vehicles. The string of data 250796 is interpreted as the date of birth, 25 July 1996, showing that the holder is only 10 years old in January 2006. The essential difference between data and information, therefore, is that data are not interpreted, whereas **information** has a meaning and use to a particular recipient in a particular context and can be used for decision making. Information comes from selecting data, summarizing it, and presenting it in such a way that it is useful to the recipient. Too often, this process is not well refined and vast amounts of data are output. This is often referred to as 'information overload', although, strictly speaking, it is not information but data, because it is not useful.

Some writers equate knowledge with information. Buckingham et al. (1987) define information as 'explicit knowledge'. In other words, information expresses what is meant clearly, with nothing left implied. Knowledge may also be seen as accumulated information. Most importantly, people with knowledge know the meaning and implications of the information presented and how to use it effectively. They should therefore be competent in completing their tasks. So, **knowledge** contains the ability to use information effectively for particular purposes.

The distinction between data, information, and knowledge is context-dependent. Let us look at another example where a line manager analyses the departmental figures and presents the results to the planning department. For the line manager, the results are an interpretation of events and are therefore information rather than data. They have meaning for the line manager. For the central planners, these figures are the raw input for their own analyses, not yet interpreted, and are therefore data rather than information. Therefore, information is such because it is relevant and understandable to some person or group. But the central planners need to have the knowledge to use this information effectively. This may be explicit, that is, communicable to others or tacit, that is, only implicitly understood.

Having given a preliminary definition of information, we need to discuss what is meant by a **system**. This is more difficult because it is a term that is used widely in many different fields of activity. Therefore, the ecological system is a view of the world that includes the relationship between flora and fauna which we call the balance of nature, and the educational system could be viewed as our understanding of the relationship between teachers, students, books, and colleges whose purpose is to pass on knowledge to all members of the community. Systems are related to each other. Telephone bills are produced by a billing system, forwarded by a postal system, and paid using a banking system. The banking system will have a customer service system, a cheque processing system, and a payroll system, among others. Smaller systems within larger systems are called **subsystems**. An information system will also have subsystems within it. An airline information system may have subsystems to report on the status of passengers, report on flights, and to analyse costs and profits. All these examples of systems include a collection of parts or subsystems that work together.

The 'system' part of 'information system' represents a way of seeing the set of interacting components, such as:

- people (for example, systems analysts, business users, line managers);
- objects (for example, computer hardware devices, a user interface, telecommunications networks, the World Wide Web);
- procedures (for example, business processes, an information systems development methodology, business rules).

All this must take place within a boundary that separates those components relevant to the system (for example, to do with purchasing a product or service online) and those concerned with the environment around the system (for example, other information systems, customers, suppliers, governments, laws, and so on).

Systems also have a purpose. For example, many information systems are designed to provide relevant information to users to help their decision making. Information needs to be presented at the right time, at the appropriate level of detail, and of sufficient accuracy to be of use to its recipient. This will help to ensure that the corporate information resource is utilized fully.

Buckingham et al. (1987) define an **information system** as:

A system which assembles, stores, processes and delivers information relevant to an organisation (or to society), in such a way that the information is accessible and useful to those who wish to use it, including managers, staff, clients and citizens. An information system is a human activity (social) system which may or may not involve the use of computer systems.

This definition is useful in that it emphasizes the human and organizational aspects of information systems. It suggests that the information system is useful, in other words, has a purpose, usually improving the effectiveness in the way that the organization does things – information systems do not exist for their own sake. The definition also makes clear that not all information systems use information technology, though this book is mainly about information systems that are computer-based – they use information technology for at least some of the work.

This book is about **information systems development**, that is, the way in which information systems are conceived, analysed, designed, and implemented. In Chapter 3 we suggest a generic approach to information systems development called the information systems development life cycle. At first, it might sound like a prescriptive, mechanistic process. In reality, however, it is very often far from that. Indeed, as we shall see later, there are different ways of developing information systems: there are many methodologies, techniques, and tools to help support the development process.

2.2 Need for a methodology

The early applications of computers – say, until the 1960s – were largely implemented without the aid of an explicit information systems development methodology. In these early days, the emphasis of computer applications was toward programming, and the skills of programmers were particularly appreciated. The systems developers were therefore technically trained but were not necessarily good communicators. This often meant that the needs of users in the application area were not well established, with the consequence that the information systems design was sometimes inappropriate for the application.

Few programmers would follow any formal methodology. Frequently they would use rule-of-thumb and rely on experience. Estimating the date on which the system would be operational was difficult, and applications were frequently behind schedule. Programmers were usually overworked, and frequently spent a very large proportion of their time correcting and enhancing the applications that were operational.

Typically, a member of the user department would come to the programmers asking for a new report or a modification of one that was already supplied. This might occur because the present system did not work as specified or because of changes in the organization and its environment. Often, implementing these changes had undesirable and unexpected effects on other parts of the system, which also had to be corrected. This vicious circle would continue, causing frustration to both programmers and users. This was not a methodology, it was only an attempt to survive the day.

As computers were used more and more and management was demanding more appropriate systems for their expensive outlay, this state of affairs could not go on. There were three main changes:

- There was a growing appreciation of that part of the development of the system that concerns analysis and design and therefore of the role of the systems analyst as well as that of the programmer.
- There was a realization that, as organizations were growing in size and complexity, it was desirable to move away from one-off solutions to a particular problem and toward a more integrated information system.
- There was an appreciation of the desirability of an accepted methodology for the development of information systems.

2.3 Information systems development methodology

It was to answer the problems discussed in the previous section that methodologies were devised and adopted by many organizations. We have already discussed the term methodology. An **information systems development methodology** can be defined as:

A collection of procedures, techniques, tools, and documentation aids which will help the systems developers in their efforts to implement a new information system. A methodology will consist of phases, themselves consisting of subphases, which will guide the systems developers in their choice of the techniques that might be appropriate at each stage of the project and also help them plan, manage, control, and evaluate information systems projects.

But a methodology is more than merely a collection of these things. It is usually based on some 'philosophical' view, otherwise it is merely a method, like a recipe. In Part VI we look in detail at 25 distinct methodologies. Methodologies may differ in the techniques recommended or the contents of each phase, but sometimes their differences are more fundamental. Some methodologies emphasize the human aspects of developing an information system, others aim to be scientific in approach, others pragmatic, and others attempt to automate as much of the work of developing a project as possible. These differences may be best illustrated by their different assumptions, stemming from their 'philosophy' which, when greatly simplified, might be that, for example:

- a system that makes most use of computers is a good solution;
- a system that produces the most appropriate documentation is a good solution;
- a system that is the cheapest to run is a good solution;
- a system that is implemented earliest is a good solution;
- a system that is the most adaptable is a good solution;
- a system that makes the best use of the techniques and tools available is a good solution;
- a system that is liked by the stakeholders is a good solution.

Techniques and tools feature in each methodology. Particular techniques and tools may feature in a number of methodologies. A **technique** is a way of doing a particular activity in the information systems development process, and any particular methodology may recommend techniques to carry out many of these activities. In Part IV we look at 29 different techniques.

Each technique may involve the use of one or more **tools** that represent some of the artefacts used in information systems development. A non-computer-oriented example may help. Two techniques used in the making of meringues are (1) separating the whites of eggs from the yolks and (2) beating the whites. The methodology may recommend the use of tools in these processes, for example, an egg separator and a whisk. In this text, tools are usually automated, that is, computer tools, normally software to help the development of an information system. In Part V we look at 8 tools. These tools might enable some development tasks to be done automatically or semi-automatically. Indeed, some tools have been designed specifically to support activities in a particular methodology. Others are more general purpose and are used in different methodologies.

This book is about methodologies, the differences between them, why these differences exist, and which methodology might be appropriate in given circumstances. As we shall see, methodologies differ greatly, often addressing different objectives. These objectives could be:

1. *To record accurately the requirements for an information system.* The methodology should help users specify their requirements or systems developers investigate and analyse user requirements, otherwise the resultant information system will not meet the needs of the users.

2. *To provide a systematic method of development so that progress can be effectively monitored.* Controlling large-scale projects is not easy, and a project that does not meet its deadlines can have serious cost and other implications for the organization. The provision of checkpoints and well-defined stages in a methodology should ensure that project-planning techniques could be applied effectively.

3. *To provide an information system within an appropriate time limit and at an acceptable cost.* Unless the time spent using some of the techniques included in methodologies is limited, it is possible to devote an enormous amount of time attempting to achieve perfection.

4. *To produce a system which is well documented and easy to maintain.* The need for future modifications to the information system is inevitable as a result of changes taking place in the organization and its environment. These modifications should be made with the least effect on the rest of the system. This requires good documentation.

5. *To provide an indication of any changes that need to be made as early as possible in the development process.* As an information system progresses from analysis through design to implementation, the costs associated with making changes increase. Therefore, the earlier changes are effected, the better.

6. *To provide a system that is liked by those people affected by that system.* The people affected by the information system, that is, the stakeholders, may include clients, managers, auditors and users. If a system is liked by the stakeholders, it is more likely that the system will be used and be successful.

An information systems development methodology, in attempting to make effective use of information technology, may also attempt to make effective use of the techniques and tools available. Information systems development methodologies are also about balancing technical aspects with behavioural (people-oriented) aspects. As we shall see in the book, there are many views as to where this balance lies and how the balance is achieved in methodologies. At one extreme are the methodologies aiming at full automation of information systems development as well as the information system itself. However, even in these systems people need to interact with the system. At the other extreme, perhaps, are attempts at full user participation in the information systems development project and user-led design. Even here, user solutions may make full use of the technology, and there are a growing number of tools designed to aid users develop their own information systems. The balance between technological aspects and people aspects is one that we will return to as it is a continual theme in information systems development.

Having stated that this book is about information systems development methodologies, not all organizations use a standard methodology. They might have developed their own or adapted one to be more appropriate for their own circumstances. Many organizations may only use some aspects of a standard methodology. Other organizations use no methodology at all. The ways that organizations use (or do not use) information systems development methodologies will be another theme of the book.

Summary

- Data represent unstructured facts about events, objects, or people.
- Information has a meaning and use to a particular recipient in a particular context and can be used for decision making.
- Knowledge is accumulated information and contains the ability to use information effectively for particular purposes.
- An information system is a system that assembles, stores, processes, and delivers information relevant to an organization (or to society), in such a way that the information is accessible and useful to those who wish to use it, including managers, staff, clients, and citizens. An information system is a human activity (social) system that may or may not involve the use of computer systems.
- An information systems development methodology is a collection of procedures, techniques, tools, and documentation aids which will help the systems developers in their efforts to implement a new information system. A methodology will consist of phases, themselves consisting of sub-phases, which will guide the systems developers in their choice of the techniques that might be appropriate at each stage of the project and also help them plan, manage, control, and evaluate information systems projects.

Questions

1 Discuss the definitions of data, information, knowledge, information system, methodology, technique, and tools, and look for alternatives in the literature. Do you think our definitions are adequate? Do they miss out on any important aspect?

2 Why do you think there is a need for a methodology to develop information systems? Why do you think that many organizations do not use them?

3 Of the six objectives for an information systems development methodology discussed above, what for you counts as most important, and why?

4 Why do you think there are so many methodologies, techniques and tools available?

Further reading

Kendall, K. and Kendall, J. (2004) *Systems Analysis and Design*, 6th edn, Prentice Hall, Englewood Cliffs, New Jersey. This is one of a large number of American texts that cover the basic field of systems analysis attractively and thoroughly. Most texts of this type tend, however, to emphasize technological aspects.

Part II: The life cycle approach

Contents

Chapter 3 – Information systems
                development life cycle   p 31

In Part II we introduce the information systems development life cycle, which is a general approach to developing systems. It has been the basis of many systems development projects since the 1970s. The phases in the approach are described along with the strengths of the approach. We use this approach as an exemplar and introduce methodologies, techniques, and tools by means of this approach. However, we also discuss its weaknesses, or at least problems relating to the way it has been used in practice, as they lead on to the alternative approaches to developing information systems, which are introduced in Part III of the book. Many of the alternative themes discussed there have in their roots attempts to address some of these weaknesses.

3 Information systems development life cycle

3.1 Information systems development life cycle (SDLC)

The SDLC has had an enormous influence as a general approach to develop information systems. Although there are many variants, it has the following basic structure:

- feasibility study;
- system investigation;
- systems analysis;
- systems design;
- implementation;
- review and maintenance.

These stages together are frequently referred to as 'conventional systems analysis', 'traditional systems analysis', 'information systems development life cycle', or, more frequently in the USA, the **waterfall model**. The term 'life cycle' indicates the staged nature of the process. Further, by the time the review stage comes, the information system may be found to be inadequate and it may not be long before the process starts again with a feasibility study to develop another information system to replace it. In the next few sections we describe each stage in turn.

1 Feasibility study

Among other reasons, a computer system might be contemplated to replace an old system because increasing workloads have overloaded the present system, suitable staff are expensive and difficult to recruit, advancing technology leads to new possibilities, there is a change in the type of work, or there are frequent errors. The next stage is to look in more detail at the present system and then to determine the requirements of the new one.

The feasibility study looks at the present system, the requirements that it was intended to meet, problems in meeting these requirements, new requirements that have come to light since it was first implemented, and briefly investigates alternative solutions. These must be within the terms of reference given to the analyst relating to the boundaries of the system and constraints, particularly those associated with resources. The alternatives suggested might include leaving things alone and improved manual as well as computer solutions.

For each of these, a description is given in terms of the technical, human, organizational, and economic costs and benefits of developing and operating the system. So, any proposed system must be feasible:

- *legally*, that is, it does not infringe any national or, if relevant, international company law;
- *organizationally and socially*, that is, it is acceptable for the organization and its staff, particularly if it involves major changes to the way in which the organization presently carries out its processing;
- *technically*, that is, it can be supported by the technology available and there is sufficient expertise to build it;
- *economically*, that is, it is financially affordable and the expense justifiable.

Of the possible alternatives, a 'recommended solution' is proposed with an outline functional specification. This information is given to management as a formal report and often through an oral presentation by the systems analysts to management who will then decide whether to accept the recommendations of the analysts. This is one of the decision points in the SDLC – whether to proceed or not.

2 Systems investigation

If management has given its approval to proceed, the next stage is a detailed fact-finding phase. This purports to be a thorough investigation of the application area. The information obtained will be much more detailed than that recorded in the feasibility report. It will look at:

- the functional requirements of the existing system (if there is one) and whether these requirements are being achieved;
- the requirements of the new system as there may be new situations or opportunities that suggest altered requirements;
- any constraints imposed;
- the range of data types and volumes that have to be processed;
- exception conditions;
- problems of the present working methods.

These facts are gained by interviewing personnel (both management and operational staff), through the use of questionnaires, by direct observation of the application area of interest, by sampling, and by looking at records and other written material related to the application area:

- *observation* can give a useful insight into the problems, work conditions, bottlenecks, and methods of work;
- *interviewing*, which may be conducted with individuals and groups of users, is usually the most helpful technique for establishing and verifying information, and provides an opportunity to meet the users and to start to overcome possible resistance to change;

- *questionnaires* are usually used where similar types of information need to be obtained from a large number of respondents or from remote locations;
- *searching records and documentation* may highlight problems, but the analyst has to be aware that the documentation may be out of date;
- *sampling* may be used but often requires specialist help from people with statistical and other skills.

A great deal of skill is required to use any of these effectively, and results need to be cross-checked by using a number of these approaches. It may be possible to find out about similar systems implemented elsewhere as the experience of others could be invaluable.

3 Systems analysis

Armed with the facts, the systems analyst proceeds to the systems analysis phase and analyses the present system by asking such questions as:

- Why do the problems exist?
- Why were certain methods of work adopted?
- Are there alternative methods?
- What are the likely growth rates of data?

In other words, it is an attempt to understand all aspects of the present system and why it developed as it did, and eventually indicate how things might be improved by any new system. So, systems analysis provides pointers to the new design. In particular, the analysts will emphasize the need to ascertain what are the requirements for the new system (see Section 5.6).

4 Systems design

Although usually modelled on the design suggested at the feasibility study stage, the new facts may lead to the analyst adopting a rather different design to that proposed at that time. Much will depend on the willingness to be thorough in the investigation phase and questioning in the analysis phase. Typically, however, the new design might be similar to the previous system, but avoiding the problems that occurred with the old system and without including any new ones. Sometimes, the new system is somewhat more radical.

This stage involves the design of both the computer and manual parts of the system. The design documentation set will contain details of:

- input data and how the data is to be captured (entered in the system);
- outputs of the system;
- processes, many carried out by computer programs, involved in converting the inputs to the outputs;
- structure of the computer and manual files that might be referenced in the system;
- security and backup provisions to be made;
- systems testing and implementation plans.

5 Implementation

Following the systems design phase are the various procedures that lead to the implementation of the new system. If the design includes computer programs, these have to be written and tested. New hardware and software systems need to be purchased and installed if they are not available in the organization at present. It is important that all aspects of the system are proven before **cutover** to the new system, otherwise failure will cause a lack of confidence in this and, possibly, future computer applications. The design and coding of the programs might be carried out by computer programmers. Alternatively, application packages might be purchased to form part of the final system. In this approach, the analysis and programming functions are separate tasks carried out by different people.

A major aspect of this phase is that of **quality control**. The manual procedures, along with the hardware and software, need to be tested to the satisfaction of users as well as analysts. The users need to be comfortable with the new methods. The departmental staff can practise using the system and any difficulties experienced should be ironed out. The **education and training** of user staff is therefore an important element of this phase. Without thorough training, users will be unfamiliar with the new system and unlikely to cope with the new approach (unless it is very similar to the old system).

Documentation, such as the operations and user manuals, will be produced and approved and the live (real, rather than test) data will be collected and validated so that the master files can be set up. **Security** procedures (see also Section 8.6) need to be tested, so that there is no unauthorized access and recovery is possible in case of failure. Once all this has been carried out, the system can be operated and the old one discontinued. If cutover to the new system is done 'overnight', then there could be problems associated with the new system. It is usually too risky an approach to cutover. Frequently, therefore, there is a period of parallel running, where old and new systems are run together, until there is complete confidence in the new system. Alternatively, parts of the new system can be implemented in turn, forming a phased implementation. If one part of the system is implemented 'to test the water' before the rest of the system is implemented, this is referred to as a pilot run. Acceptance testing comes to an end when the users feel assured that the new system is running satisfactorily and it is at this point that the new system becomes fully operational (or 'live').

6 Review and maintenance

The final stage in the system development process occurs once the system is operational. There are bound to be some changes necessary and some staff will be set aside for maintenance, which aims to ensure the continued efficient running of the system. Some of the changes will be due to changes in the organization or its environment, some to technological advances, and some to 'extras' added to the system at an agreed period following operational running. Inevitably, however, some maintenance work is associated with the correction of errors found since the system became operational.

At some stage there will also be a review of the system to ensure that it does conform to the requirements set out at the feasibility study stage, and the costs have not exceeded those predicted. The evaluation process might lead to an improvement in the way other systems are developed through the process of **organizational learning**.

As commented earlier, because there is frequently a divergence between the operational system and the requirements laid out in the feasibility study, there is sometimes a decision made to look again at the application and enhance it or even develop another new system to replace it. This could also occur for another reason. Changes in the application area could be such that the operational system is no longer appropriate and should be replaced. The SDLC then finishes and starts at the point where there is a recognition that needs are not being met efficiently and effectively and the feasibility of a replacement system is then considered and the life cycle begins again.

In the following sections we discuss the key concepts of methodologies, techniques, and tools in the particular context of the SDLC.

3.2 Methodology

There were many variants of the SDLC. The earliest in the UK was that proposed by the National Computing Centre (NCC) in the late 1960s. At the time, the NCC was sponsored by the UK government to provide a lead in computing for both government and private organizations. One of its functions, therefore, was to suggest good standards for developing computer applications. It had a great impact on the data processing community, particularly in the UK, and represented a typical methodology of the 1970s based on the SDLC. This methodology has been improved and altered since, so that a number of approaches used today can be regarded as a modern variant of this approach, the most obvious being SSADM (Section 21.1).

The use of a methodology improves the practice of information systems development. The attributes that we would expect of a methodology include:

1. A series of **phases** starting from the feasibility study through to review and maintenance. The phases are expected to be carried out as a sequential process. Each of these phases has subphases. The activities to be carried out in each of the subphases are usually spelt out clearly in the methodology documentation, usually found in manuals. The outputs (or 'deliverables') of each subphase are also detailed. Deliverables may include documents, plans, or computer programs.

2. A series of **techniques**, which include ways to evaluate the costs and benefits of different solutions and methods to formulate the detailed design necessary to develop computer applications, are also detailed. We illustrate some of these techniques in Section 3.3.

3. A series of **tools** to help the systems analysts in their work. By tools, we mean, in this book, software packages that aid some aspect of the approach. We discuss some of these tools in Section 3.4.

4. A **training** scheme so that all analysts and others new to their roles and responsibilities could adopt the standards suggested. Typically, there may be a training course offered that might take up to six weeks. A qualification might be offered.

5. A **philosophy**, perhaps implied rather than stated, which might be that 'computer systems are usually good solutions to organizational problems and processing'. We

suggest this because the assumption is that the inevitable consequence of using most methodologies is a new computer application to perform the required functions.

Of course, the NCC methodology itself became dated and was replaced by other methodologies, many of which were also based on the life cycle but incorporated the latest methods, techniques, and tools. Many more modern methodologies such as SSADM, Merise, and Yourdon Systems Methodology, which are described later in the book, are life cycle approaches.

For example, SSADM (Section 21.1) has the following stages:

- Stage 0 – feasibility;
- Stage 1 – investigation of current environment;
- Stage 2 – business systems options;
- Stage 3 – definition of requirements;
- Stage 4 – technical system options;
- Stage 5 – logical design;
- Stage 6 – physical design.

Although the terminology used for each stage seems different, we can see the progress through the life cycle from feasibility study, via investigation and requirements analysis, through to design. In fact, SSADM does not offer much specific advice regarding implementation, as this is seen as being particular to the application environment and the assumption is made that little general advice can be given. Each of these seven stages of SSADM are further detailed into several steps (the methodology documentation consists of several manuals).

Although we have argued that most methodologies follow a life cycle of sorts, some circumvent stages in the wish to develop applications quickly, such as James Martin's Rapid Application Development (Section 23.1), for example; whereas others do not assume such a logical step-by-step approach (Multiview, for example – Section 26.1). Yet others concentrate on only part of the life cycle, like Soft Systems Methodology (Section 25.1). Finally, some methodologies concentrate on particular views about the emphasis that should be placed when developing applications, such as ETHICS (Section 24.1), which emphasizes people aspects and participation.

3.3 Techniques

Most methodologies recommend a number of documentation aids to ensure that the investigation is thorough. The early methodologies might have included:

- various flowcharts, which might, for example, help the analyst trace the flow of documents through a department;
- an organization chart, showing the reporting structure of people in a company or department;
- manual document specifications, giving details of documents used in the manual system;

- grid charts, showing how different components of the system, such as people and machines, interact with each other;
- discussion records on which the notes taken at interviews could be recorded;
- file and record specifications describing, in the former case, all the data items in a record, including their names and descriptions, size, format, and possible range of values.

Many of the documentation techniques outlined above have been replaced. The most well-used techniques are probably now data flow diagramming (Section 12.1), which is important in specifying the process aspects of the system, and entity-relationship diagramming (Section 11.1), which details the data aspects. Data flow diagrams that describe the logic of the system or the physical implementation of the new design show:

- data flows into the system from the environment;
- data flows out of the system into the environment;
- processes that change the data within the system;
- data storage within the system;
- the boundary and scope of the system.

Entity-relationship diagrams, on the other hand, represent the data aspects. An entity is a data type, such as a student, lecturer, course, and classroom. These diagrams show how these entities relate to each other and they also show the detail about each entity, similar to that provided in a record specification form.

As we will see in Part IV, there are many techniques used in information systems development. Some are associated with particular methodologies, like rich pictures (Section 10.1) in Soft Systems Methodology, others are more generic, such as dataflow diagrams and entity-relationship diagrams, which are recommended in many methodologies.

3.4 Tools

When we talk of tools in this book, we are referring to software packages that aid aspects of information systems development. Most tools require significant computing power and good graphics capabilities. Most tools available now were not available in the early days, although there were project management tools (we discuss Microsoft Project in Section 18.4) and report-generating packages, such as RPGII. Users of RPGII would complete a series of forms specifying how the required report should be structured and what files contained the data necessary to be on the report. This has been incorporated as one aspect of modern toolsets (Chapter 19). Oracle, for example, will have an application generator as part of its toolset. Other tools support the production of documents, such as dataflow diagrams and entity-relationship diagrams. They include Visio, discussed in Section 18.3, but again this feature is included in many toolsets. As you will see in Part V, there are many different types of tool available, which ease much of the systems development process.

Probably the greatest benefit of using tools is that analysts and designers are not reluctant to change documents, because the change process is simple. Frequent manual redrawing is not satisfactory, because of the effort involved and the potential of introducing errors when redrawing. Without such drawing tools many a small change required by a user would not be incorporated into the documentation of the system.

3.5 Potential strengths of the SDLC

The SDLC has a number of features to commend it. Methodologies incorporating this view of applications development have been well tried and tested. The use of documentation standards in such methodologies helps to ensure that the specifications are complete, and that they are communicated to systems development staff, the users in the department, and the computer operations staff. It also ensures that these people are trained to use the system. The education of users on subjects such as the general use of computers is also recommended, and helps to dispel fears about the effects of computers. Following such a methodology also prevents, to some extent at least, missed cutover dates (the date when the system is due to become operational) and unexpectedly high costs and lower than expected benefits. At the end of each phase the technologists and the users have an opportunity to review progress. By dividing the development of a system into phases, each subdivided into more manageable tasks, along with the improved training and the techniques of communication offered, we have the opportunity for control of the applications development process.

3.6 Potential weaknesses of the SDLC

The criticisms of the systems development approach to applications development, or, to be more precise, of the way in which it was used include:

- failure to meet the needs of management;
- instability;
- inflexibility;
- user dissatisfaction;
- problems with documentation;
- lack of control;
- incomplete systems;
- application backlog;
- maintenance workload;
- problems with the 'ideal' approach;
- emphasis on 'hard' thinking;
- assumption of 'green-field' development.

We will discuss each of these in turn.

1 Failure to meet the needs of management

As can be seen in Figure 3.1, although systems developed by this approach often successfully deal with such operational processing as the various accounting routines (payroll, invoicing, etc.), the needs of middle management and top management can be ignored. Management information needs, such as that required when making decisions about where to locate a new

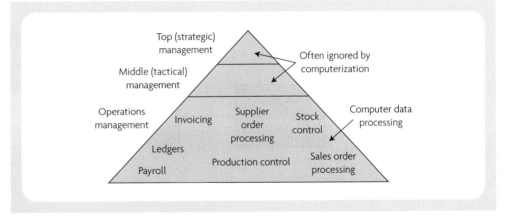

Figure 3.1: Failure to meet all the needs of management

factory, which products to stop selling, what sales or production targets to aim for and how sales can be increased, are neglected. Although some information may filter up to provide summary or exception reports, the computer is being used largely only for routine and repetitive tasks. Instead of meeting corporate objectives, computers are being used to help solve low-level operational tasks. In general, the danger is that such systems are rather limited in their scope and rather unambitious.

2 Models of processes are unstable

The conventional SDLC approach attempts to improve the way that the processes in businesses are carried out. However, businesses do change, and processes need to change frequently to adapt to new circumstances in the business environment. Because these computer systems model processes, they have to be modified or rewritten frequently. It could be said therefore that computer systems, which are to some extent 'models' of processes, are unstable because the real-world processes themselves are unstable.

3 Output-driven design leads to inflexibility

The outputs that the system is meant to produce are usually decided very early in the systems development process. Design is 'output driven' (see Figure 3.2) in that, once the output is agreed, the inputs required, the file contents, and the processes involved to convert the inputs to the outputs can all be designed. However, changes to required outputs are frequent and, because the system has been designed from the outputs backwards, changes in requirements usually necessitate a very large change to the system design and therefore cause either a delay in the implementation schedule or are left undone, leading to an unsatisfactory and inappropriate system.

4 User dissatisfaction

This is often a feature of many computer systems. Sometimes systems are rejected as soon as they are implemented. It may well be only at this time that the users see the

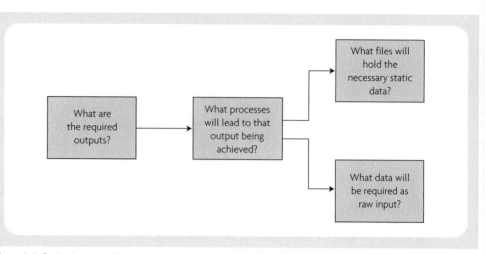

Figure 3.2: Design is output-driven

repercussions of their 'decisions'. These decisions have been assumed by the analysts to be firm ones and, as computer systems can prove inflexible, it is normally very difficult to incorporate changes in requirements once the systems development is under way. Many companies expect users to 'sign off' their requirements at an early stage when they do not have the information to agree the exact requirements of the system. The users sign documentation completed by computer-oriented people. The documents may not be designed for the users. On the contrary, they are designed for the systems analysts, operations staff, and programming staff who are involved in developing the system. Users cannot be expected to be familiar with the technology and its full potential, and therefore find it difficult to contribute to the debate to produce better systems. The period between 'sign-off' and implementation tends to be one where the users are very uncertain about the outcome and they are not involved in the development process and therefore may lose their commitment to the system. They might have become disillusioned with computer systems and, as a consequence, fail to co-operate with the systems development staff. Systems people, as they see the situation, talk about user staff as being awkward or unable to make a decision. Users may first see the system only on implementation and find it inappropriate. This alienation between technologists and users has even been known to lead to users developing their own informal systems which are used alongside the computer system, ultimately causing the latter to be superfluous.

5 Problems with documentation

We have suggested that one of the benefits of the methodology is the standardization and documentation that methodologies use. Yet this is not ideal. The orientation of the documentation is frequently toward the computer person and not the user, and this represents a potential source of problems. The main purpose of documentation is that of communication, and a technically oriented document is not ideal.

6 Lack of control

Despite the methodological approach enabling estimates of the time, people, and other resources needed, these estimates can be unreliable because of the complexity of some phases and the inexperience of the estimators. Computer people have been seen as unreliable and some disenchantment has ensued in relation to computer applications.

7 Incomplete systems

Computers are particularly good at processing a large amount of data speedily. They excel where the processing is the same for all items: where the processing is structured, stable, and routine. This often means that the unusual (exceptional conditions) is frequently ignored in the computer system. They are too expensive to cater for. If they are diagnosed in the system investigation stage, then clerical staff are often assigned to deal with these exceptions. Frequently they are not diagnosed, the exceptions being ignored or forgotten, and the system is falsely believed to be complete. These exceptions cause problems early on in the operational life of the system. These problems mark a particular technical failure, but they also indicate a systems failure, a failure to analyse and design the system properly.

8 Application backlog

A further problem is the application backlog. There may well be a number of systems waiting to be developed. The users may have to wait some years before the development process can get under way for any proposed system. The process to develop an information system will itself take many months and frequently years from feasibility study to implementation. Users may also postpone requests for systems because they know it is not worth doing because of the backlog. This phenomenon is referred to as the invisible backlog.

9 Maintenance workload

The temptation, then, is toward 'quick and dirty' solutions. The deadline for cutover may seem sacrosanct. It might be politically expedient to patch over poor design rather than spend time on good design. This is one of the factors that has led to the great problem of keeping operational systems going. With many firms, the effort given to maintenance can be as high as 60–80 per cent of the total workload. With so many resources being devoted to maintenance, which often takes priority, it is understandable that there is a long queue of applications in the pipeline. The users are discouraged by such delay in developing and implementing 'their' applications because of the necessity to maintain the legacy systems (see Section 8.1).

10 Problems with the 'ideal' approach

The SDLC model assumes a step-by-step, top-down process. This is somewhat naive. It is inevitably necessary to carry out a series of iterations when, for example, new requirements are discovered or subphases might prove unnecessary. In other words, information systems development is an iterative process, whatever textbooks say. The political dimension where, for example, people have their own 'agenda' transcends the rationale of any methodology. Users and analysts will need to interpret the methodology, and its techniques and tools, to be relevant to a particular problem situation. It also assumes a tailor-made rather than packaged solution, usually for a medium- or large-scale application using mainframe computers. With the widespread use of PC systems, such an approach may be inappropriate and unwieldy.

11 Emphasis on 'hard' thinking

The SDLC approach may make a number of simplistic assumptions. It assumes that there are 'facts' that only need to be investigated and identified; it assumes that there can be a 'best solution' identified that will 'solve the problem'; it assumes that this 'ideal solution' can be easily engineered by following a step-by-step methodology; and it assumes that the techniques offered will analyse and design all that needs to be done. But we are not engineering a simple mechanical object. The world of information systems is concerned with organizations and people as much as technology. The situations encountered are often ambiguous, issue-laden, messy, and problematical, and an alternative approach might well be more appropriate in these difficult but common situations.

12 Assumption of 'green-field' development

The SDLC has been criticized for embodying the assumption that all systems developed are new and that developers begin with a 'green field', that is, there are no existing systems that have to be taken into consideration. This may have been true when the first computerized systems replaced manual systems, but it is no longer the case. New developments are not nicely separated from each other and any new development work is more likely to be an incremental development of an existing system than a complete replacement. It is also likely to have to be integrated with many existing systems in an organization, typically having been developed at different times, with different developers, on different architectures and platforms. This makes development a much more messy activity than would otherwise be the case. This assumption is not confined to the SDLC, as many methodologies also implicitly make this assumption. One that specifically addresses the complexity of existing systems is Renaissance (see Section 25.5). This issue is also addressed elsewhere, for example, in evolutionary development (Section 7.1) and legacy systems (Section 8.1).

The above criticisms of the SDLC should be regarded as 'potential' criticisms, as many organizations using the approach do not fall into all or even most of the potential traps. However, this book concerns itself with improving information systems development by adapting the SDLC or using alternative approaches to developing information systems.

In Part III, we will look at themes regarding differing views about information systems development. They can be seen as ways in which people in the field have reacted to the potential problems of the SDLC. This has resulted in many views about what a methodology should emphasize: people aspects, planning aspects, automation . . . there are many such views. To some extent these can be seen as representing different philosophies, for example, a people-oriented approach might be seen as being based on the philosophy that people have a right to design their organizational environment. But these themes also reflect greater possibilities, for example, software tools that can both make the process easier and faster, and also make some approaches possible, which aim toward automation of the development process. Similarly, the craft of systems analysis has been improved greatly through the use of alternative techniques.

3.7 Conclusion

Before we consider approaches to systems analysis, which represent advances on the traditional SDLC approach, the reader should stop and consider that many systems developed today are still done using no real methodology at all. This is particularly true in organizations using PCs, perhaps developing websites as their first experience of computing. The SDLC has many advantages over trial and error.

The SDLC is being used successfully. In many respects, there is nothing intrinsically wrong with the SDLC. Much depends on the way in which it is used. There must be sufficient resources assigned to the process; it needs to be well managed and controlled so that any deviation from the plan is noticed and dealt with; it should be seen not as a rigid process but as a flexible and iterative one; the feasibility study needs to be seen as a way of exploring alternatives rather than as a way of advancing a limited point of view; and systems development should not be seen as a purely technological process but one involving all users and developers, indeed the organization as a whole. However, many of the alternative approaches address some of the potential weaknesses suggested above. This may mean improving one or more of the phases, including the use of an alternative technique or a new software tool.

We also do not wish to assert that any of the alternative methodologies represents a panacea. Major concerns in computing remain, for example:

- meeting project deadlines;
- system maintenance;
- staff recruitment and retention;
- user dissatisfaction;
- changing requirements.

Notwithstanding these continuing pessimistic trends, there are many successful information systems, some developed using the basic SDLC approach. In Part III we will look at the various themes in information systems development methodologies that are seen as developments or alternatives to the basic approach discussed in this chapter. These alternative approaches usually address one or more of the potential criticisms of the SDLC approach discussed in the previous section.

Summary

- The SDLC has had an enormous influence as a general approach to develop information systems. Although there are many variants, it has the following basic structure: feasibility study; system investigation; systems analysis; systems design; implementation; review and maintenance.

- An information systems development methodology is likely to include: a series of phases with subphases, each having expected outputs (or 'deliverables'); a series of techniques; a series of tools; a training scheme and some underlying philosophy.

- Although the SDLC has many strengths and is still used today, it also has many potential weaknesses, which has led to alternative methodologies, techniques, and tools being available.

Questions

1. Distinguish between techniques, tools, and methodologies in the context of the SDLC.
2. Discuss the strengths and the limitations of the traditional systems development life cycle.
3. Would you adopt the SDLC for developing a small application on a PC? Would you modify it in any way? Give reasons for your answer.

Further reading

Avison, D.E. and Shah, H.U. (1995) *The Information Systems Development Cycle: A First Course in Information Systems*, McGraw-Hill, Maidenhead, UK. Avison and Shah describe an up-to-date approach modelled on the SDLC. It can be regarded as a prequel to this present book.

Part III: **Themes in information** systems development

Contents

Chapter 4 – Organizational themes   p 51

Chapter 5 – People themes   p 79

Chapter 6 – Modelling themes   p 109

Chapter 7 – Rapid and evolutionary development   p 119

Chapter 8 – Engineering themes   p 151

Chapter 9 – External development   p 175

The structure of the book is based on separate descriptions of information systems development themes, techniques, tools and methodologies. In Part III we explore some of the themes that have evolved in relation to information systems development. A theme represents a particular foundation, school of thought or focus that has influenced information systems. Some of these themes are particular responses to one or more of the perceived limitations of the traditional systems development life cycle (SDLC) approach discussed in Chapter 3. However, none of these are panaceas and no one approach solves all the problems.

Themes may relate to, or are part of the philosophical background of, particular techniques, tools or methodologies. Thus the theme of participation relates to, and has been influential on, the ETHICS methodology. As techniques, tools and methodologies are described separately, the reader should be aware that these relationships exist and that some of them are quite complex. So, as an illustration of one such set of linkages we have produced a 'road map' (see Diagram 1). This shows, from the perspective of the theme of 'modelling', the linkages to techniques, tools and methodologies. Thus if a reader wants to explore and follow through from the theme of modelling to the related techniques, tools and methods, they should look at the highlighted sections. This is simply one example of the way techniques, tools or methodologies link, and the reader will need to explore their own road maps for the others. Potential links are cross-referenced in the text to help in this task.

In Part IV, we have attempted to group various themes into categories or groups, for convenience. However, there exist a range of alternative groupings. One reason for this is that a theme has a number of characteristics or features that might justify it being grouped in some other theme. This is inevitable and it should be realized that these groupings are not definitive. But hopefully they make some sense and prove helpful.

Chapter 4 groups together themes that address information systems for the organization as a whole: their inter-connection, strategic use, re-engineering and planning. We first look at the systems approach, which presents a holistic view of organizations and therefore addresses the piecemeal aspect of the traditional SDLC, which we criticized in Chapter 3, because the systems approach looks at the problem situation as a whole. Information systems strategy aims at developing information systems for the organization as a whole, but in particular to address the needs of top management, and could be said to be a 'head-on' attack on the emphasis placed on computer applications at the operational level of the firm in the traditional approach. We also present a discussion on the way in which IS projects are justified in organizations. Business process re-engineering is again a re-examination of the present systems, including information systems of the organization, and tends to be a radical rather than evolutionary approach. Such re-examination leads to redesign within the context of both the other processes and the needs of the organization as a whole. We then look at planning approaches which emphasize the way future information systems development can be organized and integrated so that again strategic as well as tactical and operational needs are included. The stages of growth model shows how information systems and information technology might typically develop as a whole in organizations over time. We then look at the issue of flexibility. Our last section in Chapter 4 considers managing the project in the hope that projects are delivered 'at cost and on time'.

Chapter 5 emphasizes the role of people in developing and using an information system. The importance of the people using the information systems and other interested parties, frequently called

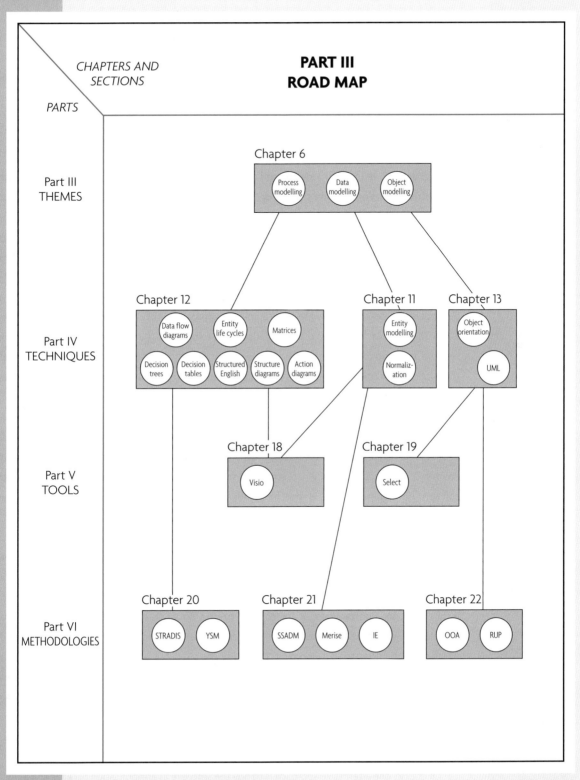

Diagram 1: Modelling Themes 'Road Map' (shows relationship between themes and techniques, tools, and methodologies for the modelling theme)

stakeholders, is addressed in the theme of participation. This participation can be strengthened so that there is joint application development, or so strong that it is the end-user who is developing the system. We then look at expertise and knowledge and include a chapter on expert systems and another on knowledge management. In expert systems one of the important areas is capturing the knowledge of people who are considered experts in the particular application domain so that users can build on this expertise. We then look at customer orientation. This suggests that the stakeholders of most importance are the customers (not so much technologists or even internal users). This is most strong, perhaps, in web applications. Finally, we consider the major theme of requirements elicitation, where we try to eke out the requirements of the system from the point of view of all stakeholders.

In Chapter 6 we look at modelling and approaches that emphasize the modelling aspects, either on the process side or on the data side. Modelling represents an important aspect of many methodologies. We first look at structured approaches, which centre on the techniques and tools for modelling processes. The next theme, data modelling, concentrates on understanding and classifying data and the relationships between data in an organization and is therefore related to the development of databases (and data warehousing and data mining applications). Object orientation is an alternative view, and in this approach we model objects of all kinds in the organization. Data, processes, people and software can all be modelled as objects. This is therefore a unifying approach within the overall theme of modelling.

In Chapter 7 we look at rapid and evolutionary development. The speedy delivery of applications has become more and more important and in many situations essential. In evolutionary development part of the application is operational quickly, but the application is improved over time so that it meets needs as they become known and can be implemented. Prototyping enables users to see and comment on the application before it is implemented. It uses system development software tools in a bid to improve requirements analysis, that is, to ensure 'what the users want is what they get'. Prototyping is also claimed to lead to more rapid application development and we discuss this as a separate issue. Agile development has many of the features of other approaches discussed in this chapter, though it is most associated with software development. Agile is also frequently used for developing web applications, but this also receives separate consideration in the final section of the chapter.

Another major area is that of engineering and this is discussed in Chapter 8. The first theme in this category concerns legacy systems, that is, what to do with the systems existing already in the organization. This is important, and yet as we shall see most methodologies tend to assume a new application for a new situation. We then consider software engineering. This concentrates on quality software, which is, of course, an important part of any computerized information system. Attempts have been made to automate all aspects of the systems development process and there are now a number of software tools which support automated development. This is discussed in detail later in Part V. We then look at the particular issues of engineering methodologies from components of other approaches, a process referred to as method engineering. We then look at building applications through the use of components. Another theme which is discussed greatly concerns security and this is followed by a section on database management. Both are concerns of much applications development work. Finally, we look at data warehouse and data mining.

Of course, one alternative approach is to buy the software as an application package and a discussion of this starts Chapter 9. Open source code may be freely available and used as components for a new application and this is considered in the next section. A larger application package, which may be appropriate for a number of applications, is a full-scale enterprise resource planning system, and this is considered next. We then look at outsourcing, which might lead to system development as well as information systems being done completely abroad (offshoring).

Most of the themes discussed in Part III have their counterparts in Part VI: actual methodologies that are used in organizations. Even if they are not directly linked, most of the methodologies used today will have been greatly influenced by aspects of these themes.

4 Organizational themes

4.1 Systems theory

In Chapter 2 we attempted to define the nature of **systems**. We saw how systems relate to each other and that they themselves consisted of subsystems. This gives rise to the definition of a system as a set of interrelated elements (Ackoff, 1971). A system will have a set of inputs going into it, a set of outputs going out of it, and a set of processes that convert the inputs to the outputs.

We define a **boundary** of a system when we describe it. This may not correspond to any physical or cultural division. A payroll system might include all the activities involved in the payment of staff in a business. These activities fall within the boundary of that system. Those systems outside it, with which it relates, are referred to as the **environment**. Systems thinking concerns itself with interactions between the system and its environment, not so much with how the system works, which can be seen as a 'black box'. The staff recruitment system and production systems within the firm will be part of the environment of the payroll system, as will the government's system to increase employment.

One of the bases of systems theory concerns Aristotle's dictum that the whole is greater than the sum of the parts. This would suggest that we must try to develop information systems for the widest possible context: an organization as a whole rather than for functions in isolation. If we fail to follow this principle then a small part of the organization may be operating to the detriment of the organization as a whole. If we do break up a complex problem into smaller manageable units, we need to keep the whole in mind. Otherwise, this may be reductionist, the process of decomposition distorting our understanding of the overall system. Users of many of the approaches discussed, in particular the structured approach, part of process modelling (described in Section 6.2), may succumb to this danger unless they use the approach with care. Decomposing complex structures is the accepted approach in a scientific discipline, but information systems concern people and organizations as well as technology, and the interactions are such that in these human activity systems it is important to see the whole picture. The human components in particular may react differently when examined singly than when they play a role in the whole system.

Organizational systems are not predictable as they concern human beings. The outputs of computer programs may be predictable. Human activity systems are less predictable because human beings may not follow instructions in the way a piece of software does, nor interpret instructions in the same way as other people do or in the way that they themselves might have done on previous occasions.

Another aspect of systems theory is that organizations are **open systems**. They are not closed and self-contained, and therefore the relationship between the organization and its environment is important. They will exchange information with the environment, both influencing the environment and being influenced by it. The system, which we call the organization, will be affected by, for example, policies of the government, competitors, suppliers, and customers, and unless these are taken into account, predictions regarding the organization will be incorrect. As organizations are complex systems, this would suggest that we require a wide range of expertise and experience to understand their nature and how they react with the outside world. Multidisciplinary teams might be needed to attempt to understand organizations and analyse and develop information systems.

Human activity systems should have a **purpose** and the interrelated elements interact to achieve this purpose. What then is the purpose of an information system? Depending on the application area, an information system may be constructed to help managers decide on where to build a new factory. It may provide information about customers so that decisions can be made about their credit rating. It might be to maximize the use of aircraft seats in an airline ticket reservation system. With this information provided, it is possible to control the environment rather than passively react to it.

Information systems usually have human and computer elements and both aspects of the system are interrelated. However, the computer aspects are closed and predictable, the human aspects are open and non-deterministic. Although not simple, the technological aspects are less complex than the human aspects in an information system, because the former are predictable in nature. However, many information systems methodologies only stress the technological aspects. This may lead to a solution that is not ideal, because the methodologies underestimate the importance and complexity of the human element.

Systems theory has had widespread influence in information systems work. It would suggest, for example, that whatever methodology is adopted, the systems analyst ought to look at the organization as a whole and also be aware of externalities beyond the obvious boundaries of the system. These include customers, competitors, governments, and so on. Systems theory would also suggest that a multidisciplinary team of analysts is much more likely to understand the organization and suggest better solutions to problems. After all, specialisms are artificial and arbitrary divisions. Such an approach should prevent the automatic assumption that computer solutions are always appropriate, as well as preventing a study of problem situations from only one narrow point of view. With this approach comes the acknowledgement that there may be a variety of possible solutions, none of them obviously 'best' perhaps, but each having some advantages. These solutions may involve structural, procedural, attitudinal, and environmental change.

Checkland (1981), developed further in Checkland and Scholes (1990), has attempted to turn the tenets of systems theory into a practical methodology, which is called Soft Systems Methodology (SSM) (see Section 25.1). Checkland argues that systems analysts apply their craft to problems that are not well defined and soft thinking attempts to understand the fuzzy world of complex organizations. By contrast, 'hard' approaches, such as the structured and data analysis methods, emphasize the certain and the precise in a particular domain and tend to look at the domain from one point of view, a major contrast compared to soft systems thinking.

Checkland's description of human activity systems also acknowledges the importance of people in organizations. It is relatively easy to model data and processes, but to understand organizations it is essential to include people in the model. This is difficult because of their unpredictable nature. The claims of the proponents of this approach are that a better understanding of these complex problem situations is more likely to result when compared to using hard approaches alone.

We will now look at one final contrast in hard and soft systems viewpoints. Analysts following hard approaches think in terms of systems as though they exist as such and can be engineered. The soft systems viewpoint is that systems do not exist but represent a way of viewing, and therefore understanding, complex real-world activities. However, the implication of this is that different analysts will see the real world in different ways, depending on their background, experience, and so on. The discussion between different analysts can therefore provide understanding of the real world as well as expose its complexity. It may lead to a completely different systems view of the organization being studied.

There are other approaches that have used systems ideas in their design. Beer's Viable Systems Model, for example (see Espejo and Harnden, 1989), provides a tool to study organizations holistically, analysing their structure and their information systems from many viewpoints. Multiview (Section 26.1) incorporates these ideas as well as 'hard' systems ideas in its approach.

4.2 Information systems strategy

The first business activities that were computerized tended to be the basic transaction processing systems such as payroll, sales order processing, stock control, and invoicing. The approach used was simple and did not involve changing the nature of the business activities performed. In the early days, change usually only concerned the means by which existing activities were undertaken. The same activities were computerized to make them more **efficient**. Payroll, for example, previously required large numbers of manual payroll clerks to perform the activity. When computerized, it did not need all these clerks. Early computerization therefore aimed to promote efficiency and in particular to reduce labour costs. This kind of computerization is termed efficiency projects, defined by Fitzgerald (1998) as:

> Projects that seek to reduce the cost of performing a particular process or task by utilising information technology. They do not seek to radically change the nature of the objectives that those tasks and processes were originally devised to fulfill, they simply seek to achieve the same objectives at lower cost, i.e. to perform existing tasks more efficiently.

Silk (1990) argues that cost displacement savings are relatively easy to quantify and that a clear financial case for investment can be demonstrated, using any of the available financial analysis techniques, such as Internal Rate of Return (IRR), Net Present Value (NPV), Return on Investment (ROI), or Payback Period. Such savings are quantifiable; for example, the cost savings of payroll clerks were the salary savings, and cost justification was easy to make.

This does not mean that **cost/benefit analysis** (or any other technique) is an effective evaluation method; indeed, it has been argued that the predicted benefits were rarely achieved. Sometimes the benefits or savings were exaggerated and the costs of the required computer staff underestimated or ignored. So, replacing relatively cheap payroll clerks with a computer system and relatively expensive systems analysts, programmers, operators, etc. did not always produce the large predicted benefits. Nevertheless, the mechanical evaluation process itself was relatively straightforward. IBM, for example, developed a method known as SESAME that used a cost/benefit technique to assess a computer project in comparison with a non-computerized system (Lincoln and Shorrock, 1992).

However, in more recent times, the opportunities for further information systems investments based on efficiency and labour displacement criteria have been limited. First, the number of opportunities is reduced, as more and more projects are implemented, and, second, the returns are declining, as the most clear-cut efficiency improving opportunities have already been addressed. In most large organizations almost all the activities are now computerized or supported by computers in some way. So, overall, the propensity for reducing traditional labour costs via information systems is declining.

The cost justification for revamping systems that have already been computerized, i.e. recomputerizing, is more difficult because the labour displacement opportunities are no longer available – those savings have already been made in the original computerization. However, more recently, there have been a number of justifications based on cost savings relating to reductions in the managerial workforce, particularly middle management levels associated in particular with enterprise resource planning systems (Section 9.3). It is this area that will probably provide the basis for most labour-saving cost displacement information systems projects in the future.

Labour costs have formed the largest and easiest costs to displace using information technology (IT), but other costs have also been amenable to displacement, at least to some degree. Paper and postage costs have been displaced, at least to some extent, by Electronic Data Interchange (EDI), electronic mail, and electronic commerce. Property costs have been displaced, or reduced, where organizations have used IT to enable the move of back-office functions away from town centres to cheaper areas (or even to home working). Inventory costs have been displaced by information systems-enabled Just In Time (JIT) systems. Such cost displacement savings can be relatively easily quantified in these systems which seek to perform essentially the same functions as before, but to perform them at less cost.

A major problem with the cost/benefit justification (based on labour displacement cost savings or other efficiencies) of new computer systems or IT investments in organizations is that it is very difficult to justify systems that do not deliver savings but may deliver benefits such as increased market share, new opportunities, or, in general, better management information. These kinds of investment are very unlikely to pass the usual efficiency-based evaluation or justification process that is used in most organizations.

Over the years computing and technology investment has moved from just being about cost savings and efficiencies to more strategic or effectiveness systems. This in particular regards the use of information systems and IT as a direct tool for obtaining competitive advantage. Information systems can be used to improve the business in the marketplace and in this way to help:

- redefine the boundaries of particular industries;
- develop new products or services;
- change the relationships between suppliers and customers;
- establish barriers to deter new entrants to marketplaces.

The basic objective of this type of information system is to identify better ways of doing things, leading to increased revenues, greater functionality, better products and services, improved presentation or image, and improvement of the organization's competitive positioning. This is usually referred to as using information systems to gain **competitive advantage**. This is IT for **effectiveness** rather than efficiency. The basic objectives of effectiveness projects are not simply to reduce the costs of performing existing tasks but to identify ways of doing different things that better achieve the required results, leading to increased revenues and better service. Effectiveness projects are not addressing efficiency criteria but seek to improve organizational effectiveness.

The distinction between efficiency and effectiveness hinges on the notion of objectives. It is possible to perform some set of tasks very well and efficiently, i.e. in the cheapest possible way, but this does not mean that performing that particular set of tasks is the best way of achieving the objectives associated with those tasks. Efficiency is broadly concerned with how we do things, whereas effectiveness is concerned with what we do.

Although making justifications based on effectiveness criteria is not impossible, it is much more difficult than making a financial case on the basis of efficiency/savings obtained. One of the reasons for this increased difficulty is because there is an extra stage of proof to go through.

For effectiveness projects it is not just necessary to identify the benefits, for example, better service, but also that the recipient of those benefits will recognize and value the improvement and change their behaviour in some positive way as a result, e.g. by existing customers buying more services, by not transferring their allegiances elsewhere, or for the improved service to attract new customers. Customers or potential customers must first be aware of the change and second perceive it as an improvement over the service they currently use (which may be no service at all or a competitor's service). So, the **benefit realization** process comprises at least two stages, first the provision or implementation of the project to provide benefit and second the effect of that benefit on the wider environment and any resultant behaviour change.

Figure 4.1 illustrates this notion; for efficiency projects the identified efficiency concept is implemented and the benefit accrues, usually as reduced costs, (or not) and that is the end of the process. It is internally controllable by the organization and usually does not depend on any interaction with the outside environment. Effectiveness projects have a similar first stage

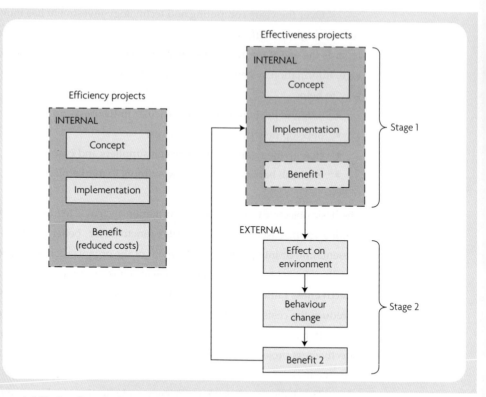

Figure 4.1: The benefits realization process

process which may result in some Stage 1 benefits. However, it is not necessary for any cost benefits to accrue at this stage. The second stage of the process is the effect of the implementation of Stage 1 on the environment and any resultant change in behaviour leading to Stage 2 benefits. Stage 2 benefits may be, for example, improved sales resulting in improved revenues.

The fact that there are two stages to the process unfortunately provides greater opportunities for the potential benefits to get lost or become dissipated and makes the evaluation of such effectiveness projects difficult. In the case of cost displacement the organization is more in control of the realization of benefits whereas in the improved services case (i.e. effectiveness justifications) the organization is in control of achieving the first stage but not the second stage which is the impact on others, for example suppliers, customers, and potential customers. This second stage has been the area where miscalculations are frequently made; for example, it is often assumed that the move from Stage 1 to Stage 2 is a logical and deterministic process: that an improved service will lead to more people buying that service, or that better information will lead to better decision making.

In some information systems development methodologies, a phase or stage relating to realizing projected benefits has been included to ensure that any measure implemented in the information system leads to the hoped-for change in behaviour. This is long term and difficult to prove and it has often been the case that organizations have abandoned any attempt at justification and instead rely on management perception of the benefits, sometimes referred to as strategic insight, intuition, or blind faith.

The use of information systems for effectiveness or competitive advantage has been shown to have the possibility of generating very significant rewards for the organization in terms of income generation rather than cost savings. There are some 'classic' examples which purport to show how some organizations 'have seized the opportunity to use information systems to gain competitive advantage' (see Eardley et al., 1996). In particular, American Hospital Supply Corporation's then 'revolutionary' 1978 order-entry system directly linked customers into their systems, which made ordering by the customer very easy. Customers had access to the order-distribution system so that, for example, customers could perform various functions, such as inventory control, for themselves using the system. This helped reduce costs for both the company and their customers, and enabled American Hospital Supply to develop and manage pricing incentives to the customer across all product lines.

There are a variety of other examples where information systems have been used to establish competitive advantage, such as DEC XCON, Xerox's customer support system, Merrill Lynch's cash management account system, American Airlines' Sabre reservation system (followed by Apollo), and the 'electronic' newspaper, USA Today. More recently there have been notable examples such as Dell Computers and Amazon who have made strategic investments in IT in attempts to change the nature of business in their particular industries, with Dell achieving significant success and Amazon less so, at least in terms of financial returns. In the UK there are also some well-known examples, such as Thompson's Tour Operator Reservation System, Sainsbury's laser scanning point of sale system, and First Direct bank.

It is from these kinds of early pioneering strategic examples that the concept of using information systems to gain competitive advantage has evolved.

Porter (1980) offers an industry analysis framework of competitive strategy to help identify the competitive forces that any company needs to consider. These five forces are illustrated diagrammatically in Figure 4.2 and Porter suggests that significant strategic advantage can be gained by diminishing supplier or customer power, holding off new entrants into the industry, lowering the possibility of substitution for its products, or gaining a competitive edge within the existing industry. This framework can be helpful in focusing the role of information systems to improve competitive positioning. Earl (1989), for example, identifies four ways in which Porter's model helps:

- it deals with the industry and competitive dynamics;
- it highlights that competition is not simply concerned with the action of rivals;
- it facilitates discussion and is based on sound principles of industrial economics;
- it focuses on the few dominant forces necessary.

Figure 4.3 shows an extension to Porter's work by Earl to illustrate the strategic role that IT can play on each of Porter's dimensions. American Hospital Supply, for example, illustrates the use of IT to address the customer competitive force. IT was used to help lock in the customers using both mechanisms. Another aspect of this case is the role of a dedicated champion to push the

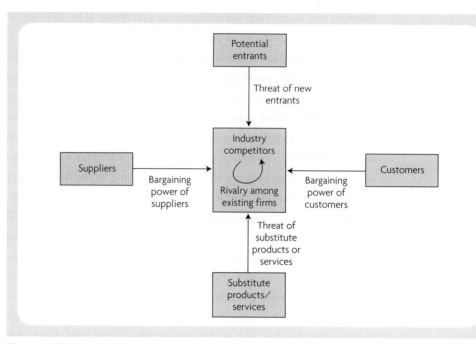

Figure 4.2: A framework of competitive strategy (adapted from Porter, 1980)

Competitive force	Potential of IT	Mechanism
New entrants	Barriers to entry	1. Erect 2. Demolish
Suppliers	Reduce bargaining power	1. Erode 2. Share
Customers	Lock in	1. Switching costs 2. Customer information
Substitute products/services	Innovation	1. New products 2. Add value
Rivalry	Change the relationship	1. Compete 2. Collaborate/alliances

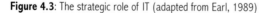

Figure 4.3: The strategic role of IT (adapted from Earl, 1989)

ideas in the organization, to overcome all the inevitable objections, to sustain the momentum over a fairly long period, and to inspire people with the vision. This person is unlikely to come from the information systems domain, although he or she may be a 'hybrid manager', that is, a person having knowledge and experience in both the IT and user domains. The role of people in successful information systems development is stressed further in the theme on partici-pation, discussed in Section 5.1.

Although many organizations now recognize the need to address effectiveness as well as efficiency in their use of IT and the resulting need for an IT strategy, the approach to developing such a strategy is by no means clear or universally agreed upon. Indeed, there has been a degree of scepticism that cast doubts on some of the more simplistic approaches to IT for competitive advantage. It has been argued that some of the successes are due to the fact that the product or service offered was good rather than having anything to do with the information system. Other criticisms relate to sustainability, i.e. even if the IT investment has generated a strategic benefit how long will it last? Can competitors copy the IT system easily and reproduce the benefits in their own organizations? Obviously not being able to sustain the advantage is problematic, and in such cases maybe all that has been achieved is an increase in the cost of doing business in that particular industry. It is clear that investing in IT to gain competitive advantage is not a simple panacea and some organizations have adopted rather simplistic approaches as follows:

- *Technology-driven model* – the reaction of some organizations has been implicitly to make the assumption that investment in IT will automatically result in business success and the achievement of competitive advantage. It is the embodiment of the view that if the technology exists it should be employed. It might be seen in the unthinking purchase of the latest IT product. This strategy is usually driven by technologists. A result of this approach is information technology that may not be appropriate for the needs of the organization and a lack of control over IT budgets. As no business benefits are stated, there is no way of measuring whether any such benefits accrue. This strategy is obviously irrational although understandable in some ways due to the rapid pace of technological change exceeding the ability of many managers to keep up. Technology adopted in this way may cause disruption in the organization and incompatibilities between information systems, rather than making for greater efficiency and effectiveness. For example, the introduction of the personal computer in many organizations was unplanned and could be described as technology-driven, and although this led to many individual benefits there were often other longer-term problems of compatibility, support, and integration. Strassmann (1990) concludes that there is little evidence that technology-based systems of information management have produced benefits that could be claimed to justify their costs. Even if the claimed (and unproved) individual gains are summed, they do not lead to an overall improved performance of the organization. It would thus appear that an approach based purely on the technology-driven model is unlikely to prove adequate. More recently a somewhat different argument has been put forward by Carr (2003) that IT is no longer strategic and that 'IT doesn't matter'. Carr argues that IT evolvement has been very similar to other earlier technologies such as the railways or electric power. For a short period when they are being developed they provide companies who are early adopters with strategic advantage but as the technology becomes cheap and ubiquitous, everyone adopts, and any strategic advantage is lost, i.e. the technology becomes a commodity. Commodities may be essential to competitiveness, they are necessary, but they provide no strategic advantage to a company and do not differentiate one company from another in any way. Carr argues that IT was no different to this and that IT has become a commodity and should be thought of in the same way as a utility such as water or electricity with the IT infrastructure just being the 'pipeline' through which information flows. The publication of Carr's paper caused a great deal of controversy with many putting counter-arguments with an ongoing debate ensuing (see for example Stewart, 2003).

- *Competitor-driven model* – an alternative model or approach that some organizations have adopted is to react to their competitors by copying them. There is evidence that this happens in some sectors rather more than others. In the UK banking sector, for example, there seems to be a 'knee-jerk' reaction to copy each other in terms of technology and services. In manufacturing, there appears to be an assumption that anything Japanese is by definition the best approach to follow. The competitor-driven model is an approach based upon the fear that an organization's competitors will use information technology to gain significant advantage over them. Therefore, they must be 'tracked' and copied at each stage of their development. The fear is that companies that do not follow the same path in information technology will ultimately be squeezed out. The competitor model says that we will not allow this to happen by matching our competitors at every point. The problems with this approach are threefold. First, that by simply following competitors, the organization will never innovate to its own particular strengths and advantages. Second, it may miss opportunities for being a leader itself. Third, it may still lose out by not itself being the first in the field, particularly in situations where being first in the field enables barriers to entry to be constructed.

A more thoughtful approach to the formulation of IT strategy and the alignment of IS development with business needs has been defined by Earl on the basis that no one-dimensional approach is likely to be successful. He suggests an approach that combines a variety of different elements and techniques, including both top-down and bottom-up analysis. The individual elements are not necessarily new but they are combined into what is argued to be a comprehensive and effective approach. Figure 4.4 illustrates this 'multiple methodology'.

The first element of the model is a top-down analysis of the business and its goals and objectives leading to an identification of the potential role that IT might play in achieving these objectives. This is a top-down business-led activity in which IT people would play a supporting role rather than the other way round. It is best achieved by the use of established techniques such as Critical Success Factors (CSFs) (see Section 15.2), SWOT analysis (see Section 15.5), and Porter's 'five forces' model (see above) to help assess competitive positioning, industry factors, competitor strengths, etc. with a view to using IT to address strengths and counter weaknesses.

The second element of Earl's model is a bottom-up analysis of the organization's current systems. This is a critical element of the model because, as in so many approaches to strategy formulation, the current systems are ignored and a 'green field' situation is assumed. This is usually quite unrealistic and leads to failure because the existing legacy systems in an organization have been ignored. The analysis consists of evaluating the strengths and weaknesses of

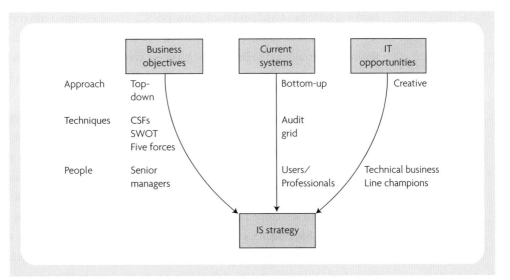

Figure 4.4: Earl's multiple methodology (adapted from Earl, 1989)

the existing IT provision and existing systems on the basis of, first, business contribution and value to users and, second, technical quality.

Figure 4.5 shows a typical systems audit grid that might be helpful in this analysis. One point of the analysis is to try to run down and exit from systems that perform poorly on both dimensions. It is frequently found that there are many such systems, particularly in organizations where IT has evolved in the organization in a somewhat *ad hoc* manner. These systems are often expensive to run and maintain, and contribute little to the organization in the direction that is required by the strategy. They should be phased out in order to free up valuable resources. On the other hand, those that are high on both dimensions should be enhanced and evolved. Earl makes the point that many strategic systems, or systems that have helped to provide competitive benefit, have in fact been based on enhancements and redirection of existing systems rather than the construction of totally new ones. Where this can be achieved it obviously provides a head start and potentially a reduction in the cost and lead times for developing strategic systems. An example is the telephone banking system of First Direct in the UK which built a strategic telephone interface system on top of the existing traditional retail

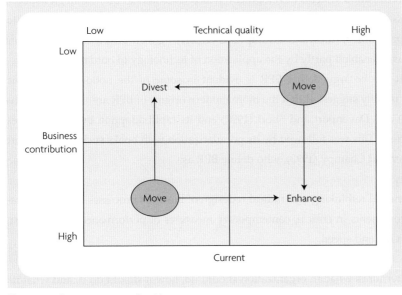

Figure 4.5: Current systems audit grid

banking systems of Midland Bank (now HSBC). Those systems with combinations of highs and lows on the grid need to be carefully evaluated and moved into one of the other boxes. Probably the systems evaluated highly on the business contribution dimension would have more potential to be moved into the 'enhance' category rather than those systems that are simply high on technical quality.

The third element of the model is the identification of IT opportunities. This is not just an across-the-board look at the potential of the latest technological innovation, as in the technology-driven model, but an attempt to assess the enabling effects of IT on the particular organization and business, given its strategic direction. Of course, in some cases IT can help to shape the strategic direction and therefore the left and right legs of the multiple methodology model must be developed iteratively. Earl suggests that this element is a creative one and that many of the best ideas for applying IT in an organization come from the people who understand the business on the ground, particularly at the customer interface, rather than the IT specialists, although the specialists can act as catalysts in this process. So, a range of people need to be involved in this part of the process. The three elements of the model thus provide a multidimensional, integrated strategy formulation approach which, it is argued, results in a coherent information systems strategy that supports and reflects the overall business objectives, addresses the potential that IT might provide, and considers the reality of existing IS provision.

The emphasis on strategic information systems and a top management view has led to a number of implications for methodologies. Many, such as Information Engineering (Section 21.3) have strategic planning as an important early phase in the development of an information system. In the next two sections we look at business process re-engineering and then at information systems planning.

4.3 Business process re-engineering (BPR)

One approach to information systems development, which takes into account strategic aspects, is business process re-engineering. It has presented organizations with the opportunity to rethink outdated procedures, rules, and assumptions underlying their business activities. This opportunity is usually enabled partly by the application of technology to outdated processes.

Although it can be argued that BPR is evident in some of the notions of Scientific Management, it is usually suggested that the more modern origins of BPR are the descriptions of Hammer (1990) and Davenport and Short (1990) and its rapid adoption by consultancies around the same time. This was followed by its popularization with books such as Davenport (1993) and Hammer and Champy (1993), who define BPR as:

> The fundamental rethinking and radical redesign of business processes to achieve dramatic improvements in critical, contemporary measures of performance, such as cost, quality, service, and speed.

Re-engineering determines what an organization should do, how it should do it, and what its concerns should be, as opposed to what they currently are. It is a radical approach and entails 'business reinvention – not business improvement, business enhancement, or business modification'. This early view of BPR is thus very different to incremental change (such as might be apparent when 'computerizing' clerical systems or 're-computerizing' computer systems), it is about fundamental change. Emphasis is placed on the business processes (and therefore information systems that reflect them and enable the change), but it also encompasses managerial behaviour, work patterns, and organizational structure. BPR is thus a total approach, involving top management, total organizational restructuring, and a change in the way people think. According to Hammer and Champy, organizations re-engineer for four main reasons:

- they face severe commercial pressures and have no choice;
- competitive forces present problems unless the organization takes radical steps to realign business processes with strategic positioning;
- management in the organization regard re-engineering as an opportunity to take a lead over the competition;
- publicity about BPR has prompted organizations to follow the lead established by others.

The essential first steps in planning a BPR programme are: to develop business vision and process objectives; to identify processes to be redesigned; to understand and measure the existing processes and to identify information systems levers that help push the changes; and finally to design and build prototypes.

The model of BPR created by Hammer and Champy describes the characteristics of re-engineered processes as follows. Several jobs are combined, performed by a 'case worker' responsible for a process. 'Case team' members are empowered to find innovative ways to improve service, quality, and reduce costs and cycle times. Due to process integration, fewer controls and checks are necessary, and defects are minimized by an entire process being followed through by those ultimately responsible for the finished product. Workers make decisions according to the requirements of the whole process. The steps in a process are performed in the order decided upon by those doing the work, rather than on the basis of fragmented and sequential tasks, and this enables the parallel processing of entire operations.

The outcomes of such BPR programmes often include:

- flatter organizational structures;
- greater focus on customers;
- improved teamwork, leading to a more widespread understanding of the roles of others.

As originally conceived in the early 1990s, BPR often necessitated the recruitment of re-engineering teams consisting of, for example, strategists in information systems, business analysts with computer skills, organizational development specialists, and organization and methods experts. Customer service teams were used to maintain and develop a focus on future business, and team 'facilitators' to coach team members, resolve conflicts, and rectify operational problems.

According to Davenport and Short (1990), information systems and BPR have a recursive relationship. On the one hand, IT usage was to be determined on the basis of how well it supported redesigned business processes. On the other, BPR was often enabled by information technology and information systems. The combination of information systems and BPR presented the key opportunity to change the way in which business was conducted radically. A successful BPR programme was fundamental enough to lead to sustainable competitive advantage for organizations brave enough to seize such opportunities.

However, the original enthusiasm for BPR has subsequently been tempered by a number of factors. The very radical approach, which often resulted in large numbers of redundancies, typified by the 'don't automate, obliterate' (Hammer, 1990) view of BPR has proved alienating. Grover (2001) has characterized this as the 'slash and burn mentality' of BPR, arguing that:

> Unfortunately, many corporations responded to reengineering by performing major work force reductions under the aegis of reengineering. Such efforts were not strategically driven, and led to firms losing vital components of the work force that reduced their ability to be creative and productive . . . Optimizing process at the cost of people has been a major problem of reengineering.

But perhaps the most significant factor is the reported failure rates of BPR initiatives. Wysocki and DeMichiell (1997) suggest that failure rates as high as 70 per cent are common. Even Hammer and Champy (1993) quote failure rates of 50–70 per cent, but in their view this is because companies and managers are not radical enough. They highlight senior managers' lack of ambition for radical change, and their unwillingness to embrace the BPR concept fully. Furthermore, they argue that managers fail to comprehend the degree of change required, not only in business processes, but also in managerial behaviour and organizational structure. Moad (1993) identifies lack of resources and senior management support as well as the time that it takes to achieve returns, which result in pressure to abandon the BPR programme.

These factors have resulted in something of a backlash against BPR. Critics have suggested that there is little new theory underpinning the ideas and that it has been usurped by consultancies for their own ends (Earl and Khan, 1994; Peppard and Preece, 1995; Grover and Malhotra, 1997).

In an IS context, BPR has been influential. It has led to a more radical look at IS as a whole in organizations, and can therefore be associated with systems thinking in this respect. Many organizations replaced a myriad of legacy systems (Section 8.1) with integrated enterprise resource planning systems (Section 9.3) as a way of 'solving' their IS problems.

BPR, according to Melao and Pidd (2000), is itself now undergoing re-engineering. They suggest that more recent literature (e.g. Davenport and Stoddard, 1994, and Burke and Peppard, 1995) has begun to 'soften' the radical approach to change associated with the early versions of BPR and they identify nine issues, or continuums, for thinking about BPR (see Figure 4.6).

In Figure 4.6, the first column indicates the issue or element of the continuum, the second column represents the traditional or classic view of BPR, and the third column is the newer, broader, and softer concept of BPR which they argue is now becoming the more

Issues	Classic	New
Novel vs Established	Conceptually new model	Linking existing models
Radical vs Incremental	Radical	Incremental
Clean slate vs Existing	Starting from scratch	Existing
Broad vs Narrow	Cross-functional	Not necessarily
IT-led vs IT-enabled	IT-led	IT-enabled
Mechanistic vs Holistic	Mechanistic	Holistic
Dramatic vs Modest	Dramatic	More modest
Top-down vs Bottom-up	Top-down	Bottom-up
Methodology vs Inspiration	Methodology	Inspiration

Figure 4.6: Framework of BPR (Melao and Pidd, 2000)

common approach. So, for example, the first identified element is 'novel vs established', where novel is the original BPR which was argued to require a 'conceptually new business model' whereas established is the newer view, which argues that it does not have to be novel but can be 'the linking together of existing approaches in a novel way'. The second element of the framework of analysis is 'radical vs incremental'. Here it is argued that in the new view of BPR, the improvements or benefits need not be completely radical but can be incremental. Third is 'clean slate vs existing' with the classic BPR typically requiring organizations to start from scratch, whereas the new BPR can build on existing initiatives. The fourth is 'broad vs narrow' which is normally interpreted as to whether the area of concern is cross-functional (broad) or whether it can be narrower, and so on. Grover (2001) adopts a similar view and states that:

> We now use the term 'business process change' to reflect the importance of process instead of radical change. The strong positions of 'radical change,' 'core processes,' 'top-down,' 'break-through performance,' and so on, are giving way to the reality that there is more than one way of conducting change. Incremental and continuous approaches with bottom-up involvement within functions might be appropriate for some companies and not for others. Classical reengineering might be appropriate for others.

Grover also relates this newer approach to BPR with Total Quality Management (TQM) initiatives and suggests that these approaches, which were diverging just a couple of years ago, now seem to be converging. The point is that BPR is changing from its original concept and is being adapted by organizations to fit new circumstances and the concept, which still has merits, is now much more flexible.

4.4 Information systems planning

Information systems planning is an essential aspect of developing successful systems. In this section we focus on strategic planning and alignment of information systems with the overall strategic direction of the organization. Rather than look at individual applications and subsystems in detail, strategic planning needs to involve the top management of the organization (the managing director, finance director, operations director, and so on) in the analysis and determination of the objectives of that organization. Management should assess the possible ways in which these objectives might be achieved utilizing the information resource. The approach therefore requires the involvement of strategic management in planning information systems.

Because of the requirement to develop an overall plan for information systems development in the organization (within organizational planning as a whole), there are obvious links with systems theory (Section 4.1), and because of its emphasis on the role of strategic management, there are also obvious links to information systems strategy (Section 4.2). Further, because of this top management involvement, there are also links with the participative approach (Section 5.1) as well as business process re-engineering (Section 4.3).

Planning approaches are designed to counteract the possibility that information systems will be implemented in a piecemeal fashion, a criticism often made of applications in the past. A narrow function-by-function approach could lead to the various subsystems failing to integrate satisfactorily. Further, it fails to align information systems with the business strategy. Both top management and IS personnel should look at organizational needs in the early stages and develop a strategic plan for information systems development as a whole so that information systems are integrated and compatible. This becomes a framework for more detailed plans. Individual information systems are then developed within the confines of these plans. Plans could be made at three levels:

1. Long-term planning of information systems considers the objectives of the information systems function and provides rough estimates of resources required to meet these needs. It will normally involve producing a 'mission statement' for the information systems group, which should reflect that of the organization as a whole. The information systems plan at this stage will be an overview document, for example, providing only prospective project titles.

2. Medium-term planning concerns itself with the ways in which the long-term plan can be put into effect. It considers the present information requirements of the organization and the information systems that need to be developed or adapted to meet these needs. Information about each potential information system will be spelt out in detail, including the ways in which they address the overall strategic objectives of the organization. The ways in which the information systems are to be integrated will be stated. Priorities for development will also be established and again these will reflect the long-term plan and mission statement. A planning document will usually will be produced which shows the current situation along with an action plan for future development.

3. Short-term planning, perhaps covering the next 12 months, will provide a further level of detail. It concerns the schedule for change, assigning resources to effect the change, and putting into place project control measures to ensure effectiveness. As well as detailing the resources required for each application in terms of personnel, hardware, and budget, it will contain details of each stage in the development process as suggested in the systems development life cycle described in Chapter 3.

As indicated there is a need for information systems to address corporate objectives directly and so planning should ensure that management needs are met by information systems. Sometimes information systems are designed around management needs, a sort of 'top-down' approach, ignoring operational needs which are assumed to be fulfilled elsewhere. Managers may also set standards for information systems, and one of these requirements will be choosing an approach and a methodology for developing information systems.

There are many general information systems planning approaches, often utilizing techniques such as Critical Success Factors (Section 15.2) and SWOT (Section 15.5). Business analysis might be the first stage of the development of an information system and involves the creation of an information systems strategy group to undertake the following tasks:

- an assessment of the strategic goals of the organization, which could be long-term survival, increasing market share, increasing profits, increasing return on capital, increasing turnover, or improving public image;
- an assessment of the medium-term objectives to be used as a basis for allocating resources, evaluating managers' performances, monitoring progress toward achieving long-term goals, establishing priorities;
- an appreciation of the activities in the organization, such as sales, purchasing, research and development, personnel, and finance;
- an appreciation of the environment of the organization, that is, customers, suppliers, government, trade unions, and financial institutions, whose actions will affect business performance;
- an appreciation of the organizational culture relating to values, networks, and 'rites and rituals';
- an appreciation of the managerial structure in terms of the layers of management or matrix structure, types of decision made, the key personnel, and types of information needed to support the key personnel in their decision making;
- an analysis of the roles of key personnel in the organization.

With this knowledge, it is possible to assess the type of information that an information system might provide. The first stages of Information Engineering (see Section 21.3) are also concerned with planning aspects.

Addressing a more detailed later stage in the planning process, Lederer and Mendelow (1989) suggest a number of guidelines which should be considered when planning:

- *develop a formal plan* – set objectives and policies in relation to the achievement of organizational goals and thereby enable the effective and efficient deployment of resources;
- *link the information systems plan to the corporate plan* – provide an 'optimal project mix' which will be consistent with and link to the corporate plan, ideally over the same time period;
- *plan for disaster* – ensure that dependencies are identified and damage likelihood identified;
- *audit new systems* – evaluate present systems to identify mistakes and hence avoid their repetition and to identify areas where a small resource input might have led to a larger benefit;

- *perform a cost/benefit analysis* – identify intangible and tangible benefits and costs before putting in the required resources;
- *develop staff* – make use of and develop the skills of staff;
- *be prepared to change* – as relationships, structures, and processes change;
- *ensure information systems development satisfies user needs* – understand the tasks and processes involved to establish the true user requirements;
- *establish credibility through success* – build up user confidence through previous success, thereby promoting co-operation and lowering barriers.

The effective strategic planning of information systems and the close alignment of IS with the overall business strategy is key to organizational success. However, many methodologies, including the traditional life cycle approach, failed to emphasize this or have a strategic planning activity in their development phases. As a result, planning sometimes got ignored and information systems failed to be as effective in organizations as they might otherwise have been.

4.5 Stages of growth

The Stages of Growth (SoG) concept has been around in economics, marketing, organizational studies, and many other areas for a long time. The concept in outline is that a country, an organization, a department, a product, an individual, or whatever, is perceived to go through a number of discernible and distinct phases, over time, before it can reach a particular point of development, or maturity (they are sometimes also called Maturity Models). Usually each of the identified stages has to be gone through in order for maturity to be achieved, although sometimes there can be shortcuts in particular circumstances. Sometimes the SoG are simply an observation based on empirical studies, i.e. it appears that these are the stages of development or growth which in general the entity goes through. But it is also frequently used to prescribe the stages that the entity *should* go through, and this can be much more contentious.

In relation to information systems and information management, there are a number of SoG models. Most of these relate to the stages that organizations go through in their use, management, and experience of IT. Probably the first was Gibson and Nolan's (1974) model, which identified four stages of growth in the use of IT before maturity was reached (see Figure 4.7).

The stages closely reflect the relative amount spent on IT in an organization, as a proportion of their size, but this also happens to reflect the development and learning path in their utilization of IT. This development and expenditure was argued to follow an 'S-shaped curve' that could be divided into four stages: initiation, contagion, control, and integration. There are four elements that have to be tracked to identify where an organization is in terms of this model:

- *Scope of the applications portfolio* – typically an organization would start its IT development by implementing mainly financial and accounting applications. This would then be increased until most major functions would be supported by IT systems, through to the development of information systems to support management and decision making in mature organizations.

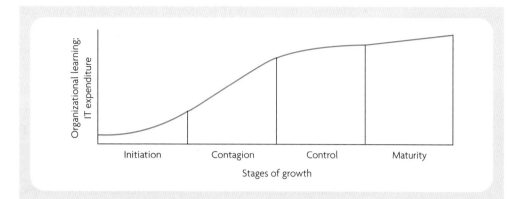

Figure 4.7: Gibson and Nolan's four stages of DP growth (Galliers and Sutherland, 1991)

- *Focus of the IT organization* – initially this would be centralized and IT-focused and move to decentralized and data resource-focused in maturity.
- *Focus of IT planning and control* – initially this would be an internal focus through to a more external focus in maturity.
- *Level of user awareness* – this concerns the source of the IT initiatives in the organization. Initially users would be reactive to what was on offer from the IT department, then user awareness would move to users becoming the drivers of IT initiatives, through to a partnership in maturity.

Subsequently, Nolan added another two stages, data administration and integration, thereby creating a 6-stage model (Nolan, 1979). Nolan's SoG model was used to identify not only where organizations were in terms of the four stages but more importantly how they should develop in order to achieve maturity. The model became very influential and a prescription for the development of the IT function in organizations in terms of structure, activities, management, etc. It was widely used by consultants and consultancy companies to drive what organizations needed to do to be mature in their use of IT.

However, the model has been criticized in a number of ways. First, it was not empirically proven, i.e. it did not originate from a quantitative study of IT organizations. It was based on what the authors had observed in organizations in their consulting roles but could not be proved. Second, not everyone agreed that this view of maturity was in fact desirable. For example, it seemed to imply ever-increasing expenditure on IT. Third, it was based on a particular view of technology at the time and a view of the IT department. This was related to 'the exploitation of a single broad strand of technology for business data processing' (Feeny et al., 1996). However, Gibson and Nolan did recognize that there would be other 'S-shaped curves' relevant to new technologies as they emerged.

Other SoG models have emerged. For example, Hirschheim et al. (1988) have evolved the Nolan model but focus on IT delivery, management, and leadership. They identify three stages:

- *Delivery* – the focus of IT is internal, reactive, and is concerned with delivering applications on time and within budget to build credibility.
- *Reorientation* – the focus moves to the exploitation of IT to support the business (what is called using IT to generate strategic advantage).
- *Reorganization* – the focus is now on relationships with the business, and ideally a partnership or coalition with the business. The application portfolio is balanced, depending on strategy and need, and the structure is probably federal with IT both in the centre and in the business units, according to what makes the most sense.

These are distinct SoG and need to be progressed through. So, for example, when IT cannot deliver (i.e. deadlines are missed, the costs run out of control, and the applications are not delivering the required functionality), there is a kind of pre-stage or Stage 0. Here, there is no possibility of IT having a relationship with the business. There is no credibility with senior business managers. Therefore, the 'delivery' stage concentrates on addressing this issue, getting to the position where the basic provision of applications can be trusted and relied upon. Once this is achieved consistently, then the IT department can be taken seriously and move to 'reorientation'. Here critical, strategic, applications can be considered and developed. IT is aligned with the business, and the IT function is now typically led by a business person who can drive the new business focus. Once these applications are achieved and successful, then the IT needs to be 'reorganized' for an overall perspective and a sustainable steady state. IT is now critical to the business and may be represented at Board level as a partner with the business. IT is integrated into the business units where appropriate, and certain functions are kept central, e.g. standards, procurement, IT careers, etc.

So, a number of different SoG models exist, and continue to be developed. Galliers and Sutherland (1991) have revisited and revised some of the earlier models and come up with the following stages:

- *ad hoc*racy;
- starting the foundations;
- centralized dictatorship;
- democratic dialectic and co-operation;
- entrepreneurial opportunity;
- integrated harmonious relationships.

Stages of growth models have also been applied to specific areas of IS; for example, IS planning and strategy, the development of end-user computing (EUC) (see Section 5.2) and information centres (IC) (also see Section 5.2). More recently SoG models have been developed to assess electronic business in organizations and the implementation and exploitation of ERP systems (see Section 9.3) (Holland and Light, 2001).

The related area of innovation and diffusion of technology (see Rogers, 1995) has also been conceptualized in terms of stages; for example, Brancheau and Wetherbe (1990) used innovation and diffusion concepts in connection with the adoption of spreadsheet software, and Cooper and Zmud (1990) in connection with the adoption and infusion of Materials

Requirement Planning. The Capability Maturity Model (CMM) of software is also essentially an SoG model and is described in Section 26.3.

4.6 Flexibility

The term 'flexibility' is used in a variety of different contexts. It is variously used to describe the inherent properties of a system, the nature of a change or implementation process, and a property of business, information, or other strategies. The term is also used in distinct senses, such as describing systems that can respond effectively to planned or unanticipated changes either where the response leads to changes in the nature of the system (flexibility is 'designed in') or where the response involves the organization changing or transforming itself (a characteristic of an 'organic' rather than a 'bureaucratic' organization). It has always been an important theme in information systems development, because computer applications have proved very poor at responding to changes in the environment that they are supposed to represent.

The flexibility of a system is typically described in terms of dimensions, such as speed of response and the range of activities that the system can perform. Implicitly, both dimensions are judged relative to some norm; for example, a *flexible* manufacturing system can do more things than 'typical' manufacturing systems. In general terms, flexible management systems may include devolved responsibility for decision making, the appropriate size of teams, a culture that can cope with broadly defined roles and *informated* team members (Zuboff, 1988).

Technical systems requisites for flexibility may include initial design, where flexibility is designed into the system, and appropriately trained operators. There may, however, be trade-offs between flexibility and other system characteristics. These trade-offs may be with political realities within an organization, system complexity, effectiveness in relation to particular objectives, the need for standardization and other factors. There may be a trade-off between organizations' needs to have uniform data standards and the need for local discretion in the face of unpredictable demands.

Flexibility is nearly always seen as a 'good thing'. It appears to have three broad advantages. First, it improves the quality of internal processes in ways that may offer a variety of performance improvements. Advantages accruing might include higher staff morale. Second, it may give firms a competitive edge, for example, through the speed of response to an unexpected increase in sales orders, which other firms could not meet. Third, it is part of the 'survival kit' of an organization. It may be that a measure of flexibility is necessary in a turbulent world: 'be flexible or cease operating'. As a number of writers have noted, the acquisition of flexibility is not without costs, and these need to be compared with the likely benefits. In terms of information systems, real costs may include hardware, software, training, reorganization, and ongoing costs.

Flexibility is a useful attribute for any organization operating in an uncertain environment. Different aspects of flexibility are apparent in related concepts such as adaptability, resilience, robustness, agility, versatility, manoeuvrability, and responsiveness. Resilience suggests an emphasis on the ability to cope with uncertainty and to withstand shocks, without fundamentally changing products or processes. However, the concept of flexibility is most closely related to, but distinguishable from, robustness.

Rosenhead (1989) has defined robustness as the extent to which future options are available. The implication is that leaving future options open is a desirable thing. He contrasts robust decision making with conventional decision making. It may be that outcomes are too far in the future to know what the aims and objectives of the decision maker may then be. Under such circumstances, it can be argued that a good decision is one that retains, as reachable, as many outcomes as possible. Concepts like adaptability, agility, and versatility suggest an ability to respond more positively when faced with a changed environment. Described in these terms, flexibility is an attribute that can enable an organization to react to developments. But flexibility and associated concepts can also describe an ability to prepare for and manage an uncertain future in a more proactive way. Therefore, flexibility can involve actions before or after events and involve either defensive or offensive measures. From this perspective, Evans (1991) identifies four archetypal manoeuvres which provide strategic flexibility:

- *pre-emptive manoeuvres* – creating options, inflicting surprise, or seizing initiatives;
- *protective manoeuvres* – insuring against losses, hedging, or creating buffers against adverse conditions;
- *corrective manoeuvres* – the ability to recover from adverse situations and learn from mistakes;
- *exploitive manoeuvres* – capitalizing on opportunities and consolidating advantages.

There are a number of circumstances in which an organization might wish to design any information system robustly or flexibly; for example, where:

- *The designers do not know what the information system is to do.* If the user/analyst team is unable to define precisely, or even imprecisely, what it is the system is likely to be called upon to do, then a flexible system, which can easily be adapted to a variety of purposes, seems potentially desirable.
- *The organization may not know how user requirements will develop after the information system has been introduced.* Traditionally, system builders believe their responsibility for the system ends with its initial implementation. However, they should be committed to developing the system beyond this time, and a flexible system, with the potential for development in various ways, is desirable because the introduction of an information system may itself be the catalyst for organizational change or development of requirements.
- *The designers may not be certain that the chosen development methodology will deliver precisely what is required.* Even if there is a precise specification of requirements, and a guarantee that these requirements will never change, there may be some doubt as to whether such a system can be delivered first time. To prevent missing the target specifications, or commencing a lengthy and costly redevelopment exercise, a flexible system, which can be altered fairly easily, is desirable.

Thus, there appear to be two essential features that make flexibility desirable: ill-defined and shifting system requirements, and mismatches between system performance and requirements. In both cases the ability to change system performance would be helpful.

Successful business strategies are often characterized by initial flexibility, vagueness, and sometimes deliberate ambiguity. In recent years, partially to cope with this, information system developers have moved away from rigid methodologies typified by the systems development life cycle (SDLC) approach, though not always successfully. For instance, Multiview (Section 26.1) emphasizes flexibility in information systems development.

Strategy needs to be most flexible when the organization operates in an unstable or turbulent environment. However, there is evidence that information technology (IT) increases environmental turbulence (indeed, it could be argued that this may be a primary motive for investment). This may, in turn, cause strategy to be couched in more nebulous terms. There is evidence both that successful firms have flexible strategies and that this stems, to some extent, from senior management preference for qualitative language which would preclude explicit statements of objectives. Flexibility is necessary in a fast-changing environment but may not be so in a stable one. The conditions under which flexibility is valued will be contingent on factors such as the rate of change of the environment and the values held by senior executives.

The need for flexibility arises partly from the nature of the strategy-making process which contains both objective and subjective elements. Such decision making has to fulfil a set of conflicting objectives, including the need to measure and combine objective and subjective aspects, to forecast future uncertain events, and reconcile different interest groups. Problems arise due to the evolutionary nature of systems. If evolutionary systems are difficult to evaluate then it seems likely that evaluation in terms of changing strategy will be even more problematic.

It may be that flexibility has a further role: as a critical success factor (Section 15.2) in its own right. The vision of senior management, not just their support, may enhance or determine success rates. Since competitive advantage comes from uniqueness, methodologies are likely to be a constraint on the identification and exploitation of such attributes.

Although some of the maintenance effort is spent on correcting systems poorly specified in the first place, enhancement maintenance, that is, adapting the systems to accommodate changes in the specification, has always taken the lion's share of total maintenance. According to Ward (1985), 98 per cent is devoted to enhancement maintenance. Software engineering techniques (Section 8.2) improve the quality of the software, but not the systems analysis. It is obvious, therefore, that time spent identifying possible changes in the investigation phase of the information systems development life cycle would be time well spent.

There are different ways to consider future requirements. They can be considered as new requirements that will arise in the future. This is the conventional way of viewing them. This is not very helpful as it implies high unpredictability, almost suggesting that it would be a waste of time to try to account for them at the design stage. More helpful, perhaps, is to consider them as existing requirements that will come to light in the future. This might suggest that if every effort was made, every technique followed, and every avenue explored, it may be possible to identify them before the design is set. In any case, new requirements that will arise in the future can also be considered as 'current requirements', because an information system should be designed so that change can be facilitated. In other words, systems should have inbuilt flexibility. Either way, the identification of future requirements and their incorporation in the new information system will reduce the maintenance burden following implementation. This suggests two things. First, that it is possible to identify future requirements at an earlier stage and, second, that systems should be designed so that it is easy to make changes to cater for requirements that are identified after implementation.

Information systems should be capable of incorporating a range of possible or most probable futures. This emphasis should diminish the maintenance problem by making change easier to incorporate. However, this may have the effect of shifting costs from maintenance to development. With these costs often coming from different budgets and development costs being, to some extent, avoidable while maintenance costs are not, the effects of making projects appear more expensive need to be carefully handled.

The difficulty of identifying future requirements discussed above and the inherent inflexibility of most information systems may lead to information systems failure or obsolescence. Later in Part IV, we discuss some techniques that might be used to help identify future requirements. In particular, we consider future analysis, risk analysis, and lateral thinking. All the techniques are concerned with reducing uncertainty.

4.7 Project management

One of the goals of an information systems development methodology is to manage information systems development projects efficiently and effectively. So, to some extent, project control is a theme that pervades methodologies themselves and the entire book. After all, methodologies were introduced largely to manage an information systems development project and to try to make the mantra 'within cost and on time' more likely, therefore reducing the risk of failure. To this we should add: 'with the required functionality and quality'.

In Chapter 14 we look at techniques, such as Gantt charts, work breakdown structures, networks (critical path or PERT analysis), CoCoMo, and function point analysis, which all attempt to help control project development. In Section 18.4, we show the features of Microsoft Project, which is a software tool for project control. A project control package can help to ensure that projects are scheduled at the earliest possible date, with the least drain on resources. Many of the methodologies discussed in Part VI, in particular those designed for large projects (e.g. SSADM, Merise and RUP), emphasize project management and control aspects. Further, PRINCE (Section 25.4) is specifically designed as a methodology supporting the management of projects. When they are used together, SSADM defines the technical work to be done and PRINCE defines the project management aspects. In this section, we look at the issues concerned with project management and control in general.

The aim of project management is to deliver the project on schedule, within budget, complete, and of the highest quality. Poor project management is a major reason for the failure of many information systems projects. Indications of poor project management might be a lack of identification and control of the activities in a project, poor estimation of the human, time, cost, and other resources required for successful completion, resistance to change, and a lack of focus on the deliverables in detail and how success can be measured.

Much of what we consider 'normal' in a methodology – for example, defining the scope, agreeing times and deadlines, allocating human resources, regular progress reporting, ensuring quality control – are within the scope of project management. They are important issues, whether the project concerns a large tailor-made system or a smaller one centred on an application package, whether the system is developed by technologists or users, indeed whatever the project. Further, project management is also closely linked with IS planning (Section 4.4).

Estimation of the resources required is difficult unless the analysis of the work involved is completed in detail. By breaking up the large project into smaller tasks and then again into activities, the process is easier. Estimation of times and costs for each activity can be based on experience of similar activities on other occasions, an average of estimates from experienced people, and by using techniques, such as CoCoMo and function point analysis (see Section 14.1).

This work breakdown structure can go down through a number of levels. PRINCE facilitates the process of breaking up the work into a series of lower levels. The analyst can better estimate the resources required for each activity because it is smaller, more tangible, and probably within the analyst's experience. The analyst also needs to work out whether some tasks need to be done on completion of other tasks or whether some tasks can be done in parallel. With all this information, and with the possible help of project management tools, it is feasible to start estimating the time and resources required for the project. Information about holidays, other interruptions such as possible sickness leave, and some slack, need to be incorporated to make the estimate more realistic. Even so, it will not be 100 per cent accurate, so the analyst may suggest a range in the estimate, from most optimistic, through most likely, through to most pessimistic.

Project management implies the existence of some sort of project manager or management team. This role requires good leadership skills, communication skills, administrative skills, and technical competence, as well as a position that is senior enough to command respect. In most organizations, there is a person allocated to this role for each project and this project leader reports to the steering committee and senior management. A member of the senior management on the steering committee may be nominated project manager and act as 'champion' of the project.

As seen by the list of requirements, the project leader role can be very demanding. Part of this role concerns people: getting people working together on the project, and getting people committed to the change, building confidence, and avoiding resistance. Change is never easy for people, so managing the change is a key aspect of the project leader's task. Much will depend on the experience that individuals have had with change in the past, which collectively is part of the organization's 'culture'.

We saw in our discussion of the SDLC in Chapter 3 that methodologies tend to break up projects into phases, subphases and tasks. Project management is partly about estimating the resources required for each (and hence compounded for the whole project) and ensuring that this forecast is achieved to the required quality as the project develops. A lot of the estimating will be done in the feasibility stage, and these estimates will form one basis for management deciding to go ahead with the project.

Identifying critical success factors and ensuring that they are thought about in every stage of the information systems project is also an important part of project management. Sometimes this aspect is distinguished through a **benefits realization programme**. People in the project team may be set aside to ensure that benefits are being realized and there may also be quality managers assigned as well, to ensure that deliverables are achieved at the right quality level. These roles add to those of the project manager, team leader, representatives from the user community, managers, directors, and others.

Of course, the project leader is concerned about ensuring all 'deliverables' are indeed delivered at the agreed levels of quality. However, the project leader should also ensure that the resources allocated to the project are realistic enough to achieve these goals. These resources include people (who have the required experience and skills for the role or training required), infrastructure (in terms of office space, hardware and software tools, training programmes) as well as an appropriate methodology.

We have mentioned the issue of quality without specifying what is meant by this potentially ambiguous term. **Quality** means that the product conforms to the requirements defined for it. An information system should have defined quality expectancies for performance (e.g. for speed of processing, the time to process 10,000 transactions), efficiency (defined levels of storage and other resources it will use), and reliability (e.g. mean time between failure of the system). Some aspects of quality may be more difficult to measure, for example, ease of use, though even here there may be some level of customer satisfaction defined through Likert scale tests (say, an average minimum of 4 on a scale of 1=bad and 5=good). 'Quality' needs to be on the agenda throughout the development of a project – a 'total quality approach'. There are quality standards, such as those of the International Standards Organization (ISO), that should be adhered to, though reaching these standards is not sufficient (nor is following a methodology, including PRINCE, a guarantee that the standard has been met).

An important aspect of project management is that of progress monitoring: the ability to reveal a separation of actual progress compared to the estimate. This will be revealed using the tools discussed in Part V (e.g. Microsoft Project). In that event, managers will be able either to make corrective actions, such as increase resources to make the estimate more likely, or to reschedule the project. If the decision is to 'accept the inevitable', at least management have been forewarned. It is obviously better than to discover there will be a delay only at the time the system was due to be operational. Further, through reporting, this is communicated to all concerned, leading to trust and understanding rather than alienation.

However, it is obviously better to have a realistic project plan and keep to that plan. In that regard, an assessment of the risks associated with the plan represents an important aspect. Risks (see Section 15.7) need to be identified, their likelihood and impact estimated, measures to counteract the risk identified, and measures to be put in place if the event occurs agreed. The latter may include risk transference procedures, such as taking out insurance or imposing penalty clauses on suppliers.

But this suggests a somewhat mechanical view of project management. Good project management has a lot to do with people management as well. It requires members of the team implementing the change to be well motivated. It also requires good management of those affected by the change. Inadequate attention to change management is one of the main reasons for project failure. Clients and users should not be disappointed, and so managing expectations so that they are realistic is important. New information systems are likely to affect the people and culture of the organization and can be gradual/incremental or sudden/transformational. The project manager needs to identify the type of change and the forces working for and against the change, listed under people, resources, time, external factors, and corporate culture. From that information it is possible to formulate an appropriate strategy corresponding to that analysis, indeed many of the arguments discussed earlier in this chapter on strategy relate also to project management. Conflicts are bound to occur, although paying heed to the above will reduce their number, and good negotiating skills could prove crucial. All this would suggest that strong top management support, good leadership and communication skills, as well as good project management processes are key to the effective management of projects.

Summary

- Systems theory has had widespread influence in information systems work. It suggests a holistic approach to viewing organizations rather than a scientific approach.

- Information systems strategy is about the way in which the organization sees the role of information systems in the company and the general attempt to identify better ways of doing things, leading to increased revenues, greater functionality, better products and services, improved presentation or image, improvement to the organization's competitive positioning, etc. to gain competitive advantage. The overall aim is to emphasize effectiveness rather than merely efficiency.

- Business process re-engineering is the fundamental rethinking and radical redesign of business processes to achieve dramatic improvements in critical, contemporary measures of performance, such as cost, quality, service, and speed. In its more recent guise, it has itself been re-engineered and it is less radical.

- Information systems planning involves top management assessing the possible ways in which their objectives might be achieved utilizing the information resource and therefore in planning information systems.

- The stages of growth model shows how information systems in an organization tend to go through a number of discernible and distinct phases, over time, before it reaches a particular point of development, or maturity. One model suggests that it develops from initiation, to contagion, control and finally to maturity.

- Flexibility in information systems is seen as an important positive attribute. It is typically described in terms of dimensions, such as speed of response and the range of activities that the system can perform.

- Project management is a theme that is important in most methodologies. It is about managing information systems development projects efficiently and effectively, delivering the project on time, within budget, complete and of the highest quality.

Questions

1. In what ways are the themes of this chapter 'organizational'? What links these themes and what separates them?

2. What is 'strategic' about strategic information systems?

3. Discuss why business process re-engineering has been softened or toned down. Do you think this change has reduced its potential?

4. For an organization of your choice, identify the 'stages of growth' that it passed through and discuss whether these are similar to any SoG model discussed in the text.

5. Discuss the difficulties related to making information systems flexible so that implementing future change is easier.

6. How are large projects controlled in your organization? Address the question in relation to the role of people, techniques and software.

Further reading

Cadle, J. and Yeates, D. (2001) *Project Management for Information Systems*, Prentice Hall, Harlow.

Checkland, P. and Scholes. J. (1990) *Soft Systems Methodology in Action*. John Wiley, Chichester.

Currie, W.L. and Galliers, R.D. (1999) *Rethinking Management Information Systems*, Oxford University Press, Oxford.

Earl, M.J. (ed.) (1996) *Information Management: The Organisational Dimension*, Oxford University Press, Oxford.

Galliers, R.D. and Sutherland, A.R. (1991) Information systems management and strategy formulation: the 'stages of growth model' revisited, *Journal of Information Systems*, Vol. 1, No. 2, 89–114.

Hammer, M. and Champy, J. (1993) *Re-engineering the Corporation: A Manifesto for Business Revolution*, Harper Business, New York.

Melao, N. and Pidd, M. (2000) A conceptual framework for understanding business processes and business process modelling, *Information Systems Journal*, Vol. 10, No. 2, 105–129.

5 People themes

5.1 Participation

We now turn to another major theme, that of the people concerned in the development of an information system, particularly users and other stakeholders; that is, people having a stake in the information systems project. In the traditional systems analysis methodology, the importance of user involvement was frequently stressed. However, the computer professional was the person who was making the real decisions and driving the development process. Systems analysts were trained in, and knowledgeable of, the technological and economic aspects of computer applications but far more rarely in the human (or behavioural) aspects which are at least as important. The end-user (the person who is going to use the system) frequently felt resentment, and top management did little more than pay lip service to computing. The systems analysts may be happy with the system when it is implemented. It may conform to what the systems analysts understand are the requirements and do so efficiently. However, this is of little significance if the users, who are the customers, are not satisfied with it.

The strategic view of information systems highlighted the necessity for top management to play a role in information systems development. The approach discussed in this section highlights the role of all users who may control and take the lead in the development process. If the users are involved in the analysis, design, and implementation of information systems relevant to their own work, particularly if this involvement has meant users being involved in the decision-making process, these users are more likely to be fully committed to the information system when operational. This will increase the likelihood of its success. Indeed, in some Scandinavian countries such a requirement may be embodied in law, with technological change needing the approval of trade unions and those who are to work with the new system.

Some information systems may 'work' in that they are technically viable, but fail because of 'people problems'. For example, users may feel that the new system will make their job more demanding, less secure, will change their relationship with others, or will lead to a loss of the independence that they previously enjoyed. As a result of these feelings, users may do their best to ensure that the computer system does not succeed. This **aggression** may show itself in attempts to 'beat the system'; for example, by 'losing' documents or even by more obvious acts of sabotage. Frequently it manifests itself in people blaming the system for causing difficulties that may well be caused by other factors. This is sometimes referred to as **projection**;

that is, they project their problems on to the system. Some people may just want to have 'nothing to do with the computer system', a kind of **avoidance** tactic. In this kind of situation, information systems are unlikely to be successful or, at the very least, fail to achieve their potential.

These reactions against a new computer system may stem from a number of factors, largely historical, but the proponents of participation would argue that they will have to be corrected if future computer applications are going to succeed and that it is important that the following views are addressed:

- users may regard the IT department as having too much power and control over other departments through the use of technology;
- users may regard computer people as having too great a status in the organization or not seeming to be governed by the same conditions of work as the rest of the organization;
- users may consider the pay scales of computer staff to be higher than their own and that the poor track record of computer applications should have led to reduced salaries and status, not the opposite.

These are only three of the arguments. Some views are valid, others less so, but the poor communications between computer people and others in the organization, symptomatic of the prevalence of computer jargon, have not helped. Training and education for both users and computer people can help address the cultural clash between them, as can time spent in user and IT departments for all. Somehow these barriers have to be broken down if computer applications are really going to succeed.

One way to help the process of breaking down barriers and to achieve more successful information systems is to involve all those affected by computer systems in the process of developing them. These are the stakeholders of the new system (see Section 16.1). This includes the top management of the organization as well as operational-level staff. Until recently, top management have avoided much direct contact with computer systems. Managers have probably sanctioned the purchase of computer hardware and software but have not involved themselves with their use. They have preferred to keep themselves at a 'safe' distance from computers. This lack of leadership by example is unlikely to lead to successful implementation of computer systems: managers need to participate in the change and this will motivate their subordinates. With the implementation of executive information systems (information systems designed for the use of top managers) and the like, this attitude is becoming less prevalent.

Attitudes are also changing because managers can see that computer systems will directly help them in their decision making. The widespread use of PCs and the information about the technology available in newspapers and other sources has also diminished the 'mystique' that used to surround computing. Earlier computing concerned itself with the operational level of the firm; modern information systems concern themselves with decision support as well, and managers are demanding sophisticated computer applications and are wishing to play a leading role in their development and implementation.

Communications between computer specialists and others within the organization also need to improve. This should establish a more mutually trusting and co-operative atmosphere. The training and educating of all staff affected by computers is therefore important. In turn, computer people should also be aware of the various operating areas of the business. This should bring down barriers caused by a lack of knowledge and technical jargon and encourage users to become involved in technological change.

Another useful way of encouraging user involvement is to improve the human–computer interface. There are a number of qualities that will help in this matter. These include visibility, simplicity, consistency, and flexibility:

1. *Visibility*. This has two aspects. First, it means that the way that the system works is seen by the users. This aspect is related to participation, because, if users understand the system, they are more likely to be able to control it. Second, visibility means providing information on the current activity through messages to the users so that they know what is happening when the system is being run.

2. *Simplicity*. This means that the presentation of information to the users should be well structured, that the range of options at each point is well presented, and that it is easy to decide on which option to choose.

3. *Consistency*. This means that the human–computer interface follows a similar pattern (sometimes referred to as 'look and feel') throughout the system. Indeed, wherever possible, all systems that are likely to be used by one set of users should follow this pattern.

4. *Flexibility*. This means that the users can adapt the interface to suit their own requirements.

User involvement should mean much more than agreeing to be interviewed by the analyst and working extra hours as the operational date for the new system nears. This is 'pseudo-participation' because users are not playing a very active role. If users participated more, even being responsible for the design, they are far more likely to be satisfied with, and committed to, the system once it is implemented. It is 'their baby' as well as that of the computer people. There is therefore every reason to suppose that the interests of the users and technologists might coincide. Both will look for the success of the new system. With a low level of participation, job satisfaction is likely to decrease, particularly if the new system reduces skilled work. The result may be absenteeism, low efficiency, a higher staff turnover, and failure of the information system.

The advocates of the participative approach recommend a working environment where the users and analysts work as a team rather than as expert and non-expert. Although the technologist might be more expert in computing matters, the user has the expertise in the application area. It can be argued that the latter is the more important when determining the success or failure of the system. An information system can make do with poor technology, but not poor knowledge of the application. Where the users and technologist work hand in hand, there is less likely to be misunderstandings by the analyst which might result in an inappropriate system.

The user will also know how the new system operates by the time it is implemented, with the result that there are likely to be fewer teething troubles with the new system.

The role of systems analysts in this scenario may be more of facilitators, implementing the choices of users. This movement can be aided by the use of application packages which the users can try out, and therefore choose what is best for them. Another possibility is the development of a prototype which users can use as a basis from which to agree final design.

Mumford (1983b) distinguishes between three levels of participation:

1. *Consultative participation* is the lowest level of participation and leaves the main design tasks to the systems analysts, but tries to ensure that all staff in the user department are consulted about the change. The systems analysts are encouraged to provide opportunity for increasing job satisfaction when redesigning the system. It may be possible to organize the users into groups to discuss aspects of the new system and make suggestions to the analysts. Most advocates of the traditional approach to system development would probably accept that there is a need for this level of participation in the design process.

2. *Representative participation* requires a higher level of involvement of user department staff. Here, the 'design group' consists of user representatives and systems analysts. No longer is it expected that the technologist dictates to the users the design of their work system. Users have an equal say in any decision. It is to be hoped that the representatives do indeed represent the interests of all the users affected by the design decisions.

3. *Consensus participation* attempts to involve all user department staff throughout the design process, indeed this process is user-driven. It may be more difficult to make quick decisions, but it has the merit of making the design decisions those of the staff as a whole. Sometimes the sets of tasks in a system can be distinguished and those people involved in each set of tasks make their own design decisions.

In Section 24.1 we discuss the ETHICS methodology which has been designed around the principles of user participation.

Of course, participation does have its problems. It might result in polarizing or fragmenting user groups, and there is a possibility of manipulating the process by selecting only those participants who are considered 'right' or by suggesting that users decide 'this . . . or there will be unhappy consequences'. Further, participation may cause resentment, either from analysts, who might see this as their own job being taken over by unskilled people, or by users, who feel that their job is accountancy, managing, or whatever, and this is being cramped by demands to participate in the development of computer systems.

One reaction to lip service participation is the growth in *end-user computing*; that is, users developing their own applications (Section 5.2). With the increase in computer literacy, software tools being designed with end-user computing in mind and low-cost hardware, end-user computing is indeed feasible. It is also one reaction to the application backlog as users see the only way that applications development will take place is to develop the applications themselves. It can be relatively unsophisticated; for example, users using menu-driven office systems, such as word processing, spreadsheet, and database packages running on the PC with Windows, through to users writing their own software using programming languages. There are programming languages such as VisualBasic available today that are easier to use and to learn when compared with Cobol, used in earlier business applications programming, to give just one example. Such end-user computing gives users control over their applications. Potential weaknesses relate to the possible inefficiency, neglect of integrity, and security issues, and the lack of an organization-wide perspective of these application 'islands'. Information centres, which are a source of advice to end-users developing their own applications and cheaper hardware and software through bulk-buying opportunities, represent one response to some of the potential disadvantages. However, this is a centralizing move which is resented by some end-user departments. Some user departments might have their own specialist computer people who are members of the department, not that of the computing department or information centre, and their role might be to provide general advice through to acting as facilitators.

Another response to the excessive power of computing people in information systems development is joint requirements planning (JRP) and joint application design (JAD) (see Section 16.2). Representatives from the user groups and computer people conduct workshops to progress the information system through the planning and design stages. The leader of the workshops has a particularly crucial role and should be trained and experienced. Executives are likely to be involved in the JRP workshops along with end-users, but they are unlikely to play a leading role in the design phase. Computer people will also be more prevalent in JAD workshops, and the use of tools and prototypes may help the process. In some organizations JAD workshops may take place in group decision support rooms consisting of linked computer systems with software, database, and other support but these can be dominated by the technology and the role of the human facilitator is again crucial.

Grundén (1986) suggests that participation implies even more fundamental changes in the organization. Figure 5.1, adapted from that paper, gives some of the characteristics of this approach when compared to the traditional approach.

The focus of the two approaches is very different. In conventional systems development, the emphasis is on the technology: computer systems, hardware, and software. The technologist drives the system. Users are given rules to follow and departures from these norms are not tolerated.

The human-oriented view focuses on the people in the organization. This may result in less complex, smaller systems which are not necessarily the most efficient from a technical point of view. Nevertheless, they are more manageable, less reliant on technology and on 'experts'. PCs are more frequently used as the technological base than mainframes. The traditional view emphasizes the technology, whereas the human-oriented view is more interested in the organization as a whole and the user as creator in that environment. Such a view is likely to lead to more effective systems. The implication is that the conventional view is more common in traditional, hierarchical, bureaucratic organizations; the human-oriented view is more common in democratic, growing, and changing organizations.

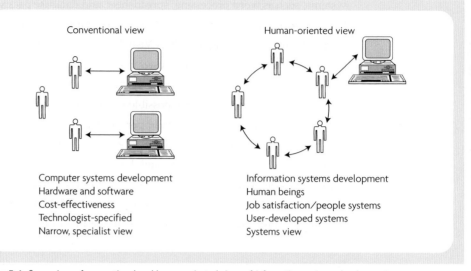

Conventional view

Computer systems development
Hardware and software
Cost-effectiveness
Technologist-specified
Narrow, specialist view

Human-oriented view

Information systems development
Human beings
Job satisfaction/people systems
User-developed systems
Systems view

Figure 5.1: Comparison of conventional and human-oriented views of information systems development

However, to some extent, both views can now be seen as distortions. In theory and practice, organizational and business needs are seen as paramount. In practice, many if not most IT staff do take account of both these views.

5.2 End-user computing

End-user computing (EUC) evolved as a concept with the development of the PC and the arrival of application packages designed for use by non-specialist IS people, particularly the spreadsheet. Prior to this, computing was seen as a black art performed by professionals using large main-frame computers hidden away in air-conditioned rooms. EUC was also driven by the backlog of applications that users required but could not get built by such professionals (see Section 3.6).

Users began to develop their own applications and some users found that they were successful and even quite liked the experience. In the early stages, they were usually self-taught enthusiasts. They had the advantage of knowing the requirements of their part of the business very well and they did not need to communicate that to a systems analyst, they just built a system themselves to meet their needs. Typically, this was either a spreadsheet model or even a small database system, and some users advanced to writing programs in languages such as VisualBasic, and even integrating their applications into the enterprise computer systems.

There are many definitions of end-user computing; some are brief, such as 'the practice of end-users developing, maintaining, and using their own information systems' (Mirani and King, 1994) through to 'the direct use, development and maintenance of information systems by those whose primary role is to achieve a business purpose rather than the indirect development and management through information systems professionals' (Smith, 1997).

As EUC developed, different kinds of user were identified. We discussed different users in Chapter 1, but in relation to EUC Rockart and Flannery (1983) identify four distinct types:

- *non-programming end-users* – users have relatively little IT skills and use limited, typically menu-driven environments or a structured set of procedures, and use software provided by others;
- *command-level users* – they perform inquiries and simple calculations and generate reports for themselves, using limited 4GLs (Fourth Generation languages) or database query languages;
- *end-user programmers* – these users can use procedural-type programming languages to develop applications for their own personal information needs;
- *functional support personnel* – these are users who have the skills to develop systems and who also become informal supporters of other end-users to develop applications within their functional areas (i.e. they provide IT support and training for other end-users).

EUC can provide an effective way of providing some of the information systems in organizations, and it results in a good deal of user satisfaction. EUC also has the benefit of taking pressure off the central IS development function, who were struggling to meet all the demands for new systems being generated in the typical business, and free up the specialist IS personnel to address those activities that EUC could not provide (e.g. the transaction processing systems, the corporate databases).

This development of EUC provided some additional benefits. It enabled the fulfilling of some needs that probably would never have been satisfied through the traditional IT department. Much of the demand for systems and applications by end-users were for things that would possibly never have passed the cost/benefit hurdles imposed by the business or been developed in the necessary time frame to be effective. But because it was the end-users developing them, they could be resourced and undertaken within the user departments. Indeed end-users often felt 'held back' and constrained by the IT department.

So, at one level the situation appeared quite healthy, the users were 'happy', so was the IT department, and the business was benefiting. However, problems began to emerge and EUC proved unsustainable in some organizations. The users needed more IT support to continue to be effective in their EUC, and the IT department was often reluctant to give such access and support. IT departments did not want these user-developed systems coming into any contact with the live-running operational systems as they tended to ignore security, back-up procedures, and recovery. They ignored IT and corporate standards, and not many user developers appreciated the need for normalized databases or referential integrity.

These end-user-developed systems, therefore, although providing benefits, were proving difficult to maintain. Sometimes they were highly dependent on one person, and there were problems when that person left or was moved, or promoted. When IT tried to take over the maintenance and upgrade of these systems, it proved extremely difficult, as there were typically no documentation and poor design. There was also some concern relating to duplication of effort and data (where the same data are held in different systems, they inevitably get out of synchronization), as local systems were produced in each unit or department. Different hardware and software proliferated throughout the organization as each department bought its own, usually in the retail market, and different contracts were entered into. An uncoordinated organizational quagmire was emerging. The situation is not dissimilar to the original *ad hoc* development of computing in organizations in the 1960s.

Corporations were justifiably beginning to worry about this situation. They could not really stop EUC because the users would revolt, and it was providing some benefits and plugging a hole that the IT department could not fill, but they could not let EUC run out of control in this way.

One 'solution' was the setting up of **Information Centres** (ICs). An IC was a unit set up by the organization, usually run by the IT department, but somewhat separate, to support and control EUC. So, the provision of a support infrastructure, the IC, was to maximize the benefits and minimize the risks of EUC.

The role of the IC varied but typically it involved some or all the following:

- defining the domains of responsibility (e.g. what could be developed using EUC and what could not);
- provision of an EUC infrastructure (e.g. PCs, communications, networks, databases, information streams, etc.);
- provision of software and tools, including evaluation, support commitments, etc.;
- support of EUC users, including education, training, help (along with development and design), help desks, etc.;
- setting of standards for EUC, including back-up, security, privacy, etc.;
- co-ordination and control of EUC in the organization, to encourage reuse and prevent duplication, encouraging cross-functionality, integration, etc.;
- procurement, to enable economies of scale to be achieved, etc.;
- dealing with external partners (e.g. contractors, suppliers, vendors, service providers, etc.).

Successful ICs were regarded by some as being critical to the success of organizational goals as a whole and their perceived importance would seem to be reflected by 78 per cent of organizations having an IC (Guimaraes, 1996). The success of ICs varied, but a number of factors have been identified as contributing to success (see Essex and Magal, 1998):

- organizational commitment to the IC, including top management support;
- rank of the IC executive;
- commitment from end-users;
- adequate budget;
- active promotion of IC services;
- quality of IC staff.

EUC and the IC were responses to developments of the 1980s and 1990s and were important concepts and solutions in large organizations trying to manage their IT, but what role do they now play? Some argue that their time has passed and that, with the increasing sophistication of users and their ability to use and develop systems for the new distributed Internet world, they have proved that they no longer need the old-style IC, and that the client part of the client–server architecture is now 'folded into mainstream applications development' (Tayntor,

1994). For others the whole notion of EUC is no longer relevant when almost everyone in business is a user and developer of some kind.

On the other hand there are those who argue that the IC is as relevant in concept today as it ever was, although perhaps its focus has changed. It is not about supporting end-users in developing their own internal systems. It is now more about supporting them in building customer-facing systems in collaboration with specialist corporate developers and developments and delivering value to the customer. In this new world the IC must transfer from a tactical problem-solving perspective to a more strategic, consultant-like role.

Whatever happens it seems that EUC is here to stay, in some form or other, although perhaps the term EUC itself may change as it looks and feels quite dated. End-users and business people (i.e. those whose prime role is not information systems development), it seems, are going to continue to have a role to play in developing systems, particularly given their increasing IT knowledge and the increasing sophistication and usability of the tools and packages available (see for example Section 18.2 on Dreamweaver). The fate of the IC is probably similar. Support for this new kind of EUC is still going to be required. Therefore, there is a need for the concept of the IC, but whether it is called that in the future is highly unlikely. What it turns out to be remains to be seen. It may, of course, just be called the IT department again!

5.3 Expert systems

The term 'expert system' derives from the fuller, and more descriptive term, 'knowledge-based expert system', and these terms tend to be used synonymously. Therefore, there are obvious links with knowledge management (Section 5.4). An expert system (ES) is a system that simulates the role of an expert. It is distinguished from other applications because its usefulness is derived from the knowledge and reasoning ability of the expert system and not from number crunching (carrying out large and complex calculations) or the repetitive processing of data, which characterize most scientific and business computing applications, respectively.

Feigenbaum (1982), one of the early pioneers in the area, defines an expert system as: 'an intelligent computer program that uses knowledge and inference procedures to solve problems that are difficult enough to require significant human expertise for their solution'. Somewhat more formally, the British Computer Society's Expert Systems Specialist Group defines an expert system as follows:

> An expert system is regarded as the embodiment within a computer of a knowledge-based component from an expert skill in such a form that the system can offer intelligent advice or take an intelligent decision about a processing function. A desirable additional characteristic, which many would consider fundamental, is the capability of the system, on demand, to justify its own line of reasoning in a manner directly intelligible to the enquirer. The style adopted to attain these characteristics is rule-based programming.

Expert systems are essentially artificial intelligence programs that contain (in some way) some of the knowledge that human specialists have (hence their inclusion in a chapter on people themes). This does not imply that an expert system builds a psychological model of how the specialist thinks, but rather that it contains a model of the expert's own model of the domain. The domain of the expert system may be a discipline or knowledge area, such as geology or medicine. Alternatively, it may be a particular narrow subset, such as 'risk factors and insurance premiums in California'.

An expert system is basically an intelligent adviser concerning one or more areas or domains of knowledge. This knowledge exists in the minds of human experts, it has been developed and evolved over time by education and experience, and it has to be captured in some way by the expert system, usually in the form of sets of rules and groups of facts. The expert system is then informed about a particular situation of concern to a user, usually achieved by the user answering questions posed by the expert system. Then the expert system comes up with intelligent advice concerning that situation. Ideally, it is able to explain to the user how it arrived at the particular advice.

In practice, expert systems vary considerably, but there are a number of common components in a structure as illustrated in Figure 5.2:

- *The knowledge base*. This contains two elements. The first is the knowledge necessary for understanding the domain, essentially the facts of the problem area. The second element is the heuristics that indicate how to use the facts to solve specific problems, essentially the rules. For example, a rule might be that 'if X is a bird then conclude that X can fly'. This rule is not true in all circumstances because emus are birds but they cannot fly. We

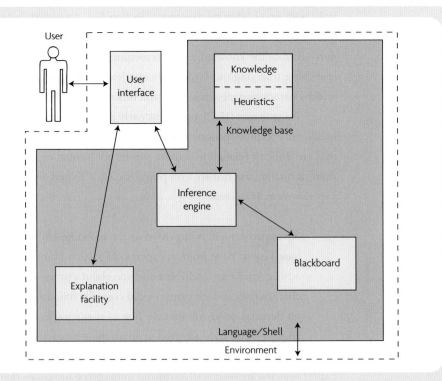

Figure 5.2: Structure of an expert system (generic)

might then introduce another rule that states that 'if X is a bird and X is an emu then conclude that X cannot fly' or we might say that 'if X is a bird then conclude with a probability of 0.99 that X can fly'. The facts contained in the knowledge base are simply assertions, for example, a robin is a bird.

- Although such rules are very common in expert systems, they are not the only way of representing knowledge. Recently, rules have been complemented by frames in some expert systems. A frame is a representation of a group of related knowledge about a particular object. These are useful in situations where there are stereotypical objects. The characteristics of a car give an example. Most cars have stereotypical characteristics, for example, they all have engines, wheels, acceleration speeds, and so on. All these characteristics are gathered together in a frame. Frames are often structured into hierarchies and are good for structuring managerial and business knowledge.

- *An inference engine.* The inference engine controls the process of invoking the rules that pertain to the solution of the problems posed to the system. It is used for reasoning about the information in the knowledge base and in the 'blackboard' (see below) and for formulating conclusions. It provides directions and organizes and controls the processing of the knowledge and the rules to make inferences based on the problems, that is, the procedures for problem solving. Three elements of the inference engine can be identified. The first is the rule interpreter which effectively executes the rules. The second is the scheduler which controls when things should be performed and in what order. The third is the consistency checker which maintains a consistent form of the process and solution. The inference engine usually keeps track of the execution steps so that the user can backtrack through the interaction as necessary.

- *A language.* This is the language in which the expert system is written. A number of specific expert system programming languages have been developed, the best known being Prolog and LISP.

- Probably the most common approach to expert systems development is to use a shell. A shell is an expert system with all the inference capability but without any domain-specific knowledge. It is thus ready for users to input their own rules and knowledge to create their own specific expert systems application. Using a shell obviously saves time as most of the system, except for the specific application knowledge, is already programmed. Many shells were originally full expert systems applications that subsequently were converted to general purpose shells by removing their knowledge component. There are a variety of different types of shell; some are PC-based, while others are for large mainframe computers. They are continuing to evolve, and some are even classed as end-user tools in which domain experts can build their own expert systems. Shells help focus the effort on the domain and knowledge acquisition rather than on the program development. They are not suitable for all types of application and can be limiting and inflexible.

- *An explanation generator or facility.* This is the part of the system that presents the reasoning behind how the system arrived at its conclusions.

- *A blackboard.* This is a temporary workspace that records intermediate hypotheses and decisions that the expert system makes as the problem and the solution evolve. It consists of three elements. The first is the plan and approach to the current problem; the second is the agenda, which records the potential actions awaiting execution, that is, the rules that appear to be relevant; and the third is the solution elements, which are the candidate hypotheses and decisions the system has generated so far. Blackboards do not exist in all expert systems, but they are particularly relevant in systems which work by generating a variety of solutions to the problem posed. They are also important in systems that integrate the knowledge of several different specialists.
- *A user interface.* This varies widely, but is a very important element in the architecture of an expert system. There will be an interface for the user and also one for entering and updating the knowledge base.
- *The environment.* This is the variety of hardware and software that surrounds the expert system. An expert system may be a stand-alone system or it may be embedded within other systems.

Expert systems emerged as an identifiable element of artificial intelligence (AI) in the late 1960s and early 1970s when it was realized that, in certain subject areas, advances could be better achieved by providing a program with substantial subject-specific knowledge and relatively simple inference capabilities rather than following the previous approach, which was to make deductions based on the general axioms of a subject area.

Expert systems have proved to be of value in certain scientific applications, particularly medical ones, but have been slower to establish themselves in commercial or business areas. Many commentators thought that business applications should be amenable to expert systems as many areas of business rely on the expertise and knowledge of specialists. Areas of finance, such as tax, credit and risk assessment, and portfolio management provide examples. A number of business-based expert systems were developed in the mid-1980s including XCON and XSEL from Digital Equipment Corporation (DEC), ExpertTax by Coopers & Lybrand, Credit Clearinghouse by Dun & Bradstreet and Authorizer's Assistant by American Express. However, although there were some successful business applications of expert systems, they were relatively few, and in general they have not lived up to their expectations. The excitement about the possibilities of expert systems in the business domain has been overshadowed somewhat by the more recent enthusiasm for knowledge management systems (see Section 5.4).

The reasons for the lack of successful business applications of expert systems are varied, but certainly one reason is the rather ill-defined and uncertain nature of most business domains. Although experts do exist, their expertise is not of the same nature as that of a chemist or a doctor. In business, there are very few hard and fast rules that everybody will agree on. One expert's views may be totally different from another. This has meant that a consensus on definitions of rules (more where probabilities are included) has proved very difficult to achieve. Most business-based expert systems have restricted themselves to areas where the domain is relatively narrow and where there exists a basic set of rules and standards. However, some applications have been successful. Many customers use an expert system on the Internet to choose the configuration for their new computer (or diagnose what was wrong with the old one). Call centre employees also use an expert system as they complete their customer profiles or give advice to customers as to how to solve their problems.

Apart from the problem of finding the right applications, the process of knowledge acquisition or knowledge elicitation has proved problematical. This is essentially the process of obtaining the expertise from human experts, although it may also involve the obtaining of knowledge from books, papers, files, systems, manuals, and so on. There are also a variety of different types of knowledge. This variety can be expressed in many different ways, but we categorize it in two types. Descriptive knowledge consists of the facts and descriptions, including the associations and relationships between them. Procedural knowledge is information about processes, procedures, and constraints involved in applying the information, for example, for decision making.

In designing expert systems, the particular level of knowledge is important. The first level is the domain knowledge. This is information specific to the domain under consideration, for example, 'knowledge of corporation tax'. A second level is also domain-specific, but at a higher level, for example, the national taxation system, in which corporation tax is a part. A final level might be the environment at large within which the taxation system operates, and relevant knowledge here might be cultural or philosophical, or it might relate to the belief and value systems. All these are potentially important to an expert system concerned with corporation tax as a way of explaining and understanding the context, constraints, ethics, history, and so on. However, the difficulties of capturing and representing the knowledge becomes progressively more difficult as the levels get higher. There are many potential levels, but, in practice, expert systems concern themselves mainly with the first level and then the rest as one general level.

The formulation of decision rules is no easy task, as we have seen above, but it is not just because business knowledge is not scientific and that the views of experts may differ considerably. It is also to do with the fact that experts are not always good at structuring or organizing their knowledge and decision-making criteria in any formal way. Experts cannot always explain why they know something or the basis for their decisions. The definition of probabilities is also extremely difficult, as is achieving consensus. Even when experts are in agreement, the knowledge still has to be formulated in ways that enable the expert systems to make use of this knowledge. The representation of knowledge, sometimes known as the construction of a knowledge map, is extremely problematical. It is these problems of knowledge acquisition and representation that are the real bottleneck in the development of expert systems for business applications.

There are also a number of other problems concerning business expert systems; one concerns the difficulties of testing and validating. It is difficult to prove that the expertise has been captured correctly and that the rules and inferences will lead to good and effective results when applied. Some companies have found this such a problem that they have abandoned their expert systems. In the USA, in particular, the fear of litigation for proffering incorrect commercial advice is very high. Yoon et al. (1995) identify a number of other characteristics that were found to be important for successful expert systems. The study was based on a study of 69 projects. The factors considered important included:

- high levels of management support for the project;
- high user participation;
- skilled expert systems developers;
- use of a systems approach to analyse the business problems;
- management of unrealistic expectations;
- use of good domain experts;
- understanding of the impact on end-user jobs.

Students of non-expert systems development will not be surprised at these findings. This raises the question of the relevance of expert systems to methodologies. Expert systems can be used for handling and applying the rules of a systems development methodology in a toolset. Another area relates to the use of methodologies, not just to develop standard information systems, but to develop expert systems. The approach to their development has been very *ad hoc*, and there is no such thing as a standard expert systems development methodology. Indeed, up until recently, there has been little interest in methodological issues in the expert systems community (but see Section 24.2). The prevailing approach to the development of expert systems may be characterized as prototyping or evolutionary. This has evolved from the trial and error approach of the early expert systems where the developers would code up a few rules and then try out the program on the users, find it was inadequate, and change or add some more rules, try it out again, and so on.

More recently, there has been an increasing focus on approaches which first attempt to acquire and structure the knowledge and then build the system. Most expert systems developments have typically been separate from information systems development methodologies, in the sense that different people are involved and the focus tends to be on solving the technical problems rather than on the process of development. For some people, expert systems have particular characteristics that mean that information systems development approaches are not relevant. They argue that the knowledge domain is more complex, and that the knowledge acquisition process and the representation of rules are the key issues in expert systems development. This complexity implies that to develop expert systems might require somebody to interface between the domain experts and the technical expert systems developers. This person is known as a knowledge engineer and typically has good cognitive and interpersonal skills.

Whether developing expert systems is radically different from developing information systems is a matter of debate. It may be that current methodologies may not need too much adapting to handle knowledge acquisition and representation as well as the acquisition and representation of data and processes. Another area of interest relates to whether any of the approaches to expert systems development have anything to offer information systems development methodologies. Some of these issues are further discussed in Section 24.2 where an expert systems development methodology (KADS) is described.

5.4 Knowledge management

We discussed the meaning of data, information, and knowledge in Chapter 2, and we explore the latter further in this section. During recent years, knowledge management has been seen to be of vital importance to organizations, as important indeed as information management. Knowledge management concerns getting information to the appropriate people, when required, helping them to share this information and experience, enabling them to use it to improve organizational performance, and putting all that in action for a specific purpose. But knowledge is more than 'information'. As Davenport and Prusak (2000) suggest:

> Knowledge is a fluid mix of framed experience, values, contextual information, expert insight, and grounded intuition that provides an environment and framework for evaluating and incorporating new experiences and information. It originates and is applied in the minds of knowers. In organizations, it often becomes embedded not only in documents or repositories but also in organizational routines, processes, practices, and norms.

If we consider the knowledge repository, for example, it can hold best practices, skills inventories, expert databases, competitive intelligence, sales presentations, and product information. It can also contain features that can be said to be a directory to help those who are searching for knowledge. Competitive advantage might be achieved through this knowledge sharing, encouraging innovation, building on past experience, and creating new capabilities within the organization. Its aim is therefore to build on organizational memory through the process of organizational learning. Its use may be apparent after a conversion process when the knowledge seekers convert the knowledge back into information relevant to their purpose.

Tiwana (2000) suggests that there are 24 key drivers that make knowledge management such an important issue for management and organizations.

1 Knowledge-centric drivers

1. The failure of companies to know what they already know.
2. The emergent need for smart knowledge distribution.
3. Knowledge velocity and sluggishness.
4. The problem of knowledge walkouts and high dependence on tacit knowledge.
5. The need to deal with knowledge-hoarding propensity among employees.
6. A need for systemic unlearning.

These knowledge-centric drivers emphasize the situation that most organizations find themselves in: important knowledge exists within the firm but it is not being used effectively. This is partly because they do not know what knowledge their employees have and the lack of motivation for knowledge sharing.

2 Technology drivers

7. The decline of technology as a viable long-term differentiator.
8. Compression of product and process life cycles.
9. The need for a perfect link between knowledge, business strategy, and information technology.

These technology drivers suggest that competitive advantage no longer can be gained from technology itself – your competitors also have this advantage. Further, the speed of change is getting even faster, not just from the point of view of technology, but also product and process life cycles. The key to competitive advantage would seem to lie in using the organization's knowledge better, and ensuring it is available for strategic decision making.

3 Organizational structure-based drivers

10. Functional convergence.
11. The emergence of project-centric organizational structures.
12. Challenges brought about by deregulation.
13. The inability of companies to keep pace with competitive changes due to globalization.
14. Convergence of products and services.

The organizational structure-based drivers emphasize the great changes that are taking place within organizations due to both internal and external drivers. When these great changes are taking place, there is a danger that the knowledge of the organization might be lost. It is therefore vital that knowledge management is seen to be as great a priority as implementing change due to the inevitable organizational changes.

4 Personnel drivers

15. Widespread functional convergence.
16. The need to support effective cross-functional collaboration.
17. Team mobility and fluidity.
18. The need to deal with complex corporate expectations.

With personnel changing roles, teams, functions, and so on, there is again a danger that the knowledge, skills, and experience are lost. People think about their new challenges rather than their knowledge of previous work.

5 Process-focused drivers

19. The need to avoid repeated and often expensive mistakes.
20. Need to avoid unnecessary reinvention.
21. The need for accurate predictive anticipation.
22. The emerging need for competitive responsiveness.

We often hear the expression, 'don't reinvent the wheel', yet it occurs regularly in practice as people do not use the knowledge gained by others or it is not available.

6 Economic drivers

23. The potential for creating extraordinary leverage through knowledge; the attractive economics of increasing returns.
24. The quest for a silver bullet for product and service differentiation.

There have been lots of 'silver bullets' claimed over the years, many relating to the use of information technology. These have provided short-term competitive advantage at best. Yet one key to competitive advantage is surely the knowledge that organizations possess. Knowledge management is about exploiting this to full effect.

Too often, organizations forget what they know, and one aspect of knowledge management is identifying its knowledge assets, and then to manage and make use of these assets by diffusion. But it is not just about forgetting knowledge already learnt. It is also about communicating quickly new knowledge just discovered in this period of rapid change. It is through the knowledge of their people that organizations have real competitive advantage, not through their technology.

Knowledge management is difficult: how can we identify and manage something that we do not know exists? In the long term, organizations try to change the culture so that knowledge sharing is the norm. In doing that, it needs to counteract the view that knowledge is power and therefore sharing it loses personal power. Reward schemes can help to make the sharing of knowledge (and also using other people's knowledge) advantageous to the individual. The theme is encouraging people to share and exploit each other's knowledge partly through the support of IT, not the use of IT to replace people. Organizations need to develop a 'culture of knowledge sharing'.

Another way to encourage knowledge sharing and knowledge management is to have an executive level role with the title of chief knowledge officer (CKO). This shows top-level commitment to the importance of knowledge. The CKO might replace the chief information officer (CIO) title discussed in Chapter 1, though the roles are very different. The CKO, for example, will not have responsibility for IT; the CIO may well have this responsibility.

An important aspect of knowledge management relates to tacit knowledge, which is knowledge that is difficult to express in words and therefore can be hidden from the rest of the organization. Among other things, a knowledge management system attempts to make this tacit knowledge explicit and public. This might be achieved through attempting to explain the knowledge to others and clarifying the knowledge through questions and answers. Further, this enables additional change. The knowledge changes from being an individual's own, to that of the community. Knowledge management is about sharing knowledge.

From the organization's point of view, knowledge management is about sharing best practice, but externally it wants to make capabilities rare, valuable, and difficult to imitate. Sun Microsystems, a company that sells hardware and software, makes available knowledge about new systems to staff out in the field for training purposes while they are at work. Staff are therefore up to date without the need to go to frequent formal training courses. Ernst & Young, a consultancy company, provides knowledge to its customers about auditing, tax updates, methods, and tools. These customers pay for the particular service, belonging to the knowledge network of Ernst & Young. 3M, a company relying on innovation, uses knowledge management systems to diffuse knowledge on best practices in their research and development and their technical communities. Their intranet site includes the ability to access library and information services, a corporate learning management system (including access to e-learning), a news management facility, a directory of e-business contacts, and communities of knowledge management and knowledge exchange.

5.5 Customer orientation

A common progression stressed in the late 20th century was the emphasis placed on the user rather than the technologist and technology. We have reflected this in a number of sections in the book and in particular in this chapter. However, toward the end of the 20th century and since, much emphasis has been placed 'further down the line' of stakeholders to the customer. Companies say they are 'customer focused'. This has been most prominent in the information systems domain in web applications (Section 7.5) where the users are likely to be customers, in data mining applications (Section 8.8) where, for example, the customers' use of the web might be traced and, more generally, in customer relationship management (CRM).

CRM is concerned with attracting valuable customers in the first place and keeping them loyal to the company. Technology is used to improve customer service and provide competitive advantage. The customer, it is hoped, is treated as an individual through this technology and finds experiences with the company very positive. Essentially the philosophy of CRM is contained in the motto 'delivering value to customers will bring value to the organization'.

Although good CRM does not necessarily mean use of the Internet, the latter has played a major part in organizations' emphasis on CRM. Internet search engines and other software have facilitated tracking customers' use of the web, and analysis of this has led to systems providing relevant information to customers. These customer data are potentially good customer knowledge, but fulfilling this potential necessitates good systems analysis and support technology.

Customers are only the final link in the supply chain and CRM also relates to other parts of the supply chain, for example, suppliers in business to business (B2B) electronic commerce.

Outside the web, similar tracking is performed through customers' use of the supermarket loyalty card. Special offers directed to the individual customer can be made based on previous purchase records, all automatically generated using computer systems. Another major application area for CRM concerns call centres where customers have their queries dealt with using on-screen tips to enable the query to be dealt with quickly and efficiently.

There are a number of implications for information systems development. One concerns ascertaining customer requirements during the investigation. A common technique is to carry out electronic surveys of customers. This is likely to achieve general information, but can be somewhat superficial. Individual customers, customer focus groups, and the like might be interviewed much as users and managers were in the past. As the customers are not part of the company and therefore not 'captive', this may require even better communication skills than those required for interviewing other employees of the organization. The development team may be somewhat different than conventionally; for example, containing marketing experts and segment managers and experts in design, data mining, and data warehousing, as well as privacy and security officers.

Emphasis is also placed on good design so that it is easy for customers of various kinds to use the systems. Some customers (like users) may be experienced and do not require nor want detailed instructions. Others may be less experienced and require detailed guidance to use the system. Further, instead of one or two user interfaces for the system, there may be a large number depending on the range of types of customer.

For help desk and call centre applications, systems analysts need to prioritize the different types of information being provided, so that particularly important information is provided first, such as information about customer name and address, product guarantee period, and other repairs made for a product, all following the customer's product serial number being input. The customers need to feel that they are getting good customer care and service. Therefore, call centre staff need to be trained to deal with customers as well as use the computer information system. Knowledge management systems may also be used here, for example, as a means of tracking down faults. The system may also keep a log of each customer transaction so that the system can 'learn' from each customer experience. Certainly sophisticated systems will analyse each set of transactions. Data mining can also be used to analyse the data gathered.

Another concern relates to security aspects (see also Section 8.6), because if customers have access so will hackers. Poor customer-oriented systems from the point of view of security are far more vulnerable than similar internal ones, although all systems are vulnerable to some extent. The company must also be seen as one that is very protective of customers' privacy. Privacy should be seen as a right, not a preference. Again, flexibility might gain greater emphasis, as these systems must keep up with competitors, and changing design, as well as content, is likely to be far more frequent than is the case with internal systems. Yet the designer has to have speed as a priority as well, for Internet users expect very good response rates. This is also an important aspect of customer satisfaction.

Although the gains of CRM are potentially high, customers might become alienated by poor systems analysis and design. The analysts' task will therefore include the ability to persuade management to bear the cost of developing effective CRM systems. Arguments such as competitive advantage, decreased customer attrition, and increased sales can help here. The impact may be great, but effective investment might include changing the business processes, even a business process re-engineering exercise, to emphasize greater attention to customer requirements, and this could be difficult. The integration necessary of the CRM system with the basic systems of the company could also be a major concern.

5.6 Requirements

The definition of requirements can be problematic, but in relation to information systems, it can be said to be everything that the set of relevant stakeholders want from a system. Relevant stakeholders encompass all those people involved with legitimate interests, including those both internal and external to the organization. It includes users, end-users, line management, senior management, customers and regulators (see Section 16.1).

Issues associated with identifying, gathering, analysing, documenting and communicating requirements have been at the heart of information systems development since the very earliest days of computing. Indeed, requirements have been more than just an issue but an ongoing problem area since this time.

Of course, defining the requirements for anything as a conceptual prelude to physical implementation has been a problem for mankind since the very beginning. Even the building of the Pyramids was beset by problems and one of the enduring riddles is why there is an unused burial chamber at the base of the Great Pyramid of Cheops! One explanation is that during its construction a new requirement arose which necessitated a much wider stone sarcophagus than the wooden one originally specified. Unfortunately, the corridor had already been cut through the bedrock and so they built a new, larger, chamber higher up. Haag (2004) reflects that the 'mystery dissolves into familiarity as you imagine the unfortunate architects pulling their hair out each time they had to redesign their building in mid-construction to suit the unpredictable desires of their clients'! It sounds familiar and yet this was around 2500 BC.

Incorrect requirements, changes to requirements, misunderstood requirements, and many other requirement problems are still with us today. So perhaps the notion of 'solving' the requirements problem is an unachievable objective, given that it has been attempted for so many thousands of years. Yet, as will be seen, many approaches, techniques and methodologies attempt to do just that. Others ignore requirements and assume that the 'requirements are a given'.

Requirements are regarded by many as the most important and crucial part of the systems development process and they are often the most misunderstood (Robertson and Robertson, 1999). They are obviously important because they determine what the system will do and to some extent how it will do it. They are also important because of the costs in both time and money of getting them wrong. According to Leffingwell (1997), requirement errors account for between 70 and 85 per cent of rework costs.

Further, there exist a number of studies that suggest that the costs of fixing errors at the requirements stage are around 80–100 times less than if an error is discovered at the implementation stage. Graphs of such relationships are frequently drawn to illustrate the point. The exact relationship should be taken with some caution as they typically are not based on any empirical data. One such is provided by McConnell (1996) and is reproduced in Figure 5.3.

Pfahl et al. (2000) distil a number of principles from software metrics and find that one of them is that, 'finding and fixing a software problem after delivery is 100 times more expensive than finding and fixing it during the requirements and early design phases'. The cost of correcting an error increases dramatically the later in the life cycle it is detected. For example if an error is detected:

- At the requirements stage, it will typically just involve a change to a document or diagram on a piece of paper to correct.
- At the design stage, then it is more expensive because it will probably require some

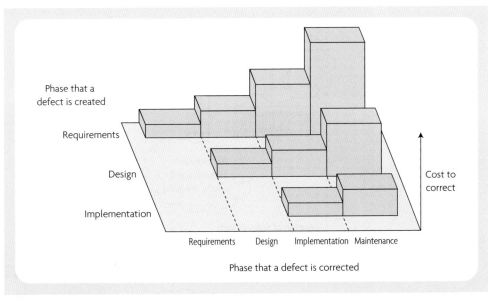

Figure 5.3: Relative costs of fixing requirements errors (adapted from McConnell, 1996)

backtracking and rechecking of assumptions and implications and may result in substantial redesign.

- At the programming stage, then previous stages may have to be revisited and possibly large chunks of code rewritten.
- At the testing stage, then previous stages such as analysis and design may have to be revisited, code rewritten and testing performed again.
- After production and release, the costs can be vast as all the above may have to be performed plus the software may have to be updated in locations throughout the world, documentation changed, users informed, processes changed, data restored, etc. And if the error has caused a user or a customer to incur costs or if the software was safety critical, lives have been put at risk or lost, then litigation may occur.

The process is often compared to the detection of errors in the automotive industry, where it can be costly in the early stages of design but once the car has entered production, and been sold to consumers, a recall may be necessary which is extremely expensive and can ruin the reputation of a company.

The traditional requirements process

Figure 5.4 illustrates a generic, and perhaps rather artificial, requirements process, but it is typical of the way that the process occurs in large organizations following a life cycle/waterfall approach (Chapter 3).

At the beginning, the stakeholders (top left) have a set of notions about what the new system should do, the way it should do it, and how it might look. These notions are typically

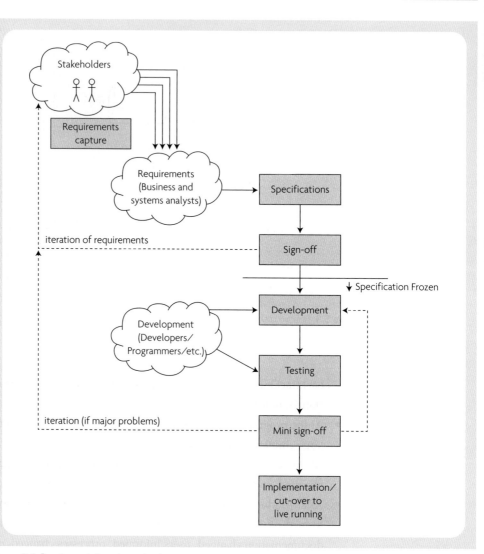

Figure 5.4: Requirements in systems development

vague and different for each stakeholder. The job of the business or systems analyst is to capture their requirements (sometimes called requirements elicitation) and organize them into a coherent whole that reflects all the needs. Any differences, duplications, etc., then need to be resolved. This process is usually time consuming and difficult, and typically involves cycles of interviews, meetings, surveys, workshops, prototypes, storyboards, etc. The analysts then produce a specification (or specifications) that documents the requirements. This specification is then fed back to the stakeholders (or their representatives) for agreement that the specification is indeed a good reflection of their needs. If so, the stakeholders are asked to sign an agreement to this effect. If not, the requirements capture process should be iterated until the stakeholders are happy. Once the specification has been signed off, the specification is forwarded to the developers as the input document to drive systems design, development and implementation. From this stage onwards the specification is 'frozen', that is, the stakeholders are not normally allowed to request any changes until after the system is implemented.

In this process (which is highly simplified here) various internal design documents are produced and towards the end of the process the systems documentation is output. This includes program structures, database designs, etc. The internal design and systems documentation is usually used when maintenance and revisions are required. Finally, a tested and working system is produced by the developers. Hopefully this meets all the stakeholders' requirements, but it needs to be tested to make sure. The specification is used as the basis for this testing and test procedures are derived from the requirements in the specification. Next the stakeholders are asked to confirm that the system meets the specification. This is in effect another mini sign-off. If significant problems are found at this late stage, iteration back to the requirements and a revised specification can occur but this will usually introduce severe delays and extra costs. Minor problems will be resolved by small-scale iteration around the development cycle. Once the stakeholders are satisfied, the system can be released for live implementation.

The process, as depicted in Figure 5.4, indicates that the stakeholders usually do not communicate directly with the developers but require the analysts as intermediaries to interpret their requirements and translate them into a written specification which the developers can understand. Thus different people attempt to elicit the requirements from the users and attempt to interpret the stakeholders' requirements such that they can be organized and communicated to those that are going to actually develop and implement the system. This is because the stakeholders are usually from a different 'world' to those who have to implement the requirements. Typically the stakeholders are from the business/organizational world with associated perspectives. But the implementers are from a somewhat different, usually more technical world. This communication problem is known as a culture 'gap' and has been much discussed in the literature (see for example, Taylor-Cummings, 1998; Saiedian and Dale, 2000; Coughlan and Macredie, 2002) (see also Section 5.1).

Many of the methodologies, techniques and tools discussed in this book are in part about making this process effective and/or changing it fundamentally (for example, see Agile Development in Section 7.4).

Real-world problems

The above is described as a generic or artificial process because, although this is in theory what happens, the practice is often somewhat different. There are many stages at which things can, and often do, go wrong. The first set of problems relate to the requirements capture process.

Requirements capture

- The analysts may not identify all the relevant stakeholders and just capture requirements from a small set of users. These may be important users but they may well ignore or misrepresent the requirements of other stakeholders. These requirements may only emerge when the system is implemented and are then very costly to add.

- The stakeholders are sometimes reluctant to participate sufficiently in the requirements gathering process, due to time or other pressures (or they may not be interested). This inevitably means that requirements will be missed.

- The analysts incorrectly capture requirements and/or misrepresent them in some way. This is very easy to do as there are so many details that must be captured. A requirement might be that a database must be updated when a transaction is complete. However, the definition of when a transaction is complete might be different for some people than others. For some, it might be when a payment is made and accepted; for others, it might be when the product or service is delivered. The success of a system can easily depend on such subtleties.

- The analysts miss some requirements, i.e. the specification is not complete. For example, analysts may forget to ask what happens in a process at the end of year. Such an omission may only be discovered after the system has been operating for nearly a year!

- Users can often over-elaborate their requirements, requesting things that are inappropriate. For example, a request for many ways of achieving the same objective may be made when one well-thought-out way might be better. Some users tend to construct what are known as 'wish lists' which are things that might be nice to have but are not really necessary. This implies that requirements need to be prioritized and costed, and that appropriate choices are made.

- Users may not perceive some of the things that they could have, possibly because nobody has discussed the range of options with them or they are not aware of what is available. A real example relates to a development where the users discovered halfway through the project that wireless technology would have been ideal as part of the solution and having discovered it demanded it be implemented, causing extra cost and delay. They accused the analysts of having hidden this information from them and a lot of bad feeling resulted. Conversely, and perhaps more frequently, the users are well aware of cutting-edge technology and solutions and demand that it is used, maybe inappropriately.

- The users disagree about some requirement. Analysts often go with the majority view or else the view of those with the most power, for example the most senior manager. However, this does not necessarily mean that they are right and that the other users did not have valid points. Sometimes, relatively low-level staff know more about the detail of their work processes than more senior people and sometimes alternative interpretations are beneficial.

- A related point is that sometimes a problem that emerges in discussing requirements reveals some deeper, more fundamental problem that needs to be resolved by the business. This may require discussions with people outside the project. For example, in a system that concerned marketing a problem was identified in the way that certain products were identified and classified. A deeper investigation revealed that there were indeed different ways in which the organization classified its products, depending on their source. The resolution of this problem would have required significant changes to other systems in the organization and it was decided that the easiest solution was to cater for both kinds of classification in the system being developed. This incurred a significant amount of redesign and extra coding and unfortunately some problems were still encountered when the system went live. The best thing in the long run could well have been to resolve these issues at an organizational level, but this was never done. In many cases such difficult problems are not resolved but are glossed over. Unfortunately, as we have seen, requirements problems missed or ignored come back to haunt developers at the implementation stage, when they are much more costly to correct.

Production of the specification

- The requirements have been captured effectively but the specification does not properly reflect them. This might be because of the inadequacies of the techniques and tools being used. Sometimes, especially in the past, the specification is written in English (or some natural language) which is not very precise and can be ambiguous. Alternatively, the specification might be diagrammatic, for example the requirements expressed as use cases (see Section 13.2), which are often not detailed enough. Whatever format the specification adopts, there are always some problems in representing all the requirements.

- In developing the specification, the analyst finds there is something they have to specify but they have not asked the users about. Often, rather than go back and ask, they make what they think is an informed guess. Almost invariably, this turns out to be wrong. This is not an uncommon occurrence because of the level of detail and precision required for processes that are to be computerized. Frequently users only express their needs in relatively high-level or overview terms and are reluctant to get into the detail that is necessary.

The sign-off

- Users may not understand the specification well enough to be able to sign it off as a correct reflection of their requirements. However, they often do sign, just hoping it is correct, usually with unfortunate results. Specifications can be expressed in terms or in forms very unfamiliar to users, for example, entity models, use cases, or even mathematical formulae.

- Users are also sometimes reluctant to question the specification for fear of delaying the project and the start of development. Often not enough time has been scheduled for this important process.

Development

- The specification may not give enough information or detail for the developers/designers to do their job and they have to make assumptions about what is needed. Assumptions, based on their own logic or what is easiest, rather than what the business actually needs, may prove incorrect. Developers are usually very reluctant to go back to users to check or confirm anything for fear of delaying the project, or revealing a larger problem.

- Users try to inject new, changing or forgotten requirements into the development process at late stages. This is known as creeping requirements. This can cause significant redevelopment work and can lead to delays in the project. It can also result in inappropriate designs, as had these requirements been known at the outset they would have been properly accommodated rather than just added on to the existing designs.

Testing

- The specification may enable the system to be tested for its basic functionality, but it often does not enable testing of the business concepts which the stakeholders' requirements usually embody. For example, it might be that the system is designed to save costs by automating certain processes, but the specification does not usually enable tests to be devised that ensure that this cost saving occurs.
- Sometimes the specification may not be a very good document to use as a basis for testing. It may be at too high a level to drive the detailed testing that is also required.
- For much the same reasons as those found in the sign-off (above), the users are sometimes reluctant to test in detail for fear of delaying the implementation.
- When problems are identified, i.e. the system does not fully meet the specification, a common reaction is not to pull together to fix the problems but rather to attempt to apportion blame. The users blame the analysts or developers for misunderstanding what they wanted and the developers blame the users for not telling them what they wanted. The specification becomes the battleground, with debates about whether a missing function was in the specification or not. If it was, the developers are responsible for the problem and must pay, and if it was not, the users must pay. This is known as the blame syndrome and is very unhealthy because it does not contribute to the overall goals of the project but pits one side against the other.

These are just some of the most common issues and problems associated with requirements. Experienced participants will have their own lists, specific to the context of the project being developed and the approaches and methods adopted. However, this is not the end of the problems. There are more, at different conceptual levels.

Changing and evolving requirements

Clearly, changing and evolving requirements create a problem for the requirements process. Requirements can and do change whilst the system is being developed and this is a major issue because the specification has been frozen. Markets and business requirements are changing all the time and information systems need to be changed as quickly. The traditional requirements process fails to cope with such real-world demands. The problem is exacerbated if the project takes many months or even years to implement because the final implemented system can be a long way from what is actually required, even if the original requirements were accurately gathered. Many approaches discussed in Chapter 7 and methodologies discussed in Chapter 23 attempt to address this problem in various ways.

Unknowable requirements

The traditional requirements process embodies an assumption that requirements are conceptually able to be discovered from stakeholders. To do this the stakeholders and the analysts need to work hard and use appropriate methods but if this is done properly, and with enough diligence, then the requirements can be discovered. There is, however, a school of thought that questions this assumption and suggests that some requirements are so complex and so obscure that they are uncapturable in this way, no matter how diligent and hard-working the participants. Development of some web-based and early e-commerce systems come into this category. Neither the companies themselves nor their customers really knew what they wanted from such systems, except in very general terms, because there was so little experience and the technology was new. In such cases organizations and developers often turned to evolutionary and agile development approaches (see Chapter 7). Maintenance, therefore, will always be with us and is a normal part of IS development and operational systems. It is not necessarily due to poor work at the requirements stage.

Non-functional requirements

Up until this point it has been assumed that requirements are the things gathered from stakeholders that they wish any new system to be able to perform, usually expressed as a set of functions or behaviours that the system should be able to do. However, there is another, often forgotten, class of requirements, known as Non-Functional Requirements (NFRs). A NFR describes not what the system will do (its functionality) but how it will do it. For example, a functional requirement in relation to an order processing system might be that the system should print a copy of the transaction. An NFR might be that this copy should be printed quickly. In other words, NFRs are characteristics of the system that the stakeholders care about and that are likely to affect their satisfaction with the system. They are sometimes referred to as *constraints* or *qualities* of the system, over and above the functional requirements. They are usually of great importance to the performance of the system. For example, if the functional requirements are all met and the transactions are correctly implemented but the printing is very slow, then the users of the system might be very dissatisfied and frustrated, even though the functions are performing perfectly. NFRs can relate to any aspect of a system but typically they relate to system performance, interfaces, designs, and software quality attributes. The printing-quickly example above is a performance NFR, an interface NFR might be that it should be 'easy' for the customer to use, a design NFR might be that the system should be 'reliable' or 'flexible'. A software quality NFR might be that the system be 'maintainable'. These are all, in some way, quality-related NFRs. Constraint-related NFRs are expressed in terms of what the system must not do or what the limits of freedom in design might be. For example, a constraint NFR might be that the system must not exceed certain limits or that a system is constrained by the hardware, or it must be written in XML.

Russell (2004) takes the categorization of NFRs further and suggests the following categories. This first group is argued to be observable, and thus easier to collect, as well as being potentially measurable, which makes them easier to test:

- **Performance** – the responsiveness of the system, for example, response time and throughput
- **Security** – the capability of securing the system from malicious or unauthorized use
- **Availability** – the proportion of time the system is available for use
- **Reliability** – the ability to perform the functionality consistently, and recover from failure as necessary
- **Capacity** – the capability of the system to handle the resources, volume, etc. and required growth
- **Usability** – the ease of use of the system commensurate with the users' abilities.

The second group are regarded as non-observable, and perhaps thus more difficult to gather and test:

- **Maintainability** – the ease with which the system can be changed
- **Portability** – the ease with which the system can be ported to alternative operating environments
- **Integrity** – the ability of the system to preserve transactions accurately and persistently
- **Scalability** – the ease with which the system can handle increased volumes
- **Manageability** – the ease with which the system can be managed and organized during its operation
- **Safety** – the capability of the system to not harm users or others
- **Efficiency** – the assessment of how well the system utilizes resources.

To help with understanding the distinction between functional requirements and NFRs it is often suggested that anything that is phrased in noun–verb form is a functional requirement, for example, 'the system prints the transaction' or 'the customer credit rating is checked'. On the other hand, NFRs are expressed with adverbs or adverbial modifying clauses, for example, 'the system prints the transaction quickly' or 'the system checks the credit rating securely'.

As has been discussed above, eliciting and specifying functional requirements is difficult but defining NFRs is sometimes considered even more difficult. Cysneiros and Leite (2001) suggest that NFRs are amongst the most difficult and most expensive to correct if they are not identified early. Indeed, they say that, 'not eliciting NFRs or dealing with them late in the process has led to a series of histories reporting failures in software development, including the deactivation of a system right after its deployment'. In the well-known case of the failure of the London Ambulance Service's Computer Aided Despatch system (Fitzgerald and Russo, 2005) the enquiry report highlighted many problems related to NFRs that contributed to the failure, such as reliability, usability, performance and cost.

Why are NFRs difficult to discover?

1. Users do not tend to think in terms of NFRs and they are usually thought to be characteristics that the developers or technical people deal with. They are often not even considered when the functional requirements are being gathered and are thus often

tagged on towards the end of the development process. This is usually a major problem as they often have an impact upon the functional requirements and may significantly influence the design decisions.

2. They are difficult to evaluate. If printing is required to be fast, it is reasonably easy to decide how fast. But how do users decide how to evaluate and quantify reliability or maintainability? A very clear understanding is needed by the users and information must be provided by the developers for informed decisions to be made.

3. They compete with the gathering of functional requirements and so usually come off second best.

4. There are tools, techniques and methods for eliciting functional requirements, as seen in this book, but there are relatively few that relate to NFRs.

5. Some NFRs are global, i.e. they arise from different parts of the system, from interactions, or from aspects not relating specifically to the system being developed but from IT or organizational strategy. For example, the system target architecture is a client–server environment running both Windows 2000 and Windows XP, or the development of the code will be outsourced to a partner in India, and therefore must be specified in a particular way.

Despite these difficulties it is usually recommended that NFRs are elicited or gathered together with the functional requirements at an early stage in the development process. A number of attempts have been made to modify use cases and UML (Section 13.2) to enable them to be used to elicit NFRs as well as functional requirements (Cysneiros and Leite, 2001).

Summary

- The movement concerned with involving all those affected by information systems in the process of developing them is usually referred to as participation.

- End-user computing is a participative approach whereby the users, not specialists, develop their own information systems.

- An expert system is an information system that simulates the role of an expert. Its usefulness is derived from its knowledge and reasoning ability.

- Knowledge management is a further development and sophistication of expert systems concerned with getting information to the appropriate people, when required, helping them to share this information and experience, enabling them to use it to improve organizational performance and putting all that in action for a specific purpose.

- Customer relationship management (CRM) is concerned with using information technology to attract customers to the organization and keeping them loyal to the company.

- Finding out what is required by the stakeholders of a new system can be problematic. Sometimes it is not until a system is implemented that the real requirements and the weaknesses of the system are revealed, when the costs to fulfil them are much greater. There are, however, a number of techniques that can capture most of the requirements.

Questions

1. A systems analyst suggests that user participation increases development times, leads to inefficient applications and makes her job less satisfying. Provide a counter-argument to this view.

2. Do you agree that knowledge management is simply a further development of expert systems or do you think there are fundamental differences?

3. Do you think the prime stakeholder of an information systems development project is the developer, user or customer? Argue a case for each. Who are the other stakeholders?

4. For a company or a university department of your choice, how would you find out the requirements of a new information system?

Further reading

Davenport, T.H. and Prusak, L. (2000) *Working Knowledge*, Harvard Business School Press, Boston.

Dyché, J. (2001) *The CRM Handbook*, Addison-Wesley, New Jersey.

Goldsmith, R.F. (2004) *Discovering Real Business Requirements for Software Project Success*, Artech House, Norwood, MA.

Jackson, P. (1999) *Introduction to Expert Systems*, Addison-Wesley, Harlow.

Mumford, E. (1995) *Effective Requirements Analysis and Systems Design: The ETHICS Method*, Macmillan, Basingstoke.

Sommerville, I. and Sawyer, P. (1997) *Requirements Engineering: A good practice guide*, 2nd edn, John Wiley, London.

Tiwana, A. (2001) *The Essential Guide to Knowledge Management*, Prentice-Hall, New Jersey.

Wiegers, K.E. (2003) *Software Requirements*, Microsoft Press, Washington.

6 Modelling themes

6.1 Modelling

We looked in Chapter 4 at approaches to information systems development that seek to address the needs of the organization as a whole and in Chapter 5, those that concern 'people aspects'. We now look at another overriding theme in modern information systems development: modelling.

A model is an abstraction, a representation of part of the real world. An abstraction is often viewed as the process of stripping an idea or a system of its concrete or physical features. Abstraction can be viewed as a simplified representation of the lower level. A benefit of abstraction is the easier development of complex applications. It provides a way of viewing what the model indicates are the important aspects of a system at various levels, so that high levels have the 'essence' of the system and low levels have the detail that does not compromise that 'essence'. The process of abstraction loses information and so a model should only lose that information which is not part of the 'essence' of the system.

It has been suggested that in information systems there are some 'natural' or 'inherent' levels, and these are the conceptual level, the logical level, and the physical level. The conceptual level is a high-level overview description of the universe of discourse (UoD), i.e. the domain of interest; this might, for example, be the overall information system, the business system, or society. The logical level is a description of the information system without any reference to the technology that could be used to implement it. Its scope is the information system itself, or part of it, and it is not concerned with modelling the UoD. The physical level is a description of the information system including the technology of a particular implementation.

The elements of the real world that the model chooses to represent are crucial. In the context of this book, it may concern a representation of one or more aspects of the current or proposed information system. Section 6.2 looks at process models in which the key element represented in the model is process. Section 6.3 looks at data models in which the key element represented in the model is data. Processes and data have traditionally been modelled separately but the object-oriented approach has modelled these elements encapsulated together (see Section 6.4) along with other elements such as people.

6.2 Process modelling

Structured methodologies use many techniques of process modelling. They have as their unifying elements an emphasis on the processes and the basic technique of **functional**

decomposition; that is, the breaking down of a complex problem into more and more detail, in a disciplined way. At the lowest level the units are simple and manageable enough so that they could be reflected in a few lines of computer program code. An example of functional decomposition, which represents a simplification of a payroll system, is seen in Figure 6.1. As well as enabling understanding of complex processes, it also enables people to view the processes at different levels. For example, systems analysts and users may wish to view the system at a high level of abstraction and programmers at a lower level.

Structured systems analysis and design has been associated with a number of consultancy houses and authors. These include Gane and Sarson (1979), DeMarco (1979) and Yourdon (1989). In Section 20.1, we look specifically at the Gane and Sarson methodology (STRADIS). This is a good example of the original structured approach. In Section 20.2 we look at a later development of the approach, Yourdon Systems Methodology (YSM). Other methodologies, for example, SSADM (Section 21.1) and Merise (Section 21.2), have been greatly influenced by this school of practice.

Some of the techniques associated with structured systems analysis and design, specifically functional decomposition, decision trees, decision tables, data flow diagrams, data structure diagrams, and structured English are looked at in Chapter 12. Many of these models are supported by tools and toolsets (Chapters 18 and 19). Many of the representations are graphical and this encourages user involvement to some extent. Although the emphasis of the structured school is on processes, there is also a consideration of the structure of the data in some of the techniques and tools.

Process modelling describes the logical (real-world) analysis of the processes and not just their physical (implementation) level designs. In other words, there is usually a clear distinction between any application logic (what a system is trying to achieve) and the computer

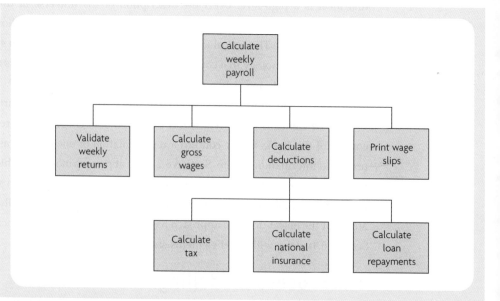

Figure 6.1: Functional decomposition

representation of that logic (how the computer system achieves it). In process-oriented approaches the documentation is produced as part of the analysis and design process as 'deliverables' and these models can be cross-checked to ensure consistency in analysis and design and hence enable quality control.

There is also a separation of data structures as seen by the user in the processing and its physical representation (a file or part of a database). This separation of logical and physical designs is an important element of a number of methodologies discussed in Part VI. It gives a level of **data independence**: in other words, the processes can change without necessarily changing the computer files. Similarly, the files can change without necessarily altering the user views of the data.

Two process modelling techniques are data flow diagramming (Section 12.1) and structured English (Section 12.4). Data flow diagrams are a particularly useful aid in communicating the analyst's understanding of the system and are a feature of many commercial methodologies. Users, both end-users and user management, of the area or department concerned can see if the data flow diagram accurately represents the system and, once there is agreement that it reflects their requirements, it can be converted to a design. Structured English is like a 'readable' computer program. It is not a programming language though it can be readily converted to a program, because it is a strict and logical form of English and the constructs reflect structured programming.

There are other variations on the 'English-like' languages used in these methodologies. Although 'English-like', they are not natural languages, which tend to be ambiguous and long-winded. Structured English and its variants, such as pseudo-code and tight English, are designed to express process logic simply, clearly, and unambiguously. This is particularly important to ensure that the system developed represents a system that is actually required.

6.3 Data modelling

Whereas structured analysis and design emphasizes processes, data analysis concentrates on understanding and documenting data which, it is argued, represents the 'fundamental building blocks of systems'. These approaches focus on the modelling of data. High-level data analysis is orientated toward modelling the 'real world' or that part of the real world of concern (e.g. organization, department, or whatever), and is implementation independent. This means that the data model, and therefore the data analysis process that derives it, is suitable whether the physical model is a database, file, or even a card index. There is therefore a separation of logical from physical data; that is, data independence. It also means that, even if applications change, the data already collected may still be relevant to the new or revised systems and therefore need not be collected and validated again.

Early database experience did not always bring about the expected flexibility of computer applications, usually because the database was not a good reflection of the organization it was supposed to represent. Modelling the organization on a computer database is not simple. It is argued that data analysis helps to achieve a data model that is independent of any database, accurate, unambiguous, and complete enough for most applications and users. Its success comes in the systematic way by which it identifies the data in organizations and, more particularly, the relationships between these data elements (i.e. the 'data structure').

Data analysis techniques attempt to identify the data elements and analyse the structure and meaning of data in the organization. This is achieved by interviewing people in the organization, studying documents, observation, and so on, and then formalizing the results through a process known as entity modelling (see Section 11.1).

Data analysis does not necessarily precede the implementation of a database or computer system; it can be an end in itself, to help in understanding aspects of a complex organization. Good models will be a fair representation of the 'real world' and can be used as a basis for discussion and understanding of the organization.

There are a number of alternative approaches to data analysis, one of the most well known being entity modelling. The basic information is obtained in a variety of ways, including the interviewing of users and management in the organization, to identify the key entities (important data groupings). Typical entities might be customer, supplier, parts (see Figure 6.2), and finished goods. Next the relationships between these entities are identified and then the entities and their relationships are expressed as a graphical model.

Data analysis is a feature of a number of methodologies. It is sometimes argued that data is more stable than processes (i.e. that processes are more susceptible to change than data). Therefore, data constitute a better, more stable, basis for the design of an information system than something that is likely to change, such as the processes. This argument suggests that in a university, for example, the data concerning students is more stable than the functions or processes that the university may perform on a student. So, there will always be students, they will have names, addresses, registration numbers, degrees they take, and so on. However, the processes, such as the detail of the registration process, may change. This argument has been questioned and some argue that data and processes are in fact equally stable (or unstable) but nevertheless data modelling is important in many methodologies. Most, including object-oriented methodologies, focus equally on the modelling of both data and processes. Additional strengths of data modelling are as follows:

PART NUMBER	COST	NAME	SUPPLIER
344	£10.00	Widgets	Smith
346	£12.00	Widgets	Jones
540	£10.00	Widget tops	Smith

Figure 6.2: Part relation

- it is a model which is readily understandable by both developers and users because of its graphical form;
- it is independent of any physical implementation (i.e. it is at a logical level);
- it does not show bias toward particular users or departmental views. The data model can reflect a variety of different views of the data.

Of course, any model can only be *a* model and not *the* model of that part of the real world being investigated. It cannot reflect reality completely and accurately for all purposes. The data model cannot be a completely objective representation of an organization, it is clearly the subjective view of those involved in its creation. Having said this, however, the data model derived from data analysis usually proves in practice to be suitable for the purpose (i.e. to represent the necessary data and to build an information system).

6.4 Object modelling

The object-oriented (OO) approach to systems development is increasingly important in the development of information systems. For some it is the means by which many of the existing problems of development, which have proved so tenacious and hard to overcome in the past, will be solved. The concepts of object-oriented modelling and object-oriented systems development are introduced in this section as a modelling theme and further developed in Chapter 13 in relation to object-oriented techniques, including UML, and then again in Chapter 22 in relation to two object-oriented methodologies.

Booch (1991) suggests that the concept of object orientation emerged simultaneously in a number of different fields in the 1970s including computer architecture and operating systems, databases, cognitive science, and artificial intelligence. However, for others its development is associated with the early programming languages, in particular, Simula and Smalltalk; in fact, according to Bahrami (1999), the term object in this context was first used in Simula. Given that the concepts of object orientation have been around since the early 1970s, it is surprising, perhaps, that they have only relatively recently become so influential. This can be said of a number of 'new' themes discussed in Part III. It may be that it takes up to 20 years for such conceptual and theoretical advances to make their impact in practice. The basic object-oriented concepts are quite different to traditional ones in systems analysis and design and this helps to explain why OO has taken some time to be widely adopted and why some systems developers have found it difficult to come to grips with in practice.

Coad and Yourdon (1991) suggest that the object-oriented concepts are based on those we first learned as children; that is, objects and attributes, wholes and parts, classes and members. We learn by identifying particular objects, such as a tree, and then identifying their component parts (i.e. their attributes), for example, branches and leaves, and then to distinguish between different classes of objects, for example, those of trees and rocks. This implies that the concepts should be simple and indeed the language Smalltalk was originally developed for children to use.

Object-oriented development is different from other models, such as process and data modelling described in the previous sections. Clearly it still involves the modelling of data and processes (or functionality), as these are the fundamentals of a system, but it does it in a completely different way. It does not treat data and processes separately but combines or encapsulates them into an object.

An object represents something in the real world, so an example might be an object representing a car, in a car rental application. It has data, such as the manufacturer, model, colour, number of seats, mileage, etc. It also has processes or actions (methods as they are termed in the object-oriented world), such as go forward, go backward, go up a gear, go down a gear, brake, accelerate, etc. But we are not particularly interested in these processes in the car rental system so we do not model them (i.e. we only model what is relevant for our application environment). In the context of a car rental application, we might be more interested in processes such as servicing a car, booking a car to a client, client returning car, checking a car for damage, etc. Some of these might be processes the car object can perform, others might be processes other objects perform on the car object. The processes that the car object can perform might be limited, but we might be interested in the car's capability of transmitting its position, its display mileage function or its self-diagnosis feature. In which case we would model them. Objects are independent of each other and exist independently (i.e. the exact way that it performs its processes and the data are unknown to other objects). This means that the internal workings or implementation of the object can be changed without it causing any problems or affecting other objects, an important benefit. So, a system is made up of a series of discrete objects that interact together by the passing of messages from one object to another, which trigger the processes of the object. So, a booking object might check the mileage of a particular car before reserving it to check if it is due for a service.

Object-oriented modelling is concerned with representing objects, including their data and processes, and the interaction of objects, in a system. We also model hierarchies of objects (called classes), so we might have a high-level car vehicle object that breaks down into various objects of cars of particular manufacturers. These concepts are described in more detail in Section 13.1.

There are a number of benefits of the object-oriented approach:

- Object-oriented concepts unify many aspects of the information systems development process. For example, the analysis of the application area can be undertaken using object analysis and object modelling, the design of the new system can use object orientation as the design approach, the human–computer interfaces can be designed using object-oriented methods, applications can be developed using object programming languages, object-oriented tools, and 'data' (using a broad definition of the term to include text, audio, pictures, video, and so on) once collected can use object-oriented multimedia databases. There is no need to transform the objects into other representations. The object-oriented theme is relevant and consistent throughout. This contrasts with other methodologies which somehow have to blend and reconcile the results of different approaches, such as that of the data and process views discussed in the previous two sections. For example, attempting to reconcile entity-relationship diagrams and data flow diagrams is not trivial because they represent different objects in different ways.

The object-oriented approach represents data, processes, people, and so on, all as objects.

- It facilitates the realistic reuse of software code and therefore makes application development quicker and more robust. In theory the organization will develop a library of object classes that deal with all the basic activities that the organization undertakes. Software development therefore becomes the selection and connection of existing classes into relevant applications, and because those classes are well tried and tested as independent classes, when they are connected together they provide immediate industrial-strength applications that run correctly in a shorter period of time. Only completely new classes will need to be developed or purchased. Proponents of object orientation believe that eventually there will exist international libraries of object classes that developers will be able to browse to find the classes they require and then simply buy them. The classes in these libraries will be guaranteed to perform as specified, and so new applications are easily developed and existing (object-oriented) applications can be modified and extended in functionality just as easily. Software development is not only quicker and cheaper but the resulting applications are robust and error-free. This still remains somewhat of a theoretical benefit as we have not yet seen very much evidence of libraries of object classes developed for reuse.

- It integrates methods of systems development with the systems context. Mathiassen et al. (2000) suggest that this is an increasingly important benefit because modern systems are not just about replacing labour-intensive operations as was the case in the past. In most organizations such labour displacement systems have all been developed, we are now at a stage of developing the second or third version of them, or developing systems with other objectives, such as supporting individuals in problem solving or communication. For such systems the method must focus on the context as well as the system. For example, in a system that dispatches ambulances to emergencies it is important that the system models the context and methods of interaction of the ambulances, the other emergency vehicles and services, the ambulance drivers and their way of working, the patients and their potential injuries and problems, and the hospitals that may receive the patients. If this context is ignored then the system may well fail, as indeed happened in the London Ambulance Service (Fitzgerald, 2000). Of course, traditional systems development has always attempted to deal with context (e.g. Checkland, 1981) but the benefit object orientation is said to provide is that it specifically addresses the modelling and understanding of context using the same methods and principles as for modelling the system itself.

Coad and Yourdon (1991) suggest a number of other motivations and benefits for object-oriented analysis (OOA), including:

1. the ability to tackle more challenging problem situations because of the understanding that the approach brings to the problem situation;

2. the improvement of analyst–user relationships, because the approach can be understood by both equally and because it is not computer-oriented;

3. the improvement in the consistency of results, because it models all aspects of the problem in the same way;

4. the ability to represent factors for change in the model so leading to a more resilient model.

These are ambitious claims, and in a later section these claims can be evaluated in the context of OOA and the RUP methodologies (see Chapter 22) and more detail on the concepts and techniques of object-oriented analysis and design are provided in Chapter 13.

Summary

- A model is an abstraction, a representation of part of the real world.
- In a three-level view, the conceptual level is a high-level overview description of the universe of discourse (UoD), i.e. the domain of interest. The logical level is a description of the information system without any reference to the technology that could be used to implement it. The physical level is a description of the information system including the technology of a particular implementation.
- Process modelling describes the logical (real-world) analysis of the processes and not just their physical (implementation) level designs. The basic technique of process modelling is functional decomposition, that is, the breaking down of a complex problem into more and more detail, in a disciplined way.
- Data analysis concentrates on understanding and documenting the data elements and their relationships.
- The object-oriented (OO) approach models objects, which represent something in the real world including people, data and processes, and the interaction of objects.

Questions

1. What are the differences between modelling an airplane and modelling an information system?
2. Why are process and data modelling separate? What else should be modelled in an information system?
3. Why do you think object modelling has been gaining more users at the expense of process and data models?
4. Do you think we can model the complexities of the real world?

Further reading

Avison, D.E. and Shah, H.U. (1995) *The Information Systems Development Cycle: A First Course in Information Systems*, McGraw-Hill, Maidenhead.

Mathiassen, L., Munk-Madsen, A., Nielsen, P.A. and Sage, J. (2000) *Object Oriented Analysis and Design*, Marko Publishing, Denmark.

Teorey, T.J. (2005) *Database Modelling and Design: the Fundamental Principles*, 4th edn, Morgan Kaufman, San Francisco.

7 Rapid and evolutionary development

7.1 Evolutionary development

Evolutionary systems development is a staged or incremental approach that periodically delivers a system that is increasingly complete (i.e. it evolves) over time. Evolutionary development is sometimes termed incremental development. The first (or even subsequent) implementation is not seen as the main objective but is just part of the continuing evolution and improvement of the system until an optimal solution to the original problem or requirements is achieved. Frequently, as the system matures only minor iterations occur until a major step change is required as a result of technological obsolescence, change of business mission, or a fundamental redesign of the required processes, in which case the system is completely thrown out and a new one begun.

When this happens the orderly, evolutionary changes are replaced by large, abrupt changes, creating shock waves throughout the organization and the process begins again. Of course, the idea that a system ever reaches an equilibrium state, where it requires relatively little maintenance and change, is not often seen because in practice the business needs and user requirements are usually evolving and changing, necessitating continual evolution.

Evolutionary systems are thus not in their final form at their first iteration or delivery. Each delivery achieves something useful and usable but is not necessarily complete. The design process stretches over the entire life of the system providing a better approximation to the ideal required product at each iteration. Figure 7.1 illustrates the concept.

It can be seen that there are a series of development efforts. The first iteration begins at a particular point in time and contains a full set of stages or phases from the identification of requirements through analysis, design, and implementation. Although not specifically mentioned here, it is frequently assumed that a prototyping approach will be used and prototyping is closely associated with evolutionary development (see Section 7.2); however, this is not strictly necessary for it to be evolutionary. The system developed in the first iteration will probably be only a subset of the total requirements, or just a first stab at them. In the bottom half of the diagram it can be seen that only a relatively small part of the total potential needs and requirements have actually been achieved at the end of the first iteration. After this, a period of learning about the system, and gaining experience of using the system, may occur. (In fact, iterations can have a time gap between them, butt up to each other, or even overlap.)

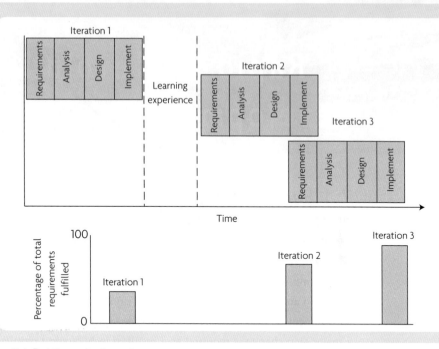

Figure 7.1: Evolutionary development

With the experience and knowledge gained, a second iteration of the system is initiated with revised and refined requirements. This second iteration essentially takes the existing system from Iteration 1 and evolves it and integrates it with the new requirements discovered in Iteration 2. This is typically a demanding task but probably involves less effort than building the original system of Iteration 1. Thus, in theory at least, the amount of effort declines at each successive iteration.

At the end of Iteration 2 a significantly greater proportion of needs should now be fulfilled. However, it is important to ensure that the environment has not changed before starting Iteration 3. For example, the system might need to reflect the introduction of some new government regulations. As well as these mandatory changes, additional improvements are also implemented resulting in a large proportion of the total requirements now being fulfilled, indeed close to 100 per cent. This might mean that the equilibrium mentioned above has been achieved; however, it is more likely that further iterations occur from time to time, as a result of further changes in the business needs and environment.

Orman (1998/9) makes the point that evolutionary development is characterized not just by its ongoing and iterative nature, but by the evolutionary nature of the system's original creation as well. Therefore, the original design, and indeed designs of subsequent iterations, should be geared to evolution. This can be achieved in a number of ways:

- *The design does not have to be perfect*. It can be just a first attempt, good enough, rather than an ideal or perfect solution. The first design often focuses on the core of the system.
- *It can allow or accommodate future change, or at least not impede such change*. This might be achieved in a number of ways, for example by building in some flexibility or redundancy, or by parameterizing as much as possible.

- *It does not have to be comprehensive*. It need only address part of the required system to begin with rather than everything that might possibly be required.

Therefore, improvements and enhancements to the system can be added later in subsequent iterations, as the system is used and experience is gained. A further benefit of this approach is that it can be quick, or at least getting to the first implementation can be relatively quick. It may not be a complete nor ideal solution but at least it is something, and it has probably been delivered more quickly than a full, traditionally developed system. Rapid application development (Section 7.3) uses many of the concepts of evolutionary development to achieve speed of delivery.

Another major benefit of evolutionary development (and something that distinguishes it from other development models) is that changing requirements over time are expected and catered for as part of the incremental development. As a change is required it is just built into the next iteration of the development. Change is seen as the norm and not a surprise or a reflection of failure in some way. What might normally be thought of as maintenance of a system, with its negative connotations, is regarded as positive (McCracken and Jackson, 1982).

Evolutionary development is regarded as highly appropriate for situations where requirements are difficult to discover in advance (or indeed where they are impossible to discover), or where the system is particularly complex. According to Orman (1998/9), it may be 'the only feasible strategy for highly unstructured or unstable systems where the traditional requirements and analysis techniques frequently fail'. He also argues that most 'strategic management applications and decision support systems fall into this category as their requirements are generally vague, and the objectives are fluid, even in the minds of end users'.

Evolutionary development differs from the SDLC (Chapter 3) in that it is not a linear development. The SDLC is designed for 'straight line development' (Pressman, 2004); that is, it assumes that a complete solution will be delivered at the end of a linear sequence. At each stage of the SDLC, as complete and comprehensive a job as possible is undertaken. For example, in the early stages, a complete and detailed set of requirements is specified in great detail. These then are typically 'frozen'; that is, they cannot be changed during the development phase. These requirements then drive the design and implementation of the complete and finished system.

This is clearly not the case in evolutionary development. With evolutionary development, just a first set of outline requirements, perhaps only for a subset of the system, would typically be identified at first, the detail of which would be assumed to be changing and changeable. These would then be used to design the core of the system that would deliver something useful but that would undoubtedly need to be changed in some way and evolve with later iterations.

Attempts have been made to combine the rigour and the management control of the SDLC with the benefits of the evolutionary approach, which has been criticized as being difficult to manage and control. Most notably, Boehm (1988) has proposed the spiral model (see Figure 7.2), which adopts the concept of a series of incremental developments or releases. As

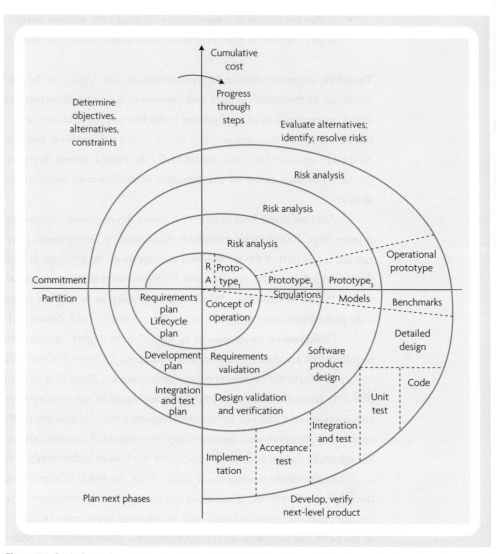

Figure 7.2: Boehm's spiral model. (Source: Boehm, B. W. (1988) A spiral model for software development and enhancement, *IEEE Computer*, Vol. 21, No. 5 © 1988 IEEE.)

can be seen, development spirals outward from the centre in a clockwise direction with each cycle of the spiral resulting in successive refinement of the system.

In each cycle of the spiral there are four main activities, represented by the four quadrants. First, there is planning (bottom left quadrant) and at each cycle round the spiral this is about planning the next part or phase, and different elements will be involved depending on what stage the project has reached. Second, there is an objectives phase (top left), then, third, a risk analysis stage (top right) concerned with the identification and resolution of risks in the system. Risks, such as design flaws, failing to meet user needs, escalating costs, losing sight of the perceived system benefits, etc. are detected early in the development process. The spiral model, apart from its evolutionary nature, is particularly focused on addressing risk. This is not generally a strength of other models of development. Indeed, it is sometimes known as the 'risk-driven model'. The fourth stage is development (bottom right). In the early stages this might concern the development of manual or paper models, it might involve the development of a prototype (see below). The prototype might be used to help define and understand the requirements or illustrate a user interface for users to react to, or whatever. These may then evolve into the delivered system, and in later cycles they become more complete systems until finally they are the fully tested and engineered releases of the system. The exact number of cycles is not defined. It depends on the nature and characteristics of the particular project and the difficulties encountered in the cycles.

The vertical axis in the model represents the cumulative cost of the development project, which increases with each cycle. Similarly, the horizontal axis represents the commitment to the project, also increasing with each cycle. Prototyping is an important element of the spiral model but some commentators regard this aspect as its main characteristic, often forgetting the importance of the other quadrants.

Although described in this section concerning evolution, it should be noted that the spiral model is not evolutionary in iterative implementations, as indicated in Figure 7.1. It is evolutionary only in its revisiting of each of the stages. The same stages are visited again and again as the project cycles around the spiral until a complete systems implementation is produced.

7.2 Prototyping

Information systems tend also to be one-offs, but the cost of building a software prototype has been difficult in the past because building a prototype was often almost as expensive and took nearly as long as developing the final system, and therefore was rarely undertaken. This has changed with the availability of software tools which have greatly reduced the costs and speeded up the process of developing a prototype. Hardgrave (1995) suggests that over 70 per cent of systems development efforts involve prototyping of some kind, and Coad and Yourdon (1991) suggest prototyping should be used for all object-oriented projects. This, together with the use of prototyping in rapid application development (RAD) (Section 7.3) and its growing use in relation to web development (Section 7.5), suggests that prototyping is here to stay and as popular as ever.

Prototyping addresses some of the problems of traditional systems analysis; in particular, the complaint that users only see their information system at implementation time when it is too late to make changes. If a prototype version of the information system is not developed first, the systems analysts are experimenting on the user. The first version of the type is also the final version and this brings about an obvious risk of failure, including outright user rejection. Therefore, the prototyping approach is a response to the user dissatisfaction found when using the traditional approach to information systems development.

According to Chen (2001), the prototyping method was formally introduced to the information systems community in the early 1980s to combat the weakness of the traditional waterfall model (Chapter 3). The early prototyping process was for developers to design and build a scaled-down functional model of the desired system and then the developers demonstrate the working model to the user. This results in feedback on its suitability and effectiveness.

The developer then continues to develop the prototype until the developers and the users agree that the prototype is satisfactory.

At that point, an important distinction is made in the literature, as to whether the prototype is then thrown away, that is, used solely to help establish and elicit the user requirements, or whether the prototype is then evolved or enhanced to form the actual system that is provided to the user for real use. So, there are two types of prototype:

- a throwaway (or expendable) prototype;
- an evolutionary prototype.

This distinction often leads to differences in the design of the prototyping tool. Some are designed to develop prototypes that are expendable only, focusing on quickly building graphical interfaces and some basic processing and functionality. Such tools develop prototypes that are:

- inefficient for operational use;
- incomplete, performing only some of the required tasks;
- inadequate, sometimes being designed for one type of user or one purpose only;
- poorly documented;
- unsuitable, or difficult to integrate with other operational systems;
- incapable of scaling to the required volumes of operational use.

It is suggested that the use of evolutionary prototyping tools leads to slightly quicker applications development, as the developers are building on something that already exists, rather than starting from scratch. However, they need to blend the power of the prototype with the ability to evolve the prototype into an effective, live running system, which can handle large volumes and has the speedy response, high functionality, security, etc. that operational systems require.

With prototyping, user acceptance of a system is regarded as far more likely. By implementing a prototype first, the analyst can show the users something tangible – inputs, intermediary stages, and outputs – before finally committing to the new design. These prototypes are not diagrammatic approximations, which tend to be looked at as abstract things, but actual physical outputs, screens, databases, etc. The formats can be changed quickly, as the users suggest amendments and enhancements, until the users see a reasonable approximation of their requirements. Further, it may only be by using this approach that the users discover exactly what they really want from the system, as well as what is feasible. It is also often possible to try out an example run using data generated by the users themselves or with real data. This helps to ensure that the system and processes can handle the inputs and data thrown at it. Without a prototype this might not happen until very late in the development process, often just prior to implementation. The key benefit of prototyping is finding out such problems as early as possible. The earlier it is discovered the easier and cheaper it is to fix.

In information systems there are different kinds of prototype with different objectives. The most common use of prototypes is to examine areas where the users and analysts are unsure of the requirements and feel they need to tease out and explore the real needs by showing and amending a physical approximation of a system. With this objective, prototyping is argued to be an improvement on the traditional form of requirements gathering. Prototyping may form part of a methodology itself, or it can be used contingently as part of an existing methodology, used when and where appropriate. It is particularly useful where:

- the application area is not well defined;
- the organization is not familiar with the technology (hardware, software, communications, designs, and so on) required for the application;
- the communications between analysts and users are not good;
- the cost of rejection by users would be very high, and it is essential to ensure that the final version has got users' needs right;
- there is a requirement to assess the impact of prospective information systems.

Alternate forms of prototyping in information systems, apart from requirements elicitation objectives, are:

- *functional prototypes* for demonstrating, testing, and evaluating the functionality of a system;
- *process prototypes* for demonstrating, testing, and evaluating the processes, sequences, responses, etc. of a system;
- *design prototypes* for demonstrating and evaluating a variety of alternate designs or solutions;
- *performance prototypes* for testing response times, loads, volumes, etc.;
- *organizational prototypes* for demonstrating and evaluating different organizational designs, cross-functional work, organizational processes, and their integration with information systems (see, e.g. Hume et al., 1999).

Often, prototyping in practice combines two or more of these objectives.

There are a variety of factors that have been suggested as important when considering whether prototyping is appropriate. Those characteristics of a project where it is normally suggested that prototyping would be particularly beneficial are as follows:

- unclear requirements;
- unstable requirements;
- high innovativeness;
- high system impact on the organization (although not safety critical);
- high system impact on users;
- relatively small project size;
- relatively small number of users;
- relatively short project duration;
- where commitment of users to a project is required.

Although these characteristics may be recommended as appropriate in the literature, of course there are always examples where they have been ignored or success has been achieved in spite of them.

Prototyping may be more than just another tool available to the analyst. It could be used as a basis for a methodology of systems development in the organization. This may have:

- *an analysis phase* designed to understand the present system and to suggest the functional requirements of an alternative system;
- *a prototyping phase* to construct a prototype for evaluation by users;
- a set of *evaluation and prototype modification stages*;
- *a phase to design and develop the target system* using the prototype as part of the specification.

Prototyping is also regarded as a way of encouraging user participation (see Section 5.1 and Chapter 16). The hands-on use of prototypes by users provides experience, understanding, and the opportunity for evaluation. Once users and managers realize that things could be changed and that they could exert influence, it can lead to improved participation and commitment to the project.

Some analysts recommend that only the most critical aspects of a new system should be prototyped. Alternatively, the prototype may be built up using the most straightforward aspects and new material added to the prototype, as users and analysts understand the application area more fully. This is somewhat akin to the evolutionary approach discussed in the previous section.

A prototype is frequently built using special tools such as screen painters and report generators which facilitate the quick design of screens and reports. The user may be able to see what the outputs will look like quickly. Whereas a hand drawing of the screen layout will need to be drawn again for each iteration that leads to a satisfactory solution, the prototype is quickly redrawn (or repainted) using the tools available. As with word processing systems, the savings come in making changes, as only these need to be drawn. Therefore, the ease and speed with which prototypes can be modified are as important as the advantages gained from building the prototype in the first place. Iteration of screens and designs becomes a practical possibility.

Frequently a prototype system can be developed in a few days, and it may not take more than 10 per cent of the time and other resources necessary compared to developing the full operational system. This can be a good investment of time and it can speed up the overall development process. Analysts can usually achieve rapid feedback from the users so that the iterative cycle can quickly work to a version acceptable to the users. It has been suggested that this typically cuts in half the time that requirements determination normally takes, with better results.

Some systems teams use the prototype itself as the artefact that the users sign off (i.e. if they are happy with the prototype and the way it performs then that is good enough). This can be a more effective approach to obtaining a user decision than signing off the traditional manual specification.

Prototyping, although having many benefits, is not without its critics. It is certainly not a panacea to all the problems of traditional development and should be used appropriately. Prototypes have been criticized on a number of fronts (see, e.g. Janson and Smith, 1985) for leading to inadequate, or partial, system designs. Designs resulting from prototyping are not properly engineered, it is argued, they just emerge, leading to poor operational systems. Some situations are just too complex for systems to be designed in this way and that more formal and deliberate design strategies are needed. This is essentially the 'quick and dirty' criticism of prototyping, which is often made. These criticisms are particularly, and probably rightly, made in connection with safety critical applications. Developers who practise prototyping have been highlighted as not having the necessary design skills. Prototyping has also been criticized for the difficulties it presents in connection with testing. Normally, systems are tested against detailed specifications but with prototyping there may not be any specification.

Prototyping is sometimes regarded as difficult to manage and control, and the danger of forever seeming to cycle around minor differences has been experienced in practice (Fitzgerald, 1999). The proliferation of requirements that prototyping leads to is sometimes regarded as a problem, as is getting users to participate. Sometimes users have questioned the long time required to develop an operational system when the time taken to develop the prototype was so short. The managing of expectations in prototyping is very important to overcome this problem. Sometimes prototypes designed to be thrown away become used as operational systems due to the pressures of managers and users who want the system immediately and cannot understand why they cannot use the prototype, especially after they have seen it 'work'. Plato (1995) states that 'prototypes have an amazing tendency to become operational systems … and they wind up becoming half-baked operational contraptions'. Some users are disappointed that the operational system is not as quick as the prototype to run. Tudhope (2000) even suggests, 'some developers go so far as to introduce code for the purpose of producing delays in the speed of operation of early prototypes to avoid unrealistic user judgements of final system response'!

There is also the risk that the system requirements may change in the meantime. Some users may argue that the time, effort, and money used to develop a prototype are 'wasted'. It is sometimes difficult to persuade busy people that this effort does lead to improved information systems.

Many of the criticisms of prototyping are really concerned with its inappropriate use. Clearly, prototyping is not a substitute for thorough analysis and design. Implementation compromises should not be made as they are likely to remain in the final system. This includes the documentation, which might be neglected as analysts argue, 'it is only a prototype'. One of the necessary components of successful prototyping is management and control so as to ensure that compromises are not made and the process of repeated iteration does not go on too long.

The information system may be implemented in stages. At each stage, the missing components are those that give the poorest ratio of benefits over costs. The analysts in this case will have to be aware of robust design and good documentation when the prototype is being developed. The prototype must be able to handle the quantities of live data that are unlikely to be incorporated when giving end-users examples of the prototype's capabilities. Otherwise, prototyping will not improve the quality of systems development. Therefore, although prototyping is frequently regarded as a 'quick and dirty' method, it need not be 'dirty'. If the prototype is well designed, the prototype can feature as part of a successful operational information system. The temptation, however, is for a quick and dirty solution because the tools can produce a quick result and analysts are tempted to move quickly on to the development phase before sufficient analysis has been carried out. The emphasis on controls in prototyping is therefore necessary. Corners should be cut only in situations where the information system will have a very short lifespan (or is set up for once-only use).

Up to this point it has been assumed that prototyping was undertaken and facilitated by professional systems developers building a prototype for users to comment upon and react to, and this is indeed the normal approach. However, sometimes prototyping is undertaken by the users themselves without specialist IS developer help. This is part of the end-user computing concept (see Section 5.2). The prototyping tools must be relatively straightforward and easy to use and the users need to be self-motivated and committed to the project. It also has the benefit of freeing up the IS specialists for other activities or projects and means that the users drive the project forward at their pace and on their terms.

As indicated above prototyping has been used for many years in systems development but recent development approaches also use prototyping (see Section 7.3). Even more recently development of web-based systems has frequently used prototyping. This may seem a little odd at first because many web-based systems do not have the internal users of traditional systems, but, for example, Chen (2001) argues that 'prototyping methods are especially suitable for web-based applications because of the ease of system delivery and updates afforded by web technology'. However, he does go on to say that 'the unique requirements of web applications require the designers to take additional considerations when using these models (i.e. *prototypes*)'.

In Chen's Modified Prototype Method (MPM) for web application development, the basic functionality of a desired system, or a component of it, is formally deployed right away, via a prototype. The maintenance phase begins right after the deployment and it continually evolves. Once a prototype is deployed, its online users' actions are monitored using web server logging facilities. The subsequent analysis of these user actions determines which parts of the application are being used the most and which are being used the least, and the information is used to evolve future development and add efficiency to the application. So, the live system is essentially a prototype and the users, out there, contribute to its development and enhancement by their actions. Another web development methodology is WISDM (Section 23.4).

7.3 Rapid application development (RAD)

Rapid Application Development (RAD) has evolved as an important theme in information systems development over the last few years. The term appears to have first been used by James Martin in 1991; the Martin RAD methodology is described in Section 23.1. As with many themes, RAD was developed as a reaction to the problems of traditional development, in particular the problems of long development lead times. It also addresses the problems associated with changing and evolving requirements during the development process. RAD has a number of general characteristics or features as follows:

1 Incremental development

An important element of the philosophy of RAD is the belief that not all a system's requirements can necessarily be identified and specified in advance. Some requirements will only emerge when the users see and experience the system in use, others may not emerge even then, particularly complex ones. Requirements are also never seen as complete but evolve and change over time with changing circumstances (Section 7.1). Therefore, trying fully to specify a system completely in advance is not only a waste of time but often impossible. So why attempt to do it? RAD starts with a high-level, rather imprecise list of requirements, which are refined and changed during the process, typically using toolsets (Chapter 19). RAD identifies the easy, obvious requirements and, in conjunction with the 80/20 rule (see the Pareto principle below), just uses these as the starting point for a development, recognizing that future iterations and timeboxes (see below) will be able to handle the evolving requirements over time. Hough (1993) suggests using the technique of functional decomposition (see Section 6.2) and each function identified and the requirements listed, but, he says, 'the precise design specifications, technical issues, and other concerns should be deferred until the function is actually to be developed'.

2 Timeboxing

The system to be developed is divided up into a number of components or timeboxes that are developed separately. The most important requirements, and those with the largest potential benefit, are developed first and delivered as quickly as possible in the first timebox. Some argue that no single component should take more than 90 days to develop, while others suggest a maximum of six months. Whichever timebox period is chosen, the point is that it is quick compared with the more traditional systems development timescale.

Systems development is sometimes argued to have three key elements. In traditional development two are typically variable: time and resources (see Figure 7.3). In traditional development when projects are in difficulty, either the delivery time is extended or more resources are allocated or both but the functionality is treated as fixed. In RAD (such as DSDM, Section 23.2) the opposite applies, resources and time are regarded as fixed (allocating more resources is viewed as counterproductive although this does sometimes happen), and so that

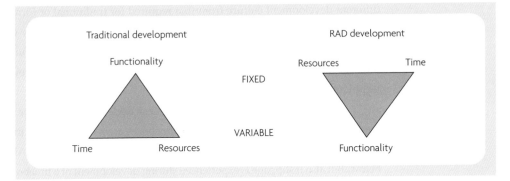

Figure 7.3: Traditional development – time and resources

only leaves functionality as a variable. So, under pressure and when projects are in difficulty, time and resources remain constant but the functionality is reduced.

RAD compartmentalizes the development and delivers quickly and often. This provides the business and the users with a quick, but it is hoped, useful part of the system in a refreshingly short timescale. The system at this stage is probably quite limited in relation to the total requirements, but at least something has been delivered. This rapid delivery of the most important requirements also helps to build credibility and enthusiasm from the users and the business. Often for the first time they experience a system that is delivered on time. This is radically different from the conventional delivery mode of most methodologies which is a long development period of often two to three years followed by the implementation of the complete system. The benefit of RAD development is that users trade off unnecessary (or at least initially unnecessary) requirements and wish lists (i.e. features that it would be 'nice to have' in an ideal world) for speed of development. This also has the benefit that, if requirements change over time, the total system has not been completed and each timebox can accommodate the changes that become necessary as requirements change and evolve during the previous timebox. It also has the advantage that the users become experienced with using and working with the system and learn what they really require from the early features that are implemented.

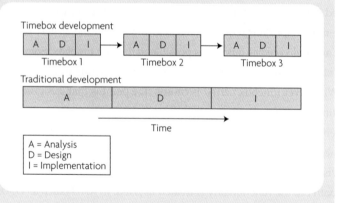

Figure 7.4: Comparison of timebox development and traditional development

Figure 7.4 illustrates three chunks of development and, although the overall time to achieve the full implementation could be the same as with a traditional development, the likelihood is that the system actually developed at the end of the three timeboxes will be radically different from that developed at the end of one large chunk as a result of the learning and evolving processes which lead to change being made to each specification at the beginning of each timebox.

Some RAD proponents argue that, if the system cannot be divided into 90 day timeboxes, then it should not be undertaken at all. Obviously such an approach requires a radically different development culture from that required for traditional or formalized methodologies. The focus is on speed of delivery, the identification of the absolutely essential requirements, implementation as a learning vehicle, and the expectation that the requirements will change in the next timebox. Clearly such radical changes are unlikely to be achieved using conventional techniques.

Once the duration of the timebox has been decided it is imperative that the system is delivered at the end of it without slippage, as timeboxes are fixed. So how is this achieved? Well, first, by hard work and long hours and, secondly, by the use of the other RAD techniques discussed below. But also if slippage is experienced during development of a timebox then the requirements are reduced still further (i.e. some of the things that the system was going to do will be jettisoned).

3 The Pareto principle

This is essentially the 80/20 rule and is thought to apply to requirements. The belief of RAD proponents is that around 80 per cent of a system's functionality can be delivered with around 20 per cent of the effort needed to complete 100 per cent of the requirements. This means that it is the last, and probably most complex, 20 per cent of requirements that take most of the effort and time. Thus why do it? – just choose as much of the 80 per cent to deliver as possible in the timebox, or at least the first timebox. The rest, if it proves necessary, can be delivered in subsequent timeboxes (or not at all).

4 MoSCoW rules

In RAD the requirements of a project are prioritized using what is termed the MoSCoW rules:

- M = 'the Must Haves'. Without these features the project is not viable (i.e. these are the minimum critical success factors fundamental to the project's success).
- S = 'the Should Haves'. To gain maximum benefit these features will be delivered but the project's success does not rely on them.
- C = 'the Could Haves'. If time and resources allow these features will be delivered but they can easily be left out without impacting on the project.
- W = 'the Won't Haves'. These features will not be delivered. They can be left out and possibly, although not necessarily, be done in a later timebox.

The MoSCoW rules ensure that a critical examination is made of requirements and that no 'wish lists' are made by users. All requirements have to be justified and categorized. Normally in a timebox all the 'must haves' and at least some of the 'should haves' and a few of the 'could haves' would be included. Of course, as has been mentioned, under pressure during the development the 'could haves' may well be dropped and even possibly the 'should haves' as well.

5 JAD workshops

RAD requires high levels of participation (see Sections 5.1 and 16.1) from all stakeholders in a project as a point of principle and achieves this partly through the JAD workshop. JAD (Joint Application Development) is a facilitated meeting designed to overcome the problems of traditional requirements gathering (see Sections 5.6 and 16.2), in particular interviewing users. It overcomes the long time that the cycle of interviews take by getting all the relevant people together in a short period to hammer out decisions. Normally, in the context of RAD, a JAD workshop will occur early on in the development process to help establish and agree the initial requirements, the length of the timebox, what should be included and what excluded from the timebox, and most importantly to manage expectations and gain commitment from the stakeholders. Sometimes a subsequent JAD workshop is used to firm up on the details of the initial requirements, etc. In some RAD approaches the whole process is driven by a series of JAD meetings that occur throughout the timebox.

Another important element is the presence of an executive sponsor. This is the person who wants the system (or whatever the focus of the meeting is), is committed to achieving it, and is prepared to fund it. This person is usually a senior executive who understands and believes in the JAD approach and who can overcome the bureaucracy and politics that tend to get in the way of fast decision making and usually bedevil traditional meetings (see Section 16.2 for further discussion of RAD).

6 Prototyping

Prototyping is an important part of RAD and is used to help establish the user requirements and in some cases the prototype evolves to become the system itself. Prototyping helps speed up the process of eliciting requirements, and speed is obviously important in RAD, but it also fits the RAD view of evolving requirements and users not knowing exactly what they want until they see or experience the system. Obviously the prototype is very helpful in this respect (Section 7.2).

7 Sponsor and champion

Having a committed sponsor and a **champion** of the systems is an important requirement for RAD and for its success. We have discussed the sponsor above. A champion is someone, often at a lower level of seniority, who is also committed to the project, who understands and believes in RAD, and is prepared to drive the project forward by their enthusiasm and hard work (overcoming bureaucracy and politics is the role of the sponsor – see 5 above).

8 Toolsets

RAD usually adopts, although not necessarily, the use of toolsets (Chapter 19) to help speed up the process and improve productivity. In general it is usually argued that the routine and time-consuming tasks can be automated as well using available tools for change control, configuration management, and even code reuse. Reuse of code (see also Section 8.5) is another way that RAD speeds the development process. However, it is not just about speed but also quality because existing code or modules have usually already been well tested, not just in development, but in real use. RAD searches for shortcuts and reuses code, maybe clones existing code and modifies it, or utilizes commercial packages, etc. where applicable. This may be code within the organization or bought from outside. Sometimes more than just a little piece of code is used; for example, complete applications may be used and the developers use this as the basis of the new system and change the interface to produce the desired results. Many 'new' e-commerce or Internet applications have been developed in this way using the legacy systems and then providing a new 'umbrella' set of applications and user interfaces on top. The idea is to leverage existing code, systems, experience, etc.

Specific RAD productivity tools have been around for some time and are developing fast, and existing tools and languages are being enhanced for RAD, particularly for rapid development of Internet and e-business-based applications.

RAD appears to be becoming relatively widely used in practice in organizations (Russo and Wynekoop, 1995; Eva and Guilford, 1996). Boehm (1999) suggests that there are 'good business reasons' why RAD has become increasingly popular. He argues that 'in general, RAD gives you earlier product payback and more payback time, before the pace of technology makes your product obsolete'. Subramanian and Zarnich (1996) suggest that the shorter elapsed time between design and implementation of RAD often results in the system being much closer to the user/business needs.

Despite the proposed advantages of the RAD approach it has been criticized as being 'quick but dirty', suggesting that some shortcuts are taken for the sake of speed, especially in relation to systems quality and documentation. One of the reasons for the establishment of the DSDM consortium was to counter such criticisms and this consortium believes that RAD can deliver quality systems (see Section 23.2).

Yourdon (2000) suggests that: 'When RAD was first introduced, developers sometimes used it as an excuse to abandon all discipline and resort to extemporaneous hacking.' He argues that this temptation is even greater in the current environment, particularly with the pressures to develop web applications rapidly.

The issue of infrastructure or underlying architecture is clearly important but often seems to be ignored or forgotten in accounts of RAD. Hough (1993) believes that the design of an application alone is not sufficient. He states:

> To realize the full potential of an application, it must be based on a sound architecture ... It is relatively easy to design a simple house (e.g. one with three bedrooms, a kitchen, etc.); however, only individuals with specific skills can properly design the architecture of a 50-storey building (how the walls are built, how big the foundations must be, etc.). Design implies caring for today's needs, whereas architecture suggests that long-term issues, including expansion, have been taken into consideration. Therefore, of key importance to Rapid Delivery is the ability to continuously add increasing functionality to the evolving application. If the overall application architecture is not sufficiently addressed early on, it will be difficult to integrate application segments over time.

Hough identifies the key architectures as: application, data, process, and technology, but he also suggests that others are important, such as presentation, control, security, and communications architectures.

Not all projects are thought to be suitable for RAD and in our opinion there are some areas in which RAD should not be used (e.g. safety-critical systems). The authors do not want the 'fly by wire' systems of the aircraft they fly to have been developed using RAD or with a fixed timebox! There may also be areas where RAD does not make sense, e.g. if the application does not need to be developed quickly, or if the requirements can be completely 'known' in advance, or if there are only 'must have' requirements, then traditional methods would still seem more appropriate. This leaves a lot of areas in-between, and we would not go as far as to advise that no large systems, or business-critical systems, or transaction processing systems, should ever be developed using RAD. Rather we would look to the maturity of the organization in using RAD and say that these areas are potentially more risky, but if the organization (developers, users, and management) are experienced and have a history of success with RAD, then they may well be able to undertake RAD in such areas.

This raises the issue of the culture change necessary for using RAD in organizations. It is clear that some developers brought up in traditional development environments find RAD very difficult to come to terms with as it goes against some of the norms that they believe are central to systems development. Others can take to it quickly but in either case a great deal of education and training in the new approach is usually required. The authors know of one organization where the culture change for existing developers was too great and they had to appoint a new team for RAD development rather than use the existing team. It is also clear that management and users need to be aware of the RAD culture. It typically requires them to have higher levels of commitment and participation throughout the process, and because of the evolutionary nature this commitment often has to be ongoing. Some of the cultural changes involved relate to the following:

- The need for interdisciplinary team working as opposed to individual specializations.
- The empowering of users and developers to make decisions as opposed to the decisions being made on high.
- The acceptance of 'good enough in the context' and fit for purpose as opposed to striving for perfection or the best possible solution.
- The notion that change is expected and not a surprise as opposed to change not being allowed once the requirements have been specified.
- The engendering of a non-blame culture and all pulling together as opposed to a search for scapegoats when things go wrong.

All these make implementing the RAD approach a challenging management of change activity.

7.4 Agile development

Development approaches known as 'agile', or representing the 'agile movement' or the 'agile school' have been around for some time. According to Highsmith (2002) the history extends over 10 to 15 years. In the UK, Tom Gilb was associated with an early evolutionary method in the 1980s with many aspects that would today be recognized as agile. Yet despite this history it has proved relatively difficult to define precisely what is meant by agile. Indeed Abrahamsson et al. (2002) categorically state that "no agreement of what the concept of 'agile' actually means exists"!

There are many different variants of the agile approach but they virtually all take as their starting point the problems and inadequacies of the traditional or waterfall approach to systems development (Chapter 3). However, these problems are particularly evident in relation to the development of web-based applications (see Section 7.5). One aspect of this relates to requirements (Section 5.6). The agile school accepts the notion that requirements are difficult for users and that often they cannot articulate or define in any detail their requirements. Users simply do not know what they want. Sometimes they may think they do but when a system is developed for them they realize that this was not what they wanted. Yet having seen the system, they now know what they do not want, and they are a little nearer to understanding what it is that they probably want.

In traditional development they would discover this at the end of a long life cycle of development, probably during user testing or even implementation. Correcting the application at this time is very late and very expensive and may, in any case, require several iterations even then.

Thus, the agile school believes that as requirements are so difficult to define, they must be evolved in some other way. This is normally achieved by adopting an evolutionary approach (Section 7.1) together with prototyping (Section 7.2) and a philosophy that embraces change as the norm, not something to be fought. Evolution and prototyping are concepts that are addressed elsewhere in this book but there is more to agile than these.

The agile school presents a relatively rare example where values, philosophies and beliefs are clearly stated (see Section 2.3). It has developed a new philosophy of development, and much like DSDM (Section 23.2), when particular agile methods or approaches are examined, they are sometimes found to contain more general discussion of values than specific detail about the approach or techniques to be adopted.

In 2001 a number of the leaders of the agile school, or proponents of what were then known as light or lightweight methods, attended a meeting, convened by Kent Beck, to form The Agile Alliance (Highsmith, 2002; Conn, 2003). This group of 17 people represented proponents of Extreme Programming (Section 23.3), SCRUM, DSDM, Adaptive Software Development, Crystal, Feature-Driven Development, Pragmatic Programming, and other likeminded people. Together they defined the philosophy of agile in a Manifesto (Beck et al., 2001). It begins by outlining what was valued in agile development, as follows:

- Individuals and interactions over processes and tools
- Working software over comprehensive documentation
- Customer collaboration over contract negotiation
- Responding to change over following a plan.

The authors of the Manifesto stated that while there is value in the items in the second part of the sentence (for example, processes and tools), agile development stresses the items in the first part more (for example, individuals and interactions).

The 'individuals and interaction' value can be interpreted as a focus on the human role and relationships of developers as opposed to the more processes and techniques focus of traditional methodologies. A community of developers should be fostered and working environments and team spirit are regarded as important.

The next value relates to working software over documentation and although most approaches focus on working software at some stage, this is about frequent delivery of working software to keep the customer involved and providing regular and rapid benefits. This is more important than producing the detailed documentation usually required in software development.

Customer collaboration is a focus on the relationship between the customer and the developers as collaborators with a common goal of producing a working system quickly rather than the more typical 'us and them' blame culture that has often existed in systems development.

The contract negotiation refers to the 'requirements definition document' or similar and the arguments over whether something was in the specification or not.

Responding to change relates to the notion, mentioned above, that change is accepted as the norm and has to be addressed not suppressed. Just because something was not in the plan does not matter. If it is now needed it should be provided.

The Manifesto went on to list 12 principles that the authors thought should be followed to reflect these values. The principles are as follows:

- The highest priority is to satisfy the customer through early and continuous delivery of valuable software.
- Changing requirements are welcome, even late in development. Agile processes harness change for the customer's competitive advantage.
- Working software is delivered frequently, from a couple of weeks to a couple of months, with a preference to the shorter timescale.
- Business people and developers must work together daily throughout the project.
- Projects should be built around motivated individuals. If the right environment and support is provided, the developers can be trusted to get the job done.
- The most efficient and effective method of conveying information to and within a development team is face-to-face conversation.
- Working software is the primary measure of progress.
- Agile processes promote sustainable development. The sponsors, developers, and users should be able to maintain a constant pace indefinitely.
- Continuous attention to technical excellence and good design enhances agility.
- Simplicity, the art of maximizing the amount of work not done, is essential.
- The best architectures, requirements, and designs emerge from self-organizing teams.
- At regular intervals, the team reflects on how to become more effective, then tunes and adjusts its behaviour accordingly.

Such a philosophy is described by Highsmith and Cockburn (2001) as 'a new combination of values and principles that define an agile world view'. The Agile Manifesto has become very influential and most discussions of agile begin with it in some form or another. But as can be seen from the number of references to other sections in this book, many of the elements are not new. Highsmith and Cockburn (2001) agree and state, 'what is new about agile methods is not the practices they use, but their recognition of people as the primary drivers of project success, coupled with an intense focus on effectiveness and manoeuvrability'.

However, the agile movement in its widest context does not just stem from a group of developers getting together and producing a manifesto. The idea of agile is an older concept than that of agile systems development and, for example, has existed for some time in the manufacturing context. Maskell (1996) suggests that agile manufacturing goes beyond the concepts of world class or lean manufacturing, which were designed to manage things that can be controlled. Agile manufacturing is designed for things we cannot control. He goes on to say that 'agility is the ability to thrive and prosper in an environment of constant and unpredictable change. Agility is not only to accommodate change but to relish the opportunities inherent within a turbulent environment'. Maskell (1996) identifies some of the axioms of agile manufacturing as follows: 'Everything is changing very fast and unpredictably. The market requires low volume, high quality, custom and specific products. These products have very short life-cycles and very short development and production lead times are required. Mass production is moribund. Customers want to be treated as individuals. This leads to a people intensive, relationship driven operation ... Products and services become information-rich.'

A number of these axioms are reflected in the software development agile movement. Baskerville and Pries-Heje (2001) suggest that they are 'conceptually and chronologically synchronous' with internet speed software development. They say that short-cycle time production is prevalent in industries 'where the length of product life cycles has been driven down by innovation in the marketplace ... and that software continues to be a member of that family of industries, which also includes such products as integrated and printed circuits'. This is interesting because software is often thought not to be like other manufacturing products. However, the association with manufacturing, and in particular, automotive manufacturing is strong in agile.

The analogy with new product development concepts has also been influential. In this environment, according to MacCormak et al. (2001), over the last twenty years research has created a good understanding of how to effectively manage and bring new products to market successfully. This understanding relates to organizational structures, teams, sequential processes, lead-times, costs, etc. However, this was true in fairly stable environments where the markets and technologies employed were well understood. But this understanding has been questioned in more uncertain and turbulent environments (Iansiti and MacCormack, 1997). Figure 7.5 illustrates the difference between the traditional model of new product development and a more flexible approach.

The traditional approach is highly structured and a new product is designed, developed, transferred to production and introduced to the market in a sequential process. The various needs and requirements of the new product are identified, the technological options are then assessed, choices are made, and designs are undertaken. This is the concept development phase, at the end of which nothing can be changed. The implementation phase involves turning the designs into a real product which can be launched into the market. However, as can be seen from Figure 7.5, the concept is frozen for a considerable time before it is introduced to the market, during which time the market may change and requirements may change, especially in dynamic and turbulent environments.

Those advocating a more flexible approach suggest a model more like the lower half of the figure where the commitment or concept freeze is delayed as long as possible. Indeed the concept development stage overlaps with implementation. As Iansiti and MacCormack (1997) argue, 'systemic changes in a product's definition and basic direction are managed proactively; designers begin this process with no precise idea of how it will end'. Implementation begins before a complete design is finished or stabilized and parts, or all of it, may well have to change as requirements evolve. This clearly requires significantly different implementation behaviour

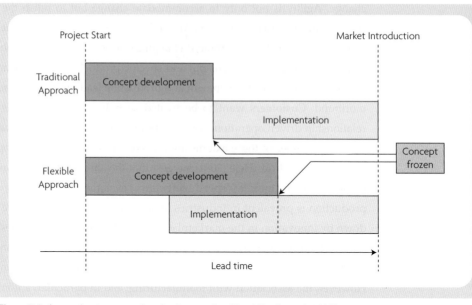

Figure 7.5: Approaches to new product development (Iansiti and MacCormack, 1997)

to the traditional approach and an understanding of things that can be easily changed and things that cannot. It requires a culture change in developers to welcome and accept such change and an ability to leave commitment to final design as long as possible. Eventually, particularly in a manufacturing environment, the concept still has to be frozen at some point to enable production, but it is later than it otherwise would be, giving more time to respond to the market. For this to be effective, an additional process of understanding the market and discovering how it is changing or evolving during the concept development process must be in place to enable an appropriate response to occur. This change is radical and runs counter to much that is traditionally held dear to designers and engineers, and in particular the lessons of the 1990s that were learnt from the Japanese in relation to quality manufacturing and especially automotive manufacturing. This is characterized by DeMarco in an introduction to the Highsmith book, as an obsession with discipline, rigour and process to achieve quality.

These changing concepts in manufacturing were very relevant to software development and reflect to some extent the agile concept. Indeed Iansiti and MacCormack use software development in some of their examples of flexible processes in action. The agile school, although adopting the concepts of flexible product development, also took things a step further. They argued that in many cases, requirements were not only difficult to understand and subject to frequent change but sometimes they were impossible to discover no matter how hard one tried. The development of early web-based systems provide an example where a more 'trial and error' form of development had to be applied (see Section 7.5).

This kind of thinking and experience implied that different ways of development had to be evolved. Thus rapid iterative and evolutionary development emerged with few detailed requirements specified, except for an understanding of the basic business objectives of the system, and little commitment to design principles, as almost everything would probably change, with a focus on feedback and rapid response.

Miller (2001, quoted in Abrahamsson et al., 2002) suggests that agile approaches have the following general characteristics:

- Modularity on the development process level
- Iterative, with short cycles enabling fast verification and corrections to be made
- Time bound, with iteration cycles from one to six weeks
- Parsimony in the development process removing all unnecessary activities
- Adaptive with possible emergent new risks
- Incremental process approach that allows functioning applications to be built in small steps
- Convergent approach to minimize risks
- People oriented, i.e. agile processes favour people over processes and technology
- Collaborative and communicative working style.

Interpreting this list, we understand the focus of agile approaches to be characterized by iteration, short life cycles, minimalism, emergence, risk acceptance, people orientation, collaboration and communication. These will be examined in turn.

Iteration

Iteration has been discussed in relation to evolutionary development in Section 7.1 and in this context and that of agile development it means the partitioning of a project into a set of parts each of which can potentially deliver working software that implements a required feature or part of a feature. Each increment thus contains some or all of the traditional phases of development, such as requirements gathering, analysis, design, build and implement. It is argued that this revisiting of these phases improves the project. For example, by revisiting the requirements in a second iteration they will often be found to have changed. Iteration also means the project is cut up into more manageable-sized chunks. Iteration thus allows for multiple passes over a project in terms of repeated implementations as each chunk or iteration is delivered.

Note this is a different form of iteration to that of DSDM where the iteration is within a phase or a number of phases (Section 23.2).

According to Szalvay (2004), this concept of iterative development comes from the 'lean development' ideas of the Japanese in automotive manufacturing. He says that they made 'tremendous efficiency and innovation increases simply by removing the phased, sequential approach and implementing an iterative approach, where prototypes were developed for short-term milestones. Each phase was actually a layer that continued throughout the entire development lifecycle; the requirements, design, and implementation cycle was revisited for each short-term milestone.' And that 'this "concurrent" development approach created an atmosphere of trial-and-error experimentation and learning that ultimately broke down the status quo and led to efficient innovation.' He also says that 'although direct analogies between industries are never seamless, the success of lean development has influenced a broad class of "iterative" software methods including the Unified Process, Evo, Spiral, and Agile methods'.

Short life cycles

Each iteration should be short such that something useful is delivered to the users in a timescale that is reasonable without them losing patience or enthusiasm or indeed without their requirements changing fundamentally. The timescales depend on the nature of the projects, but typically they are from a few days to a few weeks, but certainly not many months or years.

Minimalism

This means that developers only undertake activities that are going to help produce results in the short term. These are sometimes termed 'high value' activities. Any activity that does not lead to this goal is not undertaken. Highsmith (2002) talks about this as 'the art of maximising the amount of work not done'. This is quite different to some traditional methodologies which have lists of activities that must be performed and documents that must be produced at each stage. Many of these activities may not produce any tangible contribution to the progression of the project. Documentation is often quoted as an example, and so documentation is minimized. Jeffries, quoted by Highsmith (2002), suggests that projects should cut the numbers of documents produced and their content in half at each stage. If no problems are experienced, cut them in half again. Continue doing this until some negative effects are felt, then add a little back.

Emergence

This element of the agile approach is a result of concepts taken from complex adaptive systems (CAS) theory. CAS attempts to study and understand autonomous behaviour originally obtained from the study of living systems in nature, for example the flocking of birds, the swarming of bees, etc. It has been discovered that individual components or 'agents' in these complex activities adopt local rules, but that the outcome of the collective behaviour is characterized by an overarching order which is developed via self-organization and a collective intelligence that appears to be greater than the sum of the individual parts.

CAS has more recently been adapted and applied to human activity systems, including management. Systems development is clearly complex as are the organizations in which they operate, but the interaction of human beings involved in such development is particularly complex. Thus it was felt that maybe CAS might have something to offer in this context. CAS embodies the notion of synergy (as does systems theory, see Section 4.1) where something that is greater than the sum of the individual parts can emerge as a result of certain combinations of elements. Such emergence is difficult to create and engender but it is critical to creative thinking in groups and is particularly important where innovative thinking is required.

Typically, the agile approach involves some adoption of CAS concepts and emergence in particular, in order to try to engender such thinking and behaviour. This is done by providing a particular environment which is calculated to help emergence and by following what are termed generative rules. These are not really rules, despite the name, but are more a set of principles or guiding practices. Generative rules are the opposite of inclusive rules. Inclusive rules attempt to lay down everything and specify what must be done in all circumstances, whereas generative rules are a minimum set of principles that should be applied to generate appropriate rules in specific circumstances as and when needed. In other words, agile depends on individuals and their intelligence and creativity to discover ways of solving problems as they arise – in line with the principles. The generated rules are supposed to be simple. According to Highsmith (2002) having generative rather than inclusive rules provides three benefits:

- Flexibility, as the group can quickly adopt to changing circumstances
- Robustness, as the behaviour of a rogue individual can be dealt with by the group
- Self-organization, as typically the group does not require supervision or direction.

In the agile approach the generative rules are often, but not always, specified as the 12 Agile Principles of the Agile Alliance. Augustine and Woodcock (2003), in relation to XP (Section 23.3), suggest they are guiding vision, teamwork and collaboration, simple rules, open information, light touch and agile vigilance. Interestingly they derive these directly from CAS theory, as shown in Table 7.1.

Following these principles or practices, a manager becomes an adaptive leader responsible 'for setting the direction, establishing the simple, generative rules of the system, and encouraging constant feedback, adaptation, and collaboration' (Augustine and Woodcock, 2003). This provides teams implementing agile approaches with:

- An intrinsic ability to deal with change
- A view of organizations as fluid, adaptive and intelligent
- A recognition of the limits of external control and the importance of self-organization as a means of establishing order
- An overall problem-solving approach that is humanistic in that:
  - employees are viewed as skilled and valuable stakeholders
  - the collective ability of autonomous teams is assumed
  - up-front planning is restricted to a minimum with stress on adapting to changing conditions.

Risk acceptance

In traditional methodologies risks are to be identified, assessed and avoided, usually by applying restrictive management practices, such as bureaucratic risk assessment programmes and then opting out of any activities identified as high risk. Agile sees this as missing an opportunity, because those high risk activities have the potential of delivering large benefits. Those that return large benefits are by their nature likely to be risky. Thus, agile approaches tend to take a more proactive view of risk and see it as inevitable but also an opportunity. It is the opportunity that risk presents that is to be embraced. As Highsmith (2002) indicates, 'change generates either the risk of loss or the opportunity of gain'. In the new world of rapid change

CAS Principle	Agile Management Principle
Non-material fields exert force on material objects.	*Guiding Vision.* Recognizing vision as a non-material field rather than an elusive destination results in vision continuously guiding and influencing behaviour in positive ways.
Autonomous, intelligent agents form the basis of CAS. *Interactions* between these agents result in *self-organization* and other *emergent* phenomena.	*Teamwork and Collaboration.* Recognizing individual team members as intelligent, skilled professional agents and placing a value on their autonomy is fundamental to all other practices. Teamwork and Collaboration form the basis for rich interactions and cooperation between team members.
Local, strategic rules support complex, overlaying behaviour in a team environment.	*Simple Rules.* Simple Rules such as XP Practices support complex, overlaying team behaviour.
Information is energy that serves as an agent of change and adaptation.	*Open Information.* Open information is an organizing force that allows teams to adapt and react to changing conditions in the environment.
Emergent order is a bottom-up manifestation of order, while imposed order is a top-down manifestation.	*Light Touch.* Intelligent control of teams requires a delicate mix of imposed and emergent order.
Non-linear dynamic systems are continuously adapting when they reach a state of *dynamic equilibrium* termed the *edge of chaos*.	*Agile Vigilance.* Visionary leadership implies continuously monitoring, learning and adapting to the environment.

Table 7.1: Agile and Complex Adaptive Systems (CAS) compared (from Augustine and Woodcock, 2003)

and turbulent environments no one quite knows what will work in the market and what will not. In such environments opportunities must be explored in a quick and effective way.

It is for this reason that some agile approaches have the notion of a portfolio of projects, some of which will make it and some that may not. In the marketplace the risks relate to the product, the timing, the consumer, the technology, the costs, the benefits, etc. but those that make it can be spectacularly successful and it is argued that only agile approaches enable this. Highsmith (2002) uses an interesting analogy with oil exploration. If a company's oil exploration group was 99.5 per cent successful they would probably be thought to be too conservative because it would be assumed they were missing lots of potentially valuable oil fields, i.e. the fact that they almost always hit oil means they are not taking enough risks. The argument is that it is the same in business today. However, managing these risks is key. Agile approaches enable the potential to be explored quickly without it costing too much if they fail, and this mind-set must be adopted not just by agile practitioners but by senior strategic managers in organizations as well.

The second element of risk is about managing it within an agile project, and although risk management is an important part of agile approaches it is not always made clear exactly how this happens. Indeed, Smith and Pichler (2005) state that most agile books have 'remarkably little to say about how a development team determines the risks it faces, prioritises them, or takes action to negate their effects'. They suggest that it needs to be done by applying standard risk management processes (Section 15.7) but doing them quicker and with less effort. The agile approach helps this to be achieved. They identify five steps that are required, all traditional:

- identify the risks
- assess their severity
- prioritize the risks based on the severity
- create action plans (responses) to deal with the severe risks
- continuously monitor to ensure the actions plans are achieving their objectives.

They suggest that in agile these steps need to be undertaken but made faster and more effective. This is achieved by adopting a number of strategies. First, by building the process into each iteration as a norm and because of the nature of agile teams this will make it quick but effective. Second, by assessing risks based on experience and perceptions, rather than by long cycles of quantitative analysis. Third, by identifying underlying drivers of risk rather than superficial ones. For example, a user may not be able to make a decision about something, or identify that this is normal behaviour for this particular category of users (underlying driver), so the response might be not to try to force a decision but to try to build in a number of options that can be decided upon later.

People orientation

The team of people and how they work together is probably the most important element in the agile approach, and more important than the choice of processes, models or tools. There are of course a range of involved stakeholders, not just developers. Probably the most important are not the developers but the end-users, business users, line managers and senior business managers who are all important and must have the right 'agile-oriented' attitudes. But it is the abilities and attitudes of individual developers that are key. The agile school recognizes that people have different skills and that some developers and programmers are better than others. Good people will do a better job than not-so-good people. As Fowler (2002) states, 'if I could have a dozen top-flight people who work comfortably together, I will take them over a team of average folks any day. It doesn't matter what process is being used, because the talent will win out'. The reason for this is that good and talented people will do better than average people because they will find ways to make things work, they will not be constrained by the process when they think it is inappropriate or forcing them in directions they think are wrong. This is quite a different view than that taken by other approaches, for example, SSADM (Section 21.1).

With agile, it is not appropriate to follow a methodology slavishly and not all things will work in all circumstances. Good people provide good outputs and also apply intelligence and are able to act flexibly, which is an important aspect of agile approaches.

Collaboration and communication

Although possibly a subset of people orientation (Chapter 5), collaboration and communication between people is nevertheless also a key element of agile. Collaboration is about working together to deliver something, whereas communication refers to the sending and receiving of information. The importance given to the collaboration between stakeholders stems from the old problem of 'us and them' and the blame culture which has ruined many development projects (see above). The collaboration between developers and users with a common objective is what is argued to make a difference to projects. Nobody should try to blame others for problems, but just get on and try to solve them, in a spirit of collaboration for the good of the project (and the business). This also applies to the rapid decision making that is required for agile development, where a sharing of knowledge is required and also a recognition of what is not known nor fully understood and an acceptance that decisions will not be absolute and may have to be revised. It is understood that they were taken in good faith.

In relation to communication, advocates of agile approaches believe that the most effective communication is face to face. This is why it is usually recommended that development teams are co-located with the business. This will often be more expensive than locating the developers elsewhere, but the benefits are so great that it will pay off in the long run. If only one critical question is resolved because the developer can turn to a knowledgeable business user in seconds rather than days, or where a user can intervene when they overhear something they think is wrong, it is worthwhile. Co-location also reduces the cost of moving information between people and increases the speed at which the consequences and implications of decisions become apparent. Face-to-face collaboration also enables a large amount of documentation to be eliminated because much of it relates to communicating when you are not talking to someone face to face. It enables important discussions and decisions to be captured on bits of paper, post-it notes and whiteboards rather than formal documents (Beynon-Davis et al., 2000). Agile approaches also usually place great emphasis on team-building activities to foster collaboration and communication.

Agile methodologies

There are a number of methodologies that define themselves as agile in some way. These include Extreme Programming (XP) (Section 23.3), SCRUM (Schwaber and Beedle, 2002), Crystal (Cockburn, 2004), Agile Modelling (AM) (Ambler, 2002), Adaptive Software Development (ASD) (Highsmith, 2000), and Feature Driven Development (Palmer and Felsing, 2002. A number of authors of agile approaches do not like the use of the term methodology to describe them but prefer the term method or framework. Indeed sometimes the agile movement is perceived to be anti-methodology in general and, as discussed in Section 27.4, is part of a backlash against methodology. However, Highsmith (2002) stresses that this is not the case and argues that the agile movement in fact seeks 'to restore credibility to the concept of methodology'. This means that they feel that the concept of methodology has lost some credibility partly by the word becoming associated with the early bureaucratic methodologies of the structured school, and with notions of order, discipline and regimentation, and partly because methodologies have been thought to represent artificial notions that are not actually undertaken in practice (see Truex et al., 2000).

Limitations/Criticisms

Despite the many approaches that exist and the apparent great enthusiasm for agile development shown in recent times, it is not without its critics. Clearly, the supporters of engineering approaches and many traditional software engineers may not like some of the notions that agile embodies and there is a concern that agile might be 'a licence to hack' (Wiegers, 2001). Others question the applicability of agile in inappropriate environments and especially in larger projects. Turk et al. (2002) question some of the assumptions of agile and suggest that it is not relevant to distributed development environments with its assumptions about co-location of developers and customers. Distributed development, including offshore outsourcing (Section 9.4) and sub-contracting, is becoming an increasingly important development environment and they argue that this does not fit well with agile. They also question its relevance where there are large development teams, safety-critical domains, and large and complex systems. A concern for others is that the agile manifesto (see above) tips the balance too far to the left in its emphasis on some elements over others, for example working software over documentation, or responding to change over following a plan. Boehm (2002) for example thinks that both have importance but what is critical is getting the balance right between agility and discipline, or finding the 'sweet spot' as he terms it. Others have been more critical and suggest that the items on the right-hand side (of the agile manifesto) are essential whereas those on the left 'serve as easy excuses for hackers to keep on irresponsibly throwing code together with no regard for engineering discipline' (Rakitin, 2001).

7.5 Web-based development

Web development is sometimes argued to be different to other forms of systems development. This argument was often heard in the early days of developing systems for the web, from around 1995, because the focus was on the user interface and the multimedia nature of the web. A new breed of developers seemed to be required who were more like graphic designers than traditional developers or programmers. This changed over time as more complex web-based systems, particularly e-commerce systems, became the norm. These still required some traditional development attributes as they usually involved integration with existing and legacy information systems, a knowledge of various platforms and architectures, as well as high levels of robustness and security.

Thus web development in the context of this book is more than merely the development of an information-providing website. It involves users in symmetrical communication to exchange information and undertake tasks. This is, according to Holck (2003), a view of web information systems (WIS) as 'information system' rather than as information provider, advertisement, or community. Holck provides a discussion of whether web development is different from traditional development or not and concludes, perhaps not surprisingly, that it depends on how one defines a WIS.

One aspect that was different in the early days of web-based development of e-commerce systems was that eliciting requirements presented particular problems. Requirements were not well understood and often completely unknown because the web was so new. Further, there were no or few users, especially end-users or customers. There were people who were internal to organizations who might second guess what end-users might want, but there were few real users as yet. So how could you ask them what they wanted when you did not yet know who they were? As a result, new relationships had to be built with people and hopefully new customers found. This was quite different to traditional software development processes where the requirements were typically elicited from users in the analysis stage (see Section 5.6).

Frequently, requirements were only understood when a product or service was actually in the marketplace after its launch. The market then responded by either using the website and undertaking transactions or not. Thus the feedback was provided through the use of the site. The website itself had to be adapted again and again over time. In the early days of web-based development it was often about providing the market with something, it did not matter very much what was provided but it was important to provide something and then evolve it rapidly in response to market perceptions and feedback. It had to be a 'build it, see it, try it, change it' approach.

However, over time as more experience was gained with web-based systems and e-commerce it became somewhat easier to contact customers. Focus groups from the marketing area were often used to help elicit and confirm requirements. Nevertheless, it remained, and still remains, a somewhat difficult and essentially 'trial and error' process.

Baskerville and Pries-Heje (2001) identify ten concepts relevant to IS development for web-development (or Internet time as they term it). We will explore these concepts to gain an insight into the real-world characteristics of such projects. The first two, time pressure and vague requirements, link all the concepts:

- *Time pressure*: Competitive pressures may mean that any advantage is short-lived and will be copied quickly. Thus it is important to take advantage of any short-term gain to lead to more long-term advantages.
- *Vague requirements*: Requirements are often imprecise or not known at all and have to be created through imagination and innovation. It is not infrequent that it is only on implementation and use that the real requirements are revealed.
- *Prototyping*: The software prototype is the specification of requirements, not the thick paper report. This prototype is frequently rejected so that a more well-founded design can be constructed.
- *Release orientation*: These earlier ideas and others suggest early release and frequent re-release. Rapid application development is even more relevant to Internet projects.
- *Parallel development*: Database development can take place at the same time as the graphical design; and requirements analysis and design become hard to separate.

- *Fixed architecture*: Complexity needs to be tamed. A three-tier architecture where the business data, business logic, and the user interface are separated out allows team members to work in parallel with a degree of independence.

- *Coding your way out*: When the going gets tough, the developers need to code their way out of problems. Hacking was not originally a pejorative term, but was used to identify programmers who could write elegant and effective code quickly.

- *Quality is negotiable*: The question often arises as to whether software is developed to achieve high quality, a quick product or a cheap one. In some senses, quality has always been negotiable. In web-based projects the overriding view of quality tends to be the customer perspective and experience, rather than a defined and repeatable development process or a software product that survives an internal audit.

- *Dependence on good people*: Internet projects are completed under time pressure and typically in small teams where all members need to pull their weight. Key staff can make or break a project.

- *Need for structure*: The old structures of systems development, for example, keeping business analysts separate from software engineers, may be inappropriate to building applications in Internet time.

As Vidgen et al. (2002) show, although the ten concepts capture the emergent aspects of web IS development well, if the term 'business urgency' is substituted for the concept 'Internet time' then it is clear that the ten concepts have a more general relevance to understanding the IS development process. In situations characterized by time pressure and definitional uncertainty the response of IS developers has long been to adopt a flexible strategy to IS development using techniques such as rapid application development and prototyping. However, there are some concrete differences between Internet projects and traditional IS development:

- *Internet time.* The development time is reduced greatly – two years is unthinkable, 6 months is often unacceptable and many significant e-commerce projects are implemented in weeks. This means that the ten concepts above tend to be the norm rather than the exception.

- *Strategic implications.* The strategic implications are directly related to business goals, particularly in e-commerce projects where a revenue stream is generated.

- *Emphasis on graphical user interface.* There is a need for talented graphic designers to work with software engineers.

- *Customer-orientation.* The user is a customer rather than an employee. E-commerce applications need customer focus and marketing input. Again, these are not traditional areas of software engineering.

On the other hand, there are also similarities:

- *Databases.* Sophisticated Internet applications rely on databases and require traditional software engineering skills to implement them.

- *Integration.* Internet applications need to be integrated with enterprise applications. For example, a car ordering system that gives consumers a delivery date needs to communicate with manufacturing requirements planning (MRP) software. Enterprise Application Integration (EAI) is a challenge for Internet applications.

As Internet projects become broader in scope, requiring greater integration with front office, back office, and legacy IT systems of all sorts, then, in our view, Internet projects will become yet more difficult to distinguish from traditional IT projects. Traditional IS projects would also benefit from more attention to strategy, customers, and design aesthetics and therefore the distinctions should, over time, become less pronounced and even disappear altogether as web-based IS development becomes 'business as usual'. Table 7.2 (from Vidgen et al., 2002) shows the stereotypical differences between traditional IS and Internet development projects. Their work forms the basis of the web development methodology WISDM (Section 23.4).

Dimension	Traditional IS projects	Internet projects
Strategy	The strategic dimension is abstract The strategic dimension is addressed indirectly, through broad notions such as strategic alignment. Often the strategic dimension is not addressed at all	The strategic dimension is tangible and visible and relates closely to business goals Strategy is addressed directly, particularly for e-commerce projects in which a revenue stream is generated
User	The typical user is an employee Users can be trained and consulted directly. System use might be mandatory User needs can be understood through work studies Job satisfaction is a key aim	The typical user is a customer who makes payment for goods and services Usage is not mandatory and the customer will not attend training sessions User needs can be understood through sales and marketing methods Customer satisfaction is a key aim
Design	The development focus is on the internals of the design: the database, the programs and an architecture (e.g. three-tier) The user interface is almost an afterthought	The development focus is on the website as a visual artefact. The development cycle might start with a mock-up of the user interface Graphic design skills and a feel for web aesthetics are essential

Table 7.2: Traditional and Internet-based applications (from Vidgen et al., 2002)

Summary

- Evolutionary systems development is a staged or incremental approach that periodically delivers a system that is increasingly complete as it evolves over time.

- A prototype is an approximation of the information system to be built. Developers can design and build a scaled-down functional model of the desired system and then demonstrate this to the users to gain feedback.

- Rapid application development (RAD) follows principles and uses techniques including incremental development, timeboxing, MoSCoW rules, JAD workshops, prototyping and toolsets to achieve speedier development.

- Agile development approaches aim at flexible and quick development of software, even where requirements are difficult to define. It emphasizes interactions between people, developing software with less emphasis on documentation, collaborating with customers and responding to change in the development process.

- Web-based information systems development is just another, albeit newer, application type, but it does have some particular emphases including time pressures, design and user interface requirements, security concerns and customer orientation.

Questions

1. Compare and contrast the various approaches to information systems development discussed in this chapter with the SDLC discussed in Chapter 3.

2. What are the differences between evolutionary development, prototyping, RAD and agile development (or do you think these are very similar)?

3. Argue a case for and against that there are fundamental differences between web development and any other information systems development.

Further reading

Brinkkemper, S., Lyytinen, K. and Welke, R.J. (eds.) (1996) *Method Engineering: Principles of Method Construction and Tool Support*, Chapman and Hall, London.

Chen, J. (2001) Building web applications, *Information Systems Management*, Winter, Vol. 18, No. 1.

Martin, J. (1991) *Rapid Application Development*, Prentice Hall, Englewood Cliffs, New Jersey.

Orman, L. (1998/9) Evolutionary development of information systems, *Journal of Management Information Systems*, Vol. 5, No. 3.

Vidgen, R., Avison, D.E., Wood, R. and Wood-Harper, A.T. (2002) *Developing Web Information Systems*, Butterworth-Heinemann, London.

Warren, I. (1999) *The Renaissance of Legacy Systems*, Springer-Verlag, London.

8 Engineering themes

8.1 Legacy systems

In this book we concern ourselves in the main with developing new computer applications. However, established organizations will also be concerned with running systems that have been in operation for some time. These are its legacy systems, and some may even have been originally developed as early as the 1960s and 1970s. Because of this, some legacy systems are likely to be high-volume transactions processing applications running on mainframe computers – commercial systems frequently written in the Cobol programming language using structured files rather than database systems. With high maintenance costs, use of obsolete hardware and software, poor documentation, changes in the real-world situation and lack of people support with the knowledge required to maintain them, it is not surprising that legacy systems have become an enduring and problematic theme!

Legacy systems may well perform critical processes, a reason for their early development, and be valuable assets of the organization. However, during their lifetime they could have undergone many changes in the review and maintenance phase of the SDLC. These changes are likely to have made them more complex. Sometimes, documentation might have been neglected and is therefore not up to date. For all these reasons legacy systems will probably have become more and more expensive and difficult to maintain.

Sometimes new systems will be developed to replace legacy systems that no longer achieve their objectives, use outdated technology, or are very costly to maintain. At the other extreme are those systems that have been performing effectively with minimal maintenance necessary, sometimes for many years. However, for most organizations, many legacy systems are a major problem: both expensive to maintain and to replace.

A critical point in many legacy systems comes when a change in the environment requires a major change in the system, such as the ability to handle increasing data volumes or differing functionality (or in the recent past ensure that they will run successfully in the year 2000). The organization needs to decide whether to invest in maintenance or replacement. This is not only a question of cost. As Warren (1999) points out, if you abandon a legacy system 'you face the risk of losing vital business knowledge which is embedded in many old systems'.

If the choice is maintenance, then the organization may consider implementing new features that make the system more efficient and effective, as well as ensuring that the system incorporates new functional requirements and existing problems are corrected. If the choice is replacement, then the project is likely to be very expensive unless there is an appropriate application package (Section 9.1) that was not available previously or components available that can be conveniently reused.

One alternative to continued maintenance or replacement is to **reverse engineer** and then **forward engineer** the application. Software tools exist that will attempt to analyse the operational application programs, in terms of the rules and procedures that they embody (reverse engineering), and modify them so that they conform to new standards (forward engineering). The new standards could relate to a more recent version of the programming language or an alternative one, a restructuring of the programming code into modules that can be easier to change in the future, a restructuring of the programs (sets of modules), or a restructuring of the data source, perhaps from files to databases. Such analysis and redesign should make the application easier to maintain and possibly enable it to integrate with other applications. This is potentially a lower cost and less risky solution to a legacy problem when compared to developing a new system from scratch.

Oracle (Section 19.3) has modules designed to reverse engineer programs and the Renaissance methodology (Section 25.5) presents one approach to adapt legacy systems so that they do conform to an organization's new vision for information systems. An alternative approach is Application System Asset Management (ASAM) (see McKeen and Smith, 1996).

The implementation of an enterprise resource planning system (Section 9.3) can also address effectively the problem of legacy systems. Some modules of the ERP system may be implemented that replace legacy systems; indeed, many organizations dealt with the Year 2000 problem by replacing their legacy systems with a new ERP system. However, ERP implementation does not necessarily involve decommissioning of all legacy systems. Others might remain but be adapted so that they interface with the new ERP system.

Another way to deal with legacy systems is to manage them better. Greater resources might be given to the maintenance of such systems so that financial and other rewards make maintenance work attractive for programmers and analysts; maintenance is seen as a requirement for personnel to gain promotion; program modules are revised in turn to ensure the code is efficient; and documentation is updated to modern standards. Some organizations give their systems an annual 'service' or 'complete overhaul', rather like a piece of machinery, others have sidestepped the issue by outsourcing the maintenance of their legacy systems (Section 9.4). Whatever the choice between the above approaches, neglect is not the answer as organizations could be seriously affected by the failure of their key established applications.

Of course, today's new applications, even ERP systems, will become the legacy systems of tomorrow. Hence it is important to build into today's applications flexibility so that future change is easier to implement. This is easy to argue for, but difficult to implement, and much of the concern of good practice suggested in this book aims to ensure that the 'legacy' of today's systems are not the problems of many legacy systems implemented in the past.

8.2 Software engineering

Software forms part of any computer-based information system and software engineering concerns the use of sound engineering principles, good management practice, applicable tools, and methods for software development. This was thought to be the solution to the software crisis, that is, the ability to maintain programs, fulfil the growing demand for larger and more complex programs and the increased potential of hardware which has not been exploited fully by the software. The principles established in software engineering have now generally been accepted as a genuine advance and an improvement to programming practice, primarily by achieving better designed programs and hence making them easier to maintain and more reliable. This has led to improved software quality.

In the period before the advent of software engineering, the conventional way of developing computer programs was to pick up a pencil or to sit at a computer terminal and code the program without a thorough design phase. This *ad hoc* process was also frequently used for larger programs, but a better method for these large programs was to develop a flowchart and code from this.

The time taken to develop a fully tested program will be far greater in the long run if effort is not spent on thorough design. Without this design, it is difficult to incorporate all the necessary features required of the program. However, these omissions will only be brought to light at the program testing stage, and it will be difficult to incorporate the changes required at such a late stage. Problem solving carried out by haphazard, trial and error methods is far less successful than where good analysis and design comes before coding.

One solution to better program design was the use of flowcharting methods, but although program flowcharting does discipline the programmer to design the program, the resultant design can prove inflexible, particularly for large programs. Flowcharting usually leads to programs which have a number of branches. With this method, it is difficult to incorporate even the smallest amendments to the program in a way which does not have repercussions elsewhere. These repercussions are usually very difficult to predict, and programmers find that after making a change to correct a program, some other part of the program begins to fail.

Software engineering offers a more disciplined approach to programming which is likely to increase the time devoted to program design, but will greatly increase productivity through savings in testing and maintenance time. A good design is one achievement of the 'software engineering school' and a second is good documentation, which greatly enhances the program's 'maintainability'. A third is related to both: the regular taking of software metrics (measurements) can help general control of the project (see below).

One of the key elements of software engineering is functional decomposition. Here a complex process is broken down into increasingly smaller subsets. In Figure 8.1, 'Calculate weekly payroll' at the top level can be broken down into first-level boxes named 'validate weekly returns', 'calculate gross wages', 'calculate deductions', and 'print wage slips'. Each of these boxes is a separate task and can be altered without affecting other boxes (provided output remains unchanged). Each of these can then be further broken down. For example, 'calculate deductions' can be broken down into its constituent parts, 'calculate tax', 'calculate national insurance', and 'calculate loan payments'. Eventually this top-down approach can lead to the level where each **module** can be represented as a few simple English statements or a small

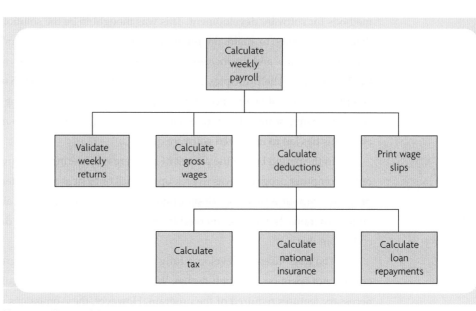

Figure 8.1: Functional decomposition

amount of programming code, the target base level. This hierarchy is sometimes described as a tree with the root at the top and the smallest leaves at the bottom. This is similar to the process referred to as stepwise (or successive) refinement (Wirth, 1971).

Figure 8.1 shows the way a process could be broken down into its constituent parts. Many programs can be broken down into the general structure shown as Figure 8.2. The top-level module controls the overall processing of the program. Separate modules at the lower level control the input and output routines (reading data and updating files), validation rou-

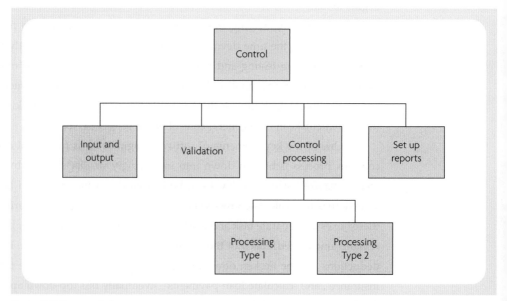

Figure 8.2: Program design

tines (checking that the data are correct), the processing routines, and the setting up of reports. Sometimes these structure charts include details of the data (called parameters) which are passed between modules. This information is written on the connecting lines in the diagram.

So far, we have equated software engineering with good practice in programming design. But this is a narrow interpretation. The term is frequently used to cover the areas of requirements definition, testing, maintenance and control of software projects, as well as the design of software.

The requirements definition must be clear and unambiguous so that the software designer knows what the piece of software should do. The designer can then devise the way that these requirements will be achieved. Even the best design will not lead to a successful implementation if the requirements are not clear. The problem with natural language is that it is frequently ambiguous, although there is a series of techniques that can be used to clarify the requirements. We may here be going beyond the boundary of software engineering into information systems development. We take the view that the requirements of the software designer will be made clear by the systems analyst. Certainly where the software is associated with a part of the overall methodology for information systems development, the likelihood of success is increased.

Supporters of software engineering practice claim that the reliability of software designed in this way is improved. These improvements do not only derive directly from better design. There is an indirect benefit. If the software is better designed, it makes adequate testing of that software much easier. By designing the program so that it is split into smaller and separate elements, it is possible to construct a series of tests for each of them. The number of tests for a larger program will be much greater. Perhaps simplistically (but it illustrates the point), a program split into four modules each containing five processing paths will require $5 + 5 + 5 + 5$ (20) tests plus one for the program as a whole once the modules are tested, whereas testing one program will require $5 \times 5 \times 5 \times 5$ (625) tests, because for each of the first five paths there will be five of the second, and so on.

Maintenance is also important because it is an activity that takes up much of the analysts' and programmers' time. Maintenance can involve the update of programs whose requirements have changed, as well as the correction of programs that do not work properly. Maintenance will be made easier by the good documentation and good design, which comes from the software engineering approach. The person making the changes will find that the software will be easier to understand because of the documentation. Further, as the program has been split into separate modules each performing a particular task, it will be easy to locate the relevant module and change it as appropriate. These changes will not affect other modules unexpectedly.

It is essential to use good control techniques to ensure that the implementation schedule for software is reasonable. Otherwise, the users and management will be unhappy about the system even if the requirements definition and software design work is first class, because expectations, which proved to be unrealistic, are dashed. A good methodology for the construction of software will be amenable to project control techniques. Breaking a system down into its parts makes estimating easier and more accurate.

As mentioned above, a major aspect of control is the ability to measure software complexity, size of task, and so on. In this way it will be possible to assess how long the software task should take and how many people should be involved and therefore be able to plan the whole process. Software metrics has become an important element of the software engineering school. Risk analysis, that is estimating the degree of uncertainty surrounding the metrics, has also become increasingly important.

Another aspect of control is ensuring the high quality of software. This implies that software is well tested; that standards, including those about documentation, have been applied; that it proves maintainable; that it is efficient; and that it meets the specification. There are now a number of official British, Australian, and international standards relating to software.

The term software engineering, therefore, is often used to cover much more than simply the design of computer programs. This is the view of many of the writers of texts on software. However, in this text the term is used to refer to the development of quality software. The wider 'environmental' issues are the concern of information systems.

In this text we are interested in producing quality information systems. We regard software engineering as a skill, which improves program design and thereby makes software, an important element of computer information systems, more effective once implemented, and easier to maintain.

A software engineering phase is usually part (though sometimes not explicitly stated) of an information system development methodology, and is explicitly part of structured approaches. Areas such as the following lie outside the scope of software engineering:

- understanding the problem area;
- understanding the needs of the user;
- looking at alternatives;
- deciding whether a computer system is needed.

If a computer system is recommended, then there will be a software engineering phase. However, the skills of software engineering should be separated from those required for information systems analysis and design. Further, it is important that the requirements of the analyst are expressed to the technologists in a way that the technologist can appreciate, and this provides one impetus in the movement toward the techniques of structured systems analysis and design which parallel, to some extent at least, those in software engineering.

8.3 Automated tools

In this section we provide some background to Part V which is devoted to a detailed look at the facilities of modern tools. The hope of automating some aspects or even all of the information systems development process is an abiding one, and has intrigued developers for many years. It has often been recognized that certain aspects of the information systems development process are repetitive or rule-based and therefore susceptible to automation. Early examples include decision table software which could generate accurate code directly from a decision table (Section 12.3), project control packages (Section 18.4) to help organize and control the development process, and toolsets (Chapter 19) to help speed up some common programming tasks. On the analysis side, software has existed for many years to help in the construction of traditional programming flowcharts.

Yet, it has often been noted by users that computer professionals have been very keen to automate everybody else but have shown a certain reticence to 'take their own medicine'. Apart from innate conservatism, there appear to have been a variety of reasons for this: some of the software was not very good or easy to use; it was often found that the benefits were outweighed by the effort and costs involved; and the technology was also a limiting factor.

Recently, some of these factors have changed. The technology is clearly more powerful, cheaper, and widely available. Improved graphics facilities have also had an impact. The quality of the software has improved, and in general there is a growing climate of opinion that automated tools may be beneficial.

There are now a number of tools that support the analysis and design process. These are tools that help the user use the techniques described in Part IV, such as data flow diagrams, entity models, and so on. They are sometimes described as documentation support tools, being designed to take the drudgery out of revising documents, because they make the implementation of changes very easy. In addition, they can contribute to the accuracy and consistency of diagrams. The diagrammer can, for example, cross-check that levels of data flow diagrams are accurate and that terminology is consistent. They can ensure that certain documentation standards are adhered to. Probably the greatest benefit is that analysts and designers are not reluctant to change diagrams, because the change process is simple. Manual redrawing is not satisfactory, not just because of the effort involved but the potential of introducing errors in redrawing. Many a small change required by a user was never incorporated into the system due to the effort required to redraw all the documentation. These kinds of documentation tool have proved both practical and useful so that the change process is not now inhibited.

Tools supporting the use of single techniques have, however, also proved limited in the sense that much of the information required for a data flow diagram, for example, would also be required, in a slightly different form, for the process logic software, and so on. It was realized that it would make more sense to have a central repository of all the information required for the development project irrespective of its graphical representation. In fact, this is the data dictionary, or perhaps as it is more correctly known, because it contains information about processes as well as data, the systems dictionary, systems encyclopaedia, or systems repository (Section 19.1).

Once most of the information concerning a development project is on a data dictionary, it is in theory only a short step to the automation, or at least the automated support, of many of the stages of the development project. Further, one of the goals of a number of methodologies is to provide automated support for all their stages. Some have the automatic generation of code as the end result of the automation of the information systems development process.

More modern software tools include:

- graphical facilities for modelling and design;
- data dictionary;

- automated documentation;
- code generation from systems specification or from the models designed using the tool;
- automatic audit trail of all changes;
- critical path scheduling with resource availability (i.e. project control);
- automatic enforcement of the standards of a chosen software development methodology.

It is important that these facilities are completely integrated (Chapter 19) so that they provide consistency in analysis and design. They are particularly effective in this regard if they are associated with a particular methodology.

Perhaps it is appropriate to provide a warning in relation to tools. They are said to reduce the skill, complexity, time, error, and maintenance associated with the development of information systems. However, many are modest in their facilities. Typically, they help people to draw diagrams, such as entity-relationship diagrams and data flow diagrams, and perform validation checks. Many of the more sophisticated tools are very costly in terms of price, but also in terms of training and support. They may also be appropriate for use with only one methodology, which ties the users to that approach and also may make updates (as the methodology is updated) both essential and expensive.

8.4 Method engineering (ME)

Method engineering (ME), sometimes referred to as methodology engineering, is the process of designing, constructing, and merging methods and techniques to support information systems development. Method engineering is often associated with the hierarchical and bureaucratic approaches of the 1980s as techniques were combined to form meta-methodologies. Normally it is concerned with the blending of methods and techniques into a methodology or framework. However, its most recent form is enterprise resource planning (ERP) systems (Section 9.3), which are combinations of application types rather than methods and techniques.

In this section, adapted from Avison and Truex (2002), we reflect on the genealogy of method engineering and on how it must evolve to meet the needs of contemporary IS development. As guides and as method engineering source books, we point the reader to Brinkkemper et al. (1996). We will discuss these challenges and point to potential directions for ME that are more likely to address the basic problems and needs of information systems development. In so doing, we briefly review five phases in the ME movement, with the fifth being ERP systems.

Early ME involved creating a standardized approach to 'engineer' systems development, usually based on process or data modelling (see Chapter 6), bringing order to work that had previously been largely trial-and-error and software-oriented. In general, this contained techniques, phases, and standards put together to form a coherent methodology to be used by systems analysts and programmers. This embodied the best practice available at the time. An example of such an approach on the process side was STRADIS (Section 20.1). We refer to this as Type I method engineering (Figure 8.3).

In Type II method engineering, ME was carried out to improve IS development methodologies so that they captured best practice by including techniques in other methodologies,

Type	Focus	Intent	Added-value	Orientation	Examples	Modelling types
Type I	Uniview	Standardize on best practice in systems development	Bring order to chaos, structured methodological approach	Technical Technologist	STRADIS (process view) D2S2 (data view)	Modelling data *or* processes
Type II	Making approach more ecumenical (filling in the gaps)	Connect univiews to engineer an overall approach	More generalizable, universal More complete	Technical Technologist	SSADM Merise	Modelling data *and* processes
Type III	Identifying and linking method fragments	Add technical richness by adding modules or method components	Broaden scope further Ability to interconnect methods in whole 'super (meta)-method'	Technical Technologist	Information Engineering with CASE	Modelling various systems components
Type IV	Contingent framework	Provide guidelines to use techniques Bring in organizational and social richness	Provides guidance about how and why we use method components Broaden scope further	Social and organizational as well as technical Middle managers and users	Multiview	Modelling the proper match of components
Type V	Enterprise resources	BPR – Re-engineer whole organizations (to fit the system?)	Provides more structure to organizational frameworks Adds BPR and workflow	Organizational and technical Top managers	SAP, BAAN, Peoplesoft	Modelling the ideal organization

Figure 8.3: Movements in method engineering (modified from Avison and Truex, 2003)

such as data flow diagrams and entity modelling, or take account of newly proposed ones, such as object-oriented methods. Many were extended to address more phases of the life cycle. This resulted in larger, generalized, perhaps bureaucratic, approaches. These widened the scope of methodologies to be more general-purpose and made them more commercially viable. Information Engineering, Merise, SSADM and Yourdon Systems Method (see Part VI) are examples of these methodologies.

But this general-purpose, universal approach did not address the problems of complexity nor the skills required to develop IS. One response was the development of tools (Part V). As a reaction to the complexity and inflexibility of methods, systems analysts frequently paid 'lip service' to a methodology, rarely following the phases exactly as described in the manual and sometimes omitting, adding, or modifying phases. Some organizations stopped using methodologies altogether.

Type III method engineering, the next development, links method fragments or components to form a 'meta-methodology'. This aimed to increase the flexibility of approaches and the customization of a 'one-off' methodology for a particular application. One major difficulty is that the analyst is confronted with a 'hodgepodge' of techniques and tools, some of which may be inappropriately linked to others; therefore, most approaches of this type provided a framework and control system to guide the analyst in choosing the appropriate method fragments.

Some of these frameworks proposed sets of situational (or contingent) factors that guided the selection of method fragments (Van Slooten, 1996). However, these frameworks tended toward a very technical view of the development of IS, indeed they have been dominated by toolsets (see Chapter 19).

Type IV method engineering recognizes that organizations are social constructs – they are about relationships between people – and artefacts do not determine how people behave. Information systems development is not just about choosing technique fragments. It was therefore broadened to include human and organization factors at least as much as technical ones. These approaches, such as Multiview (Section 26.1), recognize that IS development is constrained by organizational context while providing the means to change that context. It is not just a set of tools and techniques, however, frameworks such as Multiview provide guidance on when and where it might be appropriate to use them. We cannot assume that technical improvements will lead to organizational improvement. Any IS development endeavour can suffer the unintended consequences of action. This suggests that IS approaches applied contingently should also be applied thoughtfully.

Another concern relates to the burden on the systems analyst to choose the 'appropriate' techniques and tools for the situation, even with the guidance of such a framework. In such an informal approach, a further concern is the lack of control and application of standards in information systems development.

Avison and Truex (2003) propose, perhaps controversially, that enterprise resource planning systems (Type V in Figure 8.3) represent a type of ME approach. Enterprise resource planning (ERP) systems, such as SAP, attempt to re-engineer whole organizations and, in effect, model an ideal organization (making the actual organization fit with this ideal). Like most of its precursors to the ME movement, it became fashionable but is now being critically scrutinized.

ERP systems are built upon highly normalized, robust data models and industry-specific reference models, purported to represent 'best-of-class' process and work flow models. As such, they represent a composite of systems views and models ranging from data through process and work flow to object models. Ideally, they seek to integrate all views, such that data and process knowledge may be available to any of the application components wherever appropriate or needed.

There is a certain irony associated with ERP systems when they are seen as forms of method engineering. To some extent they hark back to the Type II ME approaches, which imply a universal view of organizational systems. The tools associated with ME approaches enforced the sequence and the description of design upon the developer. So, too, do ERP systems, by virtue of the various models and components enforcing a kind of discipline upon the organization. A final irony is that if most organizations do not take advantage of the configurability of the system, and they tend not to because of the cost, then most organizations using ERP will look very much alike in terms of business processes and organization. Moreover, legacy systems tend to incorporate aspects of organizational distinctiveness and incorporated organizational memory. Many of these legacy systems have been abandoned with the adoption of an ERP system. However, with the reduction of ERP sales growth following high demand due to the Y2K crisis, methodologists and management alike have begun to consider how to overcome the inherent problems of this universal approach to creating information systems. More recently, ERP systems have attempted to address this challenge by being more easily adapted by users as well as having modules to run e-commerce, call centre and other applications. There are also versions that are aimed at small businesses.

Method engineering will always be a theme in IS development as it is a sign of progress in the adaptation of methodologies to changing IS/IT environments and progress in methodology design itself. We return to some of these issues again in Chapter 27.

8.5 Component development

Component development is usually thought to have its origins in the object-oriented approach (see Section 6.4) because a component is often thought to be an object. Further, the objectives of the object-oriented approach, especially relating to flexibility and software reuse, overlap significantly with component development. However, others question this notion and argue that there is little evidence that in practice object-oriented principles underlie component development (Lycett, 2001).

Whatever its origins and background, the basic concept is that a system is built using already developed modules or components, combining them together in such a way that they provide the functionality desired of a new system. A component is a software element or module that provides some functionality and that has a specific and well-defined interface. Components may be relatively small chunks of code that undertake a common task or process, for example, look something up in a table, calculate an interest rate, access a database, display a message, check the validity of a product code, check that a telecommunications channel is still open, etc. Equally they may also be large chunks of code that undertake whole business functions, for example, credit card processing, checkout a product, spell-check a document, invoice a customer, etc. Or indeed anything in between.

Hopkins (2000) defines a component slightly more formally as a 'physical packaging of executable software with a well-defined and published interface'. The *published* element of the definition is interesting because it indicates that a component is somehow available in a marketplace, either to be purchased or downloaded. Thus, its interface needs to have been 'published' so that other developers can find it, will know what it does, and know how to interface with it, in other words, how to utilize it. This indicates one of the key elements of components – that they are reusable. Someone (termed the component fabricator) develops them and others (component assemblers) purchase them (or they may be freely available – see Section 9.2 on open source software) to utilize in a new development or project.

The benefit is that someone wishing to develop an application can thread or link together a series of components that will form the application they require. Further, this is likely to be quicker and cheaper than developing it from scratch (by programming the application themselves). A whole new application may be developed using only pre-programmed components, or a combination of components and new code. Assemblers and users do not have to understand the detail of the internal contents or structure of the component. They simply need to understand the functionality and the interface. Components are thus sometimes described as *black boxes*.

Component development potentially offers many benefits. Speeding the development process is of particular importance, especially with the current focus on rapid IS development (see Section 7.3). Further, the components should have all been thoroughly tested and matured in the marketplace, with any errors removed. Thus, they should be 'of industrial strength' and ready to use. Any problems or bugs in a system developed using components would in theory be due to the way in which the components have been used or combined, or due to the provision of invalid or inappropriate information in the interface, rather than the functioning and performance of the individual components. Systems developed from components should also be relatively easily changed and evolved. A component can be removed and replaced with another (or a number of others) to provide different or enhanced functionality to meet changing and evolving requirements, resulting in greater application flexibility.

The benefits are thus potentially great, indeed for some, component development heralds the death of tailor-made software development. Vitharana and Jain (2000) state that 'the traditional built from scratch software ideology is behind us, and the trend is toward CBSD (Component Based Software Development) involving component fabrication and component assembly'. Sugumaran and Storey (2003) argue that component development is likely to be more evolutionary but nevertheless it holds great promise.

There is some debate as to how widespread the uptake of CBSD actually is. It does not seem yet to have significantly penetrated the mainstream of in-house tailor-made systems development in commercial organizations, although there are some notable exceptions, for example, Dell (Lycett, 2001). Where it has made a significant impact is in the relatively new area of open source development (see Section 9.2). The reasons that it has not had the impact in the mainstream systems development arena are thought to be the following.

1. It seems there are difficulties with the commercial availability of components. According to Ravichandran and Rothenberger (2003) the market is immature and users cannot find enough components, or those that they feel they need. Obviously the lack of easily available components of the right kind, at the right price, is not going to encourage widespread take-up. There is also some evidence that companies are concerned about putting their trust, and possibly their future, into components that they do not fully know the provenance of. Developing or fabricating components is not the same as developing software for one-off applications. At least 50 per cent additional effort, some say even more, is required to make a generalized and reusable component than is required for a one-off specific application. It is more like the package market (see Section 9.1). This has discouraged the production of components as a by-product of developing in-house applications. Nevertheless, these market issues may be a somewhat temporary problem and the market will increase and mature over time.

2. It seems that it is not quite as easy to assemble components as is sometimes suggested. A good deal of experience is required and the design of a system that is composed of interrelating components is a significant task. Component-based development does not

eradicate the need for good design but it changes the nature and skills required of designers. Components have to communicate and there is a lack of models and standards to facilitate this. Sparling (2000) suggests that there is a lack of design principles for CBSD, although some have recently been proposed from Microsoft, Sun, and the Object Modelling Group (OMG). This does not mean that fully agreed standards will emerge in the near future. Further, it is not just design that has to be addressed but the whole of the development life cycle, including requirements, analysis, design and development. To make development viable, the concept of components has to be at the heart of the life cycle.

3. There are technical and technology barriers. Achieving component communication is difficult, as is getting components to function across various platforms, architectures, infrastructures, languages, versions, etc., although again there are important attempts to facilitate this from some of the large manufacturers and players in the market.

However, despite these problems, the potential of component development in reducing the costs and increasing the speed of application development is likely to make component development more and more prominent.

8.6 Security issues

Security issues are important in all information systems development. The purpose of security is to maintain business continuity, as breaches of security can damage the business in a financial sense and it can also damage its reputation. For example, if competitors have free access to your company information, then it is unlikely that your company can maintain any competitive advantage, and if you lose information on customers or products, then sales may be lost. If a breach of security at a bank is made known, for example, customers may well be concerned and take their business elsewhere. Although this is an issue previously mainly associated with large databases (Section 8.7), it is now a major issue elsewhere as well, in particular, in the context of the Internet.

Information is a major resource of the company, so there must be a high level of information security. Breaches of security may be malicious, such as fraud, espionage, vandalism and sabotage, but may also be non-malicious, such as a disk crash, fire and flood, human error and inadvertent misuse. Attempts must be made to prevent such breaches, but if they do occur, they need to be detected quickly and procedures for recovery need to be in place. This requires backup of critical data. In some circumstance it is deemed sensible to have duplicate equipment available should there be malfunction in the equipment normally used. Investment is also made in software aimed at protecting data.

Where the company uses outsourcing and offshoring (Section 9.4) security needs to be at least as good there as it is within the company, as security is only as good as its weakest link. Such considerations have been a limiting factor to outsourcing and offshoring for some companies.

Information security aims to safeguard the confidentiality, integrity and availability of written, spoken and electronically stored and communicated information. *Confidentiality* ensures that information is accessible or disclosed only to those authorized to have access; *integrity* safeguards the accuracy and completeness of information and processing methods, reflecting their trustworthiness; and *availability* ensures that authorized users have access to information and associated assets when required. One hundred per cent confidentiality, integrity and availability would be very expensive, even if feasible, so that adequate levels need to be agreed for the application area and implemented. One problem with using the Internet is that integrity in particular varies with the particular site and it is not always evident what that level might be.

Examples of how information security might be protected in an application might include:

- Data accuracy may be ensured through thorough software validation routines and testing procedures,
- Confidential data such as credit card details in an Internet application might be encrypted,
- Access to information might be password protected,
- Security software might be in place to provide a firewall and virus protection (for example, using the Norton suite of programs),
- Employee access to data is usually limited on a strict 'need to know' basis,
- Software and hardware might be used to ensure the user is who he claims to be, typically achieved through the use of passwords and fingerprint devices respectively,
- Theft of hardware is usually prevented by physical locking mechanisms and human security checks,
- Users might be required to account for their use of the system through regular reporting,
- Regular backups are usually taken and there should be clear and tested plans to recover from unexpected failure or a breach of security,
- Regular monitoring, with review and audit procedures in place, and
- Employees follow professional codes of conduct (Section 1.11).

Security needs to be seen as an important aspect of the information systems development process as well as operational systems and is part of a quality product. Thus thorough testing of programming code, the links between programs and the system as a whole is essential. The systems developers, control personnel and users should all be confident that the system is tested fully before it goes operational. All the above security procedures need to be tested fully. Developers may use live as well as test data to check procedures. Most information systems development methodologies will highlight security aspects of the development process as much as any other aspects, and include procedures for maximizing security features in the system to be developed. Of course, good project management procedures will add to this control (Chapter 14).

When accessing information, users may have different access restrictions according to authorization rules. A password may allow a user to access part information only, update part information only, create new records or have unlimited access. The latter may be a right that is allocated only to the database administrator (see Section 8.7). Highly sensitive data might be encrypted, that is coded so that it cannot be read without a decoding mechanism, to give added protection. Again, password authentication may be seen as inadequate, and to this is added biometric authentication, such as finger or voice prints, or a physical device such as a smart card.

One problem regarding investment in security is that the return on investment is not obvious, but companies can assess the costs of not having secure information. Investment tends to be justified on the basis that security levels have improved, along with a statement of the risks of security being breached and the cost-effectiveness of such investment through higher levels of protection. Security investment is high cost and coupled with a lack of awareness of these issues, it is often placed low down on investment priorities, much lower than it ought to be, and its importance only discovered when security is breached.

Good documentation forms an important part of security. Security needs to be part of each employee's role. In order to prevent collusion, the security role needs to be distributed between a number of people and this should help prevent internal fraud. There may be a procedures manual specifying the security procedures in detail and showing who has responsibility for carrying out each of these procedures. Training in these procedures is as important as that in any other aspect of an employee's role. Along with good supervision, this should ensure that procedures are carried out in a correct and consistent fashion.

There may also be legislation requiring security of information systems, for example, through requirements for personal data to be protected from misuse. For some companies, banks to give one example, security breaches can impact on the reputation of the company and therefore it has a higher profile than normal. There needs to be a regular security audit, that is, a periodical examination and check of security procedures, perhaps carried out by or with the help of outside security specialists. Some firms have made such investment to the point that they have a manager of security. But these managers often lack the resources and are not always allowed to work independently, and cannot always justify their position. In brief, they have a low status in the organization. Banks, therefore, may be exceptions to the rule and some employ 'friendly hackers' to attempt to breach security and inform on weak points so that they can be corrected.

Companies are often criticized for their lack of investment in security and its low status. Yet individuals are even more lax. In a European Password Survey (NTA, 2003), 81% of computer users had a common password for all their applications, 31% noted their passwords down, 67% of users rarely or never changed their passwords and 22% admitted that they would only ever change their password if forced to by a website or system/IT department. This is not good practice. Individuals and organizations need to take security seriously.

8.7 Database management

A large collection of books owned by the local council is not a public library. It only becomes one when, among other things, the books have been catalogued and cross-referenced so that they can be found easily and used for many purposes and by many readers. Similarly, a database is more than a collection of data. It has to be organized and integrated. It is also expected to be used by a number of users in different ways. In some companies the whole organization is modelled on a database, so that, in theory at least, users can find out any information about any aspect of the organization by making enquiries using the database.

In order to make this feasible, there needs to be a large piece of software which will handle the many accesses to the database. This software is the database management system (DBMS). The DBMS will store the data and the data relationships on storage devices. It must also provide an effective means of retrieval of that data when the applications require it, so that this important resource of the business, the data resource, is used effectively. Efficient data retrieval may be accomplished by computer programs written in conventional computer programming languages accessing the database. It can also be accomplished through the use of a query language, such as SQL, or in other ways which are more suitable to people who are not computer experts. In this section we discuss both the potential facilities of a DBMS and the data retrieval methods to access the data on the database using the DBMS.

Figure 8.4 represents a potential architecture for a database system. It has three views of the database: external, conceptual, and internal views.

The external view is the view of the data as 'seen' by application programs and users. It will be a subset of the conceptual model, which is a global or organizational view of the data. There may be a number of different user views, and different users and programs may share views. In the diagram, two users share the same view (External View 2), but their methods of

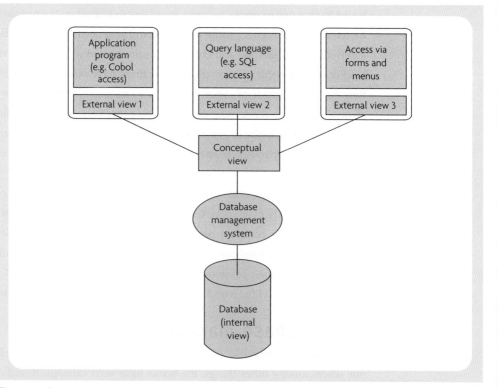

Figure 8.4: Database architecture

accessing the data are different. This arrangement enables aspects of the database in which a user may not be interested or does not have rights of access to be hidden from that user. This reduces the complexity from the users' points of view. To enable this, there is a series of mappings or transformations between the external views and the conceptual view, which is handled by the DBMS software.

The conceptual view here is different to the conceptual models referred to in Section 10.3. Readers have been warned about the non-standard terminology in information systems! Indeed, the conceptual view expressed in this model is normally that of the entity models described in Section 11.1. In these sections, we describe techniques providing a global data model. The conceptual model discussed in this section is the data model of the organization. It describes the whole database for the organization (or community of users), which could be in terms of entities, attributes, and relationships.

The internal view of the data describes how and where the data are stored. It describes the access paths to the data storage and provides details of the data storage. Again there is a mapping between the conceptual and internal views. The latter will not be of direct interest to the users, but will be of indirect interest in terms of speed of access and the general efficiency of database use. It will be of direct interest to the database administrator (see below) who is normally responsible, among other things, for the efficient organization and effective use of the database.

There is a series of mappings from the conceptual view (the overall data model) to the various external views (the subsets of the data to which applications or users have access) and to the internal view (the way in which the data are organized on computer storage devices). These mappings will be carried out by the database management system software.

The separation of different views of the data enables data independence, which is a crucial element in the database approach. This means that it is possible to change the conceptual view without changing the external views (and the application programs that use them). It is also possible to change the internal view without changing the conceptual and external views. In other words, data independence provides flexibility and efficiency.

Data independence represents only one of the hoped-for advantages of the database approach, which are outlined in the following subsections:

1 Increase data shareability

Large organizations, such as insurance companies, banks, local councils, and manufacturing companies, have for some time been putting large amounts of data onto their computer systems. Frequently the same data were being collected, validated, stored, and accessed separately for a number of purposes. For example, there could be a file of customer details for sales order processing and another for sales ledger. This 'data redundancy' is costly and can be avoided by the use of a database management system. In fact some data duplication is reasonable in a database environment, but it should be known, controlled, and be there for a purpose, such as the efficient response to some regular database queries. However, the underlying data should be collected only once, and verified only once, so that there is little chance of inconsistency. With conventional files, the data are often collected at different times and validated by different validation routines, and therefore the output produced by different applications could well be inconsistent. In such situations the data resource is not easily managed and this leads to a number of problems. With reduced redundancy, data can be managed and shared, but it is essential that good integrity and security features operate in such systems. In other words, there needs to be control of the data resource. Furthermore, each application should run 'unaware' of the existence of others using the database. Good shareability implies ready availability of the data to all users. The computer system must therefore be powerful enough so that performance is good even when a large number of users are accessing the database concurrently.

2 Increase data integrity

In a shared environment, it is crucial for the success of the database system to control the creation, deletion, and update of data and to ensure their correctness and 'up-to-dateness'; in general, to ensure the quality of the data. Furthermore, with so many users accessing the database, there must be some control to prevent failed transactions leaving the database in an inconsistent state. However, this should be easier to effect in a database environment, because of the possibilities of central management of the data resource (through the database administrator), than one where each application sets up its own files. Standards need only be agreed and set up once for all users.

3 Increase speed of implementing applications

Applications ought to be implemented in less time, since systems development staff can concentrate largely on the processes involved in the application itself, rather than on the collection, validation, sorting, and storage of data. Much of the data required for a new application may already be held in the database, put there for another application. Accessing the data will also be easier because the data manipulation features of the database management system will handle this.

4 Ease data access by programmers and users

Early database management systems used well-known programming languages, such as Cobol and Fortran, to access the database. Cobol, for example, was extended to include new instructions, which were used when it was necessary to access data on the database. These 'host language' extensions were not difficult for experienced computer programmers to learn and to use. Later came query languages and other methods to access the data, which eased the process of applications development and data access in a database environment. Once the database had been set up, applications development time should be greatly reduced.

5 Increase data independence

We have already introduced this, but it has many aspects. It is the ability to change the format of the data, the medium on which the data is held and the data structures, without having to change the programs which use the data. Conversely, it also means that it is possible to change the logic of the programs without having to change the file definitions, so that programmer productivity is increased. It also means that there can be different user views of the data even though they are stored only once. This separation of the issues concerning processes from the issues concerning data is a key reason for opting for the database solution.

6 Reduce program maintenance

Stored data will need to be changed frequently as the real world that it represents changes. New data types may be added, formats changed, or new access methods introduced. Whereas in a conventional file environment all application programs that use the data will need to be modified for each of these changes, the data independence of a database environment, discussed above, circumvents this. It is necessary only to change the database and the data dictionary (or systems repository), which will contain information about the data in the database (among other things). We discuss these in Section 19.1.

7 Provide a management view

With conventional systems, management does not get the benefits from the expensive computing resource that it has sanctioned. However, managers have become aware of the need for a corporate view of their organization. Such a view requires data from a number of sections, departments, divisions, and sometimes companies in a large organization. This corporate view cannot be obtained if files are established on an application-by-application basis and not integrated as in a database. With decision support systems using the database, it becomes possible for problems previously considered solvable only by intuition and judgement to be solved with an added ingredient, that of information, which is timely, accurate, and presented at the required level of detail. Some of this information could be provided on a regular basis, while some will be of a 'one-off' nature. Database systems should respond to both requirements.

8 Improve standards

In traditional systems development, applications are implemented by different project teams of systems analysts and programmers, but it is difficult to apply standards and conventions for all applications. Computer people are reputed to dislike following the general norms of the organization, and it is difficult to impose standards where applications are developed piecemeal. With a central database and database management systems, it is possible to impose standards for file creation, access, and update, and to impose good controls, enabling unauthorized access to be restricted and providing adequate backup and security features.

Much of the success or failure of a DBMS lies in the role of the **database administrator** (DBA), who will be responsible for ensuring that the required levels of privacy, security, and integrity of the database are maintained. The DBA could be said to be the manager of the database and, because the design of the database involves trade-offs, will have to balance conflicting requirements and make decisions on behalf of the whole organization, rather than on behalf of any particular user or according to a particular departmental objective. The role is multi-varied and is usually carried out by a database administration team. In some organizations, the information resource is regarded as one of the key elements of success (which it surely must be) and there is a high-level data administration function that includes the lower-level database administration team, responsible for the computer data. In some organizations, the board includes a chief information officer.

There are many database management systems available. These include Oracle and Access, both of which are discussed in Part V. Some of these run on large computers, others on PCs. Most DBMSs are relational or object-oriented (or a combination of the two). In a relational DBMS, the data model that the user sees is the relational one (i.e. sets of tables). In an object-oriented database management system, the data management and programming language aspects are integrated: all data is represented as objects and the programming language manipulates objects.

Of particular interest to users developing systems in a database environment is the way in which access can be made to the database. We will now turn our attention to this aspect of DBMSs. Data access may be made using soft-copy forms, menus, conventional computer languages, and query languages. Many DBMSs provide alternative ways of access. This can be a useful facility as there are many types of user. Users can be untrained and intermittent in their use of the system. These casual users should be encouraged by its ease of use. Regular users may make frequent, perhaps daily, use of the database and are usually willing to learn a simple syntax. Other users will be professional users who are computer people and will apply their long experience as computer and database users. They may be more concerned about efficiency considerations than ease of use.

As reference to Figure 8.4 makes clear, the external view of the database is that which each user 'sees'. It is derived from the conceptual view. The external view is the subset of the database which is relevant to the particular user, and, although it may be a summarized and a very restricted subset, the user may think that it represents the whole view, because it is the whole view as far as that user is concerned. The presentation and sequence of the data will also suit the context in which they are presented. The format will depend on the particular host language, query language, report writer, or other software used. Indeed, there may well be several external views, perhaps as much as one per application or user that accesses the database. But whatever the description given to the user, the underlying data will be the same.

Of particular importance in the context of this book is the use of a DBMS as a tool to develop applications. If we assume data to be key to an application, then their input, validation, storage, and retrieval form the basis of information provision to support decision making. Along with different ways to provide information (from formal languages, such as SQL, to display formats, such as query by example, through to natural language), Access, for example, enables the development of programs using VisualBasic, which can be used by 'interested users' as well as 'experts'.

8.8 Data warehouse and data mining

In Section 8.7, we discussed databases, which are organized collections of related data. Data warehouses are much larger collections of data. The data is non-volatile, that is, it is fixed and does not need updating. More up-to-date data can be added to the warehouse, but does not normally replace the old data. The data may be historical, and may not be related, apart from being relevant to the organization in some way. The data is normally collected from heterogeneous transaction systems, many legacy systems, and their related databases but organized in such a way that various other managerial uses can be derived from it. Such sources may be enterprise resource planning systems, other transaction systems, web servers, call centre systems and external providers (sources of government statistics or market research companies).

This data may be *mined* so that, for example, business trends can be identified. Data mining software picks out patterns and correlations in the data. Specialist querying and reporting tools can be used to make sense of the data through various statistical techniques. They can explain an observed event or condition, confirm a hypothesis or explore for unexpected relationships.

Data warehousing systems will extract the data, transform it into a consistent format for the system, clean the data so that it is reasonably accurate, and store, link and maintain the data in the data warehouse ready for use. Extraction of data might require 'drilling down', that is, going into more and more detail. Sometimes it involves summarizing data rather than drilling down. Data may be partitioned into sets and this process is known as 'slice and dice'. Tools such as online analytical processing (OLAP) tools are designed to do this type of analysis. All these techniques aim to focus on the data which is most relevant to the particular user requirements. Although the data in the warehouse may derive from operational processing, it is used for managerial purposes, and the data can provide a corporate-wide view.

Sometimes the term **business intelligence** is used to describe the successful use of the data for an organization. An example might be where the data provides a complete picture of the customers and this information is used to provide a better service. The software may pick out patterns related to customer demography, such as class, income, sex, age and postcode, and form part of the firm's customer relationship management. Other uses may concern using the information to develop a more profitable service, improved risk management, better cost management or even supporting new strategic options. For example, a credit card company may use data mining to detect unexpected purchases (unusual amount, place or type of purchase) or a bank may detect money laundering. Very often, the process of data mining picks up hidden or unexpected patterns of data. This may have been impossible previously, because people and smaller systems cannot work with such huge amounts of data, often many terabytes in modern data warehouses. A terabyte equals 1000 billion bytes! The data needed will normally be too complex and too large for manual analysis.

However, extracting the relevant information from such huge stores of apparently unrelated data is not a trivial task. Further, the costs of extracting, cleaning and storing such volumes of data can be very high. To this needs to be added the costs of purchasing and using tools and constructing and implementing the applications and staffing all this. The danger might be that the costs of data warehousing and data mining are borne by the business without effective use of the facility. When this is paralleled with the loss of the people expertise that the business used earlier, but laid off now to offset the costs, then the business may suffer rather than gain competitive advantage. There are also potential security and privacy problems.

As with any information system, there are different ways of developing such applications. The data in the data warehouse may be slowly built up as applications are plugged into the system in turn and corporate data is collected periodically in a piecemeal fashion. Alternatively, as particular needs are identified, applications providing related data might be hooked up to the data warehousing system. (The term data mart is usually applied to a small data warehouse that may describe its early incarnations.) A braver approach, providing potential for competitive advantage but more risky, is to have a large project aimed at creating a full data warehouse. However, in any case, the information needs of the users need to be specified, the data warehouse design needs to be agreed, the sources of data identified, the data collected, cleaned (including attempts to identify and remove duplicate information) and transferred to the data warehouse, and tools tuned to access the warehouse as appropriate for user needs. The project leaders need to ensure that users' expectations are realistic, despite the high-cost and high-profile project. Disappointment can lead to under-use of the data warehouse. On the other hand, it is also important that the project scope is not so great that it becomes unmanageable.

SPSS is a data mining tool and the company has one-quarter of a million customers and four million users. The company sees its market in predictive analytics. This studies people or company behaviours and characteristics and then generates predictive models deployed in order to optimize company relationship with customers. This includes predictive marketing, web analytics and claims (fraud detection). Amongst its toolkit are *Clementine*, a data mining workbench, *Dimensions*, a survey platform, along with the SPSS tool itself, for statistical analysis.

In the SPSS approach (Kattanjian, 2005), the following stages are perceived for the implementation of a data mining project:

1. Business understanding:
   - Statement of business objective
   - Statement of data mining objective
   - Statement of success criteria
2. Data understanding
   - Explore the data and verify the quality
   - Find outliers
3. Data preparation (this usually takes over 90% of the time)
   - Data collection and assessment
   - Consolidation and cleaning (table links, aggregation level, missing values, etc.)
   - Data selection (active role in ignoring non-contributory data, dealing with outliers, use of samples, visualization, etc.)
   - Transformations – create new variables
4. Apply models and algorithms
   - Decision trees
   - Neural networks

- Linear and logistic regressions
5. Analyse and evaluate results
    - Classifications
    - Clustering
    - Associations
    - Sequences
    - How well do models perform on test data?
    - Comparison of model results
6. Deployment
    - Determine how the results need to be utilized
    - Who needs to use them?
    - How often do they need to be used?

Let us consider a few examples in use. A leading European holiday park company wished to ensure that all interactions with customers are driven by an understanding of individual needs and preferences and has used data mining successfully. Using data from marketing, call centres, Internet use and in the parks themselves, the tools identify customers who might be most susceptible to marketing campaigns, thus reducing the costs of such campaigns and yet increasing occupancy. Further, the tools have identified cross-selling links between leisure activities and dining activities, and the tool suggests cross-selling strategies.

A company selling financial products (car, travel, health care, life insurances; consumer loans, etc.), has used the tools to change its marketing strategy from bulk campaigns to highly targeted event campaigns. It also cross-sells on the Internet and through its call centre, so that, for example, anyone purchasing insurance protecting against trip cancellation will get information about emergency medical expenses. The company argues that gains have included 35 per cent reduction in outbound marketing costs and yet a 40 per cent increase in conversion to actual sales. It also uses the tool to detect fraud and credit risk. Another company uses the tools to target high value customers in marketing campaigns and generate targeted leads for advisers.

An Internet retail company uses similar tools to model customer behaviour in the way they buy and surf the Internet site, refine marketing so that relevant contacts are targeted, improve susceptibility to buy the product and reduce the likelihood of customer unsubscription. It can also be used in industry for quality control and candidate selection, and in retail for loyalty card analysis. This is potentially positive for the company's customer relationship management (Section 5.5).

In such applications, the data mining tool may be forecasting what may happen in the future (anticipate new data using a model); classifying people or things into groups by recognizing patterns; clustering people or things into groups based on their attributes; associating what events are likely to occur together and sequencing what events are likely to lead to later events. All this may provide the companies with competitive advantage and give their customers a better service.

Summary

- Legacy systems are systems that have been in operation for some time. They may well perform critical processes, but they are often seen as a problem as they may have high maintenance costs, use obsolete hardware and software, be poorly documented, and lack support people with the knowledge required to maintain them.
- Software engineering concerns the use of sound engineering principles, good management practice, applicable tools and methods for software development.
- There are a number of tools that support tasks in the analysis and design process. There are also integrated toolsets that support many tasks.
- Method engineering (ME) is the process of designing, constructing and merging methods and techniques to support information systems development. It might be a blending of methods and techniques into a framework, methodology or mega-methodology. Its most recent form is enterprise resource planning (ERP) systems, which are combinations of application types rather than methods and techniques.
- Information systems can be developed from components. These include drivers, Internet utilities, software development software, security and database components.
- Security issues are important in all stages of IS development and operational systems. Breaches of security can be both malicious and non-malicious, but in either case they need to be prevented if possible and detected quickly otherwise.
- A database is an organized and integrated collection of data. A database management system is software that validates, stores, secures, displays and prints the data in ways that the users require.
- Data warehouses are large collections of related and unrelated data and data mining attempts to identify business trends or improve customer relationship management using software tools.

Questions

1. Provide a balanced argument to an IT manager asking, 'What should we do about our legacy systems, continue to maintain them or replace them?'.
2. What is the difference between software engineering and information systems engineering? Is one merely a subset of the other? In which case, which is the subset of the other? Give reasons for your answer.
3. Do you think the whole process of systems development could be automated?
4. Are there any problems to counteract the apparent advantages of blending techniques to engineer a best or ideal methodology?
5. Why is security such a concern for Internet users? Are they correct to be concerned?
6. What is the difference between a database, a data warehouse and data mining?

Further reading

Avison, D.E. and Cuthbertson, C. (2002) *A Management Approach to Database Applications*, Maidenhead, McGraw-Hill.

Brown, A.W. (ed.) (1996) *Component-Based Software Engineering*, IEEE Computer Society, Los Alamitos, CA.

Pressman, R.S. (2004) *Software Engineering: A Practitioner's Approach*, 6th edn, McGraw-Hill, London.

9 External development

9.1 Application packages

An enduring theme of application development concerns the wish to reduce the effort involved by purchasing a ready-made solution. Usually this means the purchase of an application package. Most packages used carry out applications that any type of business has. For example, on PCs these include word processing, spreadsheet, database, calendar management, e-mail, and so on. There are also packages designed for particular market sectors, for example, hotels, shops, and transport companies. A rather more ambitious type of application package are the enterprise resource planning systems (ERP) which are integrated packages that cover a range of applications for businesses, for example, sales order processing, invoicing, ledger systems, production control, stock control, human resource management, and the rest. We look at these separately in Section 9.3.

Although the purchase of application packages and ERP systems means that much of the development work is done externally, it does not mean that there is no systems development work in-house, as the packages need to be tailored for the company. In the case of ERP systems, this tailoring can be very substantial and implementation, a long-term exercise. However, the use of outside consultants can ease the burden on in-house staff. An alternative approach is to purchase development, implementation, and the operation of computer applications from outsourcing suppliers. Again, we look at this separately in Section 9.4.

The purchase of an application package can be a daunting task. Most businesses purchasing office applications will purchase a well-adopted package, such as those from Microsoft. However, it is much more difficult when purchasing other types of package. It is imperative that the company set up a requirements definition for the package. This will involve looking at the problems of the present system and deciding on the requirements for the new system. Very often, the old system may need replacing: the business or its environment is changing, customer service is deteriorating, or poor management decisions may be made because of inappropriate, inaccurate, or untimely information being provided. Statements about desired service levels and workload will be particularized into the provision of reports, enquiry facilities, levels of security, volumes to be processed, and timescales. There may also be constraints, such as a budget maximum, personnel restriction, and a target date for implementation.

In evaluating a particular application package, the following questions should be asked.

1 Does it meet the functional requirements?

This is of course the issue of fundamental importance. If the package does meet the functional requirements then:

- Is all the input required by the package readily available?
- Is its capacity large enough for present use or too restricting for the future?
- Does it process the data fast enough?

If only some of the requirements are fulfilled then some other questions should be asked:

- What percentage of the requirements are fulfilled without amending the application package?
- Are the limitations of the package acceptable?
- How easily can the extra requirements be fulfilled?

2 What resources are required to buy and run the package?

- What is the basic cost, maintenance cost, and the cost of extra hardware and support required?
- What labour is required to set up and run the system?
- Can the package be run on other computers (which may be important later)?

3 How many people are presently using the package?

- Is it possible to get their reactions to it?
- Were there many set-up and teething problems?
- Has it proved reliable?
- Are they presently happy with the system?
- Are they happy with the help provided by the supplier?
- What would they have done differently now?

4 What is the quality of the documentation?

This relates to any hard-copy documentation and the help and other facilities provided in the package itself:

- Is it geared to computer experts, or are the users of the system likely to understand it?
- Is it well written? The documentation should be assessed on its appropriateness for the people who are going to use the package.
- Is it good for reading, learning, teaching, referring to, reminding, and diagnosing problems?

One way of presenting the various solutions is to create a matrix, listing the requirements (and therefore ignoring irrelevant features) on the left-hand side of a table, and then listing the solutions that might be appropriate along the top.

Let us assume that the alternative solutions are three application packages. Where the package meets any criterion, put a cross at the intersection. For evaluating the package, the more crosses the better, and therefore the recommended solution will be the one with the most crosses. This is package A in Figure 9.1(a).

A package may only partly meet any requirement. The technique can therefore be improved by giving a mark out of 10 for each package/criterion. The marks for each package can be added up and the package that scores the most is chosen. This is package B in Figure 9.1(b).

The technique can be further improved by giving a weight to each criterion. Some requirements are more important than others. To emphasize the relative importance of criteria, a weight is allocated to each of them so that if a weight of 4 is assigned, then this criterion is considered to be four times as important as a criterion allocated a weight of 1. Thus, if a package scores 7 out of 10 where the weight is 3, this counts as 21 on the total. Criteria that are considered essential can be given large weights, so that packages that do not meet these requirements will be excluded. In Figure 9.1(c), the allocation of different weights to criteria has led to package C being the recommended solution. The example shows that different recommendations could be made depending on the way the technique is applied.

We will now look in more detail at some of the important considerations that need to be taken into account in evaluating and purchasing application packages and the hardware to run them.

1 A requirements shortfall

In some circumstances even the best hardware and software choice performs only some of the requirements defined. The business is then faced with a further choice in order to deal with this:

- Should it adapt the business to fit in with the requirements of the computer system?
- Should it call in the services of its computing team, consultants, or software house to adapt the package so that it does conform to the needs of the business?
- Should it decide that application packages are inappropriate and decide to develop a completely new in-house solution?

All these are feasible in certain circumstances. Whatever the choice, it is likely that some variation of the Pareto 80:20 rule will apply (see Section 7.3): about 80 per cent of the application

(a) Package/ Criterion	A	B	C	(b) Package/ Criterion	A	B	C	(c) Package/ Criterion	Weight	A	B	C
1	X	X	X	1	6	9	8	1	2	12	18	16
2	X		X	2	6		9	1	10	60		90
3	X		X	3	7		9	3	10	70		90
4	X	X		4	7	10		4	1	7	10	
5		X		5		10		5	2		20	
Total	(4)	3	3	Total	26	(29)	26	Total		149	48	(196)

Figure 9.1: Evaluating application packages

requirements will be covered by the package, but about 20 per cent will not be covered. Conversely, it is this 20 per cent of the application requirements that will absorb something like 80 per cent of the costs.

At least the prospective purchaser has had the foresight to detect the mismatch before purchasing the system. Too often the system has been purchased and the manager has discovered to his horror that some vital function is not performed. Further, as the problem was not analysed beforehand, the manager may have chosen an application package that is particularly difficult to adapt.

It may seem surprising that the extra 20 per cent of the requirements are so comparatively expensive. The reason is that the writers of an application package expect to sell many copies. This means that the initial expense of writing it is absorbed by the profit on the number of copies sold. Writing a package may cost one hundred times its selling price. However, as soon as more than one hundred copies are sold, the supplier begins to make a profit. The 'tuning' of the package for an individual user will be expensive. Although software houses design packages to be of use to as many prospective customers as possible, such a generalized package is unlikely to fulfil all the requirements of all of them. In estimating the costs and benefits of an application package, it is essential that the cost of fulfilling all the requirements are included, otherwise the justification for the purchase will be distorted.

2 Intangible costs and benefits

In evaluating systems, it is necessary to include intangible costs and benefits in the calculations. Intangible costs include: the time and effort spent looking at potential packages; and the time and effort involved in training and educating people to use the computer system and overcoming their resistance to the new technology.

Training and education are vital. The morale of staff may improve with the implementation of computer systems. But it might deteriorate if measures are not taken to maintain the confidence of employees. Staff may view computers as a threat because they may lead to a loss of work status, work satisfaction, and employment. These intangible costs are difficult to evaluate. Intangible benefits such as greater speed of data processing, improved levels of security, greater accuracy, and greater reliability are equally difficult to evaluate in money terms, but should be included in the assessment. In Section 19.5 we discuss fully the issues relating to the potential benefits and problems that might accrue from using large software packages supporting applications development known as integrated toolsets.

9.2 Open source software (OSS)

According to Feller and Fitzgerald (2002), the term open source software was coined in 1998 having previously being known as free software. It is sometimes still termed Free/Open Source Software (F/OSS). Although it is often available free, or for a relatively small fee, it is not this that makes it open source. It is more the licensing model which 'guarantees the right to copy, modify and distribute the source code of the program without discrimination' (Feller and Fitzgerald, 2000). Proprietary software does not have this kind of licence and is usually not available in a form that can be accessed and modified by others. Open source, on the other hand, is made available as source code that can be read, modified and used in other contexts.

According to Greiner and Goodhue (2005) open source communities have 'become increasingly successful, with some now threatening the market dominance of major proprietary software vendors'. Apache and Linux, for example, are two of the most popular and well-known open source software communities. Linux is now considered the most significant competitor to Microsoft's flagship operating system, Windows. Linux was started by a 21-year-old Helsinki university student, Linus Torvalds, and Linux was designed as a Unix-like operating system running on the IBM PC. Many developers responded to Torvalds' call for help in developing the system and in the ten years to 2001, 'had grown to hold 25 per cent of the general server market and 34 per cent of the Internet/Web server market' (Fitzgerald et al., 2002). Some commercial products, such as Netscape (a rival to Microsoft's Internet Explorer) became open source. Other open source products are now the backbone to many Internet transactions. Greiner and Goodhue go on to say that, 'Open source projects and their related communities come in many shapes and sizes, and the growth in the number of projects and participants has been astounding.'

According to the Open Source Initiative (OSI), a non-profit organization dedicated to managing and promoting open source software, 'the basic idea behind open source is very simple: When programmers can read, redistribute, and modify the source code for a piece of software, the software evolves. People improve it, people adapt it, people fix bugs. And this can happen at a speed that, if one is used to the slow pace of conventional software development, seems astonishing' (Open Source Initiative, 2005). They state that open source is a 'rapid evolutionary process producing better software than the traditional closed model, in which only a very few programmers can see the source code'. Open source software is thus about rapidly evolving software development but more than that it is developed and evolved by a distributed, and often global, set of talented and determined people who have no vested interests, usually work for free (or nearly free) and who are motivated by producing better software.

OSS also has something of a political, philosophical and even ideological background. For example, the Open Source Initiative, after making their case for OSS, say '… and last but not least, it's a way that the little guys can get together and have a good chance at beating a monopoly' (Open Source Initiative, 2005). The monopoly that most people would have in mind in this context is of course Microsoft. For some, the OSS movement grew out of opposition to the proprietary standards that Microsoft have exploited. For many this is regarded as against creativity and innovation. The development of the Internet itself is often seen as a triumph of open standards.

It is also interesting to note that OSS and component development have come together somewhat, and as mentioned in Section 8.5 the open source community is one of the most successful examples of the use of component development. Examples range from complete applications or products, such as the Linux operating system and the Perl programming language, through to small modules that are used to form part of an application. With open source components, people have free access to the code, and the software can be freely used by anybody as a component of their own aggregate software. Collections of routines are publicly available, typically downloadable from the web, to help build applications.

These components include drivers (for example, for networks and audio applications), system utilities (for example, emulation, file, graphics and backup components), Internet utilities (for example, file transfer and log analysis), software development tools (for example, build tools, code generators, debuggers), games, security, database components, user interfaces, internationalization functions, security, and interfaces to commercial software. There are various sources of these components and IBM, Red Hat and Cosource are example providers.

Thus, open source products tend to be of the technical and infrastructure type, such as operating systems, utilities, network technologies, etc. Fitzgerald et al. (2002) suggest that software developed by open source methods tends to be of this particular type because of the nature of the open source phenomenon. There are fewer business applications, the main focus of this book. The reason for this is that in such software the design element 'is almost a given', i.e. there is general agreement on the basic design and that the requirements are well understood and relatively stable. This has to be the case for a large number of globally distributed developers to be able to work on the code effectively. This also helps to explain why business applications, which are characterized by discussion, disagreements and changing requirements, are not really amenable to open source development, and indeed according to Fitzgerald et al. (2002) 'are not likely to be'.

Another interesting aspect of open source is that software is developed by different individuals and groups in parallel, often geographically dispersed. Even though not working on the same project with the same company, nor following a standard project management approach, the peer review taking place potentially leads to better debugging and improves the speed of development, as well as the more obvious cost advantage. Peer review is also provided by users who are seen as co-developers. Because of the independent peer review and prompt feedback, open source software can be of high quality and very robust. This is partly due to the process, but equally due to the spirit in which the process is undertaken: there is a great sense of community felt by open source software developers along with cultural norms which Raymond (2001) expresses.

Stewart and Gosain (2001) also talk about the phenomenon of evolving and self-organizing entities that surround the open source software developing process. These developers rarely receive direct financial rewards for their efforts although they gain in skills, reputation and esteem amongst their peers. Developers are usually volunteers; they work in loose teams and may work only at their personal discretion rather than as directed by anybody. They provide their own time and development resources including computers and tools. The projects they work on are not usually owned or managed by a corporation.

So how do such projects work? Scacchi (2004) has undertaken a study of a variety of open source developments and suggests that the organization and cooperation of people is a challenge because they have to do it themselves but it is one that has been addressed. Open source projects have to be well designed in terms of modularization and the interfaces between them such that they can be coded independently and in a distributed environment.

Typically, an individual, or more likely a group of people, get together to develop some software or set of software, e.g. collaborative tools or a new game. Sometimes this might be a development from scratch or sometimes from an existing 'closed' product. These people would be the core developers who then design the project and divide it up into modules and tasks. They then usually seek help, known as submissions, from other like-minded people who are prepared to give up their time and effort in this way. Like-minded is key in open source as they tend to have an interest in the area, and the same views on software development, needs and philosophies. The project may then evolve in various ways with new people joining and maybe even taking over from some of the original core developers. Task will be allocated and roles distributed. The core developers may also seek submissions from people they know, and sometimes receive spontaneous contributions to the project that they may accept or not. Gacek et al. (2004) have classified a set of roles and tasks that might be performed in an open source project, see Figure 9.2. It is noticeable that developers are also often users, indeed Gacek et al. suggest that developers are always users of the code produced, i.e. this is part of their motivation.

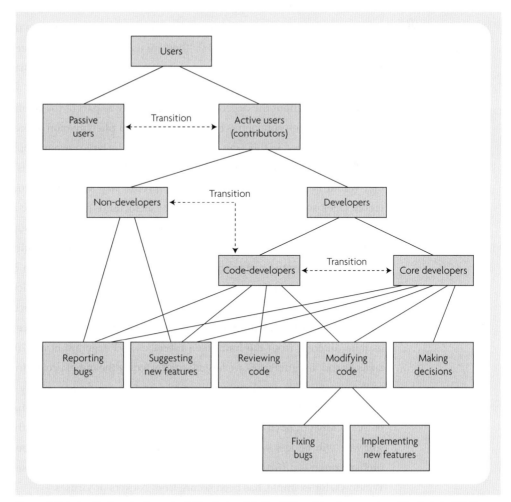

Figure 9.2: A classification of roles and activities that contributing OSS participants can perform (Gacek et al., 2004)

> This means that open source developers are a subset of the open source user community, i.e. all open source developers are users ... This characteristic explains the fact that there are normally no precise specifications or requirements documents clarifying what is to be achieved in the project. It also highlights that it is quite unrealistic to expect the open source community to start developing arbitrary kinds of software. Software developers are usually not expert users of medical systems, nuclear plant control systems, or air traffic control systems.

Code is then developed and the open source mantra is to 'release early and release often' with the objective of obtaining quick feedback from users and then identifying changes and improvements to be included in the next release. Fitzgerald et al. (2002) state that there was a new release every day in the early stages of development of Linux, and Scacchi (2004) refers to a project that had 360 release versions. Controlling releases (and indeed the whole project) is important and open source developers tend to adopt some kind of version control tool, such as CVS (Concurrent Versions System). CVS is itself an open source product and is free. According to Ximbiot who provide support for CVS it is 'the most popular version control system available today' (Ximbiot, 2005). These kinds of tools are used to manage the project and mediate control. According to Scacchi (2004), 'they are used as both a centralized mechanism for coordinating and synchronizing development, as well as a venue for mediating control over what software enhancements, extensions, or upgrades will be checked-in and made available for check-out throughout the decentralized community as part of the publicly released version'.

Whilst open source has been something of a phenomenon, there are some criticisms of software developed in this way, and not just from the large corporations that see OSS as something of a threat. For example, sometimes the level of documentation is not up to the standard that would be provided by traditional or proprietary software, particularly user documentation. Open source is also not always quite as focused on ease of use and what is sometimes called 'fit and finish', e.g. the provision of wizards, etc. The level of support is sometimes not what it might be, although of course there are some open source products where it is as good as any. Sometimes the sustainability of software has been criticized in the sense that it is difficult to guarantee that the product will be maintained over a long period (although this is also sometimes true with commercial software). The lack of discipline forced upon developers by methodologies has sometimes been cited as a downside of open source and the OSS approach is quite different from the traditional lifecycle. As Fitzgerald et al. (2002) say, 'the planning, analysis and design phases are largely conducted by the initial project founder, and are not part of the general OSS development life cycle' and so the design is only as good as the person involved and yet this is critical to the overall success of the project. Much has also been made of the fact that in open source the source code is freely available and this might mean that the software will be less secure than if it was kept secret. This might be thought to be a large problem but it is difficult to prove and certainly keeping source code secret has not meant that Windows, for example, has been secure from hackers or viruses.

9.3 Enterprise resource planning (ERP)

Enterprise resource planning (ERP) systems are much more ambitious application packages. We therefore discuss these in a separate section, although they are in essence a type of application package. The success of an organization may depend largely on integrated systems, and in particular on the effective transfer of information throughout the supply chain. ERP systems form a complex series of software modules used to integrate many business processes. Originally, these included production, inventory management, and logistics modules for manufacturing organizations. Later, they developed from materials requirement planning and manufacturing resource planning systems to encompass the capabilities of money resources planning systems, so that ERP systems supported all the basic financial applications and other organizational functions, such as human resource management. Now they include:

- Strategic planning
- Sales and distribution
- Marketing
- Financials
- Controls
- Quality management
- Supply management
- Materials management
- Plant maintenance
- Production planning
- Workflow, and
- Human resource management.

These cover all the standard business processes and functions of the organization (information, people, money, products and services, and equipment). More recently still, modules have been incorporated that provide the capability for e-commerce, customer relationship management, call centres and their implementation in small and medium-sized enterprises (SME). In this way ERP systems impact outside the organization as well as within it, as they allow for communication with suppliers and customers, all along the global supply chain. Indeed, they attempt to provide a complete IT solution for businesses.

If we just consider one of the above application areas, that of human resource management, we can illustrate the breadth and depth of the application as it covers:

- Job applications management
- Benefits management
- Personnel administration
- Incentive wages
- Payroll
- Time management
- Attendance management
- Travel expenses

- Organizational management
- Personnel development
- Room reservations
- Seminar and convention management
- Workforce planning
- Benefits administration, and
- Skill set inventories.

Many potential advantages are obvious. The business gains from a fully integrated system that enables visibility and integrity of data throughout the organization. Some of the potential disadvantages are also obvious; in particular, the complexity of integrating many or most of the organization's applications and the consequent cost in terms of money and time to achieve this. There is therefore some risk associated with the implementation of ERP systems.

As we have said, the latest versions incorporate Internet technology to pull suppliers and customers closer together in the supply chain. An example might be a customer buying a new car from a dealer. The customer may be able to find where the car is in the manufacturer's production schedule, and the suppliers of raw materials are able to predict manufacturer requirements based on dealers' forward orders.

Enterprise resource planning systems are supplied by SAP, Oracle, Baan, JD Edwards, and PeopleSoft (now both owned by Oracle), among others. The 'sales pitch' is obvious – there is support for every aspect of the business, each system is seen as the 'best of breed', top management can see the implications of one part of the business on others, and 'discipline' can be imposed on the workforce as all activities can be costed and controlled.

The market leader, SAP, has around 40 per cent of the overall market, over 15,000 installed sites worldwide and over US$6 billion in development. The temptation to join this bandwagon has been difficult to resist. It is even possible to have such software configured for certain sectors, including

- Universities
- Banks
- Hospitals
- Insurance
- Publishing
- Government
- Utilities
- Telecommunications
- Airlines, and
- Retailers.

Along with the software itself, there is a huge business in training and consulting.

However, many organizations have found that it is not that simple. The long implementation time span, the huge investment, the impact of everything changing at the same time, and the sheer direct and indirect costs are just some of the complaints offsetting some of the claims. For example, the applications are data intensive and data entry is a major conversion cost. Further, in order to make the likelihood of success greater, many firms change processes to fit the software (that is, minimal customization or 'vanilla ERP'), and this causes problems. In practice, implementation of ERP systems has often coincided with downsizing (in all but ERP expertise) as companies try to alleviate costs, and middle management in particular suffers. Thus many implementations have encountered people resistance and dependence on consultants. Nevertheless, many businesses have found that their ERP implementation has proved to be a great success.

ERP are a realization of method engineering goals (Section 8.4) in that they provide a consolidated and integrated toolkit of method fragments and tools for modelling and building applications. The ERP vendors and the systems analysts enforce a framework by which the systems are brought online. ERP are not simply technologically centric, because they incorporate business process re-engineering and modelling tools. Nor do they simply assume given process models, because, depending on the management's choice of implementation, they have a very definite focus on organizational considerations. In particular, they enable central control of organizational activities:

- How business activities are conducted
- What is measured and how
- How organizations are structured, and
- How work is managed.

ERP systems are significant breaks with the past in two ways. First, most organizations find them so complex that they forego significant customization and effectively redesign the organization to fit the software system, frequently through business process re-engineering (Section 4.3). This trend is wholly anathematic to past practice in which the goal of information systems development has been to build systems that fit organizations, like custom-tailored gloves fit the hand. A second significant difference is that the driving force behind the acquisition of these systems tends to be top corporate management rather than IT management or even business unit management. Finally, as business process re-engineering assumptions and values are a part of these approaches and systems, the focus has moved away from the integration of existing IS and the improvement or enhancement of current systems, to a philosophy of wholesale replacement and abandonment of existing systems and ways of doing work. Organizational restructuring and significant reductions in headcount throughout the organizational hierarchy generally accompany ERP. Significantly, the organizational re-engineering usually follows the introduction of the ERP system rather than preceding its adoption. Therefore, it represents the instance of the software system calling the organizational tune as compared with a more traditional development setting in which the software is constructed to meld with the organizational demands.

ERP systems are a great deal more than a set of methods and tools; they are also regarded as infrastructures in that they incorporate a type of front-end, back-end, and middle-tier architecture that runs on a host of different networks and hardware platforms. An initial appeal of these systems has been that they provided certain interoperability and sharing of enterprise data while allowing custom configuration of the applications set. This is a point that has proved to be problematic, and we return to this later. However, the sheer complexity and the cost of bringing custom solution ERP up and running has proved so high that the pendulum has swung in the direction of standard, 'no change', implementation on rapid deployment product versions using special implementation teams. Implementation is relatively quick but does not address organizational issues such as restructuring and training. We are therefore at a stage where there is some dissatisfaction with ERP systems though they have been successfully deployed in many businesses.

9.4 Outsourcing and offshoring
Outsourcing

Outsourcing activities to third party organizations has been a significant business practice for many years. For example, a manufacturing organization may outsource all its catering, security or cleaning needs to a third party company. It may be done as a way of cutting costs, but the main reason is likely to be that the organization does not want to devote its energies to these activities. Efforts are better directed towards their own manufacturing business activities, their 'core' activities, at which they are hopefully very good, and the catering, security and cleaning are left to a third party company that specializes in providing these services to a high standard. In theory at least, the client organization does not have to worry about the service provision, except to specify and set it up originally, pay the monthly charge, and renegotiate (or terminate) at the end of the contract. However, things sometimes go wrong, so they should also be able to monitor the quality of the service and negotiate rectification when any shortfalls occur.

These same principles have been used in connection with IT provision. In the early days of computing, organizations sometimes used third party 'service bureaux' to prepare input for a system (prepare punch cards, etc.), to run applications on the bureau's mainframe, or to provide a range of other computer services. More recently, it has been common to 'outsource' some or all telecommunications provision, networking, PC maintenance, help desks, training, etc.

Exactly what is meant by the term outsourcing presents problems, and a survey by Michell (1994) illustrated the wide variety of interpretations that outsourcing vendors themselves have for what constitutes outsourcing. In relation to IT, Fitzgerald and Willcocks (1994) defined outsourcing as the commissioning of a third party (or a number of third parties) to manage a client organization's IT assets, people and/or activities to required results. The focus here is on the management of the activity by the outsourcing vendor to a defined specification or service level. This can, and often does, involve a degree of transfer of assets, including staff, to the third party organization.

So defined, outsourcing does not exhaust the ways in which markets can be used. In one type of outsourcing, contracts specify a service and result which the market is to provide and manage. The focus of outsourcing is frequently on the result or outcome of the service being outsourced rather than specifying how it is to be undertaken. So an outsourcing contract might specify that a particular computer service has to be available for 24 hours a day, seven days a week, with an up-time of 99.5 per cent over the period. It does not specify how that is to be achieved. That is up to the vendor.

This contrasts with using the market to provide resources from outside to be deployed under the client organization's management and control. For example, the hiring of contract programmers into an existing development team would not be true outsourcing. To avoid confusion, Feeny et al. (1993) have termed this 'insourcing'.

In the early and mid-1990s outsourcing in IT developed a new impetus, driven by the influence of the strategic management concept of 'core competencies' or 'core business focus'. Prior to this, vertical integration of activities was in fashion. This suggested that competitive advantage could be gained by organizations seeking to undertake activities up and down their supply chain. So a manufacturing organization might also seek to own or make the constituent parts of their product, make their own packaging, own the distribution channel, do their own marketing, etc. The idea of core competencies is diametrically opposed to this. The belief here is that it is difficult, if not impossible, for organizations to be up to world class standards in everything they do. To achieve world class standards is difficult enough, but to be world class in non-core activities as well is virtually impossible. To attempt this is likely to lead to a dissipation of resources and energies, and result in failure in all activities. Thus organizations should strive to be of world class standard in their core activities only and outsource the non-core activities to third parties or vendors where those activities are the core business. They can do these to world class standards.

When organizations examined their core activities they often found that IT was not part of that core and therefore it should be outsourced. One of the most influential examples of this was Eastman Kodak in the US who decided that film was their core business and that IT was not, so they outsourced all their IT (Lacity and Hirschheim, 1993). This was a very influential decision and many other companies followed suit or at least reviewed their core business and considered outsourcing some if not all of their IT. In the UK, British Home Stores (Bhs) outsourced all their IT as a result of a 'core focus' strategic activity (Willcocks and Fitzgerald, 1993).

This also reflected a change of philosophy in relation to IT. It had previously been suggested that IT was key to competitive advantage, particularly in information-intensive industries, but even banks were not immune to outsourcing.

The outsourcing of the whole of a company's IT to a third party is still relatively rare, although these are the deals that hit the trade press headlines. The selective outsourcing of parts of a company's IT is much more common and is known as selective outsourcing.

There are other reasons for outsourcing IT (see Fitzgerald, 1994). One reason is to obtain IT services at less cost. Why might a third party be able to provide a product or service at lower cost? The answer is usually related to economies of scale which is essentially the notion that the greater the output, the less will be the costs of production per unit of output. In IT terms, there is some evidence that economies of scale are to be found in the running of mainframes, data centres, telecommunications networks, and help desks. Other reasons relate to the possibility of gaining optimal talent (because the outsourcing company are specialists), obtaining new technology quickly (for the same reason), minimizing overheads during slack times (a specialist outsourcing supplier is likely to find work between companies evens out more) and is less bureaucratic (because IT people are not in the employ of the client).

This growth in the outsourcing of IT to third party companies has resulted in the development of some very large IT outsourcing vendors. Indeed the business is dominated by a few very large companies, such as EDS and IBM, with a larger number of much smaller niche players, specializing in, for example, a particular area such as IS development or a particular sector, such as financial trading systems. The structure of the industry has certainly changed with the development of large-scale outsourcing and some large client organizations, which would previously have had a large internal IT department, may now have almost no internal IT and focus on managing and organizing suppliers rather than doing it themselves.

Choice of supplier is crucial, and consideration needs to be made about the process, technical and industry knowledge of the supplier, its track record, its flexibility regarding details of the contract, and the cultural fit between the supplier and the customer. Of course, not all outsourcing has been successful and there is evidence of some degree of dissatisfaction. Organizations have terminated and re-negotiated agreements. However, over time vendors have probably improved and there is more knowledge about the best areas to outsource and the processes to be undertaken for success.

The outsourcing of information systems development has been an important subset of IT outsourcing and the 'core competencies' argument has sometimes been the justification. For example, a motor vehicle manufacturer might say that although we need software systems, our competencies and capabilities are not in this area therefore we will outsource it. Or it might be that although a company has the competencies in IS development, and it might be core to their business, they may not be able to keep up with the demand for new systems. This was the justification made by one of the mobile telephone companies to outsource some of their IS development. For others, the continuing problems of systems development and the difficulties of delivering systems on time and within budget (see Section 3.6) have been the justification for outsourcing. The trend for downsizing has also resulted in the outsourcing of IS development. Some companies no longer have IS staff to undertake in-house development projects. Others have found that their existing IS personnel lack the skills necessary to build systems, especially those related to distributed and client–server development.

Whatever the reason, there has been a growing tendency for organizations to outsource some or all of their systems development. However, it is not an easy option. In all outsourcing the specification of the service that is being outsourced has proved difficult and problematic, but this is a particular issue with IS development. Many of the problems encountered in ISD relate to the problems of gathering requirements from users, the changing nature of requirements, and the difficulty of accurately specifying requirements, plus the 'freezing' of requirements whilst the system is constructed (see Section 5.6). If this is difficult to achieve in-house, it can only be even more difficult if it is outsourced. Nevertheless, it is frequently attempted and is sometimes successful. In outsourcing systems development the specification of what is needed must be very tight and exact. Some have argued that the use of an outside vendor to develop systems is actually helpful in this respect because users realize that the specification has to be precise and stable and they are prepared to make compromises and not demand that everything is provided. The fact that they are also paying 'real' money, rather than internal organization transfers, also tends to focus the mind on the objectives and a realistic set of requirements.

Sometimes the whole development is outsourced including the gathering of user requirements, but more commonly it is not the whole process but a subset of it. This subset may encompass the programming, or the programming and implementation. However, according to Goldsmith (1994), even if a vendor is just undertaking the building phase, the client often 'implicitly expects some help on the design', and this can lead to serious problems. Therefore he argues for expending even more effort than usual on defining requirements when outsourcing IS.

Goldsmith also finds that client companies are much more confident about selecting software packages than they are about selecting software development services. A particular problem is that the vendor cannot easily convey a tangible example of what will be provided, 'because the vendor is proposing to develop software that does not yet exist, the buyer cannot see it or try it out. Thus, the vendor essentially is asking the buyer to evaluate a promise rather than a product' (Goldsmith, 1994). A well-known example of the failure of IS development outsourcing is that of the London Ambulance Service's Computer Aided ambulance Dispatch (LASCAD) system. This failed a few days after implementation due to lack of proper specification, design and testing, together with a poor choice of vendor (Beynon-Davies, 1995).

As in all outsourcing the issue of the contract, and the contract framework, is very important. The type of contracting framework used may have a significant effect on the price paid, the business value accruing, and on the probability of completion.

The client company outsourcing its systems development has to develop skills in selecting the correct vendor, specifying requirements in detail and writing and negotiating contracts rather than thinking about system development methodologies. Again this represents a significant shift in emphasis and downplays the importance of the way the application is developed. Client companies are no longer so interested in what methodology is used – they leave that to the vendor – they just want the outcome to be successful (see Chapter 27).

There are other potential, long-term effects for the client organization. The experience and expertise of developing and running systems in-house is being lost. This skill and expertise is being transferred to the vendor with the result that the organization is increasingly dependent on outside vendors. This can be particularly problematic if IT is strategic to the business, or where it becomes strategic after outsourcing. Some organizations, because they have outsourced their IT, fail to keep track of strategy in relation to IT and may miss a trick or be overtaken by competitors who have kept up.

Outsourcing has also been driven in some sectors and organizations by a requirement to test any new work or project in the marketplace. This is seen as a way of ensuring best value for money by having in-house provision compared with the best that the market can provide. In the

UK, for example, the public sector was required to do this and choose the lowest-cost provider. This resulted in a large number of public sector IT projects being outsourced with some, like LASCAD, unsuccessful because of the requirement to choose the lowest-cost supplier.

One area which has recently seen an increase in outsourcing is that of web development (see Section 7.5) and web technology provision. In this case, the motivation is usually that the client company does not have the necessary expertise in-house. A secondary reason concerns the need to develop quickly. It might also be done in situations where a company is unsure whether the application is going to be a key area and hedging their bets by outsourcing. There has also been a significant growth in what is termed Application Service Provision (ASP) (Currie, 2001).

Offshore outsourcing (offshoring)

The practice of outsourcing to a vendor in a geographically different area is known as Offshore Outsourcing (or sometimes Global Software Development) and it has become increasingly popular. It has been made possible by advances in telecommunications technology, which effectively remove traditional geographic boundaries. Some see it as the inevitable result of globalization. Electronic transfer of data and instructions, e-mail, video conferencing, etc. can enable the development of a system in a country many thousand of miles away from the client. However, although it has been enabled by technology it is driven by cost savings due to the relative cheapness of skilled labour available in developing countries.

As long ago as 1994, Patane and Jurison identified a number of reasons for the growth in offshore IS development (Patane and Jurison, 1994). The reasons have changed a little since then and we provide an updated view as follows:

- The relatively low cost of labour in some countries, such as India and China, compared with the US and Western Europe
- A generally well-educated labour force in such countries, particularly in relation to IT and programming skills. Indeed, a recent study indicates that IT staff in major offshore locations such as India, the Philippines and Mexico are as skilled as their counterparts in the West (Moran, 2003)
- The ease of access to, and relatively low costs of, high-speed telecommunication networks
- Worldwide availability of PCs and IT infrastructure
- Relatively low entry costs in terms of capital expenditure
- The continuing trend of freer and more open market economies and free trade
- The reduction of cultural barriers, for example, different working practices and values
- The broadening of education opportunities and in particular the worldwide nature of computing knowledge
- The drive for cost-cutting in Western companies together with the economic downturn of the late 1990s.

India was one of the first in the market with many companies offering offshore services. Bangalore is the centre of the industry with large companies such as Wipro, Infosys and Tata being based there. Wipro reported a turnover of $1.35 billion in 2004, up 36 per cent from the previous year. India has the benefit of a particularly well-educated sector of the population in computing and IT together with the advantage of being able to communicate in English. The economic effect on India has already been large with Kumra and Sinha (2003), for McKinsey, forecasting that in 2008 IT services and back office work in India will have increased five times and that the industry will be worth $57 billion a year, employing 4 million people, which will be around 7 per cent of India's GDP (Gross Domestic Product). Other countries like Singapore, Malaysia, the Philippines, Hong Kong, Taiwan, and China are also in this market, as well as some in Europe, such as Ireland, Hungary and Russia.

It seems that the US has adopted offshore outsourcing in a particularly big way. According to Kripalani et al. (2004), in 2001, 125 of the top 500 US companies outsourced some of their IT activities to Indian companies, whereas by 2004 it was 285 (Computerworld, 2004). Forrester Research (2005) has estimated that the number of US jobs moving offshore is accelerating in the short term, and it expects 830,000 jobs to move offshore by the end of 2005, a 40 per cent increase from its earlier estimate, and its long-term estimate is that 3.4 million jobs will be moved offshore by 2015. These are not just IT jobs, of course, and such estimates have caused some degree of political concern in the US over the offshore phenomenon.

As mentioned, one of the main drivers is cost-saving and having a system developed (or part-developed) overseas can reduce costs significantly. Wages of programmers can be as little as 10 to 25 per cent of those in the Western world. It has been estimated that corporations in the USA can cut their labour costs by 25 to 75 per cent using workers in India, China, and the Philippines. However, such savings are not automatic (Ante, 2004). Further, cost is only one of the factors that must be considered in any outsourcing decision.

Application systems development is regarded as particularly suitable for offshore outsourcing, although those tasks requiring significant interaction with users are often still retained in-house. These are typically the front-end tasks of user requirements elicitation and outline design, and the back-end tasks such as integrated testing and implementation in the client site. This leaves detailed design, programming, and program testing as the most suitable areas to be outsourced.

Yalaho et al. (2005) agree and argue that global software development is a knowledge-intensive and complex process and that such knowledge is best developed through learning by doing. Thus the skills required in both project management and the earliest analysis and design stages of software development, 'acquired after years of practice (experiencing failure and success)' make this an area that is not advantageous to offshore.

On the other hand, according to Schware (1987), 'Coding is a relatively simple process … it does not rely on creativity, organizational understanding, or consultation with end users. Common business programming representing more than 80% of world programming requires comparatively low-level skills.' For Schware it makes sense to only outsource parts of the IS development process and this process involves the partitioning of responsibilities between the client and the offshore service provider and the definition of clear, although overlapping, knowledge boundaries.

Yalaho et al. quote Takeishi (2002) who identifies two types of dependencies: dependence for capacity and dependence for knowledge. In dependence for capacity, a client company can carry out the task, but decides to extend its capacity by means of offshore outsourcing. In dependence for knowledge the client does not know how to do the task, and relies on the offshore vendor. According to Fine (1998), this type of dependency is far more risky for the client than dependence for capacity. Thus, according to Yalaho et al., for effective outsourcing it is important to distinguish knowledge partitioning, that is who has the knowledge for the tasks among organizations, from task partitioning, that is who does the tasks among organizations.

Of course, as indicated above, in the early days the driver for offshore outsourcing was mainly related to costs in the form of labour savings and thus the main dependency was capacity. But as the offshore outsourcing market has matured and become experienced this is changing. An increasing amount of knowledge dependency outsourcing is occurring.

In such cases Yalaho et al. suggest that the client organization might begin to lose its capability to enact the practices, making it increasingly dependent on the vendor. Further, the vendor may eventually become a competitor to the client. Once the vendor has developed skills and understanding, including market and customer knowledge, from the client, the vendor may start selling its expertise and products to the end customers directly.

Maintenance of offshore-developed software can be a problem and although many firms currently perform software maintenance of their in-house systems which have been originally developed by offshore vendors, they are now more likely to be maintained offshore by the original developers. Thus it needs to be remembered that there may be a long-term commitment to an offshore outsourcing decision.

A recent paper by Dibbern et al. (2004) provides a detailed survey of the outsourcing literature, including offshore outsourcing. The authors argue that this area has grown so fast that there has been little opportunity previously 'for the research community to take a collective breath, and complete a global assessment of research activities to date'.

Summary

- An application package is a ready-made information system. Being developed externally, the package may need to be tailored for the company.
- Open source components with the source code open to change are available so that adaptability for the application is facilitated.
- Enterprise resource planning (ERP) systems are application packages, but they are integrated systems, transferring information throughout the supply chain. ERP systems form a complex series of software modules used to integrate many business processes.
- Outsourcing is the commissioning of a third party (or a number of third parties) to manage a client organization's IT assets, people and/or activities to a defined specification or service level. Offshore outsourcing (offshoring) occurs where the work is carried out overseas, possibly where costs are reduced.

Questions

1. Create a list of criteria that you might use to assess whether an information system is developed in-house, through an application package or through an outsourcing supplier.
2. What are the special features that distinguish ERP from other information systems?
3. What are the differences between software packages and open source components?
4. Evaluate an application package or packages on the basis of the weighting systems discussed in Section 9.1 and make a case for its purchase or otherwise.
5. Do you think outsourcing ISD is a short-term 'fad' or 'fashion' or a longer-term trend?

Further reading

Feller, J. and Fitzgerald, B. (2002) *Understanding Open Source Software Development*, Addison-Wesley, Harlow, UK.

Hirschheim, R. (ed.) (2002) *Information Systems Outsourcing*, Springer-Verlag, Berlin.

Janson, M.A. and Subramanian, A. (1996) Packaged software: Selection and implementation policies, *INFOR*, Vol. 34, No. 2.

Norris, G., Wright, I., Hurley, J.R., Dunleavy, J., and Gibson, A. (1998) *SAP: An Executive's Comprehensive Guide*, John Wiley, New York.

Raymond, E.S. (2001) *The Cathedral and the Bazaar: Musings on Linux and Open Source by an Accidental Revolutionary*, O'Reilly, Sebastopol, California.

Part IV: Techniques

Contents

Chapter 10 – Holistic techniques   p 199

Chapter 11 – Data techniques   p 217

Chapter 12 – Process techniques   p 243

Chapter 13 – Object-oriented techniques   p 273

Chapter 14 – Project management techniques   p 289

Chapter 15 – Organizational techniques   p 295

Chapter 16 – People techniques   p 307

Chapter 17 – Techniques in context   p 315

We have chosen to describe techniques in Part IV separately for two reasons: first, most are common to more than one methodology and therefore to leave them to Part VI (on methodologies) would lead to repetition there, and, second, so that the principles contained in the methodologies can be described without going into the techniques used in too much detail. Although the techniques described are used in a number of methodologies, this does not mean that they are interchangeable because, as used in any particular methodology, they could address different parts of the development process, be used for different purposes, or be applicable to different objects. More obviously, but less fundamentally, they often use different diagrammatic conventions to show the same things.

Again, we have attempted to group techniques into categories suggested by chapter headings. Some are straightforward, others less obvious, and could either be in another category or even a category of their own! As in Part III we show a 'road map' of the relationships between themes, techniques, tools and methodologies but this time from the perspective of a particular technique. In this example we use people techniques and show the links to themes, tools and methods (see Diagram 2).

In Chapter 10 the holistic techniques of rich pictures, root definitions, conceptual models, and cognitive mapping are described. They help to understand the problem situation being investigated by the analysts. The first three originated in Checkland's Soft Systems Methodology, but have been incorporated into other approaches, for example, Multiview. Increasingly, these techniques are being used by analysts who may be following a methodology that does not include them 'officially' as part of that approach, but nevertheless prove helpful, particularly in the early stage of a project. Rich pictures are particularly useful as a way of understanding the problem situation in general at the beginning of a project; root definitions help the analyst to identify the human activity systems they are to deal with; and conceptual models show how the various activities in the human activity system relate to each other. Cognitive mapping is used in the SODA methodology.

Entity modelling and normalization are fundamental and common to many methodologies and are described in Chapter 11. Entity modelling and normalization are techniques for analysing data. Equally fundamental, but process-oriented, are data flow diagrams, and these are introduced in Chapter 12. Many of the other techniques described in this chapter, which can be categorized as process logic, analyse processes in some respect, and these include decision trees, decision tables, structured English, and action diagrams.

Structure diagrams and matrices are used in all walks of life as well as in many information systems development methodologies and frequently at different stages in a methodology. Structure diagrams show hierarchical structures, be it a computer program, that to represent relationships between people in a department, or the structure of processing logic. Matrices show the relationship between two things, for example, entities and processes, departments and documents, or roles of people and processes. The last technique of this chapter is the entity life cycle. The technique of entity life cycle analysis is also common to a number of methodologies. The entity life cycle is not, despite its name, a technique of data analysis. We have categorized it as being a technique of process analysis, though its main aspect is that it shows how an entity changes over time. Perhaps it should have had a category of its own: time-oriented techniques!

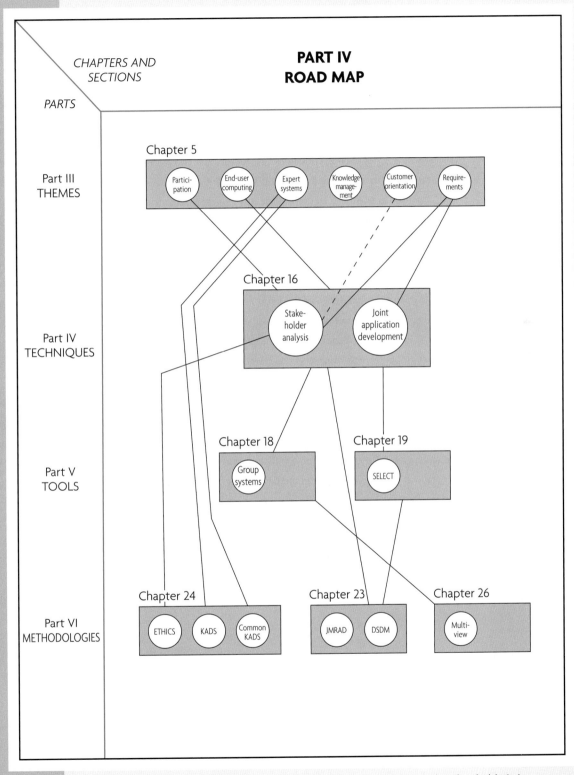

Diagram 2: People techniques 'Road Map' (shows relationships between people techniques and themes, tools and methodologies)

On the other hand, object-oriented techniques (Chapter 13) can represent data and processes, indeed the same technique can represent just about everything to do with the information system. It is not surprising, therefore, that the approach has attracted so much interest over the years, even though the more traditional data and process-oriented techniques are still widely used.

We discuss three project management techniques in Chapter 14. These are linked to the project management theme and PRINCE in the methodology section. In Chapter 15 we discuss organizational techniques. These are used by management to understand the organization: its strategy, environment, and to facilitate decision making in this context. Lateral thinking includes a series of techniques aimed at problem solving. Identifying critical success factors is a commonly used technique to ensure the most important benefits of a new information system are achieved. Scenario planning, future analysis, and SWOT analysis are techniques used to identify possible outcomes in the environment. Case-based reasoning uses the experience of previous cases to inform decision making. Risk analysis techniques attempt to make known and reduce the risks associated with investments.

People techniques include stakeholder analysis to identify all the stakeholders of an information systems project and joint application workshops, where all these stakeholders or their representatives are involved in analysis and design of the information system. We use this as the basis for our road map. We can see how the themes of participation and end-use development, for example, are connected to stakeholder analysis and joint application development. There is a connection to the tool GroupSystems. This supports group decision making, so could well enhance a JAD session. ETHICS is a methodology that has been designed to emphasize people aspects.

Finally, we argue that the use of techniques can have negative aspects. In particular, they may restrict understanding by framing the ways of thinking about the problem situation. In other words, people's understanding of a problem can be profoundly influenced by how the problem is presented to them by the technique. Different development techniques can represent the same problem situation differently, and the way in which it is represented has considerable potential for influencing problem understanding and resultant decision making. We look into these potential problems in Chapter 17.

10 Holistic techniques

10.1 Rich pictures

The analysis of such factors as interfaces, boundaries, subsystems, the control of resources, organizational structure, roles of personnel, organizational goals, employee needs, issues, problems, and concerns are not all contained in other techniques but are of importance in systems development. Understanding political aspects is essential for successful information systems. A high percentage of failure is due to ignoring these issues. When constructing a rich picture diagram such issues are taken into consideration.

An understanding of what the organization is 'about' need not take a diagrammatic form. It could be a mental map of some sort. We describe a possible diagrammatic form for rich pictures but we use the term rich picture rather than rich picture diagrams.

The technique stems from Checkland's Soft Systems Methodology (Section 25.1), but the description here is based on that used in Multiview (Avison and Wood-Harper, 1990). A rich picture diagram is a pictorial caricature of an organization and helps explain what the organization is 'about'. The rich picture should be self-explanatory and easy to understand.

One may start to construct a rich picture by looking for elements of structure in the problem area. This includes things like departmental boundaries, activity types, physical or geographical layout, and product types. Having looked for elements of structure, the next stage is to look for elements of process, that is, 'what is going on'. These include the fast-changing aspects of the situation: the information flow, the flow of goods, and so on.

The relationship between structure and process represents the 'climate' of the situation. Very often an organizational problem can be tracked down to a mismatch between an established structure and new processes formed in response to new events and pressures.

The rich picture should include all the important hard 'facts' of the organizational situation, and the examples given have been of this nature. However, these are not the only important facts. There are many soft or subjective 'facts' which should also be represented, and the process of creating the rich picture serves to tease out the concerns of the people in the situation. These soft facts include the sorts of thing that the people in the problem area are worried about, the social roles which the people within the situation think are important, and the sort of behaviour which is expected of people in these roles.

Representing the situation in terms of 'information systems needed' should be discouraged at this stage. These should come once the analysis has been carried out. Again, the question is not 'what systems does a manager think exist?', but rather 'what systems can be described in the situation?'. A 'system' in this sense is not about hardware and software but is a perceived grouping of people, objects, and activities which it is meaningful to talk about together.

Typically, a rich picture is constructed first by putting the name of the organization that is the concern of the analysis into a large 'bubble', perhaps at the centre of the page. Other symbols are sketched to represent the people and things that interrelate within and outside that organization. Arrows are included to show these relationships. Other important aspects of the human activity system can be incorporated. Any symbols can be used which are appropriate to the specific situation. We use crossed swords to indicate conflict and the 'think' bubbles indicate the worries of the major characters.

In some situations it is not possible to represent the organization in one rich picture. In this case, further detail can be shown on separate sheets. The perceived relative importance of people and things could be reflected by the size of the symbols on any one rich picture.

Figure 10.1 represents a rich picture for a professional association. We will use this case study in the description of rich pictures. The work concerning the case had three phases: an initial study, a full requirements analysis, and, finally, the development of some of the computer applications, including a computer system that handled some of the association's examinations.

The initial study was requested in correspondence from the secretary of the association. This was a top post in the organization. She felt that many of the systems ought to be computerized and she wished to know the type of computerization that would be appropriate for the situation, and whether the association should establish its own computer system or outsource (Section 9.4). The professional association is a professional body initiated for people working in or attempting to enter a particular profession. The current administrative system was purely manual. All the functions were under the control of the secretary. The education subsystem was administered by an education secretary.

The sorts of application included membership administration, examination administration, and tuition administration, requiring information about subjects, tutors, and fees. It was found that the workload at peak times of the year was becoming too demanding, membership was growing rapidly, and the administration and accounts occupied much of the time of the senior management, particularly that of the secretary.

Figure 10.1 represents an early draft of part of the rich picture of this human activity. If it has been well drawn, you should get a good idea of who and what are central to the organization and what are the important relationships. Bear in mind that there is no such thing as a 'correct' rich picture. Drawing the rich picture is a subjective process.

The act of drawing a rich picture is useful in itself because:

- lack of space on the paper forces decisions on what is really important (and what are side issues or points of detail for further layers of rich pictures);
- it helps people to visualize and discuss their own role in the organization;
- it can be used to define the aspects of the organization which are intended to be covered by the information system;

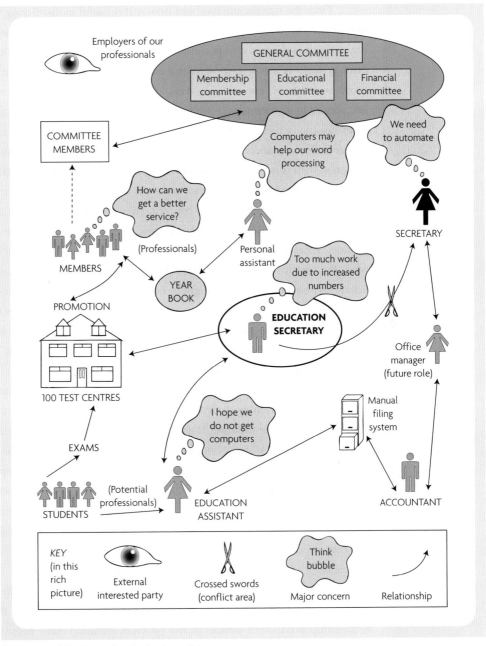

Figure 10.1: Rich picture of professional association

- it can be used to show up the worries of individuals, potential conflicts, and political issues.

Differences of opinion can be exposed, and sometimes resolved, by pointing at the picture and trying to get it changed so that it more accurately reflects people's perceptions of the organization and their roles in it.

Once the rich picture has been drawn, it is useful in identifying two main aspects of the human activity system. The first is to identify the primary tasks. These are the tasks that the organization was created to perform, or which it must perform if it is to survive. Searching for primary tasks is a way of posing and answering the question: 'What is really central to the problem situation?' For example, it could be argued that the association aims to give an excellent service to its members (which might be implied by the diagram) or increase standards in its profession (which is not implied by the diagram). Everything else is carried out to achieve that end. Primary tasks are central to the creation of information systems, because the information system is normally set up to achieve or support that primary task.

The second way that the rich picture is of particular value is in identifying issues. These are topics or matters which are of concern. They may be the subject of dispute. They represent the (often unstated) question marks hanging over the situation. In the association, they might include: 'What do we hope to achieve by installing a computer system?' This was a major issue when the systems analysis was done. This process of identifying the issues will lead to some debate on possible changes. It might be possible for these issues to be resolved at this stage, but it is essential that they are understood. Issues are important features, as the behaviour resulting from them could cause the formal information system to fail. Unless at least some of them have been resolved, the information system will have little chance of success. In some situations, the issues can be more important than the tasks.

The analyst starts by looking at an unstructured problem situation. This emphasis on the 'problem situation' as opposed to the 'problem' is important. By looking at the problem, rather than the whole situation, it will be difficult to be able to tell whether the diagnosis of the problem is correct. All too often a client will say, 'I am having a problem with X', when the problem is actually being caused by something else and X is a symptom of the overriding problem. A problem with stock control may be caused by a weak stock records system or by the fact that there is a lot of pilfering. If the analysts are limited to the official statement of the problem, then these 'real' causes may never be uncovered. It is therefore necessary for analysts to keep eyes, ears, and minds open and avoid jumping to early conclusions about the 'problem'.

There are a number of ways in which analysts get drawn into the problem situation and a number of roles that they may be called on to play. It is important that their role and their relationship with other people in the problem situation have been well defined and well explained. Whether an external consultant, a member of the internal IT department, a representative of the supplier, or a friend giving advice, it is important to think about the roles of the client, problem owner, and problem solver.

The rich picture can help the owners of the problem sort out the fundamentals of the situation, both to clarify their own thinking and decision making and also to explain these fundamentals to all the interested parties. The rich picture becomes a summary of all that is important in the situation. An analysis of the rich picture will help in the process of moving from 'thinking about the problem situation' to 'thinking about what can be done about the problem situation'.

In the example, the problem owner in the association shown in the rich picture is the general committee of the association. The analysts found that there was conflict between the secretary and the committee for whom she acted, on how best to serve the organization in its overall goal of improving the standards of the profession. The secretary, who was the client, was expected to become the system user.

As the study was under way, the role of the education secretary became more important. He was responsible for the professional examinations, and this system became the focal point of later analysis. There was a real conflict between the secretary and the education secretary regarding how and what to automate (hence the crossed swords in the rich picture). As the education secretary became a more powerful stakeholder and the analysts homed in on the examination system, he became the client. Therefore, he is central in the rich picture shown as Figure 10.1. At the time of the first part of the investigation, it was the secretary who was central.

The building shape in the left of the picture represents the one hundred test centres. It is important to draw attention to the difficulties of handling the examinations. The role of 'students', on the bottom left of the picture, changes from wanting to be professionals to 'members' who ask 'how can we get better service' once they have passed the examination at the test centre and have entered the profession.

Other stakeholders are also included in the rich picture. Developing further the theme of computerization, some actors were less positive about the prospect. The education assistant, an important actor in the examination system, thought that computerization might reduce her status and the think bubble contains 'I hope we do not get computers'. To jump a few steps, the actual system that was implemented in this situation was not a complete success, and, in retrospect, more attention should have been paid to her views. As one of the main persons involved, her misapprehension about the 'system' should have been addressed more fully.

We have included the accounting system in our rich picture. The accountant, bottom right of the picture, was satisfied with this manual filing system at the time of the investigation, but it would be looked at in the future.

In drawing the rich picture, some things have been left out which are understood by the participants but which would not be assumed by an outsider. One of these is the social roles of the people in the association and the sort of behaviour which is expected of people in these roles. Sometimes footnotes are useful to describe or list these. A second aspect is in the level of detail. The complexity of the marking system and other regulations for admission to the association, with which the stakeholders were familiar and well understood, could not be gleaned by an outsider looking at the rich picture. A second rich picture, drawn at a greater level of detail, would help here. Rich pictures can be 'decomposed' into others of greater detail. On the other hand, alternative techniques might be more appropriate at this level of detail.

With the analysts coming in as outside consultants, it was important that these 'assumptions' are drawn out. The approach adopted here was for the analyst to ask the users for a detailed explanation of a complex situation. Where a team of analysts is available, they can be divided so that different sorts of question can be asked: those relating to management strategy, those relating to data, and those relating to people's roles. The rich picture, once drawn up, proved a very useful communication tool in this situation and was refined according to new information. No thought is given at this stage to possible solutions. One of the purposes of drawing a rich picture diagram is to avoid 'design before analysis'.

The simplicity of the final rich picture is achieved by pruning the answers so that there is as much agreement as possible and so that the final picture really does represent the important people, activities, and issues of the problem situation.

Although this technique has been used by many analysts, it is not as well used as, for example, entity-relationship diagrams or data flow diagrams. The technique may not seem as formal as others and therefore may not have the credibility with managers who may also wish to avoid their political issues being disclosed and debated. Proponents of rich pictures might argue, however, that this is not because of any weakness of the approach, but partly because of a lack of knowledge about them and partly because systems analysts are not prepared to spend enough time on analysis and rush to the design and development phases.

10.2 Root definitions

The second technique originating from SSM, root definitions, can be used to define two things that are otherwise both vague and difficult. These are problems and systems. It is essential for the systems analyst to know precisely what human activity system is to be dealt with and what problem is to be tackled. The technique also originated from SSM and descriptions here are based on the Multiview approach.

The root definition is a concise verbal description of the system which captures its essential nature. Each description will derive from a particular view of reality. To ensure that each root definition is well formed, it is checked for the presence of six characteristics. Put into plain English, these are *who* is doing *what* for *whom*, and to whom are they *answerable*, what *assumptions* are being made, and in what *environment* is this happening? If these questions are answered carefully, they should tell us all we need to know.

There are technical terms for each of the six parts, the first letter of each forming the mnemonic **CATWOE.** We will change the order in which they appeared in our explanation to fit this mnemonic:

- *client* is the 'whom' (the beneficiary, or victim, affected by the activities);
- *actor* is the 'who' (the agent of change, who carries out the transformation process);
- *transformation* is the 'what' (the change taking place, the 'core of the root definition');
- *Weltanschauung* (or world view) is the 'assumptions' (the outlook which makes the root definition meaningful);
- *owner* is the 'answerable' (the sponsor or controller);
- *environment* remains the 'environment' (the wider system of which the problem situation is a part).

The word *Weltanschauung* may be new to many readers. It is a German word that has no real English equivalent. It refers to 'all the things that you take for granted' and is related to our values.

The first stage of creating the definition is to write down headings for each of the six CATWOE categories and try to fill them in. This is not always easy because we often get caught up in activities without thinking about who is really supposed to benefit or who is actually 'calling the tune'. We may question our assumptions and look around the environment even more rarely.

Even so, the difficulty for the individual in creating a root definition is less than the difficulty in getting all the individuals involved to agree on a usable root definition. Only experience of such an exercise can reveal how different are the views of individuals about the situation in which they are working together.

In trying to create the root definition for the professional association's examination system as part of the case study, the following process was followed. Initially, what were thought of as the issues and primary tasks were identified. These represented the things that the users were concerned about:

- efficient administration and management of the examinations system;
- choosing a solution which would not militate against the association's other systems, such as membership records management and accounting;
- building up a good reputation for the association.

Within this were identified three major components which are called the relevant systems. In the case study, the issues and primary tasks could largely be resolved by the following relevant systems:

- administration and management system;
- communication and motivation system;
- information provision system.

These relevant systems are subsystems to support a higher system which is to maintain and improve the reputation of the profession by ensuring high standards of entry into the profession.

The working root definition created was:

> A system owned and operated by the professional association to administer the examinations by registering, supervising, recording, and notifying students.

In the case when it was necessary to write the root definition, there was particular difficulty about the client. At first the obvious client was the secretary, but on further analysis the view was that the real client was the education secretary. Yet, as a computer solution became very likely, the person exercising power proved to be the treasurer, a member of the financial committee who would only give his consent to purchase a computer system if it was a particular brand. This happened to be that which he was experienced at using and one which the analysts felt later was inappropriate to the examinations system. There was little that the analysts could do about this political infighting, but at least they were aware of the problem.

It is sometimes difficult to produce a rigorous root definition because of these political or other problems. Sometimes it is impossible to resolve differences. However, unless they are resolved, they may be a source of difficulties later.

The CATWOE criteria were used to check and revise the above root definition as follows:

- *customers* – members of the association, the secretary, education secretary, and treasurer;
- *actors* – the association, its members, students attempting to join, and its full-time staff;
- *transformation* – to provide examinations which will ensure entry at the right level to the profession;
- *Weltanschauung* – the view that computer systems would be efficient and effective if they were used in this domain;
- *owners* – the general committee of the association (representing members of the association);
- *environment* – the particular profession.

Thus the first use for the root definition is to clarify the situation. People involved in an enterprise have very different views about that enterprise. These views are frequently at cross purposes. Not everyone, for example, thought that computer systems would be efficient and effective. This holds true even when the same words are used to describe things. This is because the differences are usually in the unstated assumptions or different perceptions of the environment. More significantly, there are sharp differences of opinion about whose problem the analysts are trying to solve, that is, who is the owner and who is the customer. It may not be possible to resolve differences of opinion and one root definition – a preferred root definition – might be chosen from the alternatives and used as a basis to develop the information system further.

Root definitions are particularly useful in exposing different views. We will look at an information system for a hospital to illustrate this. The different people involved in a hospital will look at the system from contrasting positions. Furthermore, these viewpoints in this problem situation are very emotive as they have moral and political overtones. In some situations this can lead to deliberate fudging of issues so as to avoid controversy. This is likely to cause problems in the future. Even if the differences cannot be resolved, it is useful to expose them.

Here are three different root definitions of a hospital system. They all represent extreme positions. In practice, anyone trying to start such a definition would make some attempt to encompass one or more of the other viewpoints, but any one of these could be used as the starting point for the analysis of the requirements of an information system in a hospital.

We will first look at the problem situation from the point of view of the patient, presenting a possible CATWOE and root definition (Figure 10.2).

We could therefore develop three sets of very different information systems depending on the view taken. The patient would have the system centred around patients' health records, the doctor would have the system designed around clinic sessions, and the administrator around the accounts.

These definitions have been deliberately controversial, but they attempt to show the private views of the participants as well as their publicly stated positions. There is no reason

THE PATIENT

CLIENT	Me
ACTOR	The doctor
TRANSFORMATION	Treatment
WELTANSCHAUUNG	I've paid my taxes so I'm entitled to it
OWNER	'The system' or maybe 'the taxpayer'
ENVIRONMENT	The hospital

A hospital is a place that I go to in order to get treated by a doctor. I'm entitled to this because I am a taxpayer, and the system is there to make sure that taxpayers get the treatment they need.

The perception of the doctors will be different:

THE DOCTOR

CLIENT	Patients
ACTOR	Me
TRANSFORMATION	Treatment (probably by specialized equipment, services, or nursing care)
WELTANSCHAUUNG	It is important to treat as many people as possible within a working week.
OWNER	Hospital administrators
ENVIRONMENT	National Health Service (NHS) vs Private practice. My work vs My private life

A hospital is a system designed to enable me to treat as many patients as possible with the aid of specialized equipment, nursing care, etc. Organizational decisions are made by the hospital administrators (who ought to try treating patients without the proper facilities) against a background of NHS politics and my visions of a lucrative private practice and regular weekends off with my family.

The views of the hospital administrators are likely to be different still:

THE HOSPITAL ADMINISTRATOR

CLIENT	Doctors
ACTOR	Me
TRANSFORMATION	To enable doctors to reduce waiting lists
WELTANSCHAUUNG	Create a bigger hospital within cash limits
OWNER	The government department of health
ENVIRONMENT	Politics

A hospital is an institution in which doctors (and other less expensive staff) are enabled by administrators to provide a service which balances the need to avoid long waiting lists with that to avoid excessive government spending. Ultimate responsibility rests with the government and the environment is very political.

Figure 10.2: Examples of CATWOE and root definitions from various perspectives

why definitions need to be formal and cold. Wilson (1990) carries out a similar exercise concerning the prison system. Dependent on the view taken, among other possibilities, the prison system could be seen as a:

- punishment system;
- behavioural experiment system;
- criminal training system;
- mailbag production system;
- people storage system;
- exclusive storage system.

These contribute to the eventual primary task definition:

A system for the receipt, storage and despatch of prisoners.

The alternative root definitions (briefly expressed above) indicate the difficulty of reconciling different viewpoints, and yet, if one is not agreed, it will be even more difficult to agree on final information systems needs. However, without looking at these wider views, information systems might be developed on the basis of a single (client's) view of the problem situation. Information systems are designed to serve the needs of people, and analysts are always brought directly into contact with power struggles between individuals and between viewpoints. Analysts have to make decisions, consciously or unconsciously, about which particular view of the situation or combination of views to work from. One option is to attempt to be 'scientifically detached', but this is only one of the options and is difficult to achieve in reality. In any case, such an aim would seem to be in conflict with the 'philosophy' of root definitions. We cannot be 'objective'.

In some Scandinavian countries there are laws or public agreements which state that the views of the workers or their representatives have to be sought and clearly represented at all stages in the analysis, design, and implementation of computer-based systems. In the USA the analysts usually focus on the opinion of the people who have commissioned them or are the senior people in that situation. Many analysts argue that they are making 'objective' decisions, innocent of any prejudice, but these may be based on personal and political assumptions that are never made explicit. The process described in this section should help to avoid this pitfall.

The system of communications is likely to be easier in smaller firms. The manager of a small firm contemplating a PC system may find this process rather long-winded and unnecessary. Nevertheless, the undercurrent in a small business can be just as political.

10.3 Conceptual models

Rich picture diagrams and root definitions, and the investigation and analysis preceding their construction, give an overall view of the organization. They also provide some key definitions of the purposes to be furthered by the information system. To complete the analysis of the human activity system (following SSM and Multiview), we need to build a model which shows how the various activities are related to each other, or at least how they ought logically to be arranged and connected. This is called a conceptual model. (Unfortunately, the term conceptual model is used in some other methodologies to refer to entity modelling.)

If the analysis of the human activity system is to be helpful to the organization, then it will show any discrepancies between what is happening in the real world and what ought to be happening. This may lead to changes in the organization of human activities. The purpose of the information system is to improve things, not just to 'automate or re-automate the status quo', although, as we saw in Chapter 3, many computer information systems do little more than this. So, once conceptual models are constructed, we will have a model of the required activities, which will serve as the foundation for the information model and a set of recommendations for an improved human activity system.

What do we mean by a conceptual model and what is it for? Perhaps these questions are best explained by analogy. When architects design a building they must produce two things: first, a set of artist's impressions and a scale model to show the client what is proposed and, second, a set of plans for the builder. Together these constitute the model. They will enable the builder to say how much it will cost and how long it will take and will represent all that needs to be created for the parties concerned to decide whether to go ahead with that design, modify it, or to choose an alternative.

The model serves three purposes:

- it is an essential element in the architects' design activities;
- it is a medium of communication between architects and clients to enable the right design to be selected;
- it is a set of instructions to the builders.

In computing, we also try to create models, which will serve these three purposes, but the process is not so well known, or so well tried and tested, as it is in construction. As we will see in this chapter and on looking at the various methodologies in Part VI, there are probably in use almost as many ways of describing a proposed system as there are design teams. This creates problems for users and designers as they try to understand what is being proposed.

In information systems development there is no clear-cut distinction between artist's impressions and the engineer's blueprints. There is not one version of the model for the user and another version for the computer programmer. Some may argue that this would be a valid goal, but, furthering our analogy, artist's impressions are notoriously optimistic and vague about difficulties and engineer's blueprints are very difficult to interpret by all but the trained. It is not satisfactory for the untrained to have to accept the statement: 'trust us, we're the experts'.

This means that the users and the builder of the information system must both understand the conceptual model. Of course, the information represented on a model can be complex, but the real world it represents is also complex.

Returning to the case study used for illustrative purposes in the previous two sections, the main activities of the examinations system, and consequently the information to support these activities prior to computerization, are shown in Figure 10.3.

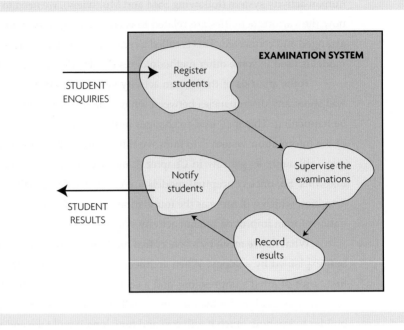

STUDENT
ENQUIRIES

STUDENT
RESULTS

Figure 10.3: Level 1 conceptual model for the professional association

The conceptual model is formed from the chosen root definition as follows:

1. Form an impression of the system to carry out a physical or abstract transformation from the root definition.
2. Assemble a small number of verbs that describe the most fundamental activities in the defined system.
3. Develop this by deciding on what the system has to do, how it would accomplish the requirement, and how it would be monitored and controlled.
4. Structure similar activities in groups together.
5. Use arrows to join the activities which are logically connected to each other by information, energy, material, or other dependency.
6. Verify the model by comparing it against the perceived reality of the problem situation.

The conceptual model shown as Figure 10.3 was derived from the root definition which was, for the professional association:

> A system owned and operated by the professional association to administer the examinations by registering, supervising, recording, and notifying students.

We start by taking significant aspects from the root definition and naming subsystems which will enable us to achieve what we require. These are the subsystems to register students, supervise the examinations, record the results, and notify the students.

So as to ensure that the most useful subsystems have been identified and understood, they are described in more detail in words and diagrammatically. In other words, there is a second layer in the conceptual model set which looks the same as the top layer, but is at a more detailed level. In other words, the technique lends itself to functional decomposition (as do many of the techniques described in this chapter).

In order to get agreement between problem solver and problem owner on these systems, it is important to ensure that there is a mutual understanding of the real-world meaning of the terms. It is necessary, for example, for the analyst to get to know what is involved in registering students (vetting enquiries and selecting potential students for registration, which is at the Level 2 conceptual model, shown as Figure 10.4) in order to understand that subsystem. As we have said above, the analyst is only concerned with 'what is conceptually necessary'. It does not matter, for example, how the enquiries are received, how the forms are sent out to be completed, or which member of staff deals with them.

The conceptual model is derived from the root definition. It is a model of the human activity system. Its elements are therefore activities and these can be found by extracting from the root definition all the verbs that are implied by it. The list of active verbs should then be arranged in a logically coherent order.

We would expect the number of activities to be somewhere in the range of five to nine. Activities should be grouped to avoid a longer list, as a long list is too complicated and messy to deal with. A shorter list suggests that the root definition is too broad to be useful. Having listed the major activities, some of these may imply secondary activities. These should also be listed and arranged in logical order around their primary activities and will form Level 2 conceptual models.

Conceptual modelling is an abstract process. Following soft systems methodology (Section 25.1) the purpose of going into this abstract world of systems thinking is to develop an

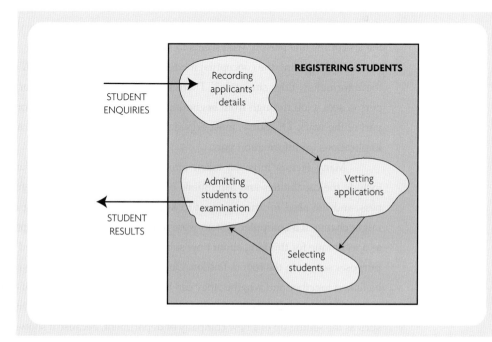

Figure 10.4: Level 2 conceptual model

alternative view of the problem situation. When this alternative view has been developed, we can return to the real world and test the model. It is constructed in terms of what must go into the system, and it can therefore be set alongside the real world. We are not concerned here with how the system will be implemented. Other techniques will be used to determine this aspect.

The conceptual model illustrates what ought to be happening to achieve the objectives specified in the root definition. There is normally more than one way of doing something and so choices will have to be made about the structure of the conceptual model. Many systems analysts adopt a very positivistic approach. An inexperienced analyst may put a flowchart in front of the user and ask: 'Is this right?' Users who may have had many years of coping with 'messy' reality may well be reluctant to answer so positively. They might wish to respond: 'Yes, but ...' or 'No, but ...'. Unfortunately the politics of the situation may be such that they may be forced to say 'yes' if they recognize some resemblance between the flowchart and reality. Alternatively they may be left to 'pick at the details'. One of the problems that the user faces is the inappropriateness of the flowcharts used in conventional systems analysis, and a strength of the conceptual model is its usefulness as a communication tool.

This conceptual model needs to be compared with reality to see whether improvements should be made to the way in which activities are organized. For example, in the second or third level detail of conceptual models, they might highlight bottlenecks, such as too many small decisions waiting for the manager or too many assistants waiting to use the same price catalogue. They may also show up circuitous routes for transferring information.

In small organizations, information handling is very informal, everyone sees what is happening or works alongside the people who know. As work diversifies and more staff are taken on, information flow is based around the experienced staff who become 'walking databases'. Such an arrangement can then ossify and become increasingly dysfunctional to new functions and new personalities. Many apparently efficient offices are thrown into disarray by the loss of the one person 'who seems to know everything'.

In comparing the conceptual model with reality the analyst will ask the question: 'Does the information flow smoothly?' There are two extreme forms of organization: where one person sees a job through from beginning to end or where each person handles a specialized part of the work. Of course, most organizations have aspects of each, and both have different implications for information flow.

Many factors must be taken into account when matching functions to staff. These include the capabilities and aspirations of staff, the demands of different aspects of the business, and the need for management to keep control of what is going on. A number of these might change if a computer system were to be introduced. The conceptual model can be used as a technique for thinking about how subsystems should be organized in order to achieve the purposes set out in the root definition. Questions can be asked about which subsystems should be linked together and whether they can be handled more efficiently if they are kept separate.

The conceptual model can also be used in the design of new human activity systems, such as the setting up of a new company or department, because it shows what activities should be carried out and how they should be related to each other.

The techniques of rich pictures, root definitions, and conceptual models will be explored further in the context of SSM (Section 25.1) and Multiview (Section 26.1).

10.4 Cognitive mapping

There are techniques and methodologies designed to make sense out of situations that are variously described as complex, uncertain, confusing, wicked, and messy. For example, rich pictures are one technique used in Soft Systems Methodology, among other approaches, to attempt to understand complexity in organizations. Cognitive mapping is a technique used in SODA, which we discuss in Section 26.2. Some of these techniques come from the management science community but are to some extent a reaction by more systemic thinkers such as Checkland, Mingers, Rosenhead, and others against the prevalent mathematical and optimizing techniques used in operations research. Our description of cognitive mapping is based on the work of Eden and Ackerman (2001).

A cognitive map is a model of the 'system of concepts' used to communicate the nature of a problem, and the concepts are related to others through an action orientation. In effect these maps show short statements (ideas, facts, circumstances, assertions, and proposals) relating to the problem situation linked by arrows showing their interrelationships; that is, how one idea might lead to or have implications for another. But it is the totality of the cognitive map that provides the most understanding. It is seen as a formal modelling technique with rules for its development. Cognitive maps can often be quite large with over 100 nodes, and they can be merged, finishing up with over 800 nodes. There is a software tool called Decision Explorer, which can be used with teams doing cognitive mapping. It draws maps, identifies clusters of nodes, and in its Group Explorer mode enables group work. Like most tools discussed in the book, it can increase productivity, but is not essential to using the technique.

The example, shown as Figure 10.5, comes from the work of Eden and Ackerman (2001), which explores the following quote about the Labour political party in the UK by Jane McLoughlin in the *Guardian* newspaper in 1986:

> The latter-day Labour party, aiming to appeal upmarket, is in a more ambivalent position. At a time when they are looking for a concordat with the unions, union opposition to profit sharing – because of the fear of collective bargaining – is hard to avoid. American experience with employee share ownership plans since the early 1980s has led to a drop in union membership in firms with these profit sharing schemes.

The cognitive map captures important phrases which represent contrasting ideas. So, at the top of the map, upmarket appeal is contrasted with concordat with the unions. The arrows show the links between the arguments. However, the cognitive map attempts to reflect the need for action (or problem-solving) orientation, so both the words might be changed and the order of representation to reflect this, but not, of course, the meaning. Figure 10.6 shows a development of this map following a merger with another.

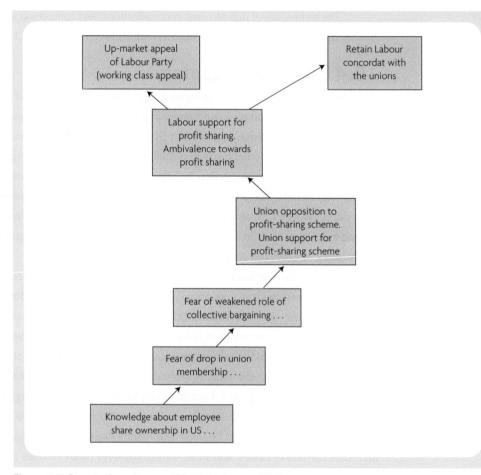

Figure 10.5: Example of cognitive map (Eden and Ackerman, 2001)

Each of the ideas expressed by individuals is converted to short statements in the cognitive map that are action or problem solving-oriented through a series of short interviews or discussions. The arrows connect options to desired outcomes, with the highest levels in the hierarchy expressing important goals. However, the map should remain in terms of the clients' thinking, not that of the consultant. The clients should still own it and regard it as a fair representation of the situation as they see it.

The process of drawing a cognitive map therefore encourages working through the detail toward the major goals through exploration or starting with the goals and, through interviews with the client or clients, elaborating on the detail that might achieve those goals (referred to as 'laddering up' or 'laddering down', respectively).

Usually, separate cognitive maps are drawn for each client, thus reducing the possibility of 'group think', and these are merged to form that of clients as a whole and tuned so that all the team feel committed to achieving the portfolio of actions, not just their particular concerns. This will require some political astuteness and conciliation skills on behalf of the consultant.

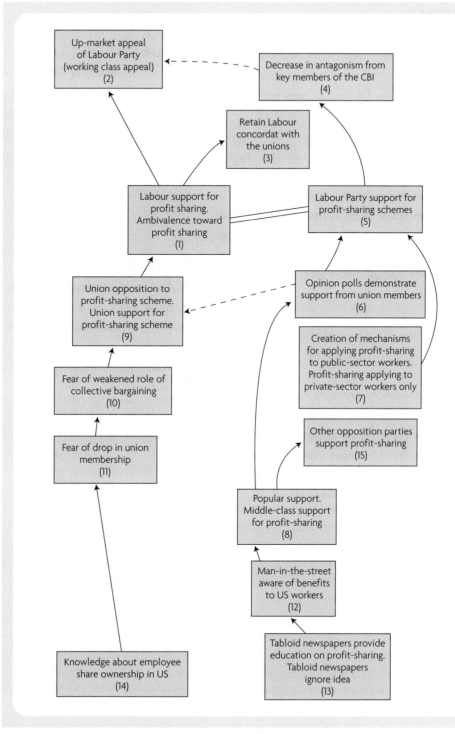

Figure 10.6: Development of cognitive map (Eden and Ackerman, 2001)

Summary

- A rich picture diagram is a pictorial caricature of an organization and helps explain what the organization is 'about'. It can represent 'soft' information representing the 'fuzziness' of many problem situations as well as 'hard' facts.

- The root definition is a concise verbal description of the system which captures its essential nature. It should reflect CATWOE (client, actor, transformation, world view, owner, and environment).

- The conceptual model is derived from the root definition. Its elements are activities and these can be found by extracting from the root definition all the verbs that are implied by it. The list of active verbs should then be arranged in logically coherent order.

- A cognitive map is a model of the 'system of concepts' used to communicate the nature of a problem, and the concepts are related to others through an action orientation. In effect these maps show short statements (ideas, facts, circumstances, assertions, and proposals) relating to the problem situation linked by arrows showing their interrelationships. But it is the totality of the cognitive map that provides the most understanding.

Questions

1. When would you use rich pictures, root definitions, conceptual models, and cognitive maps?

2. For an organization of your choice represent appropriate information in rich pictures, root definitions, conceptual models, and cognitive maps.

Further reading

Checkland, P. and Scholes, J. (1990) *Soft Systems Methodology in Action*, John Wiley & Sons, Chichester, UK.

Eden, C. and Ackerman, F. (2001) SODA – The Principles in J. Rosenhead and J. Mingers (eds) *Rational Analysis for a Problematic World Revisited*, 2nd edn, John Wiley & Sons, Chichester, UK.

11 Data techniques

11.1 Entity modelling

The theme of data modelling has been discussed in Section 6.3. Data modelling concentrates on the analysis of data in organizations and entity modelling is an important technique used to achieve this in many methodologies, including SSADM, Merise, and Information Engineering.

Just as an accountant might use a financial model, the analyst can develop an entity model. The entity model is just another view of the organization, but it is a particular perception of the organization which emphasizes data aspects. Systems analysis in general (data analysis is a branch of systems analysis) is an art or craft, not an exact science. There can be a number of ways to derive a reasonable model and there are a number of useful data models (there are, of course, an infinite number of inadequate models).

A model represents something, usually in simplified form, that highlights aspects which are of particular interest to the user, and is built so that it can be used for a specific purpose, for example, communication and testing.

An entity-relationship model views the organization as a set of data elements, known as entities, which are the things of interest to the organization, and relationships between these entities. This model helps the computer specialist to design appropriate information systems for the organization, but it also provides management with a unique tool for perceiving aspects of the business. The essence of problem solving is to be able to perceive the complex, 'messy', real world in such a manner that the solution to any problem may be easier. This model is 'simple' in that it is fairly easy to understand and to use.

Each entity can be represented diagrammatically by soft boxes (rectangles with rounded corners). Relationships between the entities are shown by lines between the soft boxes. A first approach to an entity model for an academic department of computer science is given in Figure 11.1. The entity types are student, academic staff, course, and non-academic staff. The entity type 'student' participates in a relationship with 'academic staff' and 'course'. The relationships are not named in Figure 11.1, but it might be that 'student' takes a 'course' and that 'student' has as tutors 'academic staff'. The reader will soon detect a number of important things of interest that have been omitted (room, examination, research, and so on). As the analysts find out more about the organization, entity types and relationships will be added to the model.

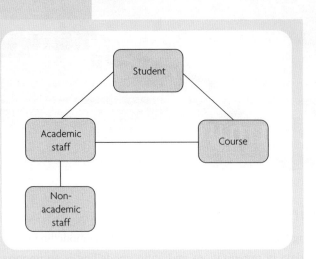

Figure 11.1: Entity modelling – a first approach

A mistake frequently made at this stage is to define the entities to reflect the processes or departments of the business, such as stock control, credit control, or sales order processing. This could be a valid model of the business but it is not an entity model and cannot be used to develop a database for the organization. Data analysis differs from conventional systems analysis in that it separates the data structures from the applications which use them. The objective of data analysis is to produce a flexible data model which can be easily adapted as the requirements of the organization change. Although the applications will need to be changed, this will not necessarily be true of the data.

The entity model is sometimes referred to as the conceptual schema or conceptual model. However, in order to avoid confusion with Checkland's conceptual models described in Section 10.3 (an entirely different model) we will use the terms 'entity model' or 'data model'.

One technique of data analysis is the entity-relationship (E-R) model, in which the real-world information is represented by entities and by relationships between entities. In a typical business, the entities could include jobs, customers, departments, and suppliers. The analyst identifies the entities and relationships between them before being immersed in the detail, in particular, the work of identifying the attributes which define the properties of entities.

Figure 11.2 relates to part of a hospital. The entities described are 'doctor', 'patient', and 'clinical session'. The relationships between the entities are also described. That between doctor and patient and between doctor and clinical session are one-to-many relationships (depicted by the crow's feet). In other words, one doctor can have many patients, but a patient is only assigned one doctor at a particular point in time. Further, a doctor can be responsible for many clinical sessions, but a clinical session is the responsibility of only one doctor. The other relationship is many-to-many. In other words, a patient can attend a number of clinical sessions and one clinical session can be attended by a number of patients. A one-to-one relationship would be shown by a line without crow's feet.

The diagram may also shows a few attributes of the entities. The particular attribute or group of attributes that uniquely identify an entity occurrence is known as the key attribute or attribute(s). The 'employee number' (Emp. No.) is the key attribute of the entity called doctor.

The technique attempts to separate the data structure from the functions for which the data may be used. This separation is a useful distinction,

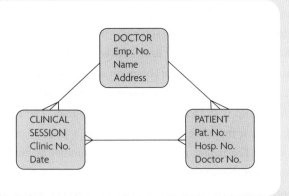

Figure 11.2: An entity set relating to part of a hospital

although it is often difficult to make in practice. In any case, it is sometimes useful to bear in mind the functions of the data analysed. A doctor and a patient are both people, but it is their role, that is what they do, that distinguishes the entities. The distinction, formed because of a knowledge of functions, is a useful one to make. However, too much regard to functions will produce a model biased toward particular applications or users and is therefore to be avoided.

Another practical problem is that organization-wide data analysis may be so costly and time consuming that it is often preferable to carry out entity analysis at a 'local' level, such as in the marketing or personnel areas. If a local entity analysis is carried out, the model can be mapped on to a database and applications applied to it before another local data analysis is started. This is far more likely to gain management approval because managers can see the expensive exercise paying dividends in a reasonable timescale. An important preliminary step is therefore to define the area for analysis and break this up into distinct subareas which can be implemented on a database in turn and merged later. Local data analysis should also be carried out in phases. The first phase is an overview which leads to the identification of the major things of interest in that area. At the end of the overview phase, it is possible to draw up a second interview plan and the next, longer, phase aims to fill in the detail.

Although it is relatively easy to illustrate the process of modelling in a book or at a lecture, in real life there are problems in deciding how far one should go and what level of detail is appropriate. The level of detail must serve two purposes:

1. It must be capable of explaining that part of the organization being examined.
2. It must be capable of being translated into a physical model, usually for mapping onto a computer database.

It is important to realize that there is no logical or natural point at which the level of detail stops. This is a pragmatic decision. Certainly, design teams can put too much effort into the development of the model.

Some decisions are based on the way in which the data is used. An example of this could be entity occurrences of persons who are female, where they relate to:

- patients in a hospital;
- students at university;
- readers in a library.

In the patient example, the fact that the person occurrence is female is important, so important that the patient entity may be split into two separate entities, male patients and female patients. In the student example, the fact that the person is female may not be of great significance and therefore there could be an attribute 'sex' of the person entity. In the reader example, the fact that the person is male or female may be of such insignificance that it is not even included as an attribute. There is a danger here, however, as the analyst must ensure that it will not be significant in all applications in the library. Otherwise the data model will not be as useful.

An entity is a thing of interest in the organization, in other words it is anything about which we want to hold data. It could include all the resources of the business, including the people of interest such as EMPLOYEE, and it can be extended to cover such things as SALES-ORDER, INVOICE, and PROFIT-CENTRE. Some entities are obvious physical things, like customers or stock. Others are transactions, like orders, sales, and hospital admissions. Some entities are more or less artificial. These are rather like catalogue entries in the library: the only reason to have them is to help people find books which would otherwise be difficult to locate. It covers concepts as well as objects. A SCHEDULE and a PLAN are concepts which can be defined as entities. An entity is not data itself, but something about which data should be kept. It is something that can have an independent existence, that is, can be distinctly identified.

In creating an entity model, the aim should be to define entities that enable the analyst to describe the organization. Such entities as STOCK, SALES-ORDER, and CUSTOMER are appropriate because they are quantifiable, whereas 'stock control', 'order processing', and 'credit control' are not appropriate because they are functions: what the organization does, and not things of interest which participate in functions. Entities will normally be displayed in small capitals in this book. Entities can also be quantified – it is reasonable to ask 'how many customers?' or 'how many orders per day?', but not 'how many credit controls?'. An entity occurrence is a particular instance of an entity which can be uniquely identified. It will have a value, for example, 'Jim Smith & Son' and this will be a particular occurrence of the entity CUSTOMER. There will be other occurrences, such as 'Plowmans PLC' and 'Archd Tower & Co.'.

An **attribute** is a descriptive value associated with an entity. It is a property of an entity. At a certain stage in the analysis it becomes necessary not only to define each entity but also to record the relevant attributes of each entity. A CUSTOMER entity may be defined and it will have a number of attributes associated with it, such as 'number', 'name', 'address', 'credit-limit', 'balance', and so on. Attributes will normally be displayed within single inverted commas in this book. The values of a set of attributes will distinguish one entity occurrence from another. Attributes are frequently identified during data analysis when entities are being identified, but most come later, particularly in detailed interviews with staff and in the analysis of documents. Many are discovered when checking the entity model with users.

An entity must be uniquely identified by one or more of its attributes, the **key attribute**(s). A <u>customer number</u> may identify an occurrence of the entity CUSTOMER. A <u>customer number</u> and a <u>product number</u> may together form the key atribute of entity SALES-ORDER. The key attribute functionally determines other attributes, because once we know the customer number we know the name, address, and other attributes of that customer. Key attributes will normally be underlined in this book.

There often arises the problem of distinguishing between an entity and an attribute. In many cases, things that can be defined as entities could also be defined as attributes, and vice versa. We have discussed one example relating to the sex of people. The entity should have importance in the context of the organization, otherwise it is an attribute.

In practice, the problem is not as important as it may seem. Most of these ambiguities are settled in the process of normalization (Section 11.2) and this often happens in database design. In any case, the analyst can change the model at a later stage, even when mapping the model onto a database, though the earlier the analyst gets it right the better. Entities are used by functions of the organization and the attributes are those data elements that are required to support the functions. The best rule of thumb is to ask whether the data has information about it; in other words, does it have attributes? Entities and attributes are further distinguished by their role in events (discussed below).

A **relationship** in an entity model normally represents an association between two entities. A SUPPLIER entity has a relationship with the PRODUCT entity through the relationship – supplies; that is, a SUPPLIER *supplies* PRODUCT. There may be more than one relationship between two entities; for example, PRODUCT is *assembled* by SUPPLIER. Relationships will normally be displayed in italics in this book. A relationship normally arises because of:

1. Association, for example CUSTOMER *places* order.
2. Structure, for example ORDER *consists of* order-line.

The association between entities has to be meaningful in the context of the organization. The relationship has information content, for example, CUSTOMER *places* ORDER. The action *'places'* describes the relationship between CUSTOMER and ORDER. The name given to the relationship also helps to make the model understandable. As will be seen, the relationship itself can have attributes.

The **cardinality** of the relationship could be one-to-one, one-to-many, or many-to-many. At any one time, a MEMBER-OF-PARLIAMENT can only represent one constituency, and one CONSTITUENCY can have only one MEMBER-OF-PARLIAMENT. A MEMBER-OF-PARLIAMENT *represents a* CONSTITUENCY. This is an example of a one-to-one (1:1) relationship. Figure 11.3 shows different conventions or notations of representing relationships found in methodologies. Very often, a one-to-one relationship can be better expressed as a single entity, with one of the old entities forming attributes of the more significant entity. For example, the entity above could be MEMBER-OF-PARLIAMENT, with CONSTITUENCY as one of the attributes.

The relationship between an entity CUSTOMER and another entity ORDER is usually of a degree one-to-many (1:m). Each CUSTOMER can have a number of ORDERS, but an ORDER can refer to only one CUSTOMER: CUSTOMER *places* ORDER.

With a many-to-many (m:n) relationship, each entity can be related to one or more occurrences of the partner entity. A STUDENT can *take* many MODULES; and one MODULE could be *taken by* a number of STUDENTS (MODULE is *taken* by STUDENT; STUDENT *takes* MODULE).

In this last example (of a many-to-many relationship), entity occurrences of the STUDENT entity could be 'Smith', 'Jones', and 'Wilson', and they could take a number of modules each. For example, in Figure 11.4 Smith might take database, IS development, and expert systems; Jones might take database and IS development, and Wilson IS development and expert systems.

Frequently there is useful information associated with many-to-many relationships and it is better to split these into two one-to-many relationships, with a third intermediate entity created to link them together. Again, this should only be done if the new entity has some meaning in itself. The relationship between COURSE and LECTURER is many-to-many, that is, one LECTURER *lectures on* many COURSES and a COURSE *is given* by many LECTURERS. But a new entity,

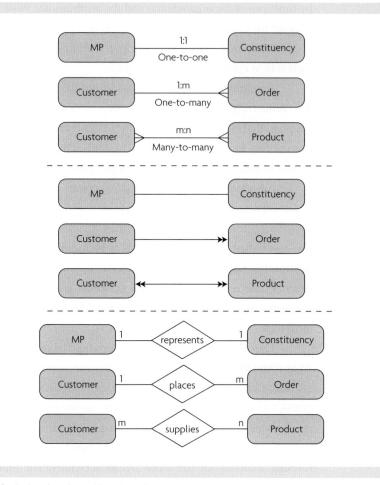

Figure 11.3: Cardinality of a relationship – three diagramming conventions

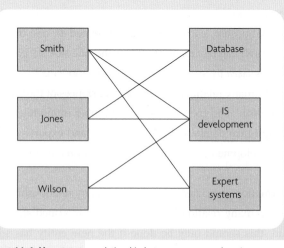

Figure 11.4: Many-to-many relationship between STUDENT and MODULE occurrences

MODULE can be described as used in the previous example which may only be given by one LECTURER and is part of only one COURSE. Therefore, a LECTURER gives a number of MODULES and a COURSE consists of a number of MODULES. But one MODULE is given by only one LECTURER and one MODULE is *offered to* only one COURSE (if these are the restrictions). This is shown in Figure 11.5.

There are other distinctions and sophistications which are often included in the model. Sometimes a 1:m or an m:n relationship is a fixed relationship. The many-to-many relationship between the entity PARENT and the entity CHILD is 2:m (that is, each child has two parents); but a PARENT *can beget* more than one CHILD.

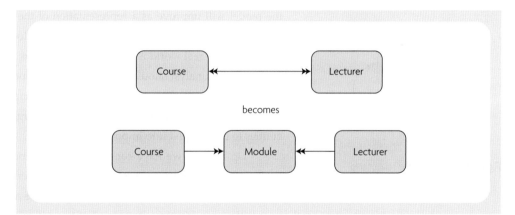

Figure 11.5: A many-to-many relationship represented as two one-to-many relationships

While some relationships are mandatory, that is, each entity occurrence must participate in the relationship, others are optional. An entity MALE and an entity FEMALE may be joined together by the optional relationship '*married to*'. Mandatory and optional relationships may be represented as shown in Figure 11.6.

Other structures are exclusive, where participation in one relationship excludes participation in another, or inclusive, where participation in one relationship automatically includes participation in another.

A relationship may also be involuted where entity occurrences relate to other occurrences of the same entity. For an EMPLOYEE entity, for example, an EMPLOYEE entity occurrence who happens to be a manager *manages* other occurrences of the entity EMPLOYEE. This can be shown diagrammatically by an involuted loop, as in Figure 11.7. Some approaches suggest that these should be eliminated by creating two entities (MANAGER and EMPLOYEE in this case).

Any relationship is necessarily linked to at least one entity. We have already looked at the involuted relationship. Where a relationship is linked to two entities (as in the case of the examples in Figure 11.4), it is said to be binary. If a relationship is linked to three entities, as in Figure 11.8, it is said to be ternary. In this example, EMPLOYEE fulfils a ROLE, EMPLOYEE fulfils a CONTRACT, and ROLE fulfils a CONTRACT. Otherwise it is *n-ary*, with the value of *n* equalling the number of entities.

A good model is one that is a good representation of the organization, department, or whatever is being depicted. The process of entity modelling is an iterative process and slowly

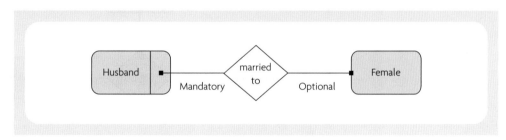

Figure 11.6: Representation of mandatory and optional relationships

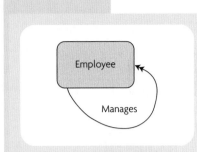

Figure 11.7: An involuted relationship

the model will improve as a representation of the perceived reality. The entity model can be looked on as a discussion document and its coincidence with the real world is verified in discussions with the various users. However, the analyst should be aware that variances between the model and a particular user's view could be due to the narrow perception of that user. If a global entity model is built for a whole organization it is usual for entities to be grouped into important clusters.

Up to now, we have considered the data-oriented aspects of data analysis, but in practice it is useful for functional considerations to be made in order to check the model. These relate to events and operations. Entities have to support the **events** that occur in the enterprise. Entities will take part in events and in the operations that follow events. Attributes are those elements which supply data to support events.

'Tom' is an occurrence of the entity EMPLOYEE. Tom's pay rise or his leaving the company are events, and attributes of the entity EMPLOYEE will be referred to following these events. Attributes such as 'pay-to-date', 'tax-to-date', 'employment status', and 'salary' will be referred to.

Operations on attributes will be necessary following the event: an event triggers an operation or a series of operations. An operation will change the state of the data. The event 'Tom gets salary increase of 10 per cent' will require access to the entity occurrence 'Tom' (or EMPLOYEE-NUMBER '756') and augmenting the attribute 'salary' by 10 per cent. Figure 11.9 shows the entity EMPLOYEE expressed as a relation with attributes. We have to check that the relation supports all the operations that follow the event mentioned.

Some methodologies, for example, Merise (Section 21.2), also define the **synchronization** of an operation (Figure 11.10). This is a condition affecting the events which trigger the operation and will enable the triggering of that operation. This condition can relate to the value of the properties carried by the events and to the number of occurrences of the events. For example, the operation 'production of pay slips' may be triggered by the event 'date' when it equals '28th day of the month'.

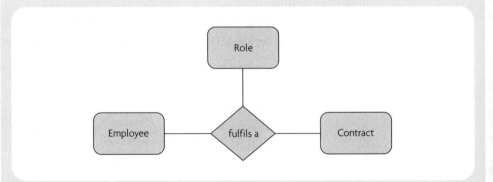

Figure 11.8: Example of a ternary relationship

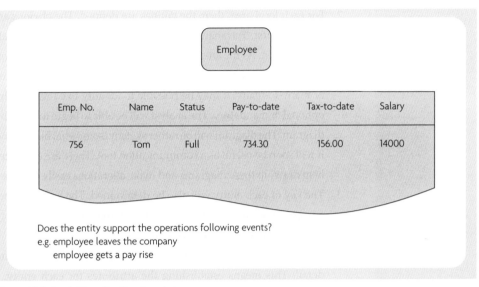

Does the entity support the operations following events?
e.g. employee leaves the company
        employee gets a pay rise

Figure 11.9: Event-driven (functional) analysis

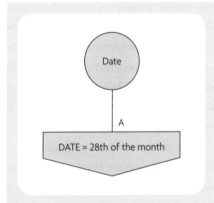

Figure 11.10: Synchronization of an operation

Some readers may be confused by the discussion of events (sometimes called **transactions**), which are function-oriented concepts, when data analysis is supposed to be function-independent. The consideration of events and operations is of interest as a checking mechanism. They are used to ensure that the entity model will support the functions that may use the data model. This consideration of the events and operations may lead to a tuning of the model, an adjustment of the entities and the attributes. We will now look at the stages of entity analysis as a whole which are:

1. Define the area for analysis.
2. Define the entities and the relationships between them.
3. Establish the key attribute(s) for each entity.
4. Complete each entity with all the attributes.
5. Normalize all the entities.
6. Ensure all events and operations are supported by the model.

We have looked at all these elements apart from normalization which is briefly described below and discussed more fully in Section 11.2.

1. The first stage of entity analysis requires the definition of the area for analysis. This is frequently referred to as the **universe of discourse**. Sometimes this will be the organization, but this is usually too ambitious for detailed study, and as we have seen the organization will normally be divided into local areas for separate analysis. Then we have the stages of entity-relationship modelling. It is a top-down approach in that the entities are identified first, followed by the relationships between them, and then more detail is filled in as the attributes and key attribute(s) of each entity are identified.

2. For each local area, then, the entities are defined. The obvious and major entities will be identified first. The analyst will attempt to name the fundamental things of interest to the organization. As the analyst is gathering these entities, the relationships between the entities can also be determined and named. Their cardinality can be one-to-one, one-to-many, or many-to-many. It may be possible to identify fixed relationships and those which are optional or mandatory. The analyst will be able to begin to assemble the entity-relationship diagram. The diagram will be rather sketchy, somewhat like a 'doodle' in the beginning, but it will soon be useful as a communication tool. There are computer software tools that can help draw up these diagrams and make alterations easily (see, for example, Section 17.3).

3. The key of each entity will also be determined. The key attributes will uniquely identify any entity occurrence. There may be alternative key attributes, in which case the most natural or concise is normally chosen.

4. The analyst has now constructed the model in outline and is in a position to fill in the detail. This means establishing the attributes for each entity. Each attribute will say something about the entity. The analyst has to ensure that any **synonyms** and **homonyms** are detected. A product could be called a part, product, or finished product depending on the department. These are all synonyms for 'product'. On the other hand, the term product may mean different things (homonyms), depending on the department. It could mean a final saleable item in the marketing department or a subassembly in the production department. These differences must be reconciled and recorded in the data dictionary (Section 19.1). The process of identifying attributes may itself reveal entities that have not been identified. Any data element in the organization must be defined as an entity, an attribute, or a relationship and recorded in the data dictionary. Entities and relationships will also be recorded in the entity-relationship diagram.

5. Each entity must be normalized once the entity occurrences have been added to the model. This process is described fully in Section 11.2. Briefly, the rules of normalization require that all entries in the entity must be completed (first normal form), all attributes of the entity must be dependent on all the key attributes (second normal form), and all non-key attributes must be independent of one another (third normal form). The normalization process may well lead to an increase in the number of entities in the model.

6. The final stage of entity analysis will be to look at all the events within the area and the operations that need to be performed following an event, and ensure that the model supports these events and operations. Events are frequently referred to as transactions. For this part of the methodology, the analyst will identify the events associated with the organization and examine the operations necessary on the trail of each of the events.

Events in many organizations could include 'customer makes an order', 'raw materials are purchased from supplier', and 'employee joins firm'. If, say, a customer makes an order, this event will be followed by a number of operations. The operations will be carried out so that it is possible to find out how much the order will cost, whether the product is in stock, and whether the customer's credit limit is OK. The entities such as PRODUCT (to look at the value of the attribute 'stock') and CUSTOMER (to look at the value of the attribute 'credit limit') must be examined. These attribute values will need to be adjusted following the event. You may notice that the 'product price' is not in either entity. To support the event, therefore, 'product price' should be included in the PRODUCT entity, or in another entity which is brought into the model.

Entity modelling has documentation like other methods of systems analysis. It is possible to obtain forms on which to specify all the elements of the data analysis process. The separate documents will enable the specification of entities, attributes, relationships, events, and operations. These forms can be pre-drawn using software tools and their contents automatically added to the data dictionary. We show the entity document as Figure 11.11.

As we have already stated, it may be possible to use completed documents directly as input to a data dictionary system so that the data are held in a readily accessible computer format as well as on paper forms. Entity modelling can be used as an aid to communication as well as a technique for finding out information. The forms discussed also help as an aid to memory, that is, communication with oneself. The entity-relationship diagrams, which are particularly useful in the initial analysis and as an overview of the data model, can prove a good

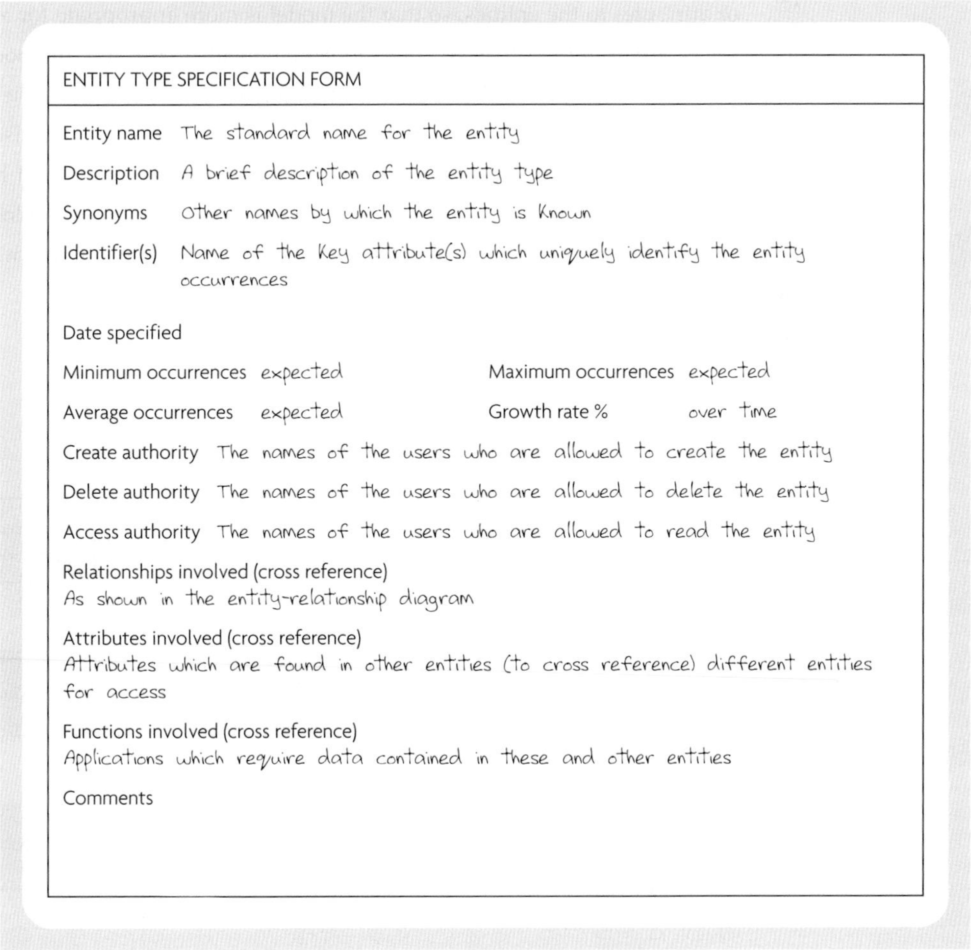

ENTITY TYPE SPECIFICATION FORM

Entity name The standard name for the entity

Description A brief description of the entity type

Synonyms Other names by which the entity is known

Identifier(s) Name of the key attribute(s) which uniquely identify the entity occurrences

Date specified

Minimum occurrences expected   Maximum occurrences expected

Average occurrences expected   Growth rate %   over time

Create authority The names of the users who are allowed to create the entity

Delete authority The names of the users who are allowed to delete the entity

Access authority The names of the users who are allowed to read the entity

Relationships involved (cross reference)
As shown in the entity-relationship diagram

Attributes involved (cross reference)
Attributes which are found in other entities (to cross reference) different entities for access

Functions involved (cross reference)
Applications which require data contained in these and other entities

Comments

Figure 11.11: Entity documentation

basis for communication with managers and users. They provide a graphical description of the business in outline, showing what the business is, not what it does. Managers and users can give 'user feedback' to the analysts and this will also help to tune the model and ensure its accuracy. A user may point out that an attribute is missing from an entity, or that a relationship between entities is one-to-many and not one-to-one as implied by the entity-relationship diagram. The manager may not use this terminology, but the analyst will be able to interpret the comments made.

Data analysis is an iterative process: the final model will not be obtained until after a number of tries and this should not be seen as slowness, but care for accuracy. If the entity model is inaccurate so will be the database and the applications that use it. On the other hand, the process should not be too long or 'diminishing returns' will set in.

The entity-relationship diagram given in Figure 11.12 shows the entities for part of a firm of wholesalers. The attributes of the entities might be as shown in Figure 11.13.

The key attributes are underlined. Perhaps you would like to verify that you can understand something of the organization using this form of documentation. It is a first sketch of the business, and you may also verify the relationships, add entities and relationships to the model, or attributes to the entities, so that the model is more appropriate for a typical firm of wholesalers. For example, we have not included payments in this interim model.

The entity-modelling approach to data analysis is often interview-driven, that is, most information is obtained through interviewing members of staff. It is also top-down, in that the entities are identified first and then more and more detail filled in. It has proved very useful and is included in many information systems development methodologies, as we shall see in Part VI. Methodologies usually have entity modelling preceding the process of normalization. This technique is described in the next section.

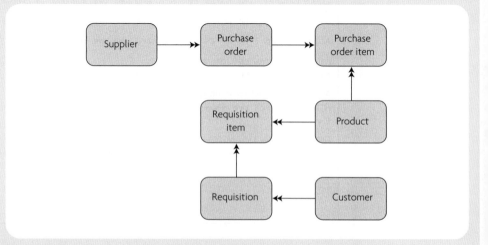

Figure 11.12: Entity-relationship diagram – a first approach for a wholesaler

SUPPLIER	<u>Supp-no</u>, Supp-name, Address, Amount owed
PURCHASE ORDER	<u>P-O-no</u>, P-O-date
PURCHASE ORDER ITEM	<u>P-O-item-no</u>, Quantity
REQUISITION	<u>Req-no</u>, Req-date
REQUISITION ITEM	<u>Req-item-no</u>, Quantity
CUSTOMER	<u>Cust-no</u>, Cust-name, Address, Amount, Credit
PRODUCT	<u>Product-no</u>, description

Figure 11.13: Entities with attributes

11.2 Normalization

Before looking at the rules of normalization, we will briefly develop a data model to use as an example. As seen in Figure 11.14, a relation is a **table** or **flat file**. This relation is called SALES-ORDER and it could show that Lee ordered 12 of 'part number' 25, Deene and Smith ordered 18 and 9, respectively, of 'part number' 38, and Williams ordered 100 of 'part number' 87.

The entities and relationships identified in the entity-relationship model can both be represented as relations in the relational model. In Figure 11.15, there are three entities in the entity model and they are expressed as three relations: COURSE, LECTURER, and TIMETABLE.

We will now introduce some of the terminology associated with the relational model. Each row in a relation is called a **tuple**. The order of tuples is immaterial (although they will normally be shown in the text in a logical sequence so that it is easier to follow their contents). No two tuples can be identical in the model. A tuple or column will have a number of attributes, and in the SALES-ORDER relation of Figure 11.14, 'name', 'part', and 'quantity' are attributes. All items in a column come from the same **domain**, that is, the domain is the set of values from which valid attributes can be drawn. There are circumstances where the contents from two or more columns come from the same domain. The relation ELECTION-RESULT (Figure 11.16)

SALES-ORDER		
CUSTOMER NAME	PART NUMBER	QUANTITY ORDERED
Lee	25	12
Deene	38	18
Smith	38	9
Williams	87	100

Figure 11.14: Sales-order relationship

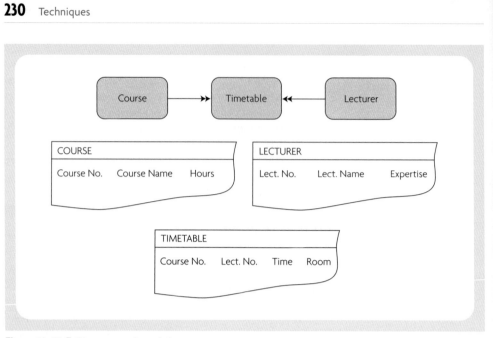

Figure 11.15: Entities expressed as relations

illustrates this possibility. Two attributes come from the same domain of political parties. The number of attributes in a relation is called the degree of the relation. The number of tuples in a relation define its cardinality.

In the SALES-ORDER relation, the key is 'customer name'. It might be better to allocate numbers to customers in case there are duplicate names. If the customer may make orders for a number of parts, then 'part' must also be a key attribute as there will be several tuples with the same 'name'. In this case, the attributes 'customer name' and 'part number' will make up the **composite key** of the SALES-ORDER relation.

What is the key for the ELECTION-RESULT relation? On first sight, 'election year' might seem appropriate, but in 1974 there were two elections in a year, and even if all three attributes

ELECTION-RESULT		
ELECTION YEAR	FIRST PARTY	SECOND PARTY
1974	Labour	Conservative
1979	Conservative	Labour
1983	Conservative	Labour
1987	Conservative	Labour
1992	Conservative	Labour
1996	Labour	Conservative
2000	Labour	Conservative
2004	Labour	Conservative

Figure 11.16: Election-result relation

formed the composite key, there are still duplicate relations. It is necessary to add another attribute which is unique, such as an election number which is incremented by one following each election. Alternatively, it would be possible to replace 'year' by 'election date' to make each tuple unique (there will not be two of these elections on the same day). Another alternative would be the composite key of 'election year' and a new attribute 'election number in year'.

There may be more than one possible key. These are known as **candidate keys**, that is, they are candidates for the **primary key**. 'Customer name' and 'customer number' could be candidate keys in a CUSTOMER relation. In this circumstance one of these is chosen as the primary key. An attribute which is a primary key in one relation and not a primary key, but included in another relation, is called a **foreign key** in that second relation.

The structure of a relation is conventionally expressed as in the following examples:

SALES-ORDER   (<u>name</u>, part, quantity)

ELECTION-RESULT   (<u>*elect-year*</u>, <u>elect-number</u>, first-party, second-party)

The process of normalization is the application of a number of rules to the relational model which will simplify the relations. This set of rules proves to be useful because we are dealing with large sets of data and the relations formed by the normalization process will make the data easier to understand and manipulate. The model so formed is appropriate for the further stages in most database methodologies and the database will be shareable, a fundamental justification for the database approach.

Normalization is a technique that is used in a variety of methodologies. For example, it is used in Gane and Sarson (STRADIS), Information Engineering, SSADM, Merise and Multiview which are all discussed in Part VI. The technique of normalization is applicable irrespective of whether a relational database is envisaged or not. It is often used in its own right as an analysis technique for the structuring of data, it can be used on its own or as a means of cross-checking or validating other models, particularly an entity model. Even in structured systems methodologies which stress processes rather than data, for example, Gane and Sarson's STRADIS, it is used to consolidate all the various data stores that have been identified in a data flow diagram into a coherent data structure.

Normalization is the process of transforming data into well-formed or natural groupings such that one fact is in one place and that the correct relationships between facts exist. As well as simplifying the relations, normalization also reduces anomalies which may otherwise occur when manipulating the relations in a relational database. In this simplifying process no data are lost or added to that provided in the original set of unnormalized relations.

Normalized data are stable and a good foundation for any future growth. It is a mechanical process, indeed the technique has been automated, but the difficult part of it lies in understanding the meaning, that is, the semantics of the data, and this is only discovered by extensive and careful data analysis.

There are three levels of normalization, and the third and final stage is known as the *third normal form (TNF)*. It is this level of normalization that is often used as the basis for the design of the data model, as an end result of data analysis, and for mapping onto a database. There are a few instances, however, when even TNF needs further simplification, and these are also looked at later in this section. TNF is usually satisfactory in practice.

There are three basic stages of normalization:

1. First normal form – ensure that all the attributes are atomic (i.e. in the smallest possible components). This means that there is only one possible value for each attribute and not a set of values. This is often expressed as the fact that relations must not contain repeating groups.

2. Second normal form – ensure that all non-key attributes are functionally dependent on (give facts about) all the key. If this is not the case, split off into a separate relation those attributes that are dependent on only part of the key together with the key.

3. Third normal form – Ensure that all non-key attributes are functionally independent of each other. If this is not the case, create new relations which do not show any non-key dependence.

A rather flippant, but more memorable, definition of normalization can be given as 'the attributes in a relation must depend on the key, the whole key, and nothing but the key'. This is an oversimplification, but it is essentially true and could be kept in mind as the normalization process is developed.

A more detailed description of normalization is now given. A key concept is **functional dependency**, which is often referred to as **determinacy**. This is defined by Cardenas (1985) as follows: Given a relation R, the attribute B is said to be functionally dependent on attribute A if at every instant of time each value of A has no more than one value of B associated with it in the relation R.

Therefore, if we know a 'customer-number', we can determine the associated 'customer-name', 'customer-address', and so on, if they are functionally dependent on 'customer-number'. Functional dependency is frequently illustrated by an arrow. The arrow will point from A to B in the functional dependency illustrated in the definition. Thus, the value of A uniquely determines the value of B.

Figure 11.17(a) is a non-normalized relation COURSE-DETAIL. Before normalizing the relation, it is necessary to analyse its meaning. Knowledge of the application area gained from entity modelling will provide this information. It is possible to make assumptions about the interrelationships between the data, but it is obviously better to base these assumptions on thorough analysis. In the relation COURSE-DETAIL, there are two occurrences of course ('COURSE'), one numbered B74 called computer science at the BSc level and the other B94 called computer applications at the MSc level. Each of these course occurrences has a number of module ('module') occurrences associated with it. Each 'module' is given a 'module-name', 'status' and 'unit-points' (which are allocated according to the status of the 'module').

(a) COURSE-DETAIL (Unnormalized)						
COURSE	COURSE-NAME	LEVEL	MODULE	MODULE-NAME	STATUS	UNIT-POINTS
B74	Computer Science	BSc	B741	Program 1	Basic	8
			B742	Hardware 1		
			B743	Data Processing 1		
			B744	Program 2	Intermediate	11
			B745	Hardware 2		
B94	Computer Applications	MSc	B951	Information	Advanced	15
			B952	Microprocessors		
			B741	Program 1	Basic	8

(b) COURSE-DETAIL						
COURSE	COURSE-NAME	LEVEL	MODULE	MODULE-NAME	STATUS	UNIT-POINTS
B74	Computer Science	BSc	B741	Program 1	Basic	8
B74	Computer Science	BSc	B742	Hardware 1	Basic	8
B74	Computer Science	BSc	B743	Data Processing 1	Basic	8
B74	Computer Science	BSc	B744	Program 2	Intermediate	11
B74	Computer Science	BSc	B745	Hardware 2	Intermediate	11
B94	Computer Applications	MSc	B951	Information	Advanced	15
B94	Computer Applications	MSc	B952	Microprocessors	Advanced	15
B94	Computer Applications	MSc	B741	Program 1	Basic	8

Figure 11.17: First normal form

1   *First Normal Form* includes the filling in of details, ensuring all attributes are in their smallest possible components. This is seen in the example in Figure 11.17(a) and is a trivial task. You may note that in Figure 11.17(a), the order of the tuples in the unnormalized relation is significant. Otherwise the content of the attributes not completed cannot be known. As we have already stated, one of the principles of the relational model is that the order of the tuples is not significant. The tuples seen in Figure 11.17(b) could be in any order in this relation. First normal form essentially converts unnormalized data or traditional file structures into fully completed relations or tables.

The key of the relation of Figure 11.17(b) is 'course' and 'module' together (a composite key) and the key attributes have been underlined. A composite key is necessary because no single attribute will uniquely identify a tuple of this relation. There were in fact a number of possible candidate keys, for example, 'module-name' and 'course-name', but we chose the primary key as above because they are numeric and unique.

Further work would have been necessary if the following was presented as the unnormalized relation:

course, course-name, level, module-details

'Module-details' has to be defined as a set of atomic attributes, not as a group item, so it has to be broken down into its constituents of 'module-name', 'status', and 'unit-points':

course, course-name, level, module-name, status, unit-points

2 *Second Normal Form* is achieved if the relations are in first normal form and all non-key attributes are fully functionally dependent on all the key attributes. The relation COURSE-DETAIL shown in Figure 11.17(b) is in first normal form. However, the attributes 'module-name', 'status', and 'unit-points' are functionally dependent on 'module'. In other words, they represent facts about 'module', which is not the whole key which is 'course' and 'module'. This is known as *partial dependency*. We may say that if the value of the module is known, we can determine the value of 'status', 'name', and 'unit-points'. For example, if 'module' is B743, then 'status' is basic, 'module-name' is data processing 1, and 'unit points' is 8. They are not dependent on the other part of the key, 'course'. So as to comply with the requirements of second normal form, two relations will be formed from the relation and this is shown as Figure 11.18. But this is only a partial advance to second normal form; there are elements in the first normal form relation not in this model.

The relation COURSE-MODULE is still not in second normal form because the attributes 'course-name' and 'level' are functionally dependent on 'course' only, and not on the whole of

(a) COURSE-MODULE

COURSE	MODULE	COURSE-NAME	LEVEL
B74	B741	Computer Science	BSc
B74	B742	Computer Science	BSc
B74	B743	Computer Science	BSc
B74	B744	Computer Science	BSc
B74	B745	Computer Science	BSc
B94	B951	Computer Applications	MSc
B94	B952	Computer Applications	MSc
B94	B741	Computer Applications	MSc

(b) MODULE

MODULE	NAME	STATUS	UNIT-POINTS
B741	Program 1	Basic	8
B742	Hardware 1	Basic	8
B743	Data Processing 1	Basic	8
B744	Program 2	Intermediate	11
B745	Hardware 2	Intermediate	11
B951	Information	Advanced	15
B952	Microprocessors	Advanced	15

Figure 11.18: First step toward second normal form

the key. A separate COURSE relation has been created in Figure 11.19. The COURSE relation has only two tuples (there are only two courses), and all duplicates are removed. Notice that we maintain the relation COURSE-MODULE. This relation is all key, and there is nothing incorrect in this. Attributes may possibly be added later which relate specifically to the *course–module* relationship, for example, the teacher or text. The relation is required because information will be lost by not including it, that is, the modules which are included in a particular course and the courses which include specific modules. The relations are now in second normal form.

3   *Third Normal Form* (TNF) is necessary because second normal form may cause problems where non-key attributes are functionally dependent on each other (that is, a non-key attribute is dependent on another non-key attribute). In the relation MODULE, the attribute 'unit-points' is functionally dependent on the 'status' (or level) of the course, that is, given 'status', we know the value of 'unit-points'. So 'unit-points' is determined by 'status' which is not a key. We therefore create a new relation STATUS and delete 'unit-points' from the relation MODULE. We check each non-key attribute and find that there are no more such dependencies. The third normal form is given in Figure 11.20.

(a) COURSE-MODULE

COURSE	MODULE
B74	B741
B74	B742
B74	B743
B74	B744
B74	B745
B94	B951
B94	B952
B94	B741

(b) COURSE

COURSE	COURSE-NAME	LEVEL
B74	Computer Science	BSc
B94	Computer Applications	MSc

(c) MODULE

MODULE	MODULE-NAME	STATUS	UNIT-POINTS
B741	Program 1	Basic	8
B742	Hardware 1	Basic	8
B743	Data Processing 1	Basic	8
B744	Program 2	Intermediate	11
B745	Hardware 2	Intermediate	11
B951	Information	Advanced	15
B952	Microprocessors	Advanced	15

Figure 11.19: Second normal form

Sometimes the term **transitive dependency** is used in this context. The dependency of the attribute 'unit-points' is transitive (via 'status') and not wholly dependent on the key attribute 'module'. This transitive dependency should not exist in third normal form.

The attribute 'status' is the primary key of the STATUS relation. It is included as an attribute in the MODULE relation, but it is not a key. This provides an example of a **foreign key**, that is, a non-key attribute of one relation which is a primary key of another. This will be useful when processing the relations, as 'status' can be used to join the STATUS and MODULE relations to

(a) COURSE-MODULE

COURSE	MODULE
B74	B741
B74	B742
B74	B743
B74	B744
B74	B745
B94	B951
B94	B952
B94	B741

(b) COURSE

COURSE	COURSE-NAME	LEVEL
B74	Computer Science	BSc
B94	Computer Applications	MSc

(c) MODULE

MODULE	MODULE-NAME	STATUS
B741	Program 1	Basic
B742	Hardware 1	Basic
B743	Data Processing 1	Basic
B744	Program 2	Intermediate
B745	Hardware 2	Intermediate
B951	Information	Advanced
B952	Microprocessors	Advanced
B741	Program 1	Basic

(d) STATUS

STATUS	UNIT-POINTS
Basic	8
Intermediate	11
Advanced	15

Figure 11.20: Third normal form

form a larger composite relation if this joint information is required by the user. The user requirements, which might include reports, are likely to contain data coming from the joining of a number of relations.

We will now consider the reasons why we normalize the relations in the first place. Unnormalized relations would have been formed by the analysts using information gained from interviews, for example, and are rough first-cut tabular representations of the data structures.

Relations are normalized because unnormalized relations prove difficult to use. This can be illustrated if we try to insert, delete, and update information from the relations not in TNF. Say we have a new 'module' numbered B985 called Artificial Intelligence which has a 'status' in the intermediate category. Looking at Figure 11.17, we cannot add this information in COURSE-DETAIL because there has been no allocation of this 'module' occurrence to any 'course'. Looking at Figure 11.18(b), it could be added to the MODULE relation, if we knew that the 'status' intermediate carried 11 unit-points. This information is not necessary in the MODULE relation seen in Figure 11.20(c), the TNF version of this relation. We simply add to the MODULE relation in Figure 11.19(c), the tuple B985, artificial intelligence, intermediate. The TNF model is therefore much more convenient for adding this new information.

If we decided to introduce a new category in the 'status' attribute, called coursework, having a 'unit-points' attached of 10, we cannot add it to the unnormalized relation MODULE (Figure 11.18(b)) because we have not decided which 'module' or modules to attach it to. But we can include this information in the TNF model by adding a tuple to the STATUS relation (Figure 11.20(d)) which is coursework, 10.

Another problem occurs when updating. Let us say that we decide to change the 'unit-points' allocated to the Basic category of 'status' in the modules from 8 to 6, it becomes a simple matter in the TNF relation. The single occurrence of the tuple with the key Basic needs to be changed from (Basic 8) to (Basic 6) in Figure 11.20(d). With the unnormalized, first normal, or second normal form relations, there will be a number of tuples to change. It means searching through every tuple of the relation COURSE-DETAIL (Figure 11.17(b)) or MODULE (Figure 11.18(b)) looking for 'status' = Basic and updating the associated 'unit-points'. All tuples have to be searched, because in the relational model the order of the tuples is of no significance. This increases the likelihood of inconsistencies and errors in the database. We have ordered them in the text only to make the normalization process easier to follow.

Another reason concerns the possible inconsistency of the data. This does not occur in the TNF relations above; but in Figure 11.17(b) the first and last tuples could have had module names 'Program 1' and 'Basic Programming', respectively, as names for the same module (B741). This would cause confusion, but the normalization process would detect the problem and the analyst will form the relation shown as Figure 11.18(b).

Deleting information will also cause problems. If it is decided to drop the B74 course, we may still wish to keep details of the modules that make up the course. Information about modules might be used at another time when designing another course. The information would be lost if we deleted the course B74 from COURSE-DETAIL (Figure 11.17(b)). The information about these modules will be retained in the module relation in TNF. The TNF relation COURSE will now consist only of one tuple relating to the 'course' B74, and the TNF relation COURSE-MODULE will consist of the three tuples relating to the COURSE B94. However, the MODULE relation (Figure 11.20(c)) will remain the same.

We have previously regarded the third normal form as the end of the normalization process and this is usually satisfactory. However, there are further levels or extensions of normalization.

Boyce-Codd Normal Form (BCNF) is one such extension. One criticism of the third normal form is that, by making reference to other normal forms, hidden dependencies may not be revealed. BCNF does not make reference to other normal forms.

In any relation there may be more than one combination of attributes which can be chosen as primary key, in other words, there are candidate keys. BCNF requires that all attribute values are fully dependent on each candidate key and not only the primary key. Put another way, it requires that each determinant (attribute or combination of attributes which determines the value of another attribute) must be a candidate key. As any primary key will be a candidate key, all relations in BCNF will satisfy the rules of the third normal form, but relations in TNF may not be in BCNF.

It is best explained by an example. In fact, the third normal form relations in Figure 11.20 are also in BCNF, so we will extend the example used so far. Assume that we have an additional relation which is also in TNF giving details about the students taking modules and the lecturers teaching on those modules. Assume also that each module is taught by several lecturers, each lecturer teaches one module, each student takes several modules, and each student has only one lecturer for a given module. This complex set of rules could produce the relation shown as Figure 11.21.

Although this relation is in TNF because the 'lecturer' is dependent on all the key (both 'student' and 'module' determine the lecturer), it is not in BCNF because the attribute 'lecturer' is a determinant but is not a candidate key. There will be some update anomalies. For example, if we wish to delete the information that Martin is studying B742, it cannot be done without deleting the information that Prof. Harris teaches the module B742. As the attribute 'lecturer' is a determinant but not a candidate key, it is necessary to create a new table containing 'lecturer' and its directly dependent attribute 'module'. This results in two relations as shown in Figure 11.22. These are in BCNF. Now, deleting the information that Prof. Harris teaches Martin (the second relation will now have three tuples) will not lose the information that Prof. Harris can teach on module B742.

Fourth Normal Form can be illustrated by looking at a relation which is in the first normal form and which contains information about modules, lecturers, and textbooks. Each tuple has a module name and a repeating group of textbook names (there could

STUDENT-MODULE		
STUDENT	MODULE	LECTURER
Bell	B741	Dr Smith
Bell	B742	Dr Jones
Martin	B741	Dr Smith
Martin	B742	Prof. Harris

Figure 11.21: Relation in TNF but not BCNF

LECTURER	
LECTURER	MODULE
Dr Smith	B741
Dr Jones	B742
Prof. Harris	B742

STUDENT-LECTURER	
STUDENT	LECTURER
Bell	Dr Smith
Bell	Dr Jones
Martin	Dr Smith
Martin	Prof. Harris

Figure 11.22: BCNF

be a number of texts recommended for each module). Any module can be taught by a number of lecturers, but each will recommend the same set of texts.

The relation seen as Figure 11.23 is in BCNF (and therefore TNF) and yet it contains considerable redundancy. If we wish to add the information that Prof. Harris can teach B742, three tuples need to be added to the relation. The problem comes about because all three attributes form the composite key: there are no functional determinants apart from this combination of all three attributes.

The problem would be eased by forming from this relation the two all-key relations shown as Figure 11.24. There is no loss of information, and there is not the evident redundancy found in Figure 11.23.

The transition to the fourth normal form has been made because of multivalued dependencies that may occur (Fagin, 1977). Although a module does not have one and only one lecturer, each module does have a pre-defined set of lecturers. Similarly, each module also has a pre-defined set of texts.

Although these examples are valid, in that they do show relations which contain redundancy and yet are in TNF and BCNF, respectively, the examples are somewhat contrived. The reader will have seen that in both examples it was necessary to make a number of special assumptions. The implication is that such problems will not be found frequently by analysts when carrying out data analysis and therefore that TNF can be a reasonable stopping point for normalization.

MODULE-LECTURER		
MODULE	LECTURER	TEXT
B741	Dr Smith	Database Fundamentals
B741	Dr Smith	Further Databases
B741	Dr Jones	Database Fundamentals
B741	Dr Jones	Further Databases
B742	Dr Smith	Database Fundamentals
B742	Dr Smith	Systems Analysis
B742	Dr Smith	Information Systems

Figure 11.23: BCNF but not fourth normal form

MODULE-LECTURER	
MODULE	LECTURER
B741	Dr Smith
B741	Dr Smith
B741	Dr Jones
B741	Dr Jones
B742	Dr Smith
B742	Dr Smith
B742	Dr Smith

MODULE-TEXT	
MODULE	TEXT
B741	Database Fundamentals
B741	Further Databases
B741	Database Fundamentals
B741	Further Databases
B742	Database Fundamentals
B742	Systems Analysis
B742	Information Systems

Figure 11.24: Fourth normal form

Summary

- An entity-relationship model views the organization as a set of data elements, known as entities, which are the things of interest to the organization, and the relationships between these entities.
- An entity will have attributes that describe the entity. The particular attribute or group of attributes that uniquely identifies an entity occurrence is known as the key attribute or attribute(s).
- The process of normalization is the application of a number of rules to the entities, now normally referred to as relations, which will simplify the model.
- For most situations a three-stage process of normalization to third normal form proves adequate.

Questions

1. Some argue that data is the 'lifeblood of the organization'. Do you agree? Provide reasons for your answer.
2. It is claimed that the entity-relationship model is easy for users to understand. Do you agree?
3. Complete the entity model shown as Figure 11.13.
4. For an organization of your choice, for example, a university library, suggest four or five entities and normalize them to the third normal form.

Further reading

Avison, D.E. and Cuthbertson, C. (2002) *A Management Approach to Database Applications*, McGraw-Hill, Maidenhead, UK.

Connolly, T. and Begg, C. (2004) *Database Systems: A Practical Approach to Design, Implementation and Management*, 4th edn, Addison-Wesley, Harlow, UK.

Date, C.J. (2003) *An Introduction to Database Systems*, 8th edn, Addison-Wesley, Reading, Massachusetts.

Lejk, M. and Deeks, D. (2002) *An Introduction to Systems Analysis Techniques*, 2nd edn, Prentice Hall, Harlow, UK.

change is thus effected at the logical level, which is the correct place, and the implications of the change are agreed, and only then the necessary design changes made. This improves and speeds up the maintenance process which, as we have seen, is a major time and resource-consuming activity.

The form of DFDs differs between the various proponents of structured systems analysis. The differences are relatively small and the basic concepts are the same. For example, the symbol used to represent a process differs. Gane and Sarson (1979) use a rectangle with rounded corners (a 'soft box') whereas many other authors use a circle. This means that superficially the DFDs look different but in practice the differences are relatively minor.

A logical DFD represents logical information, not the physical aspects. A data flow specifies what flows, for example, customer credit details. How it flows, for example, by carrier pigeon or via twisted copper wires, is immaterial and not represented in a logical DFD. A DFD is a graphical representation and is composed of four elements:

1 The data flow

Data flow is represented by an arrow and depicts the fact that some data is flowing or moving from one process to another. A number of analogies are commonly used to illustrate this. Gane and Sarson (1979) suggest that we think of the arrow as a pipeline down which 'parcels' of data are sent. Others think of it as like a conveyor belt in a factory which takes data from one 'worker' to another. Each 'worker' then performs some process on that data which may result in another data flow on the conveyor belt. These processes are the second element of the DFD.

2 The processes

The processes or tasks performed on the data flows are represented in this example by a soft box (see Figure 12.1). The process transforms the data flow by either changing the structure of the data (e.g. by sorting it) or by generating new information from the data (e.g. by merging the data with data obtained from another data source). A process might be 'validate order', which

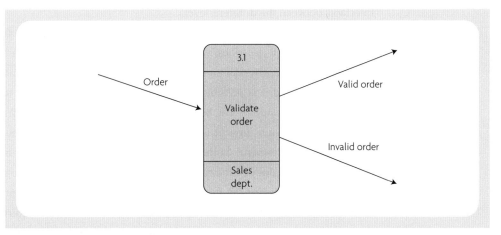

Figure 12.1: Process box and data flows

12 Process techniques

12.1 Data flow diagramming

The data flow diagram (DFD) is fundamental to structured systems methodologies and was developed as an integrated part of those methodologies. However, the DFD has been adopted and adapted by a number of other methodologies, not all structured systems type, including Multiview and ISAC. In these methodologies the DFD or similar is not the major technique of the methodology but is used in conjunction with other techniques in the analysis of processes. Like entity modelling and normalization, DFDs are an important technique in a variety of systems development methodologies.

The DFD provides the key means of achieving one of the most important requirements of structured systems, that is, the notion of structure. The DFD enables a system to be partitioned (or structured) into independent units of a desirable size so that they, and thereby the system, can be more easily understood. In addition, information is graphical and concise. The graphical aspect means it can be used both as a static piece of documentation and as a communication tool, enabling communication at all levels: between analyst and user, analyst and designer, and analyst and analyst. The graphical nature of the DFD means that it can be explained more easily to users and also means a more concise document, as it is argued that a picture can more quickly convey meaning than traditional methods, such as textual narrative. The DFD also provides the ability to abstract to the level of detail required. Therefore, it is possible to examine a system in overview and at a detailed level, while maintaining the links and interfaces between the different levels.

The DFD provides the analyst with the ability to specify a system at the logical level. This means that it describes what a system will do, rather than how it will be done. Considerations of a physical and implementation nature are not usually depicted using data flow diagrams, and it is possible for the logical DFD to be mapped to a variety of different physical implementations. The benefit of this is that it separates the tasks of analysis (what is required) from design (how it is to be achieved). This separation means that the users can specify their requirements without any restrictions being imposed of a design nature, for example, the technology or the type of access method. There exists a logical and physical independence; the hardware can be changed or upgraded without changing the functions of the system. Alternatively, if as often happens a functional change is required, the relevant part of the logical specification is changed and a new mapping to the physical system is designed. The

transforms the order data flow by adding new information to the order, that is, whether it is valid or not. It is likely that invalid orders flow out from the validation process in a different direction to valid orders. In this example the conventions used for the process symbol are as follows. The top compartment contains a reference number for the process, the middle compartment contains the description of the process, and the lower compartment indicates where the process occurs. Strictly speaking, this is not at the logical level and so would not always be used. A process must have at least one data flow coming into it and at least one leaving it. There is no concept of a process without data flows; a process cannot exist independently.

3 The data store

If a process cannot terminate a data flow because it must output something, then where do the data flows stop? There are two places. The first is the data store, which can be envisaged as a file, although it is not necessarily a computer file or even a manual record in a filing cabinet. It can be a very temporary repository of data, for example, a shopping list or a transaction record. A data store symbol is a pair of parallel lines with one end closed, and a compartment for a reference code and a compartment for the name of the data store. For example (see Figure 12.2), the process of validating the order may need to make reference to the parts data store to see if the parts specified on the order are valid parts with the correct current price associated with it. The data flow in this example has the arrow pointing toward the process which indicates that the data store is only referenced by the process and not updated or changed in any way. In this example we would expect to find another process somewhere on the DFD which maintained the parts data store. For example, in Figure 12.3, the new part data from process 6.2 is used to update the parts data store (the arrow points to the data store). The manner in which the access to the data store is made is usually regarded as irrelevant. However, it may be information which a designer needs, and therefore, in cases where it is not obvious, this information may be added

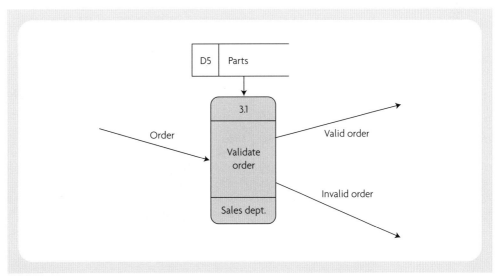

Figure 12.2: Parts data store

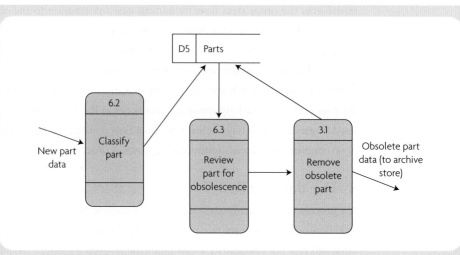

Figure 12.3: Maintenance of parts data store

to the DFD. In the example (Figure 12.2), it may be assumed that access to the parts file is via the part number. However, if a customer makes an order without specifying the part number, access via the part description may be required. This should then be specified (see Figure 12.4).

4 The source or sink (external entity)

The second way of terminating a data flow in a system is by directing the flow to a sink. The sink may, for example, be a customer to whom we send a delivery note. The customer is a sink, in the sense that the data flow does not necessarily continue. The customer is a sink down which the delivery note may fall for ever. The Department of Trade may be a sink to which a company may legally be required to provide information but never receive any in return. Sinks

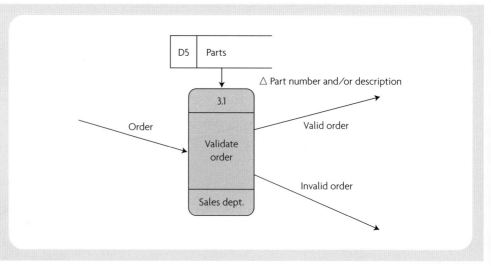

Figure 12.4: Access to data store

are usually entities that are external to the organization in question, although they need not be, another department may be a sink. It depends on where the boundaries of the system under consideration are drawn. If a DFD for the sales department is being constructed, any data flows to the production department would be represented as a sink. However, if the whole organization were being depicted, then the same data flows would go to a process within the production department. The original source of a data flow is the opposite to a sink, although it may be the same entity. For example, a customer is the source of an order and a sink for a dispatch note. Sinks and sources are represented by the same symbol which is a shadowed rectangle (see, for example, Figure 12.5). Sources and sinks are often termed 'external entities'. Figure 12.5 is an example of a data flow diagram that illustrates the combination of the four elements discussed above.

Mason and Willcocks (1994) suggest the following rules for drawing data flow diagrams assuming the analyst has described the whole logical system in narrative form:

- read the whole process a few times to get a clear picture of the system being described;
- identify the sources and sinks by circling a keyword for each of them;
- identify data stores (perhaps through the use of verbs such as 'store' or 'check' and underline them);
- draw a source entity box for the external entity which seems to start the process off and name it (entities can only link with processes);
- to its right, draw the first process box;
- from that, draw an arrow representing the primary data flow and name it;
- name the process;
- link it with any data stores and name these as well as the data flow entering or exiting it to and from the process (data cannot go from store to store, there must be an intervening process);
- link the next process and so on until all the external entities are drawn.

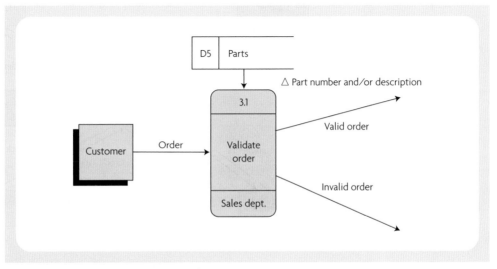

Figure 12.5: External entity customer (source of order)

We would add that all these steps will be repeated as we iterate to improve our model following discussions of it with the users.

One of the most important features of the DFD is the ability to construct a variety of different levels of DFD according to the degree of abstraction required. This means that a high-level diagram (containing the main processes) can be consulted in order to obtain a high-level (overview) understanding of the system. When a particular area of interest has been identified, then this area can be examined in more detail with a lower-level DFD. The different levels of diagram must be consistent with each other in that the data flows present on the higher levels should exist on the lower levels as well. In essence it is the processes that are expanded at a greater level of detail as we move down the levels of the diagram. This 'levelling' process gives the DFD its top-down characteristic (DFD levelling is also described in the context of YSM in Section 20.2).

Usually the very top level of a DFD is known as the Context Diagram. All the processes, data stores and internal data flows are consolidated into a single process and only the external entities and the data flows between this single process are shown. Figure 12.6 is an example of a Context Diagram for an imaginary 'Sales order processing' system showing its interactions with various external entities such as Customer, Supplier, etc.

The next level down would be the Level 0 diagram showing the major processes of the 'Sales order processing' system. These major processes might be Credit Checking, Order Handling, Stock Control, Invoicing, Packaging and Delivery, etc. This level is not illustrated. The next level down would be the Level 1 diagram and each major process identified in Level 0 would be decomposed into a Level 1 DFD. We have taken the 'Order Handling' major process (Reference 3.0 say) and illustrated this in Figure 12.7. So 'Order Handling' is now broken down into nine subprocesses and the figure shows the associated data flows and data stores.

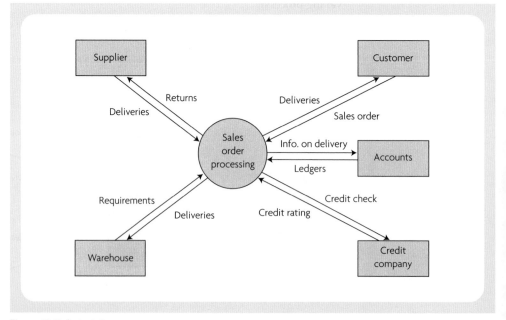

Figure 12.6: Context diagram

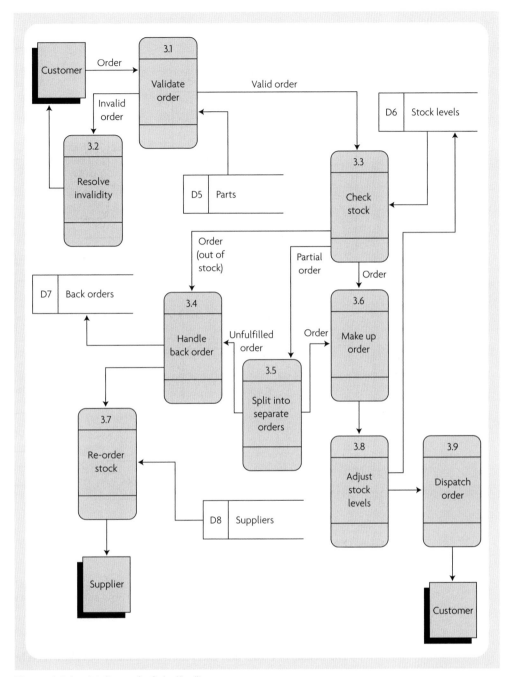

Figure 12.7: Level 1 diagram for Order Handling

Each of these nine processes can, if necessary, be expanded to still lower levels of detail. Figure 12.8 is the next level down (or an *explosion*) of the Validate Order process (Reference 3.1). This process is now expanded into five tasks with the various data flows between them. All the data flows (four in this case) in and out of Validate Order (Reference 3.1) in Figure 12.7 can be found on Figure 12.8. Any new data flows are either flows that only exist within the Validate

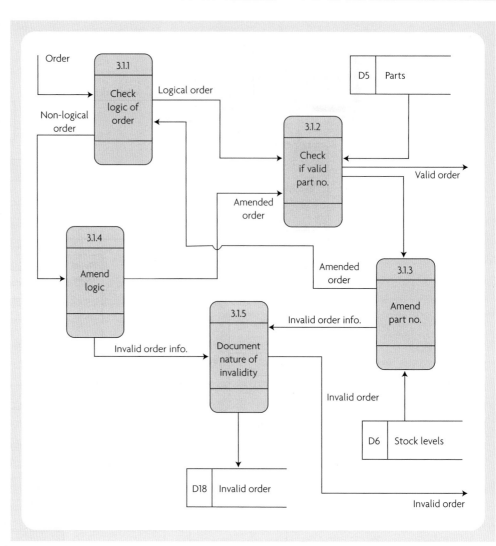

Figure 12.8: Lower level of detail for 'validate order' (Reference 3.1)

Order process, that is they are internal to it and are now shown because we have split this down into separate parts (for example, Amended Order), or because they are concerned with errors and exceptions.

The details of errors and exceptions are usually not shown on high-level diagrams as it would confuse the picture with detail that is not required at this level. What is required at the high level is 'normal' processing and data flows. To include errors and exceptions might double the size of the diagram and remove its high-level characteristics.

For example, Figure 12.8 shows new processing concerned with amending an order and a new data store which did not appear on the higher-level diagram (the Parts data store). The problem is that it is sometimes difficult to decide what constitutes an error or an exception, and what is normal. Some common guidelines suggest that if an occurrence of a process or data flow is relatively rare, then it should be regarded as an exception. However, if it is financially significant, it should be taken as part of normal processing. Overall, it depends on the audience or use to be made of the DFD as to exactly what is included. At the lowest level, all the detail, including all errors and exceptions, should be shown.

The question arises as to how many levels a DFD should be decomposed. The answer is that a DFD should be decomposed to the level that is meaningful for the purpose for which the DFD is required. There comes a level, however, when each process is elementary and cannot be decomposed any further. No further internal data flows can be identified. At this point each elementary process is described using a form of process logic. The techniques of representing process logic are described in the next four sections.

Data flow diagrams are generally logical and not physical, but they can be used to represent a physical implementation. For example, the process 'Document nature of invalidity' (Figure 12.8) might be 'type nature of invalidity on form B36 in sales office' with the B36 forms being output as part of the data flow and 'click on reason for invalidity using screen 5'. The physical DFD may then describe the present system or a proposed system.

Although it is a fundamental technique of some information systems development methodologies, e.g. Stradis, YSM and SSADM (see Part VI), DFDs have limitations even for addressing information about processes, such as how long data take to get from one process to another and the detail about the data that pass between processes in terms of peaks and troughs. Also little detail is provided about decision aspects, that is, why data flow in one way and not another. However, other techniques can be used to provide this information although none of the process techniques are as well used.

Techniques and tools closely associated with DFDs include data dictionaries as well as process logic descriptions. The data dictionary (see Section 19.1) is the central place where all the details of data flows, data stores, and processes are stored in an ordered and logical fashion. The dictionary may be manual or computerized, but the important thing is that it must be the centralized resource of the structured analysis project. It is not the same thing as the data dictionary of a database which is concerned with physical aspects of the data, although the two may very well be integrated.

The way in which process logic is described is by the use of decision trees, decision tables, structured English, and action diagrams (Sections 12.2–12.4 and 12.8). They are not complementary. A particular process is not described using all four techniques but by using whichever is the most appropriate, given the characteristics of the process concerned. So, each technique has particular strengths and weaknesses, but they all provide simple, clear, and unambiguous ways of describing the logic of what happens in the elementary processes identified in the DFD.

12.2 Decision trees

Decision trees and decision tables are tools which aim to facilitate the documentation of process logic, particularly where there are many decision alternatives. A decision tree illustrates the actions to be taken at each decision point. Each condition will determine the particular branch to be followed. At the end of each branch there will either be the action to be taken or further decision points. Any number of decision points can be represented, though the greater the complexity, then the more difficult the set of rules will be to follow.

When constructing a decision tree, the problem must be stated in terms of conditions (possible alternative situations) and actions (things to do). It is often convenient to follow a stepwise refinement process when constructing the tree, breaking up the largest condition to basic conditions, until the complete tree is formulated. The general format is shown in Figure 12.9. An example of a decision tree is given in Figure 12.10. At the first decision point (or node), the customer is classified into one of two types, private or trade. If the customer is trade, then a second decision point is reached. Has the customer been trading with us for less than five years, or five years or more? If the customer has been trading for five years or more, then the customer can obtain up to £5,000 credit, otherwise up to £1,000 credit can be given. If the customer was deemed private at the first decision point, then the action is to offer no credit at all.

Decision trees are constructed by first identifying the conditions, actions, and 'unless/however/but' structures of the situation being analysed. This can be obtained from a written statement or during interviews with users. Each sentence may form a 'mini' decision

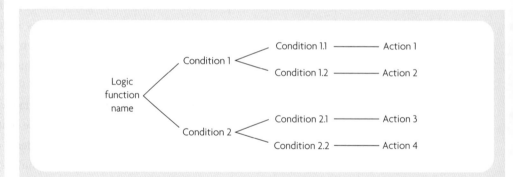

Figure 12.9: General format of decision tree

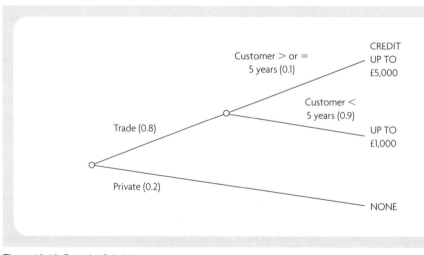

Figure 12.10: Example of decision tree

tree which combined with others could be joined together to form the version which will be verified by the users.

Sometimes it is not possible to complete the decision tree because full information has not been given in the statement. For example, one branch of the decision tree may be identified, but no indication is given on the action to take on this branch. In such cases the analyst has to carry out a further investigation by interviewing staff or by using another method of systems investigation. The technique therefore helps validate the information as well as proving a useful documentation aid.

Decision trees prove to be a good method of showing the basics of a decision, that is, the possible actions that might be taken at a particular decision point and the set of values that lead to each of these actions. It is easy for the user to verify whether the analyst has understood the procedures.

Sometimes it is possible to associate probability scores for each branch once the decision tree is drawn. With this information it will be possible to compute expected values of the various outcomes and hence to evaluate alternative strategies. For example, in Figure 12.10, research may have indicated that trade customers outnumber private customers by 4:1 and that 90 per cent of trade customers are of less than five years standing. By extrapolating these figures with the total number of customers, it may be possible to evaluate the amount of total credit to be made available to all customers.

Where the number of decision nodes becomes greater than, say, 10, drawing the decision tree begins to be overly complex and a decision table is usually chosen as it can more readily represent processes with very complex decision structures. People are usually very familiar with the meaning of decision trees, and these are more graphical; therefore, decision trees are often preferred for communicating less complex logic.

12.3 Decision tables

Decision tables are less graphical, when compared to decision trees, but are concise and have an inbuilt verification mechanism so that it is possible to check that all the conditions have been catered for. Again, conditions and actions are analysed from the procedural aspects of the problem situation, usually expressed in narrative form.

When constructing a decision table, the narrative is analysed to identify the various conditions and the various actions. The various actions to be executed are then listed in the bottom left-hand part of the table known as the action stub. In the top left-hand part, the conditions that can arise are listed in the condition stub. Each condition is expressed as a question to which the answer will be 'yes' or 'no'. There can be no ambiguity. All the possible combinations of yes and no responses can be recorded in the upper right-hand part of the table, known as the condition entry. Each possible combination of responses is known as a rule. In the corresponding parts of the lower right-hand quadrant, an X is placed for each action to be taken, depending on the rule associated with that column.

Figure 12.11 shows the decisions that have to be made by drivers in the UK at traffic lights. The condition stub (upper left section) has all the possible conditions which are 'red',

CONDITION STUB	CONDITION ENTRY
Red	Y Y Y Y N N N N
Amber	Y Y N N Y Y N N
Green	Y N Y N Y N Y N
Go with caution	X
Stop	X X X X X X X
Call police	X X X X
ACTION STUB	ACTION ENTRY

Figure 12.11: Decision table for UK traffic lights

'amber', and 'green'. Condition entries (upper right quadrant) are either Y for yes if the light is on (this condition is satisfied) or N for no if the light is off (this condition is not satisfied). Having three conditions in the condition stub, there will be 2 to the power of 3 ($2 \times 2 \times 2 = 8$) columns in the condition entry (and action entry). The easiest way of proceeding is to have the first row in the condition entry as YYYYNNNN, the second row as YYNNYYNN, and the final row as YNYNYNYN. If there were four conditions, we would start with eight Ys, eight Ns, and so on, giving a total of 2^4 ($2 \times 2 \times 2 \times 2 = 16$) columns.

All the possible actions are listed in a concise form in the Action Stub (bottom left). An X placed on a row/column coincidence in the Action Entry means that the action in the condition stub should be taken. A blank will mean that the action should not be taken. Therefore, if a driver is faced with Red (Y), Amber (Y), and Green (Y), the first column indicates that the driver should stop and call the police (a particular combination of conditions may lead to a number of actions to be taken). All combinations, even invalid ones, should be considered. The next column Red (Y), Amber (Y), and Green (N) informs the driver to stop. Only the Red (N), Amber (N), and Green (Y) combination permits the driver to go with caution.

Once the table is completed, rules which result in the same actions can be joined together and represented by dashes, that is, 'it does not matter'. The result of this is a consolidated decision table. Figure 12.12 illustrates an example decision table before and after the consolidation process. In the decision table represented in Figure 12.12(b), Rules 3 and 4 of the decision table seen in Figure 12.12(a) have been merged into a consolidated Rule 2, expressing a 'doesn't matter' condition. This is because the same processing needs to be executed whether or not condition 2 is 'yes' or 'no'.

In systems analysis, there are likely to be requirements to specify actions where there are a large number of conditions. A set of decision tables is appropriate here. The first will have actions such as 'go to decision table 2' or 'go to decision table 3'. Each of these may themselves be reduced to a further level of decision tables. The technique therefore lends itself to functional decomposition.

Sometimes the values of conditions are not restricted to 'yes' or 'no', as defined in the limited entry tables described. There can be more than two possible entries, and extended entry tables are appropriate in this case. For example, the credit allowable to a customer could vary according to whether the customer had been dealing with the firm for 'up to 5 years', 'over 5 and up to 10 years', 'over 10 and up to 15 years', and so on. The rule for obtaining the right number of combinations will need to be modified. If Condition 1 has two possibilities and Condition 2 five possibilities, then the number of columns will be 2×5, that is, 10 columns.

(a)

Invoice > £300	Y	Y	Y	Y	N	N	N	N
Account overdue > 3 months	Y	Y	N	N	Y	Y	N	N
New customer	Y	N	Y	N	Y	N	Y	N
Put in hands of solicitor	X		X	X	X	X		X
Write first reminder letter		X	X	X		X		X
Write second reminder letter					X			
Cancel credit limit			X	X		X	X	

(b)

Invoice > £300	Y	Y	Y	N	N	N	N
Account overdue > 3 months	Y	Y	N	Y	Y	N	N
New customer	Y	N	–	Y	N	Y	N
Inform solicitor	X		X	X	X		X
Write first reminder letter		X	X		X		X
Write second reminder letter				X			
Cancel credit limit			X		X	X	

Figure 12.12: Consolidating decision tables

Whereas decision trees are particularly appropriate where the number of actions is small (although it is possible to have large decision trees), decision tables are more appropriate where there is a large number of actions as they can be decomposed into sets conveniently; in other words, decision tables can better handle complexity. However, decision trees are easier to construct and give an easily assimilated graphical account of the decision structure. Decision tables have good validation procedures and, further, can be used to generate programs which carry out the actions according to the rules. Here the processing is specified by the analyst in terms of decision tables. These are transferred to computer readable format, and the programs generated automatically. Decision tables do, however, suffer from the disadvantage that no indication is given regarding the sequence that the actions are to follow (it cannot be assumed that they are to be followed in the sequence given in the decision table itself).

12.4 Structured English

Structured English is very like a 'readable' computer program. It aims to produce unambiguous logic which is easy to understand and not open to misinterpretation. It is not a natural language like English, which is ambiguous and therefore unsuitable. Nor is it a programming language, though it can be readily converted to a computer program. It is a strict and logical form of English and the constructs reflect structured programming. Like a conventional programming language, the sequence of the commands expressed in structured English is important. It reflects the sequence in which the instructions should be followed. Although decision trees or decision tables are more suitable tools to document aspects where the system has many decision points, structured English proves to be a very useful technique to express logic in a system.

Structured English is a precise way of specifying a process, and is readily understandable by a trained systems designer as well as being readily converted to a computer program. An example is given in Figure 12.13. Structured English uses only a limited subset of English and this vocabulary is exact. This ensures less ambiguity in the use of 'English' by the analyst. Further, by the use of text indentation, the structure of the process can be shown more clearly. As with all these techniques, however, although the logic can be formally expressed, there is no guarantee that the expression in syntactically correct structured English is semantically correct. That will depend on the systems investigation in the first place.

Structured English (see the example in Figure 12.13) has the following general construct:

```
CREDIT RATING POLICY

IF the customer is a trade customer
  and IF the customer is customer for 5 or more years
    THEN credit is accepted up to £5000
    ELSE credit is accepted up to £1000
ELSE the customer is a private customer
  SO no credit given
ENDIF
```

Figure 12.13: Structured English example

```
IF condition 1 (is true)
THEN action 1 (is to be carried out)
ELSE (not condition 1)
SO action 2 (is to be carried out)
ENDIF
```

Functional decomposition can be supported in structured English by a construct using the IF ... THEN construct (Figure 12.14).

Conditions can include equal, not equal, greater than, less than, and so on. The words in capitals are keywords in structured English and have an unambiguous meaning in this context.

```
IF condition
  THEN do a named set of operations
  (specified at a lower level)
```

Figure 12.14: IF ... THEN construct

The logic of a structured English construct is expressed as a combination of sequence, selection, case, and repetition structures. Any logical specification can be written using these four basic structures (Figure 12.15):

- *sequencing* shows the order of processing of a group of instructions, but has no repetition or branching built into it;
- *selection* facilitates the choice of those conditions where a particular action or set of actions (or another decision and selection) are to be carried out;

- *cases* represent a special type of decision structure (a special kind of selection), where there are several possibilities, but they never occur in combination (in other words, they are mutually exclusive);
- *repetition* or loop instructions facilitate the same action or set of actions to be carried out a number of times, depending on a conditional statement.

The actual keywords and their number will vary according to the particular conventions used; indeed, structured English is not a 'standard', but the basic structures of sequence, selection, and repetition will be common to all.

The layout of structured English is as follows:

- the use of capital letters indicates a structured English reserved word such as IF, THEN, ELSE, and GET or the operators ADD, DIVIDE, and so on, which have particular meanings in the context of structured English;
- any data elements which are included in a data dictionary are normally underlined and these will include those items associated with the particular application, such as <u>credit</u> or <u>customer</u> in the credit rating example (Figure 12.13);
- indentation is used to indicate blocks of sequential instructions to be created together, and hierarchical structures can be built by indenting these blocks;
- blocks of instructions can be named and this name quoted in capital letters to refer to the block of instructions elsewhere in the code, and this will be particularly necessary where there are a number of places in the logic where this set of instructions needs to be performed.

```
Sequencing:   Statement-1
              Statement-2
                              Statement-2-1
                              Statement-2-2
              Statement-3
              ...
              ...
              ...
              Statement-n

Selection:    IF      Condition
              THEN    Statements
              ELSE    (not condition)
              SO      Statements

Repetition:   REPEAT
                      Statements
              UNTIL
                      Condition

Case:         CASE    expression
              OF      Condition-1: Statements
                      Condition-2: Statements
                      ...
                      ...
                      ...
                      Condition-n: Statements
                      OTHERWISE: Statements
              ENDCASE
```

Figure 12.15: The basic structures of structured English

When creating structured English statements, it is best to break down complex statements into a number of simple ones. Named blocks of these simple statements can be thought of as a way of effecting functional decomposition because the block can be performed in a REPEAT statement a number of times.

Structured English has many advantages, in particular its ability to describe many aspects of analysis, its conciseness, precision, and readability, and the speed with which it can be written. However, it takes time to build up skills in its use, and it is alien to many users (despite being English-like). Indeed, the terminology might be misleading to users because the structured English meanings are not exactly the same as their natural language counterparts.

There are alternative languages to structured English, such as 'pseudo code' and 'tight English'. These vary on their nearness to programming language and readability to users. For example, pseudo code has a DO-WHILE and END-DO loop structure which is similar to constructs of some conventional computer programming languages, and is obviously more programming-oriented than structured English. Tight English code seems nearer natural language than computer programming language, though it can also be interpreted by programs. When there are a number of decision points, in tight English it is usual to use a decision table or decision tree.

12.5 Structure diagrams

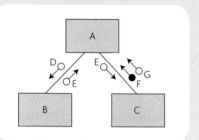

Figure 12.16: Structure diagram

Like other techniques described in this chapter, structure diagrams are used by a number of methodologies. The structure diagram is a series of boxes (representing processes or parts of computer programs, usually referred to as modules) and connecting lines (representing links to subordinate processes which show the way that data and control can be passed between processes) which are arranged in a hierarchy. Each module should be small and manageable. Structure charts therefore exemplify the functional decomposition aspect of many of these process techniques. The basic diagram of a structure chart is shown in Figure 12.16:

- Module A can call module B and also module C. This is shown by lines joining the boxes, which represent modules. No sequencing for these calls nor whether they actually occur is implied by the diagramming notation. When the subordinate process terminates, control goes back to the calling process.
- When A calls B, it sends data of type D to B. When B terminates, it returns data of type E to A. Similarly, A communicates with C using data of types E and G.
- When C terminates, it sends a flag of type F to A. A flag is used as a flow of control data. The difference in the symbol between data types and flags lies in the circle being filled in or not.

Figure 12.17 shows the structure chart for part of the processing of applications to undergraduate courses. It concerns those students that we propose to interview. CD refers to the candidate document and TR to the test result. Other charts in the set will be constructed for the processing of examination results and final assessment, among others. The top-level diagram (A in Figure 12.17) will show each of these processes as a box and refer to the lower-level structure charts.

The structure of a program should be designed to ensure that it minimizes the interdependence between modules (known as **coupling**), it encourages module reuse (known as **cohesion**), and eases the programming task and later maintenance. By minimizing the connections between modules and therefore their independence, coupling reduces the risk of errors or changes in one module affecting another. But cohesion, that is, the way the activities

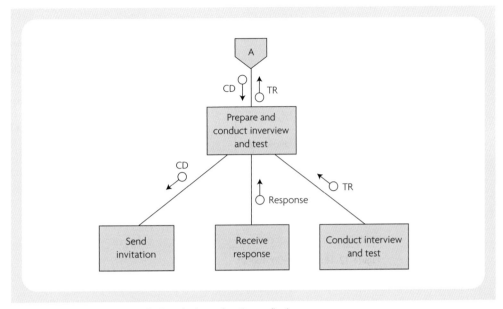

Figure 12.17: Structure diagram for interviewing and testing applications

within a module are related, should be maximized as it helps ease of understanding the module and encourages reuse because the elements of the module concern similar activities which may be used in other applications. The designer should be aware of the possibilities of code reuse, both in terms of designing modules that may be reused and using modules already written.

We have so far suggested that the structure diagram relates to computer program design. But, of course, the basic technique applies to the overall design of information systems, of which the design of computer programs plays a part. This sequence is not a coincidence, for many of these process techniques were used first in the structured programming movement which influenced many systems analysts. Structure diagrams are used in SSADM, Multiview, and elsewhere, but they are described further in Section 20.3 when discussing the entity structure step, a phase in Jackson Systems Development (JSD). Jackson has been a notable contributor to both the software world (through Jackson Systems Programming) and the technical aspects of information systems.

A further variation on the same structure chart theme, but looking somewhat different, is the Warnier–Orr diagram which has been well used in the USA. As shown in Figure 12.18, the diagrams use curly brackets to show the hierarchical structure, and they are read from left to right rather than from top to bottom. Sequential processes are therefore shown within one bracket. There is a selection construct, where a choice is made between mutually exclusive possibilities, according to a condition. This is denoted by the symbol which has a plus sign within a circle. The example provided reflects a low-level decision-making process, but Warnier–Orr diagrams can be used and are used to reflect the higher-level processing of the structure charts of the above examples.

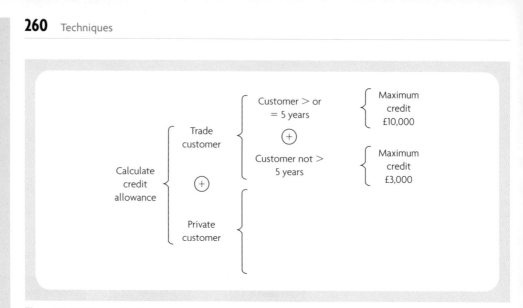

Figure 12.18: Warnier–Orr diagram

12.6 Structured walkthroughs

Structured walkthroughs are a series of formal reviews of a system or a program, held at various stages of the life cycle. This is an idea which has developed around structured systems analysis and design approaches, where the opportunities for such review are clearly identifiable. These are intended to be team-based reviews of a product, such as a program or design component and are not intended to be management reviews of individuals or their performance.

The basic idea behind structured walkthroughs is that potential problems can be identified as early as possible so that their effect can be minimized. The benefits of this approach are:

- The overall quality of the systems analysis and design of the information system is improved, since more than one person is responsible for it and the analysis and design are exposed to the scrutiny of others at every appropriate opportunity.
- There is the opportunity to detect errors earlier in the development cycle than might otherwise be possible, avoiding the errors propagating throughout the rest of the systems development process.
- All team members have the opportunity to be 'educated' in the total system, resulting in a much better understanding of the total system by a greater number of organizational personnel. This means that team members can more easily take over work from each other. Other personnel, outside the development team, such as production and operational systems staff also have the opportunity to familiarize themselves with the overall system as well as particular components.
- Technical expertise is communicated through discussion that is often generated as a result of a structured walkthrough. More experienced staff will spot common sources of potential problems and discuss these with other staff, thereby transferring their own knowledge and skills. This means that the technical knowledge is dispersed more widely than would otherwise be the case.

- Inherent in structured methodologies is that technical progress can be more readily and easily assessed, and the walkthroughs provide ideal milestones and opportunities to do this.
- If carried out in the correct spirit and atmosphere, structured walkthroughs can provide an opportunity for trainee analysts and programmers to gain experience and enable them to work on complex problems more quickly, due to their participation in walk-throughs with other more experienced team members and also because of the opportunity of having walkthroughs on their work where they receive specific comments and feedback in a non-threatening environment.

Structured walkthroughs have been identified as being of considerable value in the development of information systems, and they should be held on completion of certain phases of the development as indicated. It is impractical to hold formal walkthroughs too often, as they cause unnecessary administrative overheads. The best approach is to maintain the spirit of the concept by team members discussing all decisions with others without necessarily calling formal meetings. It is intended that the approach will normally promote discussion and exchange of ideas within the team. As stated previously, formal reviews should be held on the completion of stages and certain phases within each of the stages of the life cycle, and it is suggested that walkthroughs might be held at the end of the following:

- feasibility study;
- system investigation;
- systems analysis;
- systems design;
- program specification;
- program logic design;
- implementation;
- test plan;
- implementation plan;
- operational system plan;
- user manuals;
- review and maintenance.

Formal walkthroughs should be attended by a number of team members because responsibility for the system is then placed on the whole team. All members of the team should be given the opportunity to contribute, from the most junior to the most senior. It might be appropriate to limit the number of people attending some walkthroughs to around four to six.

For maximum benefit to be derived from the walkthrough, it is important that appropriate documentation is circulated well in advance of the walkthrough and that:

- everyone attending is familiar with the subject to be reviewed;
- each attendee has studied it carefully;
- minor points of detail are discussed before the walkthrough, so that valuable time is not wasted on trivial points.

During the walkthrough it is important that:

- all errors, discrepancies, inconsistencies, omissions, and points for further action are recorded so that this can form an action list;
- one person should be allocated the responsibility of ensuring that all points from the action list are dealt with.

It should not normally be necessary to hold another walkthrough for the same activity. Walkthroughs are a very powerful technique and are most successful where they are carried out in an 'ego-less' environment, that is, one in which the individual concerned with the particular activity does not feel solely responsible for it. It is important that all team members have responsibility for the system.

Even down to the level of individual programs, it is important that all team members have access to any code produced and feel responsible for the system as a whole rather than just that program. In practice, this means that a programmer should be able to accept criticism of the program design, code, or test plan that has been produced. Equally, the programmer should not feel afraid to discuss code produced by someone else. This type of environment is encouraged by the use of structured walkthroughs because they are a formal introduction to an approach of communication. In some organizations structured walkthroughs are used extensively during coding. The idea is not to fault-find or criticize any individuals but to identify any potential problems and resolve them as early as possible in the information systems development process.

There are two specific types of walkthrough that are commonly used in programming: code reading and dry running. These are not formalized procedures but entail team members critically examining each other's work:

1 Code reading

This is performed both before testing and on completion of testing. It is normally carried out by someone other than the developer of the code being looked at. Again the idea is that work is exposed to outside scrutiny and comment as soon as possible. The aims of initial code reading are to:

- detect any coding errors in going from the program design to the code of the programming language used;
- ensure that appropriate coding standards of the organization have been adhered to;
- check that the coding is efficient in terms of the performance that the system will produce;
- cross-educate team members, as they all have the opportunity to learn from the code developed by others.

The aims of the final code reading, a task which is usually performed by the team leader, are to:

- check that the code is of good quality and adheres to organizational standards;

- confirm that testing of the code has been completed according to the appropriate test plan;
- establish that the code is consistent with the specification that it was produced from.

Code reading is a difficult and time-consuming activity but it does have a number of benefits:

- most minor errors are discovered before testing begins, preventing more time being wasted later;
- coding standards are maintained as programmers know that their code will be specifically checked for this and therefore they are more likely to conform in the first place;
- technical expertise is communicated by the code reader passing on his or her knowledge to the programmer as a result of the code reading activity;
- an ego-less environment is created where the emphasis is continually on reviewing products rather than personnel.

2 Dry running

- This involves manually passing test data through the code. The start of the process involves the programmer listing all the variables and noting any initial values. As the test data are processed, variables are tested and updated as appropriate. This may seem a tedious task, but it has the advantage that errors and discrepancies are highlighted and it can also verify that the test data are adequate. However, if previous reviews have been performed adequately, it is only necessary to dry-run complex programs. This is because logical design errors will have been removed during the structured walkthrough of the design and clerical errors in the translation of design to code will have been found during code reading.

12.7 Matrices

One of the most common of techniques in all walks of life is the matrix, a simple tabular expression of a relationship, usually between two things (three things would require a 3-dimensional set of matrices). We find them in many methodologies. In this section we give a few examples of their use.

A common matrix is that showing the relationship between functions and events. In Figure 12.19, which is a part-formed matrix relating to the acceptance of students on courses, the Xs at the intersections between an event row and a function column shows that an event triggers a particular function. Therefore, the event 'student enquiry' triggered the 'process enquiry' and 'selection' functions. Every event should trigger at least one function and every function should have a triggering event.

A second matrix used by many methodologies is that associating functions with entities, that is, what entities are used by each function to enable that function to be carried out? Figure 12.20 shows the entities used by functions carried out in a hospital. Frequently in methodologies, the entries reveal more than this, that is, the manner of access. Therefore, C (create), R (read), U (update), and D (delete) entries will give more information about the relationship

Function name Event name	Process enquiry	Selection	Enrolment	Accounts
Student enquiry	X	X		
Student accepted		X	X	
Student registers				X

Figure 12.19: Function/event matrix

Entity name Function name	Staff member	Group session	Location	Program type	Patient
Group session attendances	R	U	R		C/U
Programs	R			R	R
Contact traced	R				U
Assessments	R				R/U/D
Program costing	R			U	R

Figure 12.20: Entity/function matrix (CRUD matrix)

between the two than a mere X. This is often referred to as a CRUD matrix. There should be represented in the full set of matrices all the entities contained in the entity model and all the functions contained in the data flow diagrams. The CRUD matrix can also be used for the relationship between entities and events. Some cells may contain more than one entity, for example, C and R. This will show that a particular event leads to an entity occurrence either being created or simply accessed.

The systems designer may well also create a key attribute and entity matrix. This will ensure that all entities have a key. Where there are two or more Xs, this will show that an entity has a composite key. Other matrices commonly found are those associating entities and attributes and that showing the relationship between user roles and functions.

A matrix showing the document flows through the system, such as that relating the data elements in the data dictionary system with the various documents that record their existence,

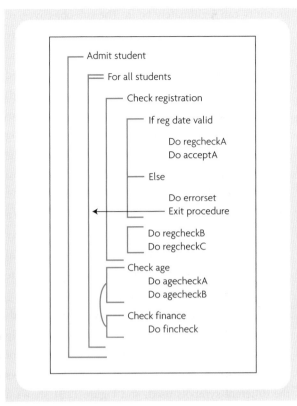

Figure 12.27: Action diagram with a number of constructs

Many argue that action diagrams are easy to construct and utilize, both by analysts and users. As well as being able to represent and communicate logic in the traditional systems development process, they are also advocated as being useful to end-users when developing their own systems using various software tools and also by information centre staff when working with users. A particular benefit of using action diagrams is that it is possible to use the same technique for representing high-level functions right through to low-level process logic.

12.9 Entity life cycle

This technique also varies slightly from methodology to methodology and is called by different names, but in essence what is being achieved is substantially the same. The entity life cycle is used at a variety of stages in a number of methodologies and is one of the few attempts to address changes that happen over **time** (most of the other techniques represent static views of a system).

The entity life cycle is not, despite its name, a technique of data analysis, but more a technique of process analysis, though we were tempted to have a chapter on time techniques with this section on its own! It does show the changes of state that an entity goes through, but the things that cause the state of the entity to change are processes and events, and it is these that are being analysed. The objective of entity life cycle analysis is to identify the various possible states that

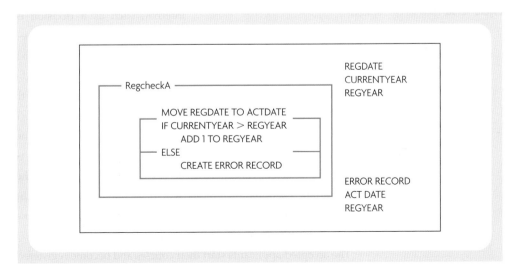

Figure 12.28: Action diagram including inputs and outputs

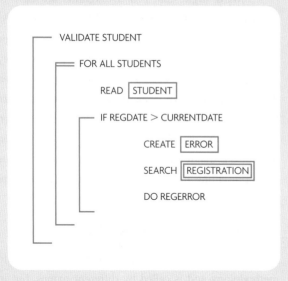

Figure 12.29: Action diagram incorporating database operations

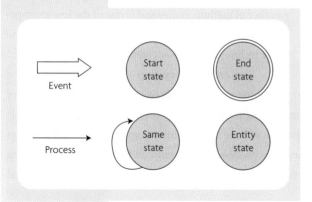

Figure 12.30: Entity life cycle symbols

an entity can legitimately be in. The sub-objectives, or by-products, of entity life cycle analysis are to identify the processes in which the entity type is involved and to discover any processes that have not been identified elsewhere. It may also identify valid (and invalid) process sequences, not identified previously, and it can form the outline design for transaction processing systems. So, as well as being a useful analysis technique in its own right, it is a useful exercise to perform as a validation of other process analysis techniques.

The documentation of the states of an entity in a diagram is one of the most powerful features of entity life cycle analysis. The diagram provides a pictorial way of communication that enables users to validate the accuracy or otherwise of the analysis easily.

The modelling conventions differ from methodology to methodology but the concepts are fairly consistent. In the following example the conventions of Information Engineering are used. Figure 12.30 shows the different symbols used, and Figure 12.31 shows a simplified example of an entity life cycle for the entity 'student' in the context of a university environment.

There is always a starting point, an event, which sets the entity into its initial state. There are three types of event. Events can be caused by external factors, such as a prospective student applies for admission into a degree programme (the starting point in the example); internal, such as a decision is made to accept the student for admission; or time-related, for example a candidate has not replied to our offer one month following our letter and is deemed to have withdrawn. There is also always (or should always be) at least one end or terminating point to finish the life cycle. In-between, there may be many different states of the entity.

In the example, the initial state of the entity is as 'applicant'. This is triggered by an event, which is the receipt of an application for admission for one of the courses at the university. The entity changes state as a result of the admissions function, which either causes the applicant to be rejected, conditionally accepted (which means that the applicant is accepted provided certain examination grades will be achieved), or unconditionally accepted. The resultant entity states are rejected, conditional, or accepted. At any time, both conditional and accepted applicants may withdraw their applications.

The accepted applicants start their courses and become registered. They may or may not graduate. Registered students may suspend their registration for a wide variety of reasons at

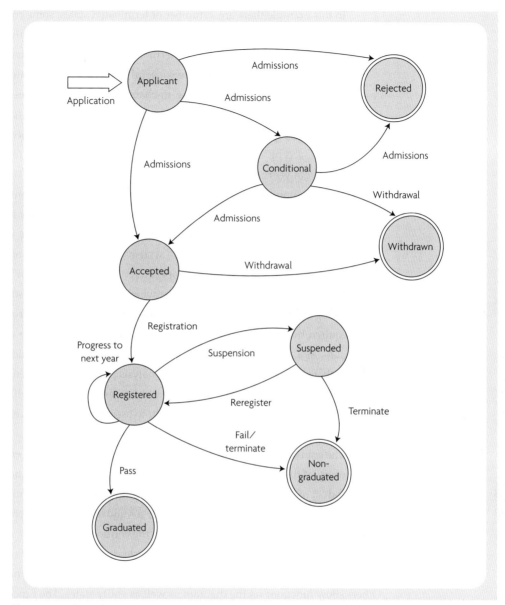

Figure 12.31: Entity life cycle for the entity student

any time, for example, ill-health, and may either return as registered or terminate as non-grad-uated. It should be noted that a function can be depicted that does not change the state of the entity. In this case the arrow points back to the same entity state. 'Progress to next year' is an example of this; it is a function that does not change the state of the entity, the student is still registered. An alternative perspective, that of three entity states named first-year students, second-year students, and final-year students, would have led to a change of state. Again, we see that systems analysis is not a science where all analysts following the same approach will result in exactly the same diagram.

In this example there are a number of terminated states; some conventions suggest that there should only be one. In this case we would add an extra state, called, for example, archived, and draw arrows from all our terminated states to this archived state.

It can be seen that the technique is useful in identifying the states of an entity, the processes that cause the states of an entity to change, and any sequences that are implied. Like many other of these process techniques, the key aspects of sequence (of business rules as the applicant proceeds through the university system to graduation), selection (between alternatives, such as, accept, conditionally accept, or reject), and iteration (e.g. registered students progressing from year to year) can be expressed clearly.

It is also important to identify the terminating states of the entity. Some information systems have not always done this and found that at a later date they have no way of getting rid of entity occurrences, which leads to obvious inefficiencies. One example found by the authors concerned the vehicle spare parts system of a large organization.

In this system vehicles require specific parts, but what is not known is which parts support specific vehicles. The result is that when vehicles become obsolete there is no way of withdrawing the parts that support that vehicle only. It is too dangerous to withdraw all the parts required by the obsolete vehicle, as many of these parts will be used by other vehicles. This results in a database which is continually growing as new vehicles and their parts are added. If there had been an entity life cycle analysis performed on the entity part, it would have been discovered that the entity occurrence did not terminate. The likelihood is that the organization would then have designed a function to associate parts with vehicles and thus be able to terminate the entity and not have these ensuing problems.

The entity life cycle diagram is a good communication tool that enables users to validate the accuracy of the analysis. It can form an outline design for transaction processing systems. A by-product of the analysis process is that functions in which the entity type is involved are identified. The process is therefore useful as a validation of the other process analysis techniques. These charts should be drawn for all the entities.

Some approaches develop entity life cycle diagrams further. For example, they might use separate symbols for various states of entities, such as set-up, amended and deleted states. Others show functions on the diagram as well as entity states. The functions implied in Figure 12.31 might include 'reject student', 'conditionally accept student', and 'student withdraws'. Their explicit inclusion, by labelling arrows in more detail, might help full understanding of the overall documentation set and also might identify functions which have been omitted. Events occurring in the problem situation which have not been modelled anywhere else in the documentation may also be identified.

In SSADM the entity life cycle is termed an entity life history. The diagram looks more like a hierarchical structure with the entity under consideration as the root or parent of the tree. The resulting diagrams are rather more complex than that shown in this section because they express more. These are described, with an example, in Section 21.1. The entity life history diagram is also very similar to the structure diagram of Jackson Systems Development (JSD).

In JSD this is the central modelling technique used in the methodology and there are also extensions to the entity life cycle technique described in JSD (Section 20.3).

In some approaches, entity life cycles are developed to provide information related to events, which are not possible to provide on standard entity life cycle diagrams. For example:

- If incorrect information has been added to an entity at what stages can it be corrected?
- What effect does premature termination have on the life of an entity?
- Can events happen out of sequence and, if so, which ones are permissible?

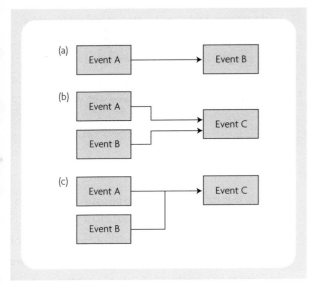

Figure 12.32: Dependencies between events

Unless such questions are considered, the analysis work is not complete and this may lead to an information system unable to deal with all possible situations. The extra information can be captured on these **state dependency diagrams** which can be created alongside entity life cycle diagrams as they show the dependencies between events. In Figure 12.32(a) event A must take place before event B; in Figure 12.32(b) event A and event B must take place before event C; in Figure 12.32(c) event A or event B must take place before event C. Of course, the interdependencies need to be thought out first and therefore the technique encourages proper analysis. Such analysis can be a vital input to the design of the information system in terms of the application logic following events.

Summary

- Data flow diagrams provide structure to a system (or part of a system) showing independent units in a graphical and concise manner. Through functional decomposition, it is possible to examine a system in overview and at a detailed level, while maintaining the links and interfaces between the different levels. The system can be described as a logical or physical model. There are four aspects: data flows, processes, data stores, and sources/sinks.

- Decision trees and decision tables aim to facilitate the documentation of process logic, particularly where there are many decision alternatives. A decision tree illustrates the actions to be taken at each decision point. Each condition will determine the particular branch to be followed. At the end of each branch there will either be the action to be taken or further decision points.

- Decision tables are less graphical, when compared to decision trees, but are concise and have an inbuilt verification mechanism so that it is possible to check that all the conditions have been catered for. Again, conditions and actions are analysed from the procedural aspects of the problem situation expressed in narrative form. They have four elements: action stub, condition stub, condition entry, and action entry.

- Structured English is very like a 'readable' computer program. It aims to produce unambiguous logic, which is easy to understand and not open to misinterpretation. It is not a natural language like English, which is ambiguous and therefore unsuitable. Nor is it a programming language. It is a strict and logical form of English and the constructs reflect structured programming. The sequence of the commands reflects the application logic.

- The structure chart is another functional decomposition technique with a series of boxes (representing processes or parts of computer programs, usually referred to as modules) and connecting lines (representing links to subordinate processes that show the way that data and control can be passed between processes) which are arranged in a hierarchy.

- Structured walkthroughs are a series of formal reviews of a system, held at various stages of the life cycle. They are intended to be team-based reviews of a product, such as a design component.

- A matrix is a simple tabular expression of a relationship, usually between two things such as entities and database files, or data items and processes.

- An action diagram represents an alternative way of representing the details of process logic (the business rules) and is specified in the Information Engineering methodology among others.

- An entity life cycle addresses changes that happen over time. It shows the changes of state that an entity goes through from when it enters the system to when it is discarded.

Questions

In relation to a video rental chain of shops:

1. Draw a decision tree and a decision table to express the logic related to the decision whether to accept an applicant as a new employee or not.
2. Draw a data flow diagram (DFD) showing how a shop might process a registered customer wishing to rent a video.
3. Draw an entity life cycle for the entity customer.
4. Perform structured walkthroughs of your colleagues' efforts and ask them to do the same with yours.

Further reading

Avison, D.E. and Shah, H.U. (1995) *The Information Systems Development Cycle: A First Course in Information Systems*, McGraw-Hill, Maidenhead, UK.

Lejk, M. and Deeks, D. (2002) *An Introduction to Systems Analysis Techniques*, 2nd edn, Prentice Hall, Harlow, UK.

13 Object-oriented techniques

13.1 Object orientation

In Section 6.4 the basic concepts and a number of advantages of the object-oriented approach were identified. In this section we examine the concepts and some of the techniques further. In Chapter 22 we look at two methodologies for information systems development that utilize the object-oriented approach. Readers interested in object orientation should look at all these sections.

We begin by looking at a definition of **object-oriented programming** (Booch, 1991) and then examining the object-oriented components of that definition:

> Object Oriented Programming is a method of implementation in which programs are organised as a co-operative collection of objects, each of which represents an instance of some class, and whose classes are all members of a hierarchy of classes united via inheritance relationships.

Such definitions are not always very clear or meaningful at first sight but this definition is in fact quite comprehensive. We begin by looking at some of the components of the definition. First, we examine the concept of an object. An **object** is something to which actions are directed; it has an identity, a state, and exhibits behaviour. The identity enables it to be distinguished from other objects; the state is the current value of the dynamic properties of the object; and the behaviour is the actions that the object can itself undertake.

Formal definitions of an object or objects are not very helpful because basically an object can be anything, so we add that it should also be an abstraction of something in the problem domain, that is, it is of interest to us (in this case, as developers), given the context of our current objectives (in this case, to develop a particular system).

Yourdon and Argila (1996) informally define an object as, 'an independent, asynchronous, concurrent entity which "knows things" (i.e. stores data), "does work" (i.e. encapsulates services), and "collaborates with other objects" (by exchanging messages), to perform the overall functions of the system being modelled'.

Strictly speaking, an object is an instance of a class of objects (i.e. a group of objects together make up a class of objects). All the objects in the class exhibit a common set of object attributes, such as structure and behaviour. Unfortunately the term 'object' is often used synonymously with class. So, strictly, the very term 'object oriented' itself should really be 'class oriented'.

A class is often illustrated by an example, which is often customer or client in many books. However, this does not really distinguish it from an entity in entity-relationship modelling for example. Therefore, a class can be a person or a place or an invoice or whatever, but it is not just the data as it would be if it was an entity. Additionally, it has actions or services that it can undertake and it can pass messages to other objects. So, it is more than just the data associated with say a customer. The customer class may include processing or code, for example, to change the status of the customer, and pass the new status to another class, such as a billing class.

Additionally, classes may be structured in a hierarchy based on inherited properties. Inheritance is a relationship among classes, such that one or more classes share the structure or behaviour of another class. The identification of classes in the design of a system is a key part. Classes are not necessarily the obvious ones we might first think of, or be the same as the entities we might identify.

Coad and Argila (1996) tell a story of a small European country that built a new social security system which kept track of pension and benefit payments, and much more besides, at great cost. They adopted an object-oriented approach and identified objects (classes) such as citizen, pension, benefit, etc. The system was not a success and was very difficult to maintain as local and European legislation changed frequently. As Yourdon and Argila say, 'nowhere was there a legislative rule object or anything similar. The secrets of legislative rules were embedded throughout the system . . . and whenever one changed, it usually required that very significant system-wide changes be made.'

Therefore, objects have to be properly identified and designed. Only if this is done correctly will the benefits of object orientation be obtained. If it is not done well then there are no particular benefits to object orientation. This means that the object-oriented methodology is of key importance. However, methodologies are not dealt with until Chapter 22.

To illustrate the general benefits and concepts of object orientation we will use an example. The object we identify is customer update. Let us assume the object is an update transaction on a customer file and, further, that it is a particular transaction to update a specific record, in a specific way, made at a particular time. The object has a unique identity, for example, 16249, the number of the transaction on the transaction log, which is unique to this transaction. The object has a state, in this case the state is 'successfully completed', that is, the update has been properly made. Other possible states of the object might be 'unsuccessfully completed', 'in-process', or 'awaiting processing'. In practice, the object may have many potential states combining a variety of attributes.

The behaviour of the object is the actions (or operations) that it can undertake: it can trigger an error message, it can change the contents of a field, it can access the status of the record, it can update the log file, and so on. The behaviour of the object is completely defined by its actions.

A class is the group of objects that share a common structure and behaviour. In this example the class might be that of 'customer record update', that is, the general class of which our object 16249 is an instance (the terms 'object' and 'instance' are in fact the same and are used interchangeably). So, the class is the general form of the object.

If we compare the concepts used in the entity-relationship model, we can make an analogy in terms of the relationship between entity types and classes, and entity occurrences and objects. The entity type is the general form and an entity occurrence is a specific instance of the entity type. For those with a programming background, class is analogous to type.

Classes are often structured in a hierarchy, which shares certain properties. Our class of 'customer record update' might be a subset of a class called 'record update', which might have more than one subclass, for example, inventory record update. Thus, there is a hierarchy of classes. This is illustrated in Figure 13.1.

Hierarchy is not simply about classes and subsets or supersets of classes but also includes the object-oriented notion of **inheritance** to define the relationships in the hierarchy. Inheritance implies that the relationship is such that the hierarchy goes from classes of a general type down to classes of a more specific type, or from classes that exhibit commonality down to those with differences at the lower levels. The classes at the top are general and are then extended with more detail at the lower level in the hierarchy. In an object-oriented programming language, a lower class in the hierarchy can be produced from the higher class 'inheriting' all the higher class's structure and behaviour. The programmer can then add some more specifics and detail to the lower-level class to make it perform more specifically than the higher class from which it was derived.

In our example, the class 'record update' would contain all the general features required to update a record. The class 'customer record update' would have inherited all these and then extended the class specifically to update the customer file. The benefits of inheritance include the fact that code can be reused; instead of separately writing code for 'customer record update' and 'inventory record update' we just inherit most of it from the higher class of 'record update'. This saves time and has other efficiency benefits, and also ensures that the processing (or rather behaviour in object-oriented terminology) in both 'customer record update' and 'inventory record update' is the same, that is, there is a standard approach.

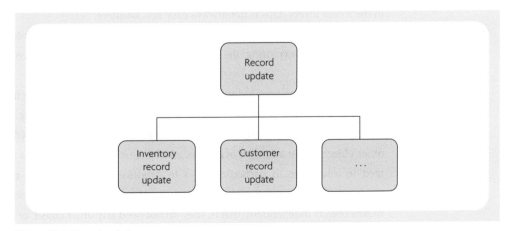

Figure 13.1: Hierarchy of classes

MS Windows-based programs, for example, use a hierarchy of windows classes to ensure that each window behaves in the same way no matter which application is running in order to achieve a consistent windowing environment. Inheritance is central to the object-oriented concept and provides much of its benefit. Indeed Booch (1991) goes as far to suggest that 'if a language does not provide direct support for inheritance, then it is not object oriented'.

It needs to be pointed out that the object-oriented notion of code reuse is different from that in traditional programming, where code is often shared and borrowed, which is a form of reuse. There are program libraries in organizations to enable code to be shared and reused. However, in this case the code is copied or cloned and used as a basis for a new program, which is then amended and extended as necessary for the new application. This may indeed save time and effort but it is very unlikely that the original code can be returned to again because there are now two separate entities: the original program and the new program, and they develop along quite separate paths in different ways. If an error is found in the original code or a change is required to the original code, such change has to be made in both programs as the new program has changed and evolved so much it is usually impossible to go back to the original and start again.

With object orientation, and true inheritance, the higher-level class could be changed and the lower-level classes would inherit that new behaviour by a simple recompilation. The processing specifically required for the lower level, in this case the code to make it specific to customers, would not be affected, because the object-oriented inheritance has ensured that they are treated and organized separately. In the example, we could change the class 'record update' and then recompile the classes 'customer record update' and 'inventory record update', and this effectively means that they then inherit the changes. So, the change is made only once and is applied to all the subclasses, and in this way really effective reuse is potentially achieved.

It should be obvious that inheritance does not just happen, it needs careful design. We will examine this in more detail later, but at the very least the decision as to the classes and their hierarchical structure needs to be thought about very carefully, such that the common aspects are included in the higher-level objects and the requirements of the special cases are contained in the lower levels. This is not always easy nor intuitive.

There is a further aspect relating to the definition of object-oriented programming made above that needs exploring: the organization of the program as '. . . a co-operative collection of objects'.

In the object-oriented world, objects sit around waiting to be activated, and that activation happens when the object receives a message from another object. Messages are the only way that objects communicate, and we may think of objects being fired off by messages from other objects or by an initial event. So, an object-oriented program is a series of objects organized to interrelate in particular ways to produce the functionality that is required of the program. This is sometimes referred to as co-operage. Communicating through messages makes objects independent; that is, they do not need any other object, or knowledge, to perform their job. They are triggered by a message from another object, they perform their job, which may itself involve the triggering of other objects, and when complete they usually return a message to that effect to the triggering object. In traditional programming this might be thought of as a bit like a subroutine, but a subroutine is not always independent nor is it triggered by the passing of a message.

For example, we might have an object 'menu-selection' which displays a menu and obtains a selection of an option from the user. The object is fired or invoked by another object passing a message to 'menu-selection'; part of the message might be the menu that should be used in 'menu-selection' itself. Let us assume that the object 'menu-selection' is the method for selection from a menu and all the associated error checking, and the message is the menu text to be used. On completion, 'menu-selection' returns the option that has been selected by the user as a message or it may invoke another object. In this way the object is independent of other objects and exists as an independent entity. The exact way that it performs its task and the data used by the object are unknown to other objects, indeed it may even be written in a different programming language. This also means that the internal workings or implementation of the object can be changed without it causing any problems or affecting the other objects.

What is important is the external interface of the object, that is, its messages. Furthermore, data and procedures are not allowed to exist externally or independently of an object, unlike in most other programming approaches. The fact that the internal processing and the details of the data are hidden (or private) is known as **encapsulation** and this is described by Booch (1991) as another of the fundamental elements of the object model.

Daniels and Cook (1992) use the analogy of an egg to illustrate the concept (see Figure 13.2). An object is an egg, the yolk is the data surrounded by the white which is the processing or operations that act on the data. The shell of the egg surrounds the whole thing and keeps it all together. It effectively hides the data and processes from the outside world; the shell is the interface and the only thing that is seen. The data and processes are said to be **encapsulated** in the object. The analogy breaks down a little because the shell of the egg, although nicely encapsulating the contents and hiding them from the outside world; does not interface very well with other eggs, nor does it send messages, or have an identity. In other approaches and methodologies, for example, SSADM, Merise, and Information Engineering, the data and processes exist, but are analysed and handled separately. Encapsulation is very different.

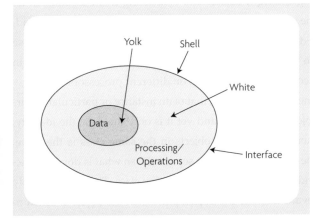

Figure 13.2: Objects represented by an egg (modified from Daniels and Cook, 1992)

An object-oriented program is simply a collection of interrelated objects where the connections are unidirectional paths along which messages are sent. The program begins with an initiation from outside, often an event, which triggers an object, from then on that object initiates others and so on. The activations form a network of objects that together make up the program (see Figure 13.3).

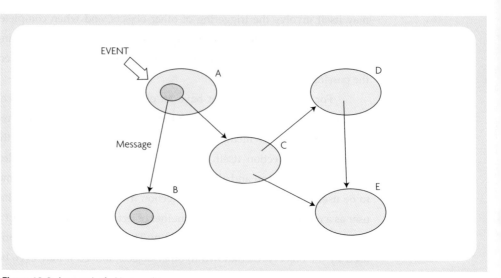

Figure 13.3: A network of objects make up a program

Programming an application consists of developing the objects and specifying the network or the co-operation between objects. It is one of the objectives and benefits of the object-oriented approach that in the future most of the objects needed will already exist somewhere, either within the organization or available commercially, and then development will essentially involve selecting or buying the right objects and connecting them together in an appropriate network. The path that the program takes through the network is called the thread, or the thread of control, and it can be difficult to identify once the program is invoked, because of course the objects are interacting in ways dependent on events and user responses. Also, in theory, the objects may be operating concurrently, and even on different processors.

In the discussion above the concept of an object is that of an instance or particular occurrence of a class, which means that objects are unique, and yet it is only the data, the identity, and state that are different. The general behaviour of objects in the same class is the same. Clearly it would be foolish to write the code for each object separately, so what is done in practice is that the code is written once and shared by objects of the same class. The object is really just a data structure holding the relevant data for the object, and when the object is asked to perform an operation the run-time system ensures that the data are connected with the relevant code. This means that in practice the implementation of object-oriented programs differs somewhat from the theory.

Having discussed the theory, we will now summarize the benefits that the proponents of the approach argue it generates (see also the benefits discussed in Section 6.4). As discussed above, the concept of inheritance leads to the controlled reuse of code. In theory, the organization will develop a library of object classes that deal with all the basic activities that the organization undertakes. Programming becomes the selection and connection of existing classes into relevant applications, and because those classes are well tried and tested as independent classes, when they are connected they provide immediate industrial-strength applications that run correctly the first time. Only completely new classes will need to be developed or perhaps purchased. Proponents of object orientation believe that, eventually, there will exist international libraries of object classes that developers will be able to browse to find the classes they require and buy them. The classes in these libraries will be guaranteed to perform as specified, and so new applications are easily developed. Further, existing (object-oriented) applications are modified and extended in functionality just as easily. Software development is not only quicker and cheaper but the resulting applications are robust and error-free. This attacks the problems that have bedevilled the software industry for so long, such as projects being delivered late, over budget, and full of errors. The implication is that the information systems developed using object-oriented techniques will be robust and error-free and quicker and cheaper to achieve. They should also be easier to maintain and hence address another problem that has bedevilled traditional software development.

13.2 Unified Modelling Language (UML)

The Unified Modelling Language (UML) is a graphical language, or a notation, for modelling systems analysis and design concepts in an object-oriented fashion. It is a set of rules and semantics that can be used to specify the structure and logic of a system. Although it has the term language, this refers to UML being a graphical language rather than a programming language.

In the early days of object-oriented development each author or vendor seemed to define their own notation for their own variant of language or object-oriented approach. As a result there was a great deal of diversity and some confusion. The same concepts had a number of different notations. This was a barrier to object-oriented use. There was concern that, although the theoretical concepts of the object-oriented approach were well understood and generally well established in education, their practical use was worryingly low. So, attempts were made to standardize notations in the hope that this would remove some of the confusion and smooth the path to increased adoption of object-oriented techniques in practice.

Therefore, UML was developed, with the first standard being made available in 1997. Booch, Jacobson, and Rumbaugh set themselves seven goals as follows (Booch et al., 1997):

1. Provide users a ready-to-use, expressive, visual modelling language so they can develop and exchange meaningful models.
2. Provide extensibility and specialization mechanisms to extend the core concepts.
3. Be independent of particular programming languages and development processes.
4. Provide a formal basis for understanding the modelling language.
5. Encourage the growth of the object-oriented tools market.
6. Support higher-level development concepts.
7. Integrate best practices and methodologies.

UML is now managed and maintained by the Object Management Group (OMG), an independent and 'vendor-neutral' organization. According to Smith (2000) UML has 'broken free from its proprietary roots to become an evolving public standard.'

The first version of UML was 1.1 and at the time of writing the latest version is 1.4, but there is also a version which includes UML facilities for the design of web applications, enterprise application integration, real-time systems, and distributed platforms. Exactly where UML will go in the future is not clear. As new developments and technologies emerge it must evolve and this means that it will grow. The problem is that as it grows in size it becomes more complex, difficult to understand, and of less use as a standard. It may well be already. Most people only use a subset of UML and as Booch et al. (1997) says, 'you can model 80 percent of most problems by using about 20 percent of the UML.' As a response, OMG is attempting to define a UML kernel.

The UML defines 12 types of diagram, divided into three categories (OMG, 2002). The first category is static application diagrams comprising Class Diagrams, Object Diagrams, Component Diagrams, and Deployment Diagrams. The second is dynamic behaviour diagrams comprising Use Case Diagrams, Sequence Diagrams, Activity Diagrams, Collaboration Diagrams, and Statechart Diagrams. The third category represents diagrams that help to organize and manage applications, composed of Packages, Subsystems, and Models.

Like other authors, we choose to illustrate only a small subset of UML, but it is enough to understand the essence of the language and prove useful enough without being overburdensome. Six diagrams are described below (we concentrate on the dynamic). The notation with examples is described and illustrated using the Select Enterprise case tool (see Section 19.4), except for the Activity Diagram, which was developed using the Rational Rose CASE tool. Of course, although the notation of UML is standard, the way the diagrams are used is a different matter and depends on what a developer is doing or what methodology is being followed.

1 Class diagram

The class diagram is used for modelling static structure in UML, and a class is represented as a rectangle separated into three parts. The top part depicts the class name, the middle part the

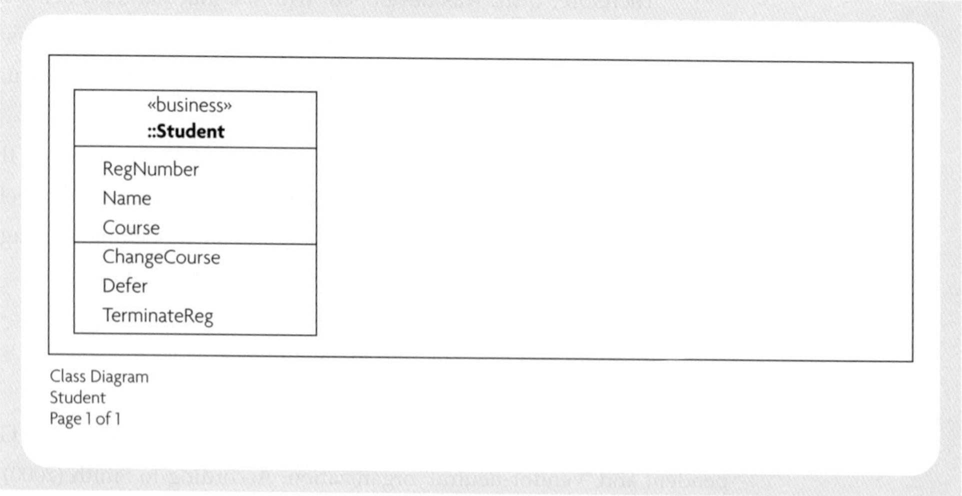

Class Diagram
Student
Page 1 of 1

Figure 13.4: Class diagram

attributes, and the lower part the operations. Figure 13.4 shows a class diagram consisting of one class named **Student**, with attributes of RegNumber, Name and Course, and operations of ChangeCourse, Defer and TerminateReg. This tells us that a class of **Student** exists with attributes as shown and operations that the class may participate in (i.e. the common behaviour of the class). Obviously in this example things are simplified; for example, the attributes of class **Student** are likely to be rather more than three in number, and probably have more than three operations, but UML allows us to use the diagrams in the way we wish for the purpose. So, in a high-level analysis activity we would not include the detail that might appear when we are conducting a low-level design activity. In fact UML allows us to omit the attributes and/or the operations from the class diagram. So, the rectangle may consist only of the class name (for example see Figure 13.5). It is thus quite flexible. Usually the class name is in bold and additionally in italics if it is an abstract class.

An object is an instance of a class. In the example of Figure 13.4, an instance of **Student** might be Smith (i.e. an instance of class **Student**). When an object is spawned from a class it has the same structure as the class, therefore the notation is the same. But in order to differentiate an object from a class, the object name is usually underlined rather than being in bold.

Figure 13.5 shows a slightly more complex class diagram. It illustrates an association between the class of **Student** and the class of **Degree** (i.e. a student is registered for a degree). An association represents a relationship between instances of classes, similar to the relationships between entities (see Section 11.1). Unfortunately the notation in UML is different, including that to represent multiplicity, which is the term given to indicate how many objects may participate in the association. In Figure 13.5 the multiplicity is shown at either end of the association line. The * represents the range zero to infinity (*0 . . . infinity*) and the 1 indicates one and only one (*1. . .1*). Therefore, a student is registered for one and only one degree (at a

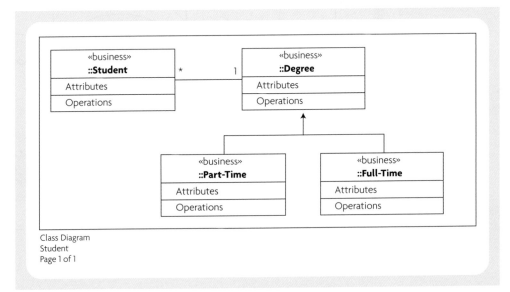

Class Diagram
Student
Page 1 of 1

Figure 13.5: Class diagram

particular time) and a degree can be taken by an infinite number of students. If there were a maximum, say 250, this would be represented as (0. . . 250).

The class **Degree** is shown as having a single-inheritance hierarchy. The superclass **Degree** is composed of two subclasses, **Part-time** and **Full-time** degrees. Actually it is not really 'composed of', rather the subclasses inherit features (attributes and operations) from the superclass. However, in UML the inheritance arrow points upward. It is single inheritance because one subclass inherits from one superclass. Multiple inheritance is where a subclass inherits from more than one superclass. For more details of inheritance see Section 13.1.

2 Use case diagram

Use cases describe the functionality of a system from the perspective of users or actors in the system. It is a high-level, user-oriented diagram. A use case is usually described as a high-level function of a system, and in UML the use case diagram shows the associations or interactions (as lines) between actors (stick figures) in the system and the use cases (ellipses). The diagram also shows the boundary of the system as a rectangle with the use cases within.

Figure 13.6 shows an example of a use case diagram for a student registration in a university. These are the interactions, as the users see them, which are essentially the users' requirements. The use case diagram does not depict the internal processes or workings of the system but only the flow of events as actors interact with the system. The total functionality of

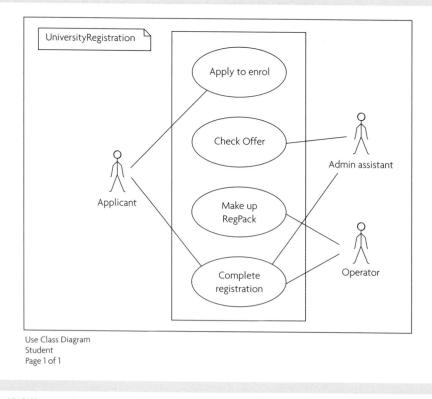

Figure 13.6: Use case diagram

a system would be represented by a number of use case diagrams. The diagrams are essentially non-technical and are supposed to represent things that yield value to the actors. For this reason it is recommended that the user should participate in the construction of the diagram.

Actors although usually depicted as stick figures need not be. They can be represented as a class rectangle. Further, an actor might not be a human actor in the system. For example, it could be a non-human piece of hardware, or another system, or perhaps even a methodology. In fact it can be anything that interacts with the system.

Along with the use case diagram a use case textual description or specification is usually added. This is often brief, but can be a fully detailed breakdown of each interaction. Some methodologies or OO approaches recommend that the use case diagrams be the starting point for the identification of classes (e.g. Bennett et al., 1999). See also Section 22.2 on the Rational Unified Process (RUP).

3 Interaction diagrams

Unlike the previous diagrams, which were basically static, the interaction diagrams attempt to model dynamic behaviour. There are two interaction diagrams: the sequence diagram and the collaboration diagram. They both are used to describe the interaction between objects and messages in a system within a single use case. We shall only describe the sequence diagram here because the two diagrams are essentially alternative ways of modelling the same thing – interactions – and because people often seem to prefer sequence diagrams. Page-Jones (2000), for example, states that they enable him to 'clearly see who says what to whom and when . . . and no other diagram illuminates timing so well'. We provide examples of interaction diagrams in the following subsection.

4 The sequence diagram

The sequence diagram shows the interaction between objects and messages over time. Time is normally represented on the vertical axis with the objects (or classes) depicted from left to right on the horizontal. The order of objects in the diagram is immaterial although they are usually drawn so that the message arrows point from left to right as far as possible, just to make it more readable and less cluttered. Each object is depicted as a rectangle (i.e. by its object symbol), usually at the top of the diagram with a dotted line (though not dotted in Figure 13.7 when produced by the select CASE tool) running from the rectangle down the page, known as the object *lifeline*, which shows the object's existence during the interaction. Messages are shown by labelled arrows between the lifelines. When a message is sent from one object to another it sets off an operation of the object. A sequence diagram is often used to expand a use case to a lower level of detail. Figure 13.7 shows an example of a sequence diagram which is an expansion of part of the Check Offer use case of Figure 13.6.

We will follow what happens when the object ValidApp receives a message, CheckApp, sent by the actor StudentApplicant (not shown in this diagram, but see Figure 13.6). The object ValidApp is then invoked which means that an operation of that object is fired. This operation (or set of operations as this is still a fairly high-level diagram) confirms that a valid application exists for this person by sending a message, GenerateOffer, to Offer. The Offer object is then

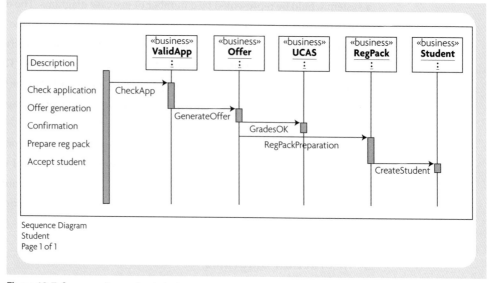

Sequence Diagram
Student
Page 1 of 1

Figure 13.7: Sequence diagram for student

triggered and an offer, in the form of the required grades at the advanced-level examination for a place on the particular course, is produced. When this has been done the Offer object sends a message, GradesOK, to the UCAS object (this is the central university admissions body in the UK) and also a message to prepare the registration pack (RegPackPreparation) is sent to the objec RegPack. This then triggers a CreateStudent message which is sent to the Student object which creates a new Student. The dotted lifeline starts lower down in the diagram because it is only at this time that the object has been created and so its lifeline starts. The other objects already exist (i.e. they have been created elsewhere), and their lifelines start at the top of the diagram. The activation period of an object is indicated by the grey rectangle on the lifeline which represents the period that the object is executing.

This is again quite a simplified example, it does not indicate what happens if errors are encountered or if the application is not valid. It should also be stated that no university functions in exactly this way, but it illustrates the principles of the sequence diagram. The strength of the sequence diagram is its simplicity and visual appeal. It is relatively easy to get a quick understanding of the flow of control and certainly much easier than looking at program code. It is also a good way of building up the specification of that flow in the first place and is relatively easy for users to understand. Other details can be included in the diagrams as necessary. For example, iteration of a message can be indicated with an asterisk.

5 Statechart diagram

A statechart diagram shows the various permitted states that an object may be in, and then transfer to, as events occur and messages are received. In concept, it is similar to the entity life cycle described in Section 12.9 except that statechart diagrams model an object rather than an entity. The general concepts will not be repeated here so readers should review Section 12.9.

In statechart diagrams (usually just called state diagrams) the state is a particular set of values of the attributes of an object at a particular time. When the value of an attribute changes, the object changes state. Obviously not all changes of attribute values are of interest. We are only interested in modelling those that are deemed to make important changes to the state of the object. For example, if a bank customer cashes a cheque the attribute 'balance', of object BankAccount, will change, but we do not usually consider the object to have changed state (unless perhaps the balance goes negative). However, if a customer tells us that their cash card has been stolen, the attribute 'suspend' will be set and this would probably be important enough for us to consider the object to have changed state.

States of an object are represented by rectangles with rounded corners (state symbols) and usually labelled with a unique name for that state. Transitions from one state to another are represented by arrows with the name of the event that triggers the change and/or the name of the attribute that changes and the condition that triggers it. Events are usually separated from attributes by a / character. A solid black dot represents the starting point for reading the diagram (i.e. the initial state). This starting point is purely notional, and an object cannot be at that point but is immediately in the first state. Similarly, the end point of the diagram is represented by a bull's eye although the object cannot leave its final state. Figure 13.8 is a statechart for object student.

The diagram starts at the black dot and the object student is in state *current* (there has already been some activity for creating the object). Three possible transitions may occur. First, a student may suspend their studies and the state of the object changes to suspend. Then a student may reinstate themselves and the state returns to current or the student withdraws completely and the state changes to *archive*. Second, the student may successfully complete their studies and the state changes to *graduate*. After a specific period the state then changes to

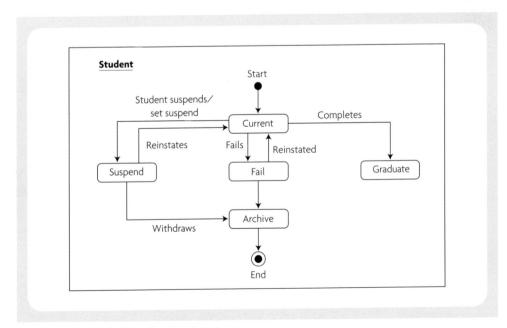

Figure 13.8: Statechart diagram for object students

archive. Third, the student may fail and the state of the object changes to *fail*. If the student re-takes exams they may be reinstated, or after a specific period the state then changes to *archive*. The bull's eye indicates that *archive* is the final state of the object.

This is the basic statechart diagram. As with other diagrams there are various extensions. For example, hierarchies of states can be nested and concurrency can be represented by splitting a state symbol into two nested halves separated by a dashed line, each half running concurrently.

6 Activity diagram

An activity diagram is described as a variation of a statechart that replaces events with activities, and focuses on internal flows and activities rather than events. Statecharts are concerned with the states of a single object, whereas an activity diagram can model the internal state of an object or a set of objects or a whole use case. An activity (or action state) is represented by a box (sometimes with curved sides).

Like statecharts they begin with a black dot (the initial state) and move directly to the first state. Figure 13.9 is an activity diagram for an operation called ProcessStudent in our student example. The first state is Check status. This is an action state because it has two possible outgoing transitions. Which one is taken depends on the 'guard condition'. In this case the applicant is classed as either EU (European Union) or Overseas. If EU the state of EU status is set. If Overseas, the quota is checked. If the quota is full, then the application is rejected. If the quota is not full, then the Overseas quota is incremented by one. The condition is in this case represented by a decision diamond after the Check Overseas quota box, with again two mutually exclusive outcomes.

Activity diagrams can indicate activities that can be performed in parallel or that must wait for another activity to complete before starting. This is represented on the diagrams by use

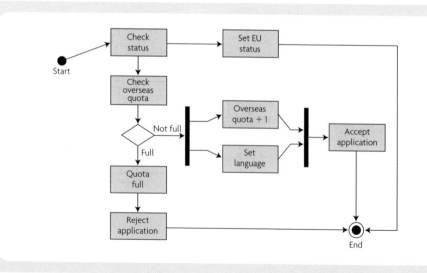

Figure 13.9: Activity diagram

of a synchronization bar, which is a short thick line. In Figure 13.9 the synchronization line after the condition quota 'not full' indicates that Set language and Overseas quota + 1 can be performed in parallel. The synchronization bar after these two states indicates that they must both be completed before the next activity.

Thus the UML is a standard modelling language, based on object-oriented concepts, for visualizing, specifying, constructing, and documenting the various models or artefacts required in systems development. A methodology utilizing the UML standard, known as the Rational Unified Process, is described in Section 22.2.

Summary

- An object is something to which actions are directed; it has an identity, a state, and exhibits behaviour. The identity enables it to be distinguished from other objects; the state is the current value of the dynamic properties of the object; and the behaviour is the actions that the object can itself undertake.

- An object is an instance of a class of objects (i.e. a group of objects together make up a class of objects). Inheritance is a relationship among classes, such that one or more classes share the structure or behaviour of another class.

- The Unified Modelling Language (UML) is a graphical language, or a notation, for modelling systems analysis and design concepts in an object-oriented fashion. It is a set of rules and semantics that can be used to specify the structure and logic of a system.

Questions

1. What is different about the object model when compared to the entity-relationship model?

2. Why have many organizations turned to the object-oriented approach? Why have many organizations not changed?

3. What are the advantages of graphical languages, like UML, over notational languages?

Further reading

Fowler, M. (with Scott, K.) (2003) *UML Distilled: A Brief Guide to the Standard Object Modelling Language*, 3rd edn, Addison-Wesley, Reading, Mass., USA.

14 Project management techniques

14.1 Estimation techniques

In this section we look at three estimation techniques for project management: **CoCoMo**, function point analysis (FPA), and work breakdown structure. A major problem of software development is estimating the effort required to develop programs. CoCoMo is an approximation of the effort needed based on experience of past projects. This is a formula for estimation based on the amount of program code required, that is, lines of code (measured in thousands of source instructions or KDSI). The formula, evolved from experience, is:

$$PM = 2.4(KDSI)^{1.05}$$

where PM is effort in people-months. So, if 1,000 instructions are required, then:

$$PM = 2.4(1,000)^{1.05}$$
$$= 2.692 \text{ (or nearly three people-months)}$$

There are further sophistications of CoCoMo taking into account other factors such as product reliability and complexity, execution time and storage requirements, experience and capability of programmers, use and experience with software tools. Such sophistications aim to increase the reliability of the formula.

Function point analysis, like CoCoMo, is an estimation technique, but somewhat more sophisticated. It tries to estimate the functionality of the system being delivered to the end-user. It does this by analysing the system in terms of information systems requirements, based on the inputs, outputs, files, updates, interfaces, reports, and enquiries, each being assigned a number of function points. An estimate of the technical complexity and such considerations as staff experience and deadline pressure can also be calibrated in the formula to make allowance for these differences. The weights for each criterion are based on measurements from previously developed systems.

This approach is used for software development estimation in methodologies such as SSADM and Merise. The criteria used are easier for analysts to estimate than lines of program code. Even so, the complexity ratios and other adjustment factors require an experienced analyst to estimate. Tables have been created which help determine the effort and elapsed time required to complete the project. At its most sophisticated, this approach will take into consideration supervision levels, documentation, quality levels required, training required, familiarization required, data conversion, reviewing required, technical support required, staff experience, and user involvement among other factors.

One of the major tasks in estimation is breaking down the project into its basic work elements. The first-level decomposition will be based on the various phases and subphases, and each is broken down into its technical, management, user liaison, administrative, quality assurance, and other tasks. At the lowest level, the tasks cannot realistically be broken down further and are small enough to make estimation fairly accurate. For each task, analysts may use experience from past projects or seek advice from others for estimation. Work breakdown structure is a first phase in developing PERT charts below.

14.2 PERT charts

PERT (Project Evaluation and Review Technique) is based on project network diagrams or charts with a particular feature for estimating the elapsed time of activities.

In a PERT chart the activities are represented by arrows, which join the nodes (circles). The latter represent events, that is, the completion of activities. Figure 14.1 shows a network. The arrows represent the activities, though the length of the arrow does not indicate the time taken for each task. Arrows drawn from the same node indicate tasks that can be carried out simultaneously. Arrows following others indicate tasks that are dependent on the completion of these other tasks.

The manual development of networks is lengthy, and project control software makes the task much easier. Such a software tool can draw the network and highlight critical activities on which any slippage of time will cause the whole project timescale to suffer. The path of the critical activities joined together forms the critical path, and it is useful for the package to highlight these activities. In Figure 14.1, the activities A-B-C-E are on the **critical path**. Each activity

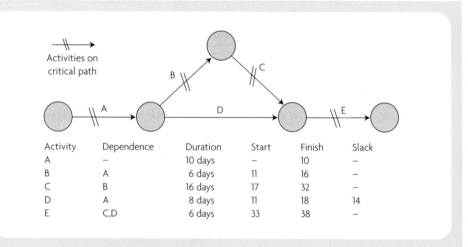

Activity	Dependence	Duration	Start	Finish	Slack
A	–	10 days	–	10	–
B	A	6 days	11	16	–
C	B	16 days	17	32	–
D	A	8 days	11	18	14
E	C,D	6 days	33	38	–

Figure 14.1: Project control – the network and critical path

on the critical path has been marked with short parallel lines. If it is possible to reduce the time of these activities, perhaps by moving resources allocated from other activities to them, then the overall project time should decrease. Activity D is not on the critical path, and there is a slack of 14 days on this activity. In other words, there can be a delay of up to 14 days on D without delaying the overall project. If feasible, it might be expedient, therefore, to move resources from activity D to an activity on the critical path. This change can be entered into the package and the results recalculated.

PERT tools such as Microsoft Project (Section 18.4) can aggregate the various resources, such as the number of people working on the activity, and attempt to smooth their use throughout the project. This can be particularly useful as management reviews and when approving plans.

It is usually better to use resources as smoothly as possible in the lifetime of the project; otherwise staff will be used efficiently for only part of the project. However, unless the resources are taken out of non-critical activities, this process is likely to lead to an increase in the overall project time. Once smoothing has been done, a bar chart showing the resource allocation over time can be produced (see Figure 14.2).

Normally there is a trade-off between time and cost (assuming the same quality); in other words, the more resources allocated (and the more costly the project), the quicker it can be finished. The user may like to input various estimates of resource availability, basing them on past experience in terms of minimum, most likely, and maximum figures. This will give three different results for time/cost comparisons.

Many project control packages (see Section 18.4) will report on inconsistencies within the network, such as the same resource being used at the same time. Although the plan should allow for minor deviations, the package may permit the user to ask 'what if?' questions so that the consequences of more major deviations can be seen, for example:

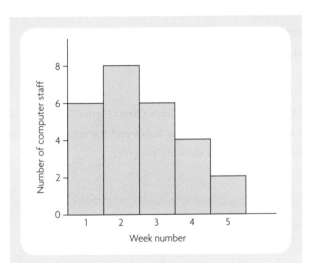

Figure 14.2: Project control – resource schedule

- reallocating staff;
- unexpected staff leave;
- machine breakdown.

Useful reports from a package might include a list of activities presented in order of:

- latest starting date;
- earliest starting date;
- by department;
- by resource;
- by responsibility.

Packages can simulate the effects of:

- prolonging an activity;
- reducing resources applied to it;
- adding new activities.

Similarly, it can be used to show the effects of changing these parameters on project costs. The manager may be faced with two alternatives: a resource-limited schedule, where the project end date is put back to reflect resource constraints, or a time-limited schedule, where a fixed project end date leads to an increase in other resources used, such as people and equipment.

Once a project is under way progress can be monitored to:

- compare the time schedule with the actual progress made;
- compare the cost schedule with the actual costs;
- maintain the involvement of users and clients;
- detect problem areas;
- replan and reschedule as a result;
- provide a historical record, which can be used for future project planning.

14.3 Gantt charts

Another way of displaying activity information is to use a Gantt chart (Figure 14.3). In this chart, the estimated duration for each activity is shown in clear boxes and the actual duration in shaded boxes. The Gantt chart is particularly good at showing progress.

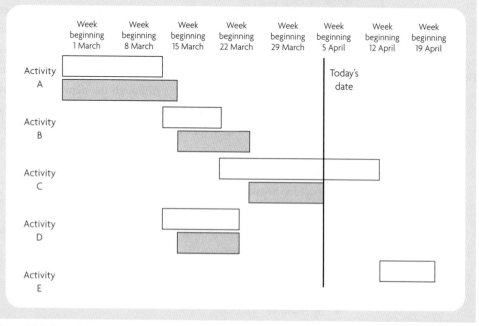

Figure 14.3: Gantt chart

In the example, a delay in activity A by three days is recorded, and this has set the start of activities B and C back by three days. Unlike the losses caused by delaying activity A, the gains made in reducing activity D by three days shown in the chart will not make any difference to the overall project time. In the event of delays impacting the overall duration the analyst can replan and mitigate the delay as best he or she can. Identifying and reporting delays is particularly important in organizations with many projects as the delay of one project may lead to delays in others.

A Gantt chart is often an integral part of a project management tool (Section 18.4) that may also provide guidance on how to put right any deviation from the schedule. This may be achieved by, for example, increasing the resources on some activities and rescheduling others. Goal-seeking analysis is a feature of a number of project management tools utilizing Gantt charts. See Section 18.4 for examples of their use using a project management tool.

Summary

- CoCoMo, function point analysis (FPA), and work breakdown structure are estimation techniques for project management.
- CoCoMo is a formula, based on experience of past projects, which approximates the effort needed in terms of program code required.
- Function point analysis is a more sophisticated technique. It tries to estimate the functionality of the system being delivered to the end-user by analysing the system in terms of information systems requirements, based on inputs, outputs, files, updates, interfaces, reports, and enquiries, each being assigned a number of function points, plus an estimate of the technical complexity and other considerations.
- Work breakdown structure is a decomposition of work into its technical, management, user liaison, administrative, quality assurance, and other tasks. For each task, analysts may use experience from past projects or seek advice from others for estimation of work. Work breakdown structure is a first phase in PERT or network analysis.
- PERT (Project Evaluation and Review Technique) is based on project network diagrams for estimating the elapsed time of activities. In a PERT chart the activities are represented by arrows, which join the nodes (circles). The latter represent events, that is, the completion of activities. It is used for estimating the time taken to complete a project and highlighting those activities where delays can be critical.
- In a Gantt chart, the estimated duration for each activity can be compared to the actual duration, and is therefore good at showing the progress of a project in terms of time.

Questions

1. Complete a work breakdown structure for an information systems development project following the SDLC (use the text in Chapter 3 to guide you).
2. Construct a PERT chart showing the interdependencies of the tasks and estimate the time for each activity. Use a project management tool if one is available.

Further reading

Cadle, J. and Yeates, D. (2001) *Project Management for Information Systems*, Prentice Hall, Harlow, UK.

15 Organizational techniques

15.1 Lateral thinking

A further way to identify areas of change is to use some of the techniques used in lateral thinking. De Bono (1990) identified many techniques to try to restructure 'patterns', where patterns are people's internal models of a particular situation. The techniques range from lateral thinking 'games' to practical problem restructuring techniques. Most of these techniques, which include generating alternatives, challenging assumptions, fractionation, and brainstorming, are well known, and here we give only brief overviews. However, brainstorming is used most in systems analysis, so we will discuss this more fully.

Although we often assume that we 'naturally' search for alternatives, this is limited in practice, and the **lateral search** is aimed at looking for as many alternatives as possible. De Bono argues that the search should not be limited to obvious candidates as 'unlikely' alternatives may reveal one that will have the most impact. This is related to the second technique, that of **challenging assumptions**. But this may also be applied to asking questions about whether our basic ideas are sound. Restructuring our views and patterns may give more insight into the problem. The third technique, **fractionation**, based on functional decomposition, entails breaking things down into their constituent parts. However, in the lateral thinking context, it also involves reassembling the parts in a different way, and this may reveal a useful new view of the situation.

Brainstorming is a team activity aimed at generating a cross-stimulation of ideas. It is used in a semi-formal setting to generate ideas, where the ideas of one person serve as a stimulus to generate further ideas from other people, which in turn serve as a stimulus for further ideas, and so on. Judgement on the usefulness or validity of the ideas is 'suspended' until the brainstorming session is completed. The aim is to get a free flow of ideas.

Brainstorming is the most used of the lateral thinking techniques in systems analysis and is, perhaps, the least structured of the fact-finding techniques used in the systems development process. Different people can express their ideas together, and it therefore allows many views and opinions to be considered.

There might be problems with using lateral thinking techniques. For example, they rely on the skill, or arguably the art, of practitioners, and hence there is a lack of consistency; the techniques are not very formal or controllable; different people may be more inclined to lateral thinking than others; there is no definite stopping point in some of the techniques nor is there a definite result at the end; and there are costs involved.

Furthermore, brainstorming sessions may be difficult to arrange:

- Who does the systems analyst involve and from what business areas and should there be both management and shop-floor participation?
- How many people should be invited – the larger it is then the more ideas that can be expressed, but too large a group becomes difficult to handle?
- How can the subject matter be limited – we need a wide-ranging discussion to enable full expression of opinions, but this may lead to little being decided on the most pressing decisions?

Brainstorming sessions are therefore potentially chaotic situations, as well as potentially very enlightening, so care must be taken on their planning. But, whereas most of the methods we have discussed concern the analyst interpreting the results of interviewing or observing through the production of diagrams, for example, data flow diagrams, entity-relationship models, and the like, brainstorming provides the opportunity for the group sharing of ideas and perceptions, which can be very helpful as well. It also provides an opportunity to gauge reactions to suggestions and proposals. It forms, therefore, part of the change management process of an information system.

Some lateral thinking techniques could be used to move out of the 'current time' constraint and help identify possible areas of future changes. They are probably better used as a toolset within more formal techniques, such as future analysis or risk analysis, though they could be used on their own as quick alternatives. They could also be used throughout the development process, for example, through stating and challenging the assumptions made at each stage.

In Figure 15.1 we show how future analysis, risk analysis, and lateral thinking might be used in the requirements analysis phase of the development of an information system in terms of change identification inputs, processing, and outputs.

15.2 Critical success factors (CSFs)

Critical success factors were popularized in the information systems arena by Rockart (1979, 1982). CSFs are usually understood to be the set of factors that can be considered critical to the continued success of an organization or a business. These factors may be 'skills, tasks, or behaviours' according to Bisp et al. (1998). They can operate at a number of levels and be used for a variety of purposes. They can, for example, be used at a macro-level and used to define the critical factors for a particular industry or even for an overall economy. More usually, however, they are used at the micro-level and applied to a business or company, or a part of a company. In the context of strategic IS planning the unit of analysis is the business unit, because, according to Ward and Griffiths (1996), this is the 'practical level to determine strategy'. Sometimes CSFs are used at the individual level, so, for example, the CSFs for a Chief Executive Officer (CEO) might be identified.

Only a limited number of factors should be identified as critical, and the limited number is an important aspect of the analysis. If too many are identified, they are probably not all

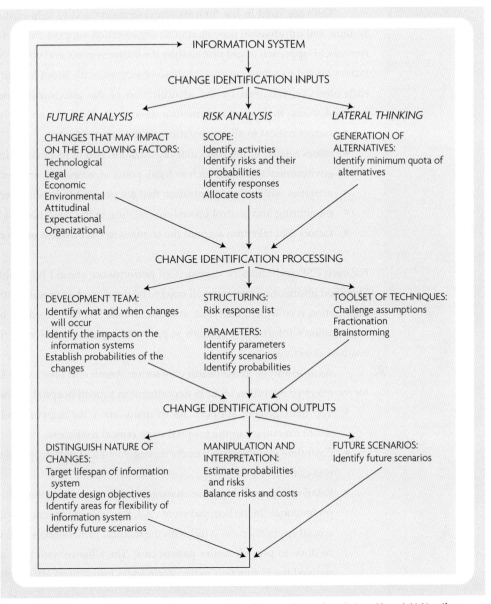

INFORMATION SYSTEM

CHANGE IDENTIFICATION INPUTS

FUTURE ANALYSIS

CHANGES THAT MAY IMPACT
ON THE FOLLOWING FACTORS:
Technological
Legal
Economic
Environmental
Attitudinal
Expectational
Organizational

RISK ANALYSIS

SCOPE:
Identify activities
Identify risks and their
  probabilities
Identify responses
Allocate costs

LATERAL THINKING

GENERATION OF
ALTERNATIVES:
Identify minimum quota of
  alternatives

CHANGE IDENTIFICATION PROCESSING

DEVELOPMENT TEAM:
Identify what and when changes
  will occur
Identify the impacts on the
  information systems
Establish probabilities of the
  changes

STRUCTURING:
Risk response list

PARAMETERS:
Identify parameters
Identify scenarios
Identify probabilities

TOOLSET OF TECHNIQUES:
Challenge assumptions
Fractionation
Brainstorming

CHANGE IDENTIFICATION OUTPUTS

DISTINGUISH NATURE OF
CHANGES:
Target lifespan of information
  system
Update design objectives
Identify areas for flexibility of
  information system
Identify future scenarios

MANIPULATION AND
INTERPRETATION:
Estimate probabilities
  and risks
Balance risks and costs

FUTURE SCENARIOS:
Identify future scenarios

Figure 15.1: Techniques for improved requirements analysis – future analysis, risk analysis and lateral thinking (from Avison et al., 1994)

critical and at too low a level. The focus is on the relatively few areas where things must go absolutely right to ensure business success. The process thus includes a fundamental assessment and prioritization of factors (i.e. those factors that are critical and those that are not). It is usually suggested that CSFs should be measurable. Indeed, not only measurable but actionable, and linked to perceived value in the market. Strictly speaking CSFs themselves should not be prioritized or ranked because, as the term implies, they are all critical and thus equally important. If they can be ranked in terms of importance, as is sometimes seen, it implies that they are probably not all critical.

CSFs are used in the information systems context to help ensure that the information systems and information provision in an organization support the overall business strategy. A typical CSF approach might first analyse the business goals and objectives and then identify the factors critical to achieving each of those objectives, with about four to six CSFs identified for each objective, followed by the identification of the information and information systems required, if any, to support and monitor these CSFs. The CSFs are likely to include:

- factors critical to all organizations in the same industry;
- issues related to the particular organization and to its position in the industry;
- environmental factors, such as legal, political, economic, and social aspects;
- activities within the organization that are proving to be short-term problems;
- monitoring and control procedures relating to the operations of the firm;
- factors that take into account the changes in the business environment.

For each CSF, indicators or measures of performance should be established and trailed. The identified information systems will need to be developed or modified to ensure that the critical information is collected, analysed, and distributed. In other words, it helps to ensure that the organization's information systems support the agreed critical activities and thus the wider business objectives.

As an example, in the health-care sector, Arnett and Litecky (1992) identified three CSFs for the effective operation of an IS department in a small hospital. These were:

- Top management support. As in many areas the support and understanding of the senior executives in the hospital were critical to success.
- Cash-flow maintenance. Timely management and monitoring of the hospital's cash flow was critical.
- Relationship with key user managers/medical professionals. Relationships with other professionals in the hospital were vital. Key user managers provided a means to identify crucial applications as it did with the medical professionals who understood what could be done to provide better patient care. The alliance with both groups of professionals enabled the IS function to transform ideas into actions effectively.

They then went on to identify the specific skills needed by IS professionals such that the CSFs could be met and help ensure the organization's success. This is an example of cascading the CSF analysis downward to ensure that all the elements in the organization contribute to addressing the identified CSFs. In this case the critical skills required of the IS professional were identified as follows:

- ability to wear several hats (i.e. be a 'jack of all trades');
- knowledge and appreciation of the work of hospital administration (particularly regarding cash flow and accounting issues);
- knowledge and appreciation of the work of medical staff (including knowledge of medical terminology and medical software);
- negotiation skills (particularly with software vendors);

- PC and networking skills;
- a specific programming skill.

So, using the CSF technique, it is argued, helps identify those relatively few things that must be achieved to ensure success at a strategic level. Then it can be cascaded down to specific lower-level activities or elements that contribute to the achieving of the overall CSFs. In this example, the skills required of the IS professional are specifically tailored to the needs of the health-care sector and the characteristics of a small hospital. It might be worth noting that the particular hospital found it very difficult to find IS staff with all the necessary skills, but at least they knew as a result of the CSF analysis what they were looking for.

15.3 Scenario planning

One of the main concerns of the systems analyst is to ensure that the system they are designing will be suitable for the organization in the long term. We know that we cannot assume that the conditions of today will prevail in the future. Therefore we cannot assume that the system we build today for today's situation will be appropriate for the future. There are a number of techniques that can help us plan for future change. An obvious technique is forecasting, where we base our understanding of the future on present trends. However, sometimes the future cannot be planned in this way – it is much more uncertain. One technique for such circumstances is scenario planning.

Porter (1985) defines scenario planning as 'an internally consistent view of what the future might turn out to be – not a forecast, but one possible future outcome', whereas Ringland (1998) suggests that it is 'that part of strategic planning which relates to the tools and technologies for managing the uncertainties of the future', Assuming, then, that we cannot easily predict the future, how can we identify possible future outcomes? Three ways of identifying future scenarios are expert scenarios, morphological approaches, and cross-impact approaches.

In the case of **expert scenarios**, people outside the company deemed as experts in the problem area are asked for their views on possible outcomes. A consensus may emerge from the views. This might be difficult to manage, so there are skills needed from the internal people to liaise with experts. A **Delphi** approach might be used so that experts see the views of other experts and some general consensus is achieved. In the Delphi approach, people in turn give their views so that the next person can build on the previous assertions. Sometimes experts might suggest two or three possible outcomes. Some of these cannot be said to be 'expected', but they might be plausible and worth planning for. In this case the systems analyst may incorporate in the design the possibility of each of the emerging scenarios.

Morphological approaches identify a number of future states built on different assumptions. These assumptions might relate to expected states for the economy, depletions or findings of natural resources, changes in people's values or lifestyles, or changes in the political persuasion of a new government. Scenario analysis might consider different combinations of values for these key values. The spreadsheet package Excel has a scenario manager tool that can show these different scenarios side by side. For each of these different scenarios, their implications are then discussed.

Cross-impact approaches identify potential events, trends, and conditions that impact on the decision and on each other. Sometimes probabilities are assigned to each of them so the likelihood of these impact factors can be estimated.

Other techniques, such as brainstorming (as well as Delphi), may be used with these approaches to eke out possible scenarios or a consensus. Further, the approach might be used in a number of methodologies in a requirements analysis phase or, in general, to create a more robust strategy for systems design.

As Ringland (1988) shows, a number of good business choices have been based on scenario planning. A particular example is British Airways' decision following the Gulf War not to reduce capacity, a reaction of other airlines to the fall in demand. British Airways' scenario assumed that demand would go back to 'normal' levels and capacity would pick up. This proved a realistic scenario, and BA was able to provide the necessary capacity when demand did pick up and gain competitive advantage. Following 11 September 2001, however, BA had to reduce capacity in the face of large losses. Scenarios may have again suggested that demand would pick up and airlines' capacity become stretched, but such losses are unsustainable even in the short to medium term.

The UK National Health Service also used scenarios to plan their strategy in 1994. The future requirements of the NHS were uncertain and 30 experts provided opinions on the range of NHS services and perceptions of them. These included clinical practice, public values, the socio-political context, demography, and disease trends. Following this, 12 people participated in a weekend workshop. Such things as an ageing population, increased cost of medicines, and technology (and its greater sophistication) were predictable, but the scenarios concerned people's attitudes toward the NHS and their willingness to pay through taxes.

Having agreed possible scenarios, the organization then needs to assess the implications of present actions and alternative future decisions based on these scenarios. Scenario planning may also provide early warning and guidance so that potential problems can be detected and avoided before they occur. Further, strategy formulation can be proactively devised by considering the present implications of possible future events. Aspects of possible or desired future scenarios can also be seen and appropriately dealt with. Of course, in a methodology context, such scenario planning can ensure a greater likelihood of the information system being appropriate in the longer term despite environmental change.

15.4 Future analysis

Future analysis (Land, 1982) is one of a series of techniques aimed at predicting potential change in the environment of the information system so that it can be designed to cope with that change when and if it occurs. It has two stages. The first attempts to discover the changes that may impact on the information system by classifying them into major categories. These include changes in technology, legal requirements, economic and other environmental factors, attitudes, expectations, tastes or in climates of opinion, and within the organization. Simulation, statistical forecasting, and economic modelling can all be used to help identify the changes. These changes are examined in terms of effect on the logic of the information system, on the 'traffic' of the information system, that is, changes in the volume or frequency, and on the timescale – short term or long term.

The second stage assesses the kind of future the information system under development will have to face, addressing issues such as what and when conceivable changes are likely to occur, the impact that they will have, and the probabilities of the changes. It might be necessary at this point to distinguish the nature of the changes – 'mandatory' or 'desirable'. The outcome of this stage is a target lifespan for the information system, an updated list of design objectives, identification of sections of the information system which may need extra flexibility built into them, and a number of future scenarios with which the design will have to cope.

A similar approach is Fitzgerald's (1990) flexibility analysis. Fitzgerald argues that enhancements to information systems tend to fall into three broad categories: environmental, technical, and organizational. Environmental changes might include government legislation or changes in other external agencies; organizational change may refer to changes to strategy or organizational structure; while technical change refers to technical developments or constraints. Although some changes did not fall easily into any one category, Fitzgerald suggests that 70 per cent of enhancements are attributed to organizational changes. More importantly in this context, this study suggests that 45 per cent of the changes analysed were known to the department concerned, and all but 5 per cent known by some member of the organization. But they were not allowed for in the initial designs. Flexibility analysis is about identifying these change factors.

Flexibility analysis needs to be carried out by a team that includes senior functional and strategic management and representatives of other groups in the business. This 'interdisciplinary' team is likely to have more success than a team of systems analysts alone. The latter will be able to assess how difficult and expensive it might be to incorporate the changes. Such an investigation will increase the development time of the project, but is worthwhile in the long term. The investigation and analysis phases are likely to be affected most, though such a process can work with prototyping, and some support software for flexibility analysis is available.

15.5 Strengths, Weaknesses, Opportunities and Threats analysis (SWOT)

SWOT is an acronym for Strengths, Weaknesses, Opportunities, and Threats and is perhaps the most well-known approach to defining strategy, having influenced both practice and research for over 30 years (Zach, 1999). It is used to identify and analyse the strengths, weaknesses, opportunities, and threats that apply to a business or organization. The unit of analysis can be the overall organization, the business unit, the department, or even the individual. It can be used for a quick, 'back of the envelope' assessment or an in-depth, highly researched analysis. It is usually used as a group technique, rather than something performed by an individual, and frequently is employed initially in high-level brainstorming sessions, although it may be subsequently iterated and refined.

Business strategy has been equated with crafting and maintaining a profitable fit between a commercial venture and its environment, and SWOT analysis is the traditional means of searching for insights into ways of realizing the desired alignment. Undertaking a

SWOT analysis involves describing and analysing a company's internal capabilities (i.e. its strengths and weaknesses) in relation to the competitive environment (i.e. the opportunities and threats it faces). Strategy is then formulated as a balancing act between the internal and external factors as it attempts to sustain the company's strengths, overcome its weaknesses, avert or mitigate the threats, and exploit the opportunities discovered in the SWOT analysis. SWOT is often used as a technique within some broader method or process for developing business strategy (e.g. Strategic Auditing, Balanced Scorecard, Strategic Analysis, Five Forces).

For example, as part of something called the Marketing Strategy Worksheet (MSW), SWOT is used at the organization level, first, to identify the internal strengths and weaknesses of the corporate set-up, for example, by looking at strengths and weaknesses of the marketing, finance, operations, and human resources functions. This is followed by an analysis of the external opportunities and threats at both a micro- and macro-level. At the micro-level this focuses on customers, competition, intermediaries, resources, etc. and at the macro-level on legal, technological, social, economic, and environmental opportunities and threats. This helps to steer the SWOT technique in particular directions. So, for example, it forces one to think about environmental threats and opportunities which otherwise might get forgotten. However, as in most descriptions of SWOT, these are only offered as guidelines. The outcomes of the SWOT are then documented in the worksheet and issues prioritized. MSW then goes on to use the SWOT analysis to define a Mission Statement, Objectives and Strategies, a Capsule (overview) Marketing Strategy, and finally a Budgeted Marketing Mix. As can be seen this is a marketing strategy method utilizing SWOT (Buttle, 1992).

SWOT is not specifically a technique related to information systems, but it can be, and has been, used to think through how IT could enhance the strengths and opportunities revealed by the SWOT and help counter the weaknesses and threats. For example, Zach (1999) advocates its use in the context of Knowledge Management. He argues that knowledge is an organization's most valuable resource, and that to remain competitive it must be effectively managed. However, many of the programmes to develop knowledge management do not explicitly link it or frame it within the organization's business strategy:

> In fact, most knowledge management initiatives are viewed primarily as information systems projects. While many managers intuitively believe that strategic advantage can come from knowing more than competitors, they are unable to explicitly articulate the link between knowledge and strategy.
>
> (Zach, 1999)

He argues for a framework that helps make the link between an organization's competitive position and its intellectual resources, and capabilities based on a SWOT analysis of their knowledge-based capabilities, comparing their knowledge to that of the competitor's and to the knowledge required to execute their own strategy and the information systems to support them.

SWOT, although a very popular technique in practice, has been criticized. Valentin (2001) suggests that the technique is made to look much too easy and is simply seen as answering a few questions from a checklist or filling in the quadrants of a one page worksheet. He argues that, actually, strategically significant SWOTs are not apparent at a glance and that a much more in-depth analysis is required. The examples provided, such as 'attractive customer base' or 'likely entry of potent new competitors', '. . . seldom reveals which factors are pivotal and which are just peripheral and they do not shed much light on the sustainability of advantages and the persistence of disadvantages'.

Hill and Westbrook (1997) have also been critical of SWOT's 'shallow' use by consultants with outcomes that are as likely to be as misleading as illuminating. Nevertheless SWOT is a well-established technique, widely used in business, and likely to remain so, partly in fact because of its simplicity but mainly because of its perceived benefits and rapid insights. In the information systems context it is typically used to help identify strategy and ensure that the business strategy is aligned with the IS strategy and IS provision.

15.6 Case-based reasoning

A **case** reveals knowledge in its natural context. It represents an experience that teaches a lesson relating to the goals of the practitioner. This lesson can be useful in understanding a new situation. Therefore, we may use this learning from previous cases to solve a new problem, adapt a solution that does not quite fit, warn of possible failures, and interpret a situation. Although it is knowledge, it is specific rather than general knowledge.

Case-based reasoning is not new, even though it is relatively recent as an explicit technique in business situations, having derived from the cognitive psychology literature of the 1980s. But the principles are not new. English law is largely case-based. A lawyer will draw on previous cases that are similar to the present case that will support the client (the other lawyer will seek similar cases that will have the opposite effect). Mediators and arbitrators will also use cases represented as precedents. When doctors diagnose a patient's illness, they look for similar cases, and usually reduce the likely possibilities through the process of eliminating other possibilities. Mechanics repairing a car will often go through a similar process.

Similarly, managers making decisions might base their decision on previous experience in other cases. The manager might see particular similarities and if the case is appropriate, it should bring light to the new situation. They might be most appropriate where situations recur. But situations are rarely the same. Nevertheless, it might be feasible to adapt the old case to the new situation. Drawing on previous experiences might help the practitioner even where the situations are incomplete or fuzzy.

A reflective practitioner will learn as a consequence of reasoning on the basis of case knowledge. But such a practitioner needs to have had a number of experiences to draw on and needs to integrate these experiences in memory. The practitioner has to have the ability to understand new situations in terms of these old experiences. Further, as no two cases are alike, the ability to think about a new situation in terms of old cases is required.

The use of case-based reasoning (CBR) can therefore help solve what might be perceived as very new and difficult problems. It may speed up problem solving by reducing areas of difficulty and help in new domains. Experience gained in previous cases can help in evaluating solutions, interpreting open-ended and ill-defined aspects of the problem. Further, by drawing on less successful cases, the practitioner may prevent repeating mistakes of the past.

On the other hand, there is a danger that practitioners try to 'fit' an inappropriate case to a new situation or blindly depend on all aspects of a related case. Of course, the approach is less relevant to analysts who are less experienced and do not have the cases to draw on for support.

In information systems, CBR stemmed from artificial intelligence as expert systems software has been developed to support such reasoning. For example, expert systems (see Section 5.3) can represent the rules that formed decisions in previous cases. They can reflect the attributes of each situation for comparison purposes. Most software systems that support CBR for various types of application – Chef, Julia, Casey, Hypo, Protus, and Clavier – are therefore based on expert systems.

Such systems may use keywords relating to previous cases in new cases to use as a basis for comparison. They may ask the user questions to confirm that the cases do have significant aspects in common. Unlike humans, they do not 'forget' the experience that has been entered into the system. On the basis of all this information they look for the best match case and reveal the solution used then and possibly use experience in other cases to adapt the solution better to the new case. Those based on expert systems ought to offer the user the reasoning that lies behind the solution offered.

Watson (1997) provides a full account of numerous applications of CBR in industry and commerce. Most are used to help provide customer support for hardware and software suppliers, retailers and bankers, and for outsourcing suppliers. It can also be used to help store corporate memory supporting knowledge management systems (Section 5.4).

CBR can also be used for programmers developing new systems or maintaining old ones, as well as systems analysts designing aspects of systems used before – even identifying components used in previous systems. For example, an analyst might use CBR to identify different decision points (goods out of stock, customer exceeds credit limit) and identify other cases where these situations occur and repeat the rule (and programmers reuse the code).

15.7 Risk analysis

Risk analysis (or risk engineering) is another approach that helps to manage uncertainty and its effects. It consists of identifying areas of possible risk, estimating and allocating probabilities to the risks, identifying possible responses (which may be pre-emptive or after the fact) and allocating costs to the risks and actions. The result is a trade-off between expected risk and expected cost for different alternatives. In principle, in-depth risk analysis ought to lead to the formulation of a risk management strategy consisting of a set of response options aimed at dealing with specific sources of risks.

Various methods for risk analysis exist, for example, SCERT (Synergistic Contingency Evaluation and Review Technique). This has been used for large engineering projects but contains principles relevant to a wide variety of applications. Indeed, it is potentially invaluable in formulating corporate strategy (Cooper and Chapman, 1987). One application in the context of information systems development is in project planning. In this context, the basic risk engin-eering notion of alternative views and representations of any given situation applies. There are a variety of associated models, and the need to select that view which is the most appropriate to the particular circumstance is important.

SCERT consists of four stages. The first, the scope phase, identifies aspects of interest, in terms of activities, associated risks, and responses to the risks. The approach uses precedence and bar chart representations and extensive, structured, verbal documentation about the activi-ties, risks, and responses. The second phase structures the risks and responses, identifying specific and general responses, and identifies decision rules. This leads to a risk–response list which can be represented in diagrammatic form. The parameter phase identifies parameters with which the outcomes are to be judged, and scenarios and their probability. Such parameters will include money, safety, and timescale. The final manipulation and interpretation phase esti-mates the probabilities along with the associated risks within an activity and attempts to strike a good balance between risks and costs. Usually an allowance is made in the budget for contin-gencies.

Throughout the whole process there is feedback, and this continues until the problem description, structuring, probabilities, decision rules, and their schedule implications have been agreed. More formally, the structure and parameter phases are first performed with 'primary' risks and then again for 'secondary' risks (i.e. those that are due to the responses of the primary risks). There may also be a case for looking at tertiary risks and responses as well. Generally available software supporting simulation and PERT may also be useful.

Risk analysis need not be solely concerned with identifying risk. It can also be concerned with identifying opportunities. The costs allocated to each risk need not be represented in terms of money alone and could be represented in terms of time, social, reliability, and safety metrics. The general outline of risk analysis as stated here may not be completely applicable for pro-ducing information systems or smaller systems generally, but the 'methodology' may be tailored to match the problem area. Ideally such an analysis would be undertaken at a very early stage in a project.

There are potential problems with risk analysis. For example, it will be difficult to identify all the activities and risks, and estimate (accurately) the probabilities of risks. However, there is no limit to the amount of time that could be spent attempting to analyse risk and plan reactions to it. Indeed, complexity and uncertainty may be so great that any analysis of risk must be greatly simplified. Analysis consumes resources, and this may lead to choosing the option that identifies 'general responses' to several problems rather than identify in detail every source of risk. This reduces effort in dealing with uncertainty, and general responses are a natural first line of defence in coping with 'unforeseeable' threats or opportunities.

An important result of more detailed risk analysis is that decision makers can gain an understanding of the trade-off between expected risks and costs of different alternatives, giving a firm basis on which to make and compare decisions. Risk analysis is likely to be more useful at the start of information systems development, though some of the principles can be carried through to further stages.

Summary

- Lateral thinking consists of a number of techniques, from lateral thinking 'games' to practical problem restructuring techniques, including generating alternatives, challenging assumptions, fractionation, and brainstorming.

- Critical success factors are those factors – skills, tasks, or behaviours – that can be considered critical to the continued success of an organization. For a project, they will represent those elements that are crucial to its success.

- Scenario planning is an internally consistent view of what the future might turn out to be. Plans can be made on the basis of these scenarios.

- Future analysis is a technique aimed at predicting potential change in the environment of the information system so that it can be designed to cope with that change when and if it occurs.

- Strengths, Weaknesses, Opportunities, and Threats (SWOT) analysis is used to identify and analyse four crucial factors that apply to an organization. It is usually carried out by a group and used to develop a strategy with greater knowledge of the organization and its environment.

- A case reveals knowledge in its natural context. It represents an experience that teaches a lesson relating to the the goals of the practitioner. Case-based reasoning is about using these lessons in understanding a new situation. So, we may use this learning from previous cases to solve a new problem, adapt a solution that does not quite fit, warn of possible failures, and interpret a situation.

- Risk analysis (or risk engineering) consists of identifying areas of possible risk, estimating and allocating probabilities to the risks, identifying possible responses, and allocating costs to the risks and actions. The result is a trade-off between expected risk and expected cost for different alternatives, which might lead to a risk management strategy consisting of a set of response options aimed at dealing with specific sources of risks.

Questions

1. Undertake a CSF analysis of Amazon.com.
2. Use SWOT to look at a university department with which you are familiar.
3. What might be the future factors to be considered in the development of a new banking application?
4. What 'basket of techniques' might be used to estimate the risk of a student not passing the end-of-year examinations? How might this inform a strategy to make success most likely?

Further reading

De Bono, E. (1992) *Serious Creativity: Using the Power of Lateral Thinking to Create New Ideas*, Advanced Practical Thinking, Harper Collins, New York.

Fitzgerald, G. (1990) Achieving flexible information systems: The case for improved analysis, *Journal of Information Technology*, Vol. 5, No. 1, 5–11.

Moynihan, T. (2002) *Coping with IS/IT Risk Management: The Recipes of Experienced Project Managers*, Springer-Verlag, London, Practitioner Series.

Porter, M.E. (1991) Toward a dynamic theory of strategy, *Strategic Management Journal*, Vol. 12, 95–117.

Ringland, G. (1998) *Scenario Planning: Managing for the Future*, John Wiley & Sons, Chichester, UK.

Watson, I.D. (1997) *Applying Case-based Reasoning: Techniques for Enterprise Systems*, Morgan Kaufmann, San Francisco.

16 People techniques

16.1 Stakeholder analysis

Stakeholders are those people or groups of people with a stake in an information system. Traditionally in information systems, this has meant users of various kinds. But views have broadened more recently. In this section, we first look at users and then develop the concept.

In Section 1.9 we discussed the various types of user. Very often in practice, however, these all get subsumed into the one term 'user', though it is common to distinguish between user and end-user, where the end-user is a person who actually interacts with the system rather than necessarily utilizing the output or outcome of the interaction.

Each type of user may be a single individual or a group of people, or indeed diverse groups of people, internal and/or external, to the organization owning or responsible for the information system. An individual or group may undertake more than one of the roles identified in Section 1.9. In a small organization one might possibly find that a single individual might play all roles.

Therefore, there are different types of user in any system, and they need to be recognized and addressed, particularly in the design and development of a system, as they are likely to have fundamentally different requirements. Of course, there are other important ways of distinguishing users – other than by the function or role they perform in relation to the system. One distinction that is frequently made is that of skill and knowledge level. This can be an influential factor particularly in terms of the design of a system. End-users may be very knowledgeable in terms of their understanding of the application itself; that is, they are knowledgeable about banking and the current account application, or they may be knowledgeable about IT and software, or both. Equally, end-users may have little knowledge of either of these domains. The frequency with which an end-user uses a system is also important. Someone who uses the system on a regular basis, with a lot of knowledge of the system and the application, needs a very different interface to someone who uses the system irregularly and perhaps does not have as much knowledge. The occasional user will need more prompts, explanations, and guiding; for the experienced regular user this will just drive them mad. Regular professional travel agent users enter a series of short obscure codes to access and navigate their airline booking systems whereas the Internet interface for occasional users (i.e. the general public) is very different, with detailed guidance and the ability to enter destinations, dates, etc. in full or by selecting from pull-down menus, in a very laborious fashion compared to travel agents.

The term 'stakeholders' is sometimes used as a surrogate for users, but it was introduced into the information systems literature to represent a broader set of people who have involvement, influence in, or are affected by the development, use, implementation, and impact of information systems (see also Section 5.1). The term user was felt to be too narrow and limited to represent the wide range of people who are increasingly involved with information systems and their impact, especially with the ever more distributed and inter-organizational nature of today's information systems. It was not just users who need to be involved, consulted, considered, and to express their opinions, but also those outside that category, such as government, society, environment, shareholder, employee, customer, supplier, patient, politician, lawyer, regulator, citizen, subject, etc. So, in IS the term 'stakeholders' was adopted to represent this broader, more diverse group.

The stakeholder concept has been used extensively and in a variety of contexts outside information systems, particularly in the strategic management area in relation to the corporation or the firm. Various stakeholder theories have been defined. Donaldson and Preston (1995) have proposed a stakeholder theory framework for 'examining the connections, if any, between the practice of stakeholder management and the achievement of various corporate performance goals'.

A stakeholder might be defined as anyone who has a stake in, or claim on, or can affect or is affected by the organization. However, Smith and Hasnas (1999) argue that it is currently understood in a narrower sense as referring only to those groups that are either 'vital to the survival and success of the corporation or whose interests are vitally affected by the corporation'. In this context they identify seven primary stakeholders:

- shareholders and investors;
- employees;
- customers;
- suppliers;
- trade associations;
- environmental groups;
- public stakeholder group, that is, the government and communities that provide infrastructures and markets (regulation and taxation agencies).

In the context of information systems, the analysis of stakeholders usually involves a rather unstructured identification of a set of relatively narrow stakeholders. It is often done in a kind of brainstorming session (Section 15.1) and then documented as a list or a set of interconnecting circles, sometimes known as a **stakeholder map**. These stakeholders are then considered as having some relevance or potential input to a system under development who then might be consulted and involved. Usually each stakeholder group is considered as having some specific requirement that needs to be considered and addressed in the system. They are seen as groups who have diverse requirements that need to be addressed by the system for it to be successful.

Kambil and van Heck (1998), for example, have used stakeholder analysis in the analysis of successes and failures in the introduction of a new IT-based trading mechanism in the Dutch flower markets. The new system and its effects on the market had the potential to lead to differential levels and allocations of costs across stakeholders. These costs and their distribution had to be considered by designers of new IT-enabled market mechanisms for it to be successful. However, the concept of stakeholder is quite limited. It appears to consist of buyers, sellers, and intermediaries, and there is no description of how the stakeholders were identified. Similarly, in relation to minimizing the risks concerned with the introduction of new IT systems, McManus (2001) suggests that: 'at the beginning of a project the manager should try to hold a risk workshop with the key stakeholder groups', but again no detail is provided as to how to do this nor how to identify the stakeholders, or any notion of a stakeholder theory.

Sometimes the stakeholders might be considered as having rights, which the system should not violate (e.g. privacy). Even more unusually, stakeholders might be identified in an attempt to minimize some detrimental effect of the system on the stakeholders. Usually the consideration of stakeholders in IS is undertaken because it is felt that it is a 'good thing' to involve a range of stakeholders. This is seen as more likely to lead to a successful information system. There is a managerial imperative underlying stakeholder analysis which sees it as 'good for business'.

However, much of the literature is not like this, but is normative and advocates that firms should address and reflect the interests of all their stakeholders, not just that of shareholders and investors, on moral or philosophical grounds. Stakeholder theory is frequently discussed in relation to business ethics and corporate social responsibility.

Introna and Pouloudi (1999) adopt this view in connection with a discussion of privacy and the possible effects of IT. They identify the following fundamental aspects of stakeholder theory (generalized):

- Stakeholders are persons or groups with legitimate interests in procedural and/or substantive aspects of the domain of concern. Furthermore the interests of all stakeholders in the domain are of intrinsic value.
- The ability, or influence, of the different stakeholders is unequal and the weaker should not be subsumed to the stronger.

They also argue that Stakeholder Analysis provides a way to make explicit, or give a voice to, the claims of all those stakeholders involved. Smith and Hasnas (1999) think that Stakeholder Analysis of this kind is problematic when applied to IS. It is not an easily applied 'cookbook' but requires 'substantial and non-trivial evaluation'. He identifies four problems. First, it is often unclear who constitutes the stakeholders for an IS initiative. The primary stakeholders are probably relevant, but it is difficult to determine all the others, particularly as interorganizational systems are becoming so common. 'Managers confront many new challenges in identifying all the stakeholders in such complicated webs.' Second, it is difficult to establish not only legal but also the moral rights of stakeholders in an IS context. Third, Stakeholder Theory requires managers to consider stakeholders' interests, which are often ill defined and subjective. For example, should customers be informed of potential data uses and should they be given an opportunity to 'opt out'? Fourth, the balancing of stakeholder interests is very difficult, for 'managers must contemplate a myriad of possibilities, calculate a number of cost–benefit analyses (with attention to each stakeholder's gain or loss), and balance these in some rational fashion'. This particular challenge is regarded as a significant weakness of Stakeholder Theory.

16.2 Joint application development (JAD)

JAD (Joint Application Development, or sometimes Joint Application Design) is a facilitated meeting designed to overcome the problems of traditional requirements gathering (see Sections 3.1, 5.6 and 7.3), in particular interviewing users. It overcomes the long time that the cycle of interviews typically takes by getting relevant stakeholders together in a meeting of a defined length, usually away from the office, which is highly focused on outcomes and making decisions.

The underlying concepts of JAD as a facilitated meeting process are not new (Andrews, 1991). According to Wood and Silver (1989) they were first developed by IBM in 1977 to help elicit requirements. However, it was not picked up by many other organizations until around the mid-1980s when it became popular. Carmel and Nunamaker (1992) estimated that there had been over 10,000 JAD-type meetings to help design and define information systems. Exactly how one knows how many such meetings there have been is difficult to imagine; nevertheless, the point is that JAD has become increasingly popular as a technique in IS development used to make development decisions and define requirements. Although used in a number of different approaches and methodologies JAD workshops are particularly associated with RAD (Rapid Application Development) (see Section 23.1).

The typical characteristics of a JAD workshop are as follows:

- *An intensive meeting of business users (managers and end-users) and information systems people*. There should be specific objectives and a structured agenda, including rules of behaviour and protocols. The information systems people are usually there to provide assistance on technical matters, for example, implications, possibilities, and constraints, rather than decision making in terms of requirements. Non-participating observers may also be present. The number of people involved in the workshops varies and this will depend on the type of system, its complexity, and its reach. Fifteen participants has been suggested as the ideal number. One of the most important people is the executive owner or executive sponsor of the system.
- *A defined length of meeting*. This is typically one or two days, but can be up to five. The location is usually away from the home base of the users, but most importantly away from interruptions. Telephones and e-mail to the outside world are usually banned. The participants are expected to attend full-time and cannot drop in and out of the meeting.
- *A structured meeting room*. The layout of the room is regarded as important in helping to achieve the meeting objectives. The round table principle is usually employed, and the walls of the room are typically covered in whiteboards and pinboards. When tools and toolsets are employed (see Section 18.1), these are usually placed at the side with the ability to display output on large screens and print when necessary.
- *A facilitator*. This is a person who leads and manages the meeting. He or she is inde-

pendent of the participants and specializes in facilitation. This person may be internal to the organization or brought in from outside and will understand the psychology of group dynamics and the tasks that the participants are undertaking. A facilitator is responsible for the process and outcomes in terms of documentation and deliverables. He or she will control the objectives, agenda, process, and discussion, using a variety of techniques to help move the meeting forward and achieve the objectives. Techniques such as brainstorming, reflection exercises, and cooling breaks will be used.

- *A scribe.* This is a person (or persons) responsible for documenting the discussions and outcomes of the meeting (including the use of tools and toolsets).

From these characteristics it can be seen that there are a number of principles underlying JAD. First, the user design should be moved forward as quickly as possible. There may be a series of JAD meetings which either address different parts of the design area or more commonly take the design from overview to more detailed levels. Often further work is carried out between the meetings, such as the preparation of more sophisticated prototypes, but decisions are only taken at the meetings. The proponents of JAD argue that it replaces cycles of interviews and meetings on an individual basis that normally take many months. This can significantly reduce the elapsed time required to achieve the design goals. In the traditional approach, meetings usually consist of a small group or are held on a one-to-one basis. When analysts find a conflict or discrepancy between users as to requirements or interpretations, they have to reschedule all these meetings again to try to resolve things. It may be necessary to cycle round the groups more than once. Typically, this takes a great deal of time, because setting up meetings is notoriously difficult in most organizations. JAD seeks to overcome these kinds of problem with one or two major workshops.

The second key element is getting the right people together for the workshop. The right people are all those with stakes in the proposed system, including end-users, and those with the authority to make binding decisions in the area. This avoids all the time-consuming cycles that are encountered with traditional methods. Rapid applications development (RAD), which advocates JAD, argues for the use of small, empowered teams giving the participants the power to make decisions that may commit other colleagues and parts of the business. This empowerment is important to successful RAD outcomes.

A third element is the commitment that the JAD meeting engenders. With traditional meetings, commitment is often dissipated over time, and decisions may be taken off the cuff in small meetings where all information is not available and implications are not fully understood. With JAD, it is all out in the open and high profile. Decisions tend not to be taken lightly, but when they are made, they are made with conviction and commitment. In particular, because JAD focuses upon the benefits of the system for the business and users, the commitment is more marked and visible.

Perhaps the most important single aspect of JAD is the facilitator. This person can make or break a workshop and is critical to determining whether the objectives are achieved. Apart from skills in handling JAD workshops, along with an understanding of group dynamics, it is the independence or neutrality of the facilitator which is crucial. This enables facilitators to achieve more than any other stakeholder who might be regarded with suspicion by others. A facilitator is able to avoid, and smooth, many of the hierarchical and political issues that frequently cause problems and will be free from the taint of organizational history and past battles.

There have been a number of examples of JAD in action with seemingly beneficial results. For example, a 60 project study by Jones (1986) showed that projects that did not use JAD missed up to 35 per cent of required functionality which resulted in 50 per cent more code being needed in the system. Whereas when JAD was used only 10 per cent of functionality was missed with minimal impact on the code. However, Davidson (1999) in a study of JAD concluded that although the participants believed that its use resulted in favourable outcomes there was little evidence of substantive change in the ISD process (i.e. the ISD status quo was not radically changed). However, these examples were not, it seems, used in the context of the kind of cultural change required for successful RAD. There was also some evidence that JAD was sometimes used prematurely in large projects.

Summary

- Stakeholders are those people or groups of people with a stake in the project. They include users of various kinds: government, society, shareholders, employees, customers, suppliers, patients, politicians, lawyers, regulators, and citizens.
- Joint Application Development (JAD) is a facilitated meeting or workshop designed to overcome the problems of traditional requirements gathering to agree a design for the information system that fully takes into account the views of users and other stakeholders.

Questions

1. Identify the stakeholders in a student administration system in a university.
2. Do you think that the views of all stakeholders should be taken into account in designing a system?
3. How can the views of such stakeholders as 'society' be ascertained?
4. In what ways does a JAD meeting overcome the problems of traditional requirements gathering approaches?
5. Why is a facilitator usually considered a necessary element of a JAD meeting?
6. What elements in the development of an information systems can JAD meetings be used for?

Further reading

Vidgen, R. (1997) Stakeholders, soft systems and technology: Separation and mediation in the analysis of information system requirements, *Information Systems Journal*, Vol. 7, No. 1.

Wood, J. and Silver, D. (1995) *Joint Application Development*, John Wiley & Sons, New York.

17 Techniques in context

17.1 Introduction

This book focuses on three aspects of information systems development: methodologies, techniques and tools. Having looked at various techniques in Part IV, we now devote this chapter to commenting on the use of techniques in practice, and the way that the adoption of techniques is not necessarily neutral. Similarly at the end of Part V, we discuss issues associated with the adoption of tools and toolsets, and in Part VII, we discuss issues concerned with the use of methodologies.

Although, as will be seen in Part VII, there is debate about the possible negative aspects of using methodologies and tools (along with that discussing their potential benefits), techniques, on the other hand, are seen largely as benign, very often as simple aids to help carry out a task and are used and recommended in many methodologies. They might be seen as supporting the collection, collation, analysis, representation or communication of information about system requirements and attributes (or a combination of these). However, following Adams and Avison (2003) we argue in this chapter that techniques can also have negative aspects and there are as many dangers in their use as there are in using methodologies and tools. In particular, techniques may restrict understanding by **framing** the ways of thinking about the problem situation. In other words, people's understanding of a problem can be profoundly influenced by the technique or techniques that they have available. Different development techniques can represent the same problem situation differently, and the way in which it is represented has considerable potential for influencing problem understanding and resultant decision-making. The common quote that suggests that 'a man with only a hammer in his toolkit sees every problem as a nail' is a good example of framing.

In this chapter, we develop a classification that attempts to separate the objective and subjective characteristics of techniques. The classification may be used to provide guidance in selecting appropriate combinations of techniques to limit any negative framing influences and support an appropriate problem-learning environment. We propose this as only the beginning of a full classification of techniques (a larger study is provided in Adams and Avison, 2003). We hope that this will lead to the adoption of techniques based on appropriateness to the problem situation rather than because it just happens to be part of a methodology being used or the habits of the individual or organization.

The structure of the chapter is as follows. Firstly, there is a discussion on techniques, what they offer developers and an examination of their main characteristics. There follows an examination of how particular aspects of techniques may influence problem cognition. These are then used to inform and develop a two-dimensional classification of techniques used in IS, based on final presentation attributes and the underlying technique paradigm. Finally, some implications of these framing influences are discussed.

17.2 Techniques – potential benefits of their use and characteristics

Given that development techniques play such an influential role in how an information system is developed, it would be useful to consider what is gained through using development techniques. The use of a technique may offer one or more of the following advantages:

- Reduces the solution of a problem to a manageable set of tasks
- Provides guidance on addressing the problem situation
- Adds structure and order to tasks
- Provides focus and direction to tasks
- Provides cognitive tools to address, describe and represent the problem situation
- Provides the basis for further analysis or work
- Provides a communication medium between interested parties
- Provides an output of the problem-solving activity
- Provides support to the practitioner.

To give some specifics, techniques may be used to:

- Help understand the problem situation (rich pictures – Section 10.1)
- Identify human activity systems (root definitions – Section 10.2)
- Analyse data aspects (normalization – Section 11.2)
- Show data relationships (entity-relationship models – Section 11.1)
- Show how data changes over time (entity life histories – Section 12.9)
- Analyse processes (data flow diagrams – Section 12.1)
- Understand the decision-making process (decision trees – Section 12.3)
- Show how things interact (matrices – Section 12.7)
- Analyse hierarchical structures (structure diagrams – Section 12.5)
- Analyse projects and their resource requirements (PERT – Section 14.2)
- Analyse possible outcomes (SWOT – Section 15.5)
- See how previous experiences can inform new ones (case-based reasoning – Section 15.6)
- Understand the roles of people (stakeholder analysis – Section 16.1).

We might classify techniques according to their major characteristics:

- Visual attributes, for example, visual representation and structure of technique output
- Linguistic attributes, for example, terminology and language used – not just English language, but also others such as mathematical and diagrammatical

- Genealogy attributes, for example, history of a technique or related technique
- Process/procedure attributes, for example, description and order of tasks
- People attributes, for example, roles of people involved in tasks
- Goal attributes, for example, aims and focus of technique
- Paradigm attributes, for example, discourse, taken-for-granted elements, cultural elements
- Biases, for example, particular emphasis, items to consider, items not considered
- Other attributes, for example, techniques or application-specific attributes.

However, the selection of appropriate techniques is not simple or straightforward. The same techniques may be advocated in several domains. Thus, PERT charts (Section 14.2) can be useful in controlling any project, not just information systems projects. Techniques play an influential role in how an information system is developed and, to some extent, the selection and use of techniques distinguish one development methodology or approach from another. For example, process-oriented methodologies emphasize data flow diagrams (Section 12.1), data-oriented ones emphasize entity-relationship diagramming (Section 11.1) and many people-oriented approaches emphasize stakeholder analysis (Section 16.1). These are very different techniques, aimed at emphasizing particular aspects of the problem situation, which methodology authors have deemed important. New technologies and applications might give rise to new techniques and new tools to support development. Thus, computer-produced mind-maps (Section 15.1), rich pictures (Section 10.1) and charts might be used as aids to creativity. As we have argued, techniques may be similar: structured English and pseudo code, for example, are similar ways of representing process logic though the latter is nearer machine language and the former nearer natural language. Indeed, a range of techniques might be used for a particular stage of information systems development. Some techniques are called different names depending on the methodology. Thus, entity life histories or entity life cycles are different names for the same basic technique. To make the situation even more confusing, different methodologies use different representational symbols for the same technique. There are several variants of data flow diagrams, for example. A process in a data flow diagram, for example, is sometimes represented as a circle and at other times by a rectangle. These confusions may cause minor problems when using techniques, but this chapter addresses much more fundamental dangers.

17.3 Techniques impact on problem understanding: Potential blocks to problem cognition

To help understand the way that techniques can impact a problem and influence its solution, some problem cognition issues are discussed. James Adams (1987) in his book *Conceptual Blockbusting* identifies a number of conceptual blocks as follows:

- *Perceptual blocks*

   Seeing what you expect to see – stereotyping

Difficulty in isolating the problem

Tendency to delimit the problem area too closely (that is, imposing too many constraints on the problem)

Inability to see the problem from various viewpoints

Saturation (for example, disregarding seemingly unimportant or less 'visible' aspects)

Failure to utilize all sensory inputs

- *Emotional blocks*

Fear of taking risks

No appetite for chaos

Judging rather than generating ideas

Inability to incubate (ideas)

Lack of challenge and excessive zeal

Lack of imagination

- *Cultural and environmental blocks*

Cultural blocks could include:

Taboos

Fantasy and reflection seen as a waste of time

Reasons, logic, numbers, utility, practicality seen as *good*

Feelings, intuition, qualitative judgments seen as *bad*

Tradition being preferable to change

Environmental blocks could include:

Lack of cooperation and trust among colleagues

Autocratic boss

Distractions.

These 'blocks' indicate that techniques could have a variety of adverse influences on problem cognition. The influences could derive from each of the areas identified (for example, 'blinkered' perception from a particular perspective, a failure to provide emotional support as a transitional object, and suggestions of a flawed approach and logic.

17.4 Techniques impact on problem understanding: Visual and linguistic influences on problem cognition

The Gestalt psychologists revealed a number of relevant concerns relating to the way we use techniques. As Mayer (1996) argues: '*In Gestalt theory, problem representation rests at the heart of problem solving – the way you look at the problem can affect the way you solve the problem. ... The Gestalt approach to problem solving has fostered numerous attempts to improve creative problem solving by helping people represent problems in useful ways.*' The key element here is that the way in which a problem is represented will affect the understanding of the problem. Relating this to techniques, we can deduce that the visual, linguistic and other representation imposed by a technique will impact problem cognition. The Gestalt movement has led to various strands of techniques, such as lateral thinking (Section 15.1) and other creative techniques. The Gestalt psychologists indicate a potential strong influence on problem understanding, that of functional fixedness: *'prior experience can have negative effects in certain new problem-solving situations … the reproductive application of past habits inhibits problem solving'* (Mayer, 1996). The implication is that habits 'learnt' using previous techniques and problems would bias the application of new techniques and problems. This could explain the dominance of certain techniques used in IS development, such as data flow diagrams and entity-relationship or object models. It may also explain why many 'new' techniques are often rehashes of older ones. It might also explain why relatively new techniques take so long to get established.

Another aspect relates to framing situations in either a negative or a positive perspective. Such bias is likely to influence problem cognition, particularly on estimating the likelihood of events. Milburn (1978) identified a *time influence* on viewing problem situations, particularly with positive and negative events: negative events are seen more likely in the short term while positive events are seen as increasingly likely over time. This, Milburn argues, *'might help explain why often so little long-range planning is done: If one feels that things are bound to get better later on, no urgent pressure is felt to plan for problems which might occur later.'* The implications are that the way in which a technique frames a problem situation, in both a positive/negative way and in a time perspective, will influence problem cognition, particularly in regard to the perception of the likelihood of events.

Another major area in which a technique can influence cognition can be deduced from support theory (Tversky and Koehler, 1994), which indicates that support for an option will increase the more that the option is broken down into smaller component parts with each part being considered separately. The more specific the description of an event, the more likely the event will seem. The implications are that the more a technique breaks down a situation into component parts or alternatives, the more likely the situation will seem. In addition, a technique's underlying paradigm is likely to dictate whether or not a problem situation is broken down into increasingly smaller component parts or if a situation is represented in a negative or positive light. This would seem to explain the predominance of techniques used in IS that use functional decomposition as the basis of analysis (for example, data flow diagrams and other techniques of structured approaches). Alternative techniques might look into problem situations holistically, such as rich pictures (Checkland and Scholes, 1999; Avison and Wood-Harper, 1990). These are less well used in the IS domain.

Another aspect relates to structure influences. Prescriptive structures are also likely to exert influence on problem cognition. For instance, hierarchy and tree structures are likely to do this by binding attributes together (for example, on the same part of a tree structure) and limiting items to the confines of the imposed structure. The implication is that techniques dictating hierarchical structures will force a (self-perpetuating) category inclusion bias. An element in one branch of a hierarchical structure will automatically have different properties to an element in another branch of the hierarchical structure. For instance, take a functional breakdown of an organization; we might conclude from category inclusion that a task in an accounting department will always be different to a task in a personnel department, which clearly may not be the case as both departments will have some similar tasks, such as ordering stationery.

The discourse and language used to describe a problem are also likely to play a role in problem understanding. Adams (1987) discusses various different types of 'languages of thought' used in problem representing and solving. People can view problems using mathematical symbols and notation, drawings, charts, pictures and a variety of natural verbal language constructs such as analogies and scenarios. Further, people switch consciously and unconsciously between different modes of thought using the different languages of thought. The information systems development environment is awash with technical jargon and language constructs.

Perceptual processing is profoundly influenced by the sequence of information provided and the relational constructs of information. The sequence and number of items in a list will influence how people will understand (and recall) the items and how people will categorize them. The implications of this are that the language and sequence of describing a problem situation, the questions asked and how they are asked and the implied relationships (all of which are usually prescribed by a technique) will bias problem understanding, for example, by forcing 'leading questions' or 'leading processes'.

Language aspects highlight another set of possible influences, that of communication between different groups of people (for example, such as that between analysts and users). Differences of perspective between different groups of people in the development process have been discussed within the IS field under the heading of the 'softer' aspects or as the organizational or people issues (for example, Checkland, 1981). Identifying differences and inconsistencies can be classed as a useful task identifying and dealing with requirements.

There are also likely to be individual preferences, and corresponding biases for some techniques or specific tasks within techniques. Couger (1995) notes that: '*It is not surprising that technical people are predisposed towards the use of analytical techniques and behaviorally orientated people towards the intuitive techniques.*' In addition, there may be some biases between group and individual tasks. This might suggest that the make-up of IS development teams should be mixed in all sorts of ways.

Goal aspects also profoundly influence problem understanding by providing direction and focus for knowledge compilation. Goals influence the strategies people undertake to acquire information and solve problems. Further, when there is a lack of clear goals, people are likely to take support from a particular learning strategy, which will typically be prescribed by the technique. The implications are that techniques with clear task goals will impact the focus and form of information collection (for example, what information is required and where from, along with what information is not deemed relevant) and how the information is to be processed. Further, if there are no clear goals, then people are likely to rely more heavily on the learning method prescribed by the technique.

The framing influences discussed above indicate that any *framing effect* due to the characteristics of a technique is likely to be complex and interwoven. However, some main themes emerge. It is clear that the visual, structure and linguistic aspects are related and could be classified as 'representational' influences. Equally, several aspects can be classed collectively as 'paradigm/process' influences. Arguably, the more prescribed and structured a technique is, then the more likely that 'predictable' framing influences can be detected.

17.5 Applying lessons from cognitive psychology: A macro analysis of techniques

As the previous discussion shows, the literature from cognitive psychology indicates two main types of influences: representational influences (for example, prescribing certain visual and other 'language' representations) and paradigm/process influences (for example, underlying approach, prescribed processes and tasks utilizing specific language). The 'visual and language' characteristics imposed by techniques are likely to be explicit, in that specific representations will be prescribed, such as distinct diagrams, tables and other visual outputs in the final representation. The process characteristics may also be explicit (for example, list and order of tasks to be completed). The underlying paradigms for techniques are likely to be far less explicit. However, there is likely to be some relationship between paradigm and process since an underlying paradigm is likely to dictate the processes and activities to be undertaken. As such, a classification by representational influences and paradigm/process influences might provide a reasonably objective metric with which to view influences on cognition.

Representational influences

The final representation of techniques, consisting mainly of visual characteristics and some language constructs, fall into three distinct categories: a matrix/table structure, a hierarchical structure and a non-hierarchical structure. In addition, there is a fourth category which accounts for techniques that do not prescribe a particular presentation. This could be called *structure free*. The characteristics of each are detailed below:

- *Matrix/Table structure*
  lists, tables or matrix format
- *Hierarchical structure*
  functional breakdowns
  similar visual representations to a hierarchical organization chart
- *Hierarchy-free structure*
  clear relationships
  network structures
  non-hierarchical structures
- *Structure free (other, non-prescribed)*
  non-prescribed structures
  non-diagrammatical structures (for example, verbal, written structures)
  other, freer structures.

Paradigm/process influences

Examining different categories for the underlying paradigm and processes of techniques, one could group by the qualitative/quantitative (or by subjective/objective, judgmental/rational) characteristics. Objective techniques, that is, those relying on 'hard' data, use rigid scientific or mathematical 'rules'. They are aimed at situations where following these rules is appropriate. Subjective techniques are those that rely more on judgement and interpretation of complex situations. However, few decisions (if any) are made completely on quantitative or qualitative information, but techniques could be classed on some form of qualitative/quantitative scale. However, deciding which is the predominant characteristic is itself subjective: There is no clear boundary line between qualitative and quantitative techniques and sometimes one technique is placed in a different category by different authors.

Another approach to categorize the paradigm/process elements of techniques is to consider how innovative or creative they are. Techniques may be described as dominated by analytical thinking or intuitive thinking, similar to de Bono's 'vertical' and 'lateral' thinking classification (de Bono, 1977). As with the qualitative/quantitative categories discussed above, this grouping is open to considerable interpretation as to where a technique would fit.

Another approach at categorization might be exploratory techniques, which start by assessing the present situation and move towards the future, and normative techniques, which start by assessing future goals, aims and desires, then working back towards the present situation.

There are similarities with each of these classifications. They seem to have fairly 'closed' techniques at one end of a continuum and fairly 'open' techniques at the other. Of course, placing techniques on this closed/open continuum is likely to be subjective (as with the previous classifications), but attempting to do this may provide a useful perspective:

- *Closed paradigm/process*
  stays within defined scope
  closed set of rules and language
  prescriptive processes and tasks
  prescriptive representations
  mostly objective
- *Open paradigm/process*
  open, less defined scope
  more open set of rules and language
  less prescriptive processes and tasks
  less prescriptive representations
  mostly subjective.

17.6 A two-dimensional classification: Visual/language and paradigm/process influences

There are likely to be some correlations between the 'visual/language representation' and the 'paradigm/process representation'. For instance, techniques that impose strong hierarchical structures are likely to have 'closed' dispositions (for example, more formal and structured, restrictive scope and a more objective paradigm). Grouping according to visual/language and paradigm/process attributes can be represented two-dimensionally, as in Figure 17.1.

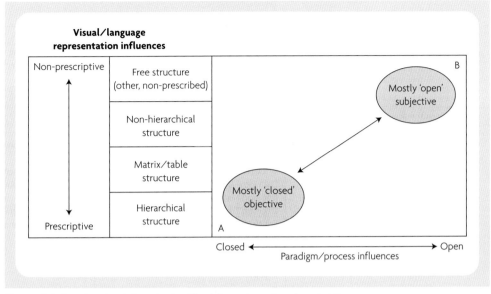

Figure 17.1: Grouping of techniques by visual/linguistics and paradigm characteristics

Looking at Figure 17.1, the closer to 'A', the more closed and 'objective' a technique will be. It is likely to have the following cognitive influences:

- Reduced, but defined, problem/solution spaces
- More 'social defence' attributes providing cognitive and emotional support, but also more 'rituals' of technique detracting attention from the actual 'problem situation'
- More functional fixedness and category inclusion biases (or similarly, item ordering and relational/proximity biases)
- More normative representation and correspondingly, more acceptance of results
- More detailed description of items, resulting in increased perception of likelihood of items
- More defined goals, resulting in more rule-induced learning activity.

In addition, there are likely to be further influences, such as individual biases towards different types of techniques (or tasks within them) and negative versus positive framing. It could also be argued that techniques closer to 'A' would provide more support for novice developers, whereas techniques closer to 'B' would be more suited to experienced developers (i.e. novices are likely to need more guidance and cognitive support in a development). However, most people use those techniques with which they are familiar, and this suggests a general bias towards 'A'.

We have attempted to allocate each technique to one of the four quadrants: non-prescriptive structure/closed paradigm, non-prescriptive structure/open paradigm, prescriptive structure/closed paradigm and prescriptive structure/open paradigm. Some techniques have aspects that relate strongly to more than one quadrant, and we have therefore assigned these

Non-prescriptive structure	Cognitive Mapping Critical Path Analysis/Method (CPA/CPM) Entity-Relationship Models Program Evaluation and Review Technique (PERT) Scenario Planning Stakeholder Analysis	Brainstorming Cognitive Mapping Lateral Thinking Techniques Rich Pictures Root Definitions Scenario Planning
Prescriptive structure	Action Diagrams CoCoMo Critical Success Factors (CSF) Data Flow Diagram Decision Matrices Decision Trees Entity Life Histories (Cycles) Future Analysis Normalization Program Evaluation and Review Technique (PERT) Requirements, Needs and Priorities (RNP) Risk Assessment (RA)/ Engineering (RE)/ Management Stakeholder Analysis Strategic Options Development and Analysis (SODA) Structured English Structured Walkthroughs Synergistic Contingency Evaluation and Review Technique (SCERT)	Case-based Reasoning Future Analysis Lateral thinking techniques Risk Assessment (RA)/ Engineering (RE)/ Management SWOT analysis (Strengths, Weaknesses, Opportunities and Threats)
	Closed paradigm	**Open paradigm**

Table 17.1: IS techniques grouped into structure and paradigm

techniques to these quadrants. Readers may disagree with aspects of our allocation. However, as Table 17.1 evidences, there is a major bias in information systems towards closed techniques and, within the closed techniques, prescriptive over non-prescriptive structures. A similar bias was found in Adams and Avison (2003) in their study of over 80 techniques.

17.7 Conclusion

There are three major supports to information systems development: methodologies, techniques and tools. Techniques are usually seen as benign, supporting 'objectively' the work of users and developers, and providing them and their work with many benefits. We have argued that the choice of techniques used in information systems development influences problem understanding, and hence influences the analysis and design of proposed solutions and their eventual implementation. Biases are likely to be perpetuated or even worsen as one proceeds through the development stages of a project.

The influences can be considered under certain visual and other 'language' representation characteristics and under paradigm/process characteristics. Works from the cognitive psychology literature indicate how these characteristics are likely to affect problem under-standing. This is known as the framing effect. Techniques may provide barriers to problem cognition rather than enlighten, and visual and linguistic influences may blinker perception in one direction.

By classifying the characteristics of techniques, we have tried to indicate how different types of technique are likely to influence problem cognition, and in doing so have tried to map the framing effect of techniques. Our classification places techniques into one of four quadrants: prescriptive/closed; prescriptive/open; non-prescriptive/closed; non-prescriptive/open.

It is a concern that in our classification (and the larger study of over 80 techniques), over three-quarters are found in the closed rather than open paradigm, suggesting that the vast majority of techniques used in systems work are limited understanding techniques. Further, the majority are also prescriptive, further limiting perception by restricting the user.

Summary

- The use of techniques can blinker or frame the way people think about a project, blocking full understanding, and prevent the best analysis, design and/or implementation.
- Techniques may be classified according to their visual, linguistic, genealogy and other attributes.
- Techniques can be prescriptive or non-prescriptive and open or closed (or somewhere along each axis) and systems developers might use a mix of techniques to provide the widest understanding of a situation.

Questions

1 For techniques not included in Table 17.1, allocate them to a quadrant in the table.

2 Discuss whether we have placed each technique in the correct quadrant of Table 17.1. For those in more than one quadrant, attempt to allocate them in one only. Why do you think we have allocated them to more than one?

3 What techniques were used in a project with which you are familiar? Was a balance struck between prescriptive and non-prescriptive techniques and open and closed techniques?

Further reading

Adams, C. and Avison, D. (2003) Dangers inherent in the use of techniques: Identifying framing influences, *Information Technology & People*, Vol. 16, No. 2.

Part V: **Tools and toolsets**

Contents

Chapter 18 – Tools   p 331

Chapter 19 – Toolsets   p 361

A s in Part IV we again show a 'road map' of the relationships between themes, techniques, tools, and methodologies (see Diagram 3). This time the perspective is that of a tool and a toolset. We use Microsoft Project and Select in this example and show the links to themes, techniques, and methodologies.

We looked at different aspects of software in Chapters 8 and 9. By tools we mean software packages that support the analysts and users in particular aspects of the application development process. Toolsets are integrated software environments. Technology has continued to develop greatly throughout the computer era, becoming much more powerful, cheaper, and more widely available. Improved graphics facilities and windowing environments have also had a major impact on tools. Indeed, some people have suggested that the latest generation of automated tools is a panacea for the problems of systems development: improving systems quality and eliminating the development and maintenance backlog. Although we do not share this belief, the quality of software has greatly improved over time so that automated tools are potentially very beneficial. In Part V we look at the whole range of tools that support the information systems development process.

We first look in Chapter 18 at a number of software tools: groupware (with GroupSystems as exemplar), website development (with Dreamweaver as exemplar), drawing tools (with Visio as exemplar), project management tools (with Microsoft Project as exemplar), and database management systems (with Access as exemplar). In all these cases we describe briefly how the tool is used and show the screenshots generated. However, in all these cases a full description of the tool would require a book (or books), so we can only provide a flavour of their potential in this chapter.

Although the main focus of Chapter 19 is on toolsets, we start with a discussion of systems repositories as these are an essential ingredient of any toolset. Toolsets are developments on what used to be called CASE tools and are whole software environments supporting the information systems development process. Some toolsets address the early stages of systems development (strategy, planning, and analysis); others address physical design, programming, and implementation stages; yet others integrate the two into a single, fully integrated development and support facility. We look at three exemplars. The first is Information Engineering Facility, a toolset supporting users of the Information Engineering methodology and one of the first of these toolsets. We then look at Oracle, at present the most widely used and one of the most powerful and inclusive toolsets. Finally we discuss Select, which is used in many universities, and we illustrate its use through screenshots.

Following this, we look in detail at the potential benefits and provide an overall evaluation of such tools and toolsets. An opportunity is taken to discuss the human, social, and organizational aspects. The likelihood of successful adoption of these tools (or, indeed, information systems in general) is as much to do with people and organizational factors as with the qualities of the software itself. We have therefore taken the opportunity in the chapter to look at a number of wider considerations in adopting software. Some of these debates are developed further in Part VII, which follows the descriptions of a number of information systems development methodologies in Part VI.

Descriptions of software packages are likely to become dated more quickly than other aspects described in this book. Therefore the descriptions of tools provided in the next two chapters show the reader the purpose, capabilities and scope, and enable him/her to get a 'look and feel' for each. Those readers who wish to use the software need, of course, to follow a dedicated text, online tutorial or similar, for each tool.

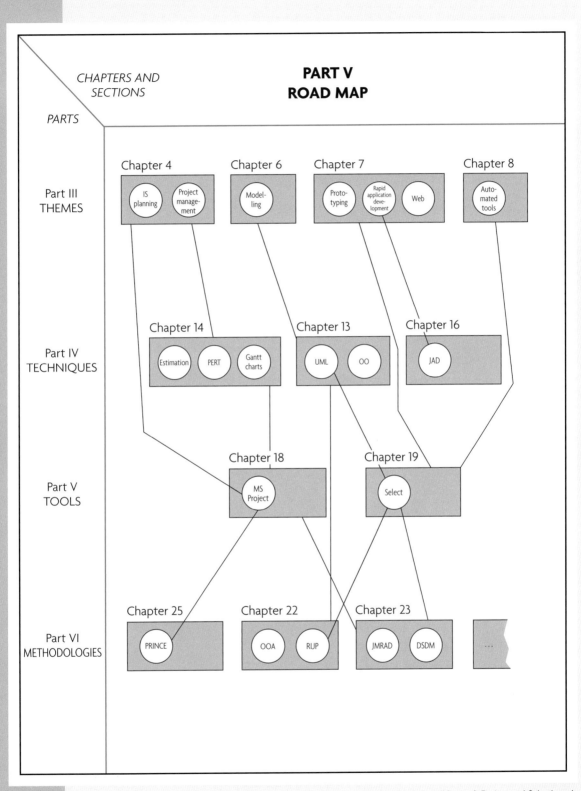

Diagram 3: MS Project tool and Select toolset 'Road Map' (shows relationships between a tool and a toolset (Microsoft Project and Select) *and* themes, tools and methodologies)

18 Tools

18.1 Groupware: GroupSystems

Much, indeed most, information systems development work is carried out by a project team, that is, a group of individuals working in collaboration. Software tools are now available to support group work. It takes many different forms. Some people might include e-mail in this category, for example, it can be used for information dissemination, though this has very limited groupware facilities. The most popular software of this type is Lotus Notes or Microsoft Exchange. There is also specialist software available, such as GroupSystems and others for videoconferencing. These software tools can help group work in many ways: for instance, sharing knowledge, discussing problems, co-ordinating effort, and collective decision making.

Groupware can be effective where the group is working together in a room or when the team is dispersed. For example, GroupSystems is particularly suitable for groups working together in a room and supports, for example, brainstorming, categorization, electronic whiteboards, and voting. It can be used with other software so that databases, spreadsheets, websites, and so on can inform the decision-making process. Video conferencing facilities offer people the opportunity to see as well as talk to others in another site or sites.

Lotus Notes is a well-used group support system. Whereas the emphasis in GroupSystems is on group brainstorming and decision support, Lotus places emphasis on communications between group members. It enables the documentation, sharing, storage, and access of information. It can be used to display the historical conversation over time between team members that led to a particular decision. 'Notes', for example, on viruses can be displayed on a bulletin board for all the team (or the company). The calendar can be used to co-ordinate schedules of physical and virtual meetings. In general, it helps to inform members of the project team about progress in real time as well as stimulate that progress.

The use of such systems may change the nature of group work from being largely face to face to largely online. Trust becomes a major factor as online work lacks the usual signals gained from face-to-face contact. Videoconferencing can have an obvious role here. Even web cameras used with e-mail and Internet conversations can also provide support, as can voice-mail. Such systems also change roles, responsibilities, interactions, and the way work is carried out. Indeed, they may be introduced to achieve this result as well as encourage and support teamwork in general. Of course, there are also privacy and security implications as well as individuals' concerns about sharing their knowledge with others and perhaps losing their individual competitive advantage. The way such systems are introduced will therefore be a key factor in its potential for goodwill and better teamwork.

Failla (1996) provides a useful case study relating to how software is developed at IBM's international network of laboratories. Teams of work groups and managers can exist 'virtually', in many different locations. Developers write code usually in small groups while managers divide the work between group members and groups. Electronic mail, forums (shared files), conference call systems, faxes, and videoconferencing are all cited as tools supporting group work. IBM's commitment to technology supporting group work is exemplified by its purchase of Lotus for its Lotus Notes software, which it now supplies as well as uses internally.

We will look at GroupSystems software, which developed from the work of Jay Nunamaker and others at the University of Arizona (Nunamaker et al., 1991), see www.groupsystems.com for the latest version. Although it can be used for remote decision making and at different times, its most popular use is in special meeting rooms, sometimes called 'pods', where each user has a workstation with the software installed. The brainstorming feature enables each member of the team to create ideas and comment on them. These are usually expressed anonymously so that it is the ideas that have force (or not), and not who is expressing those ideas. Brainstorming should encourage unusual thinking and ideas and innovation, and the anonymity of expressing these ideas reduces the inhibiting factors. Some ideas will be rejected but others will be kept and organized into separate categories for further evaluation. At the end of the meeting there is a complete record available.

Figure 18.1 shows the basic elements of the software tool. An agenda has been agreed for the morning meeting. This starts by generating ideas on the subjects for the meeting and cat-

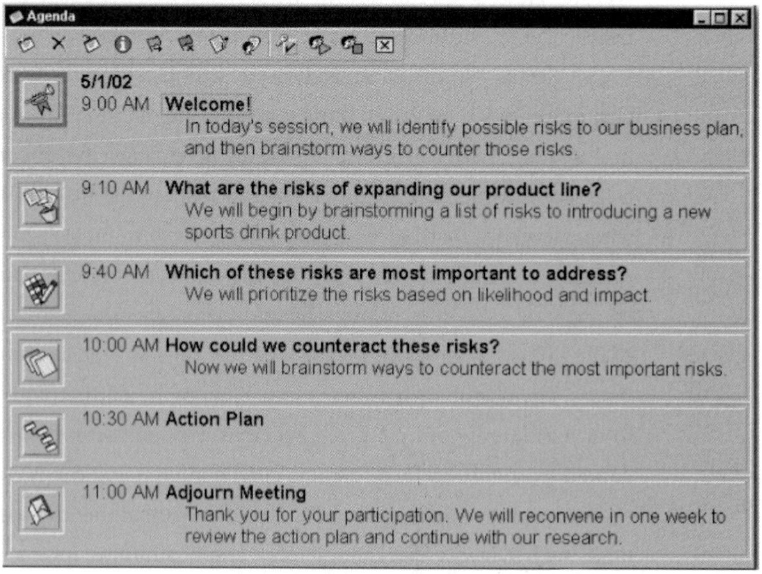

Figure 18.1: GroupSystems

egorizing them. These are explored in more detail through the expression of ideas in the brainstorming part. Figure 18.2 shows how ideas are brainstormed and organized. These may then be further commented on by other members of the group with an action plan agreed.

There may be some agreements without a vote, but there is an electronic voting system that can also be used. Figure 18.3 shows how opinion can be expressed by use of a 'barometer'

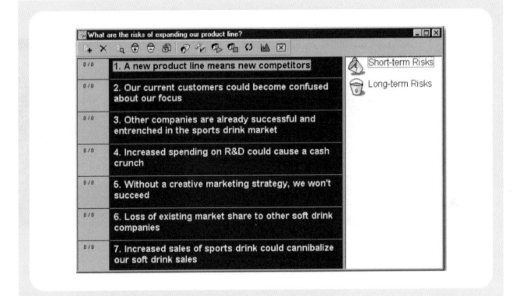

Figure 18.2: GroupSystems brainstorming

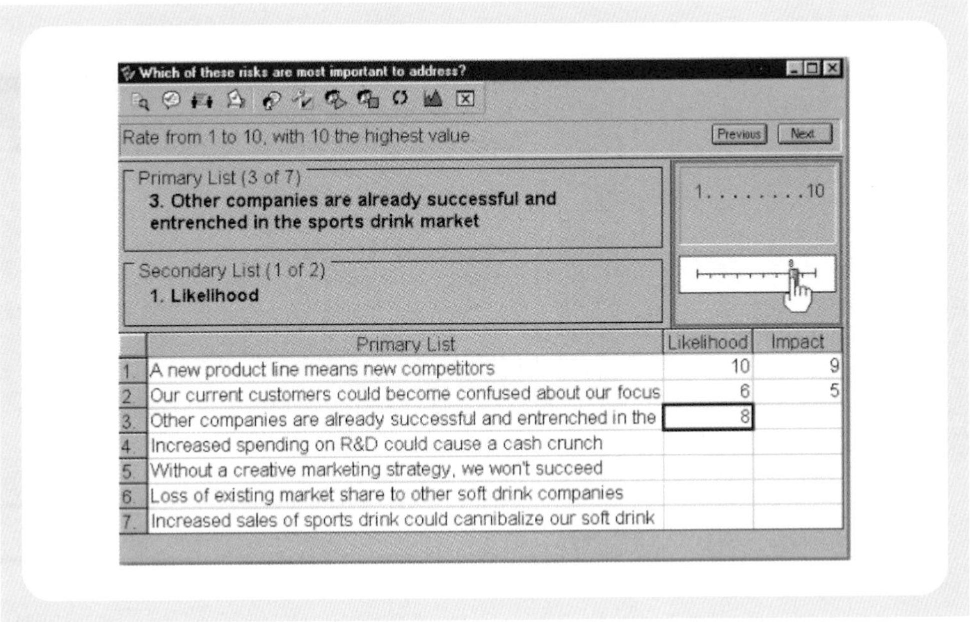

Figure 18.3: Expressing opinion

and Figure 18.4 how the opinions of everyone in the group can be aggregated. Such statistical information as standard deviation and mean can also be provided. Group members can vote or express their opinions in terms of rank order, true/false, yes/no, etc. Results can be displayed through a variety of charts and tables.

The system provides an electronic whiteboard; electronic handouts and information can

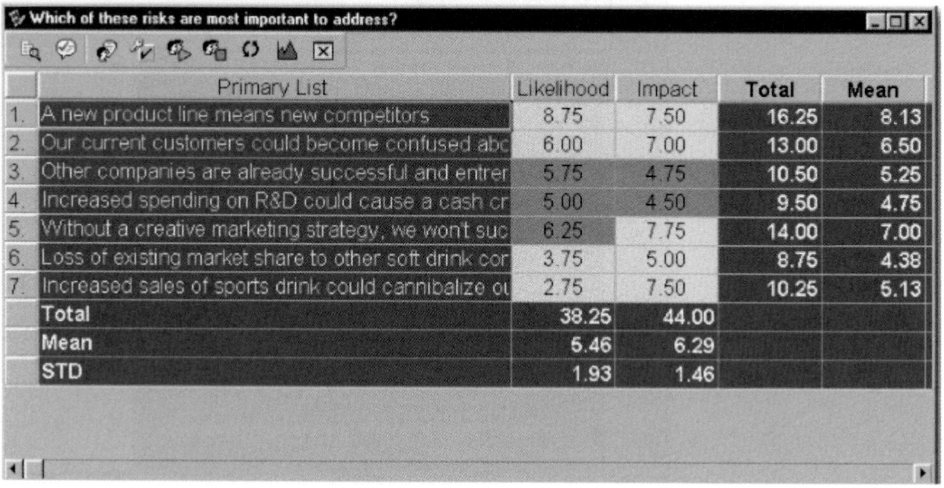

Figure 18.4: Aggregating opinions

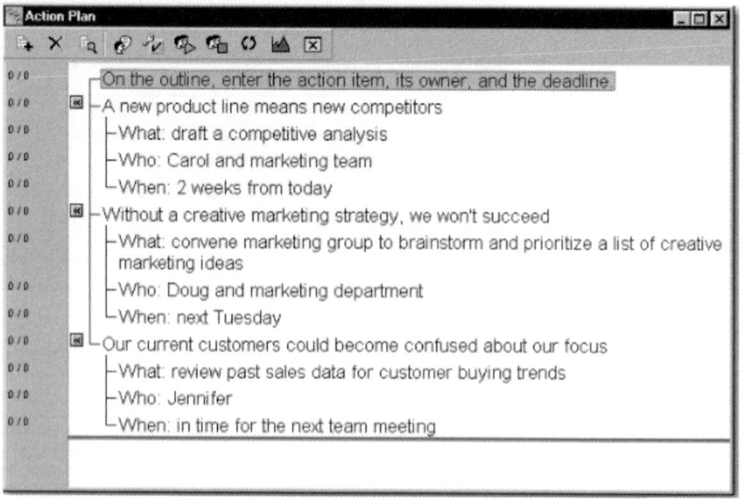

Figure 18.5: Agreeing an action plan

be obtained from other sources, such as spreadsheets, databases, and web pages. We see in Figure 18.5 how an action plan can be expressed following the group discussion.

Bikson (1996) provides a very interesting case study of GroupSystems use at the World Bank. Its use needs a facilitator who provides the technological support when needed. This may be a technical role only, or some 'project leadership' role may be included alongside the technical role, for example, to ensure that the schedule and agenda are kept. The website (Groupsystems, 2005) enables testing of the application through a group meeting using the software hosted by the supplier.

18.2 Website development: Dreamweaver

Dreamweaver (version 8 was released in 2005) is one of a number of software tools that support the production of websites. Alternatives include Microsoft's FrontPage. It is not, however, the basis for an e-commerce site. Here, a more sophisticated tool such as ColdFusion is necessary.

As Figure 18.6 shows, when opening Dreamweaver, you are confronted with a document window and several floating windows called palettes. The document window will form what will be seen in the browser window (e.g. Internet Explorer), that is, the web pages themselves.

The package comes with a tutorial, and we will use some examples from this. Figure 18.7 shows one of the web pages created alongside the HyperText Mark-up Language (HTML) code that the design reflects. HTML contains the text plus all the instructions for displaying the text (colour, bold, italic, spacing, font size, font type, pagination, etc.). These instructions reflect the mark-up code that used to be added by proofreaders for printers of newspapers and books. In this case, the mark-up code is produced by Dreamweaver itself. Users do not have to work with HTML code, but they can if that is their preference (usually computer-experienced people such as programmers might prefer to write or edit code). The designers will specify what they want in the document window and this is translated to HTML.

Many of the commands used will be familiar to users of conventional word processors and products of the Microsoft Office suite (Ctrl + C for copy; Ctrl + V for paste; Ctrl + F for find) and there is also a spellchecker provided. Emphasis is placed on enabling the analyst to create an attractive style for the pages (some example style sheets are provided in the package), and there can be styles defined for the site so that the user sees a similar 'look and feel' for the site as a whole. Style definitions can include type (e.g. font formatting), background (e.g. colour), block (e.g. alignment and indentation), box (e.g. margin controls), and lists (e.g. numbered or bulleted). Text, images, icons, and other material can be cut and pasted from other sources into the web pages.

A particularly important page is the home page. This will normally be the users' first encounter with the website. It has therefore to be attractively designed, informative, and also enable the users to go quickly to the sections of the site that are of most or immediate interest.

The web page shows the content and also the hyperlinks to other web pages. Therefore, if the user double-clicks on products or special promotions (Figure 18.7), this will bring in these related pages. *Our story, products,* and *this week* are also links, but this time in the form of icons that are double-clicked rather than words. They change shade when the user moves the cursor

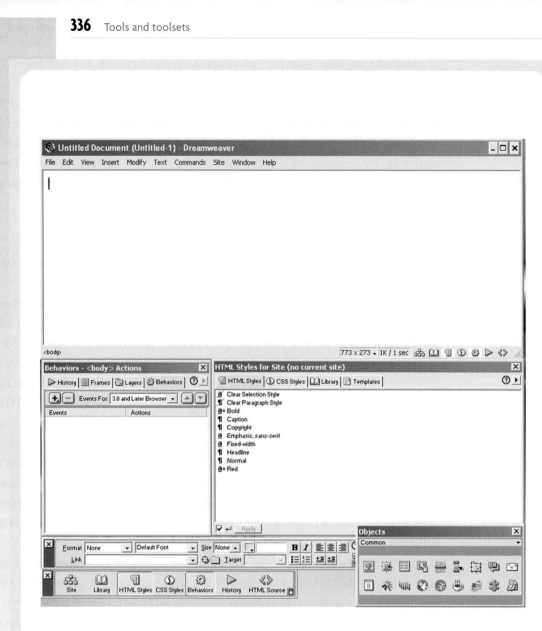

Figure 18.6: Dreamweaver document window and palettes

over them, and this signals to the user that they represent links. The objects palette seen towards the bottom of Figure 18.6 can enable the use of images, tables, layers, movies, Applets, Active X controls, as well as links of various kinds.

Of course, good design is particularly important for web applications as it could be the customer who is using the application, rather than a captive user (an employee) as in most information systems. Happily, Dreamweaver, as with similar tools, provides analysts with a number of templates that can be used as a basis for their particular web pages. The page prop-

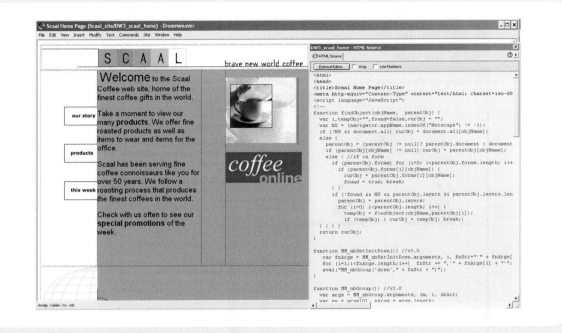

Figure 18.7: Tutorial webpage and HTML code

erties dialogue box applies to the entire page, and enables the definition of margins, and so on, while the visual properties dialogue box enables the definition of background colour, text colour, background image, etc. At first all these floating boxes can be daunting to the new user, but they separate the various requirements, and the experienced user works well in this mode.

Dreamweaver, like all modern software of this type, offers WYSIWYG (what you see is what you get), so that as analysts design the page they can preview it as it will eventually appear in a browser. Eventually the site will contain lots of pages and links between them and also links to external sites, designated universal resource locators (URLs). Figure 18.8 shows some of the linked pages for the tutorial site.

The ability to show images has, perhaps, been one of the key selling features of web pages. Dreamweaver enables images to be pasted and edited into the web page. Alternatively it can be used in conjunction with full image editors such as PaintShopPro or Photoshop. The two main image formats of the web, GIF and JPEG, among others, are supported by Dreamweaver. However, the designer has to be aware of potential downloading time problems, so images, animation, and other moving images need to be used with care. The Dreamweaver libraries can be used to store images, animation, sound files, Shockwave and Flash files, Java Applets, ActiveX controls, frames, and other components that might be used in other websites.

Another key feature of the web is the ability to link with other pages in the site and externally with only a click of the mouse. In Dreamweaver a link can be achieved by high-lighting the text or image that is to be the link (for example, products in Figure 18.7), and in the link text box (bottom left in Figure 18.6) typing the location of the document to which it is

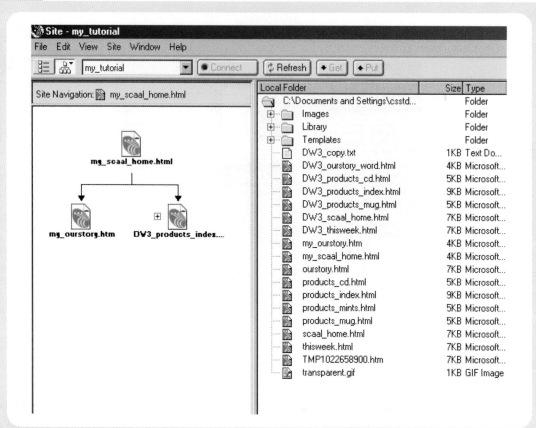

Figure 18.8: Site map

linked. Other links might be to e-mail addresses. Navigation bars (list of related links) can be created, and this is another way to help guide users through the site.

Many pages on the web consist of several frames, containing text, graphics, tables, images, navigation bars, and other distinct displays. It might be useful to display the navigation bar on the page whatever scrolling the user is doing. This is made possible because each frame is independent of the others on the same page as it is an individual HTML document. Nevertheless, from the designer's point of view, it is possible to drag and drop objects from one page to another in the process of creating the web pages.

Forms used for surveys, for example, are a common feature in websites, but they can be notoriously slow and difficult to create. However, their creation is also facilitated using Dreamweaver. Again, in the object palette (see Figure 18.6), forms can be chosen as one of the options. Forms can be created using tick boxes, radio buttons, lists, drop-down menus, jump menus, and text fields. A password field enables the user's typing to be replaced by asterisks when displayed. Another option is a multi-line feedback box to be used for the user's general unstructured comments. The Macromedia website (Dreamweaver, 2005) contains a demonstration called a 'virtual tour', which highlights the newer features of the package.

18.3 Drawing: Microsoft Visio

Drawing tools help developers to create and maintain the various diagrams required in systems development. Most are designed to support the drawing of one or more of the common diagramming techniques and do not usually support any particular methodology. However, sometimes they offer the same technique drawn under the conventions of different methodologies. For example, they may give the options of drawing data flow diagrams following the Gane and Sarson (STRADIS), Yourdon, or other conventions.

Drawing tools help draw many of the diagramming techniques described in Chapter 6, such as data flow diagrams, entity models, process logic techniques, entity life cycles, and so on. They are sometimes described as documentation support tools, being designed to take the drudgery out of drawing and revising documents, and thus making the implementation of changes easier.

Additionally, they contribute to the accuracy and consistency of diagrams. For example, in a data flow diagram, a drawing tool can check that levels of the diagram are consistent with each other. This will include ensuring that all data flows on a higher-level diagram appear on the lower-level ones and that the descriptions are consistent across a set of diagrams. Drawing tools can also be used to ensure that certain documentation standards and conventions are adhered to. For example, processes must have inputs and outputs, and data stores must be specified in terms of contents and flows. Most drawing tools will prevent users from doing things that are not permitted in that diagrammatic convention. With an online reference manual and context-sensitive 'help' facilities available in some tools, users can be guided through the technique.

Probably the greatest benefit of their use is that analysts and designers are not reluctant to change diagrams, because the change process is simple. Frequent manual redrawing is not satisfactory, because of the effort involved and the potential of introducing errors when redrawing. Without such drawing tools many a small change required by a user would not be incorporated into the documentation of the system.

Such drawing tools usually support only a single technique or a few basic techniques (most commonly, data flow diagrams and entity-relationship diagrams). Although useful, these are somewhat limited in the sense that, for example, much of the information required for a data flow diagram is also required, in a slightly different form, for other process logic representations and elsewhere. It makes more sense to have a central repository of all the information required for the development project. This will ensure consistency between techniques as well as within a diagramming technique. This provides the potential for integrated, co-ordinated, and consistent support throughout all aspects of the life cycle, ensuring a smooth transition from one phase to the next. With these ambitions the simple drawing tools began to evolve into more sophisticated products originally known as CASE tools but more commonly now seen as application support toolsets (see Chapter 19).

Visio is a well-used drawing tool for PCs and workstations. As can be seen from Figure 18.9, it has the same 'look and feel' as other Microsoft products. There are a number of templates

Figure 18.9: Visio 'look and feel'

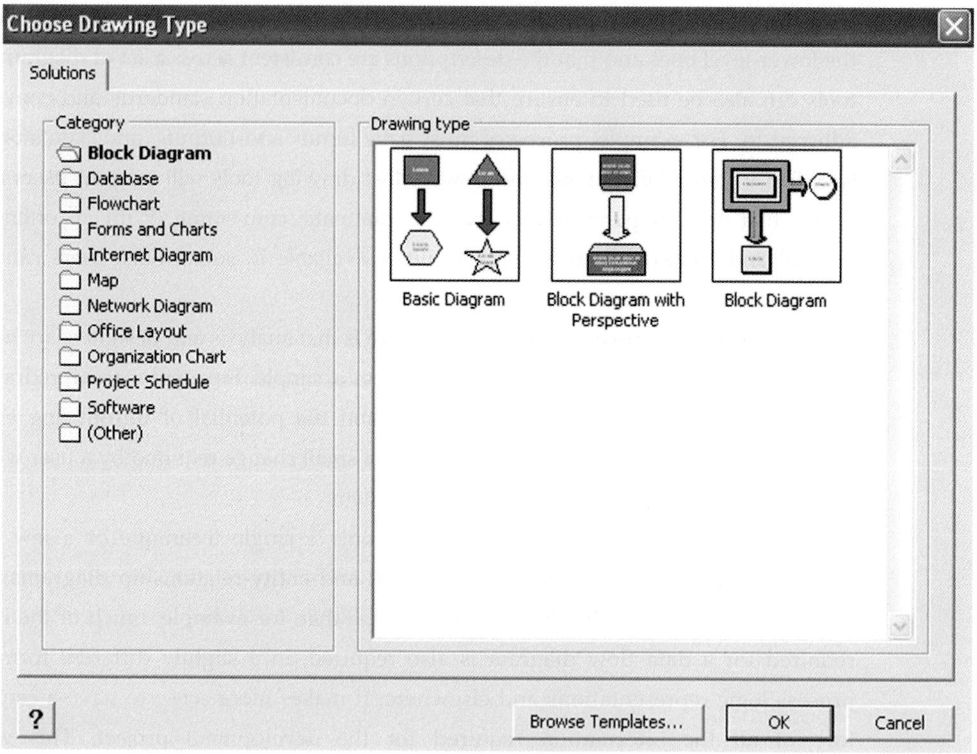

Figure 18.10: Drawing types

available, depending on the type of diagram to be drawn. Figure 18.10 shows some database drawing types including diagrams based on Chen's entity-relationship models (ERMs) or James Martin's approach; Figure 18.11 shows some of the flowcharts which include mind mapping diagrams (see lateral thinking, Section 15.1) and total quality management; Figure 18.12 shows website diagrams; and Figure 18.13 shows data modelling diagrams, including object diagrams,

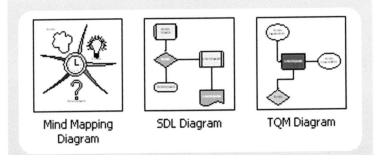

Figure 18.11: Flowcharting types

data flow diagrams, UML diagrams, Jackson diagrams, Coad and Yourdon, and SSADM, all discussed in this book. Figure 18.14 shows the stencil set of all the symbols for SSADM diagrams; Figure 18.15 shows the four basic shapes of data flow diagrams; and Figure 18.16 shows a data flow diagram being constructed using these shapes.

Although not a sophisticated integrated toolset such as the systems described in Chapter 19, Visio enables the easy creation and update of most basic diagrams used in many of the techniques and methodologies discussed in this book. It is possible to download a copy of Visio Professional 2003 as a trial version for one month (Microsoft, 2005), and such possibilities may be available for other tools as well.

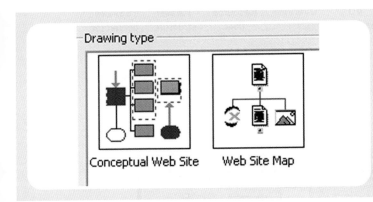

Figure 18.12: Website design

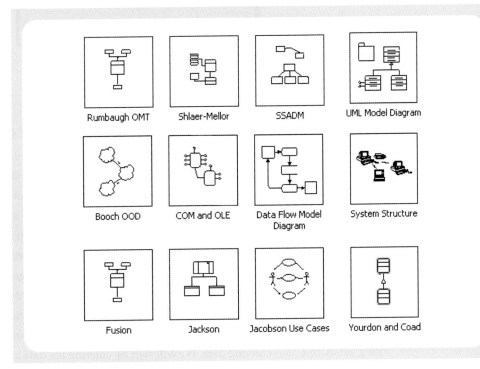

Figure 18.13: Methodology choices

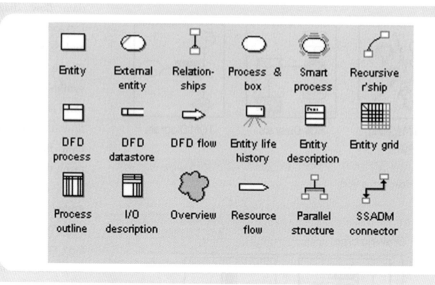

Figure 18.14: Visio – SSADM symbols

Figure 18.15: Visio – Gane and Sarson (STRADIS) data flow diagram symbols

18.4 Project management: Microsoft Project

We looked at a number of project management and control techniques in Chapter 14, and we look at a methodology, PRINCE, in Section 25.4, which is specifically designed for the project control aspects of developing projects. Here we look at the output of software tools, which provide speed and accuracy to the project planning process and help to achieve the matching of project plans with their execution. In particular, we look at Microsoft's planning tool Project. It features techniques we looked at such as work breakdown structure, critical path analysis, and Gantt charts in Chapter 14, plus some others like calendar management. Although essentially a package for workstations and PCs, Project will manage thousands of tasks and resources for each project, create all sorts of diagrams, ways to present information, track progress, and anticipate problems. On the other hand, no software package can eliminate the work of planning, estimating, and entering data. We will start by entering the data as shown here in Figure 18.17.

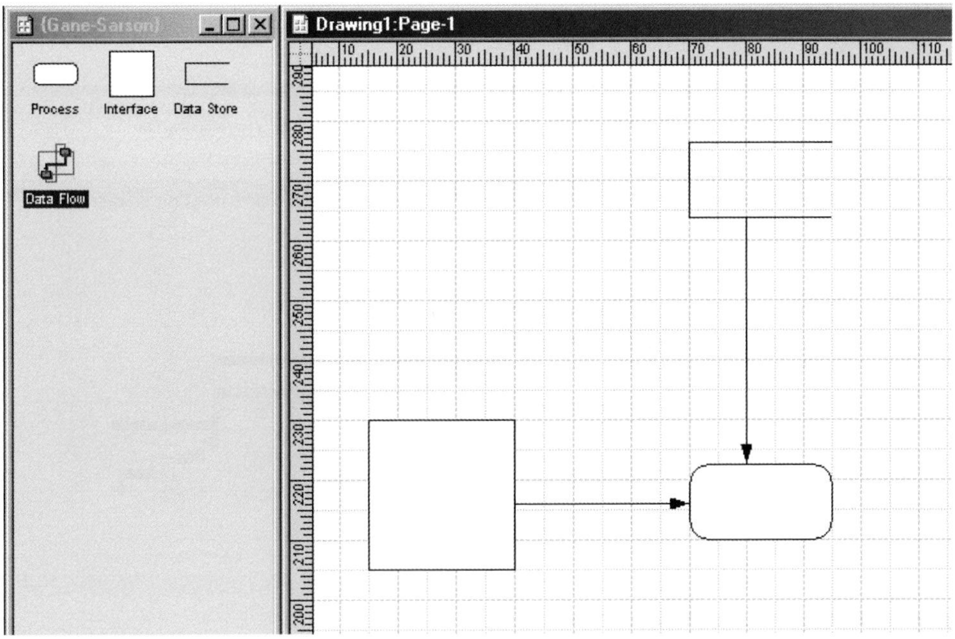

Figure 18.16: Visio – Developing the data flow diagram

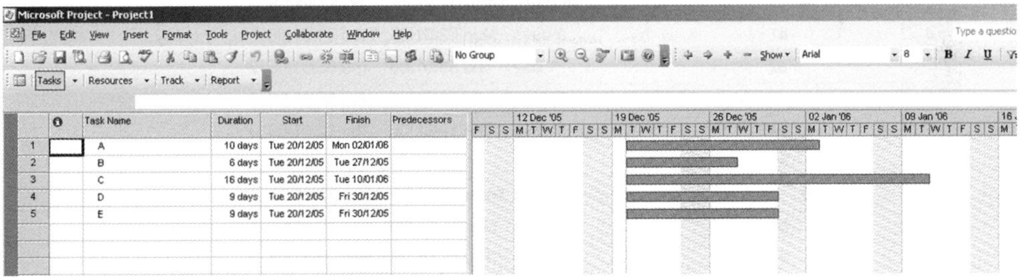

Figure 18.17: Project – Entering the work breakdown structure activities and durations

In Figure 18.18 we have added some subtasks and also shown their interdependencies as well as those between the main tasks. Note that most activities are completed sequentially, but task D, for instance, can be done in parallel with other tasks.

Figure 18.19 shows how these interdependencies can be displayed, along with the start and end dates.

In Figure 18.20 we show the calendar display and in Figure 18.21 part of the PERT network itself.

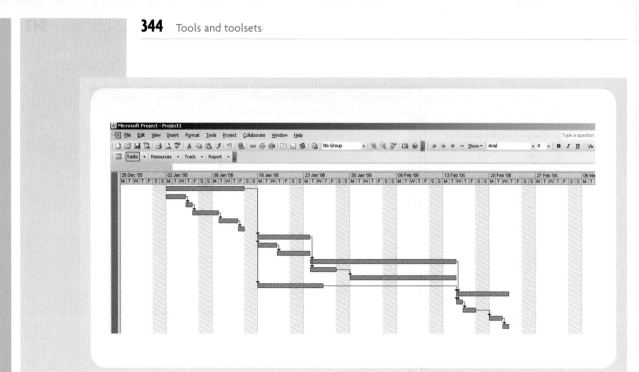

Figure 18.18: Project – Interdependence of tasks

	❶	Task Name	Duration	Start	Finish	Predecessors	Resource Names
1		**A**	**10 days**	**Mon 02/01/06**	**Fri 13/01/06**		
2		a1	3 days	Mon 02/01/06	Wed 04/01/06		
3		a2	1 day	Thu 05/01/06	Thu 05/01/06	2	
4		a3	2 days	Fri 06/01/06	Mon 09/01/06	3	
5		a4	3 days	Tue 10/01/06	Thu 12/01/06	4	
6		a5	1 day	Fri 13/01/06	Fri 13/01/06	5	
7		**B**	**6 days**	**Mon 16/01/06**	**Mon 23/01/06**	1	
8		b1	3 days	Mon 16/01/06	Wed 18/01/06	1	
9		b2	3 days	Thu 19/01/06	Mon 23/01/06	8	
10		**C**	**16 days**	**Tue 24/01/06**	**Tue 14/02/06**	7	
11		c1	4 days	Tue 24/01/06	Fri 27/01/06	7	
12		c2	12 days	Mon 30/01/06	Tue 14/02/06	11	
13		**D**	**8 days**	**Mon 16/01/06**	**Wed 25/01/06**	1	
14		**E**	**6 days**	**Wed 15/02/06**	**Wed 22/02/06**	10,13	
15		e1	1 day	Wed 15/02/06	Wed 15/02/06	10,13	
16		e2	2 days	Thu 16/02/06	Fri 17/02/06	15	
17		e3	2 days	Mon 20/02/06	Tue 21/02/06	16	
18		e4	1 day	Wed 22/02/06	Wed 22/02/06	17	

Figure 18.19: Project – Start and end dates

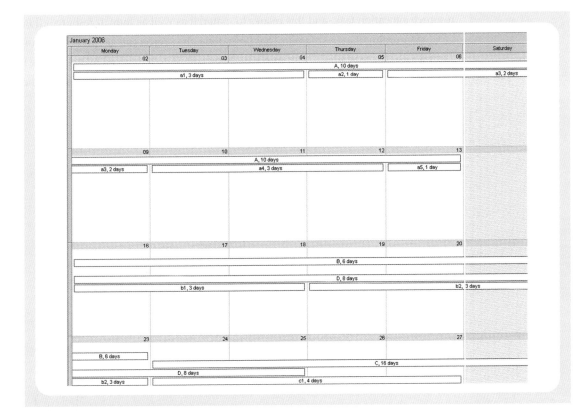

Figure 18.20: Project – Calendar display

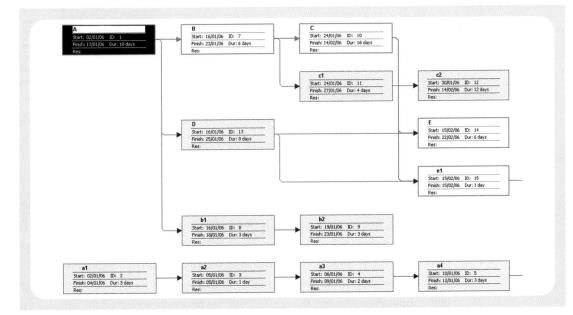

Figure 18.21: Project – PERT chart

So far, we have not input resource usage. Figure 18.22 shows the beginnings of the resource allocation task, and this relates to material as well as human resources (Figure 18.23).

Assuming that the planning has been done, the project can get under way, and in Figure 18.24 we can see that all the tasks in A have been completed (denoted by ticks in the left column and a black line through their respective bars on the Gantt chart). Note that some of the tasks in B and D (done in parallel) have only partially been completed.

Resource Name	Work	Details	02 Jan '06 M	T	W
⊟ Joe	10.67 hrs	Work	8h	2.67h	
a1	2.67 hrs	Work	2.67h		
a2	8 hrs	Work	5.33h	2.67h	
⊟ Bill	2.67 hrs	Work	2.67h		
a1	2.67 hrs	Work	2.67h		
⊟ Ted	2.67 hrs	Work	2.67h		
a1	2.67 hrs	Work	2.67h		
⊟ Bill (50%)	8 hrs	Work		5.33h	2.67h
a3	8 hrs	Work		5.33h	2.67h
⊟ Ted (50%)	8 hrs	Work		5.33h	2.67h
a3	8 hrs	Work		5.33h	2.67h

Figure 18.22: Project – Adding resource names

	❶	Resource Name	Type	Material Label	Initials	Group	Std. Rate	Cost/Use	Accrue At	Code
1		Joe	Work		J		€25.00/hr	€0.00	Prorated	
2		Bill	Work		B		€25.00/hr	€0.00	Prorated	
3		Ted	Work		T		€15.00/hr	€0.00	Prorated	
4		Bill (50%)	Work		B		€15.00/hr	€0.00	Prorated	
5		Ted (50%)	Work		T		€20.00/hr	€0.00	Prorated	
6		David	Work		D		€20.00/hr	€0.00	Prorated	
7		Postage	Material		P		€0.60	€0.00	Prorated	

Figure 18.23: Project – Resource details

There are many ways the data can be reported. In Figure 18.25 we display the main types (many graphical); Figure 18.26 shows the various current activity reports; Figure 18.27 shows one of these reports showing completed tasks; Figure 18.28 shows a resource allocation report ('who does what?'); and Figure 18.29 shows a resource usage graph.'

As for all the tools outlined in Part V, we hope that we have given the reader some idea of the application package, though, in truth, we have only scratched the surface of the possibilities provided by Project. At the time of writing, the latest version available was Project 2003 (Microsoft, 2005).

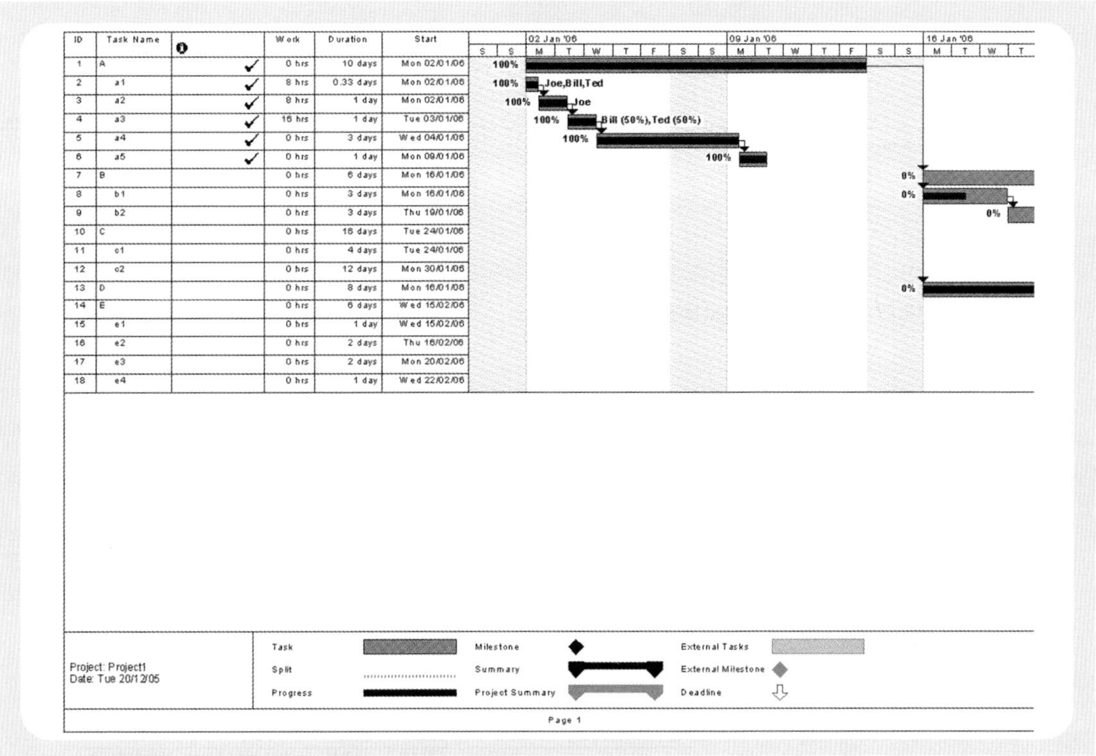

Figure 18.24: Project – Project development through time

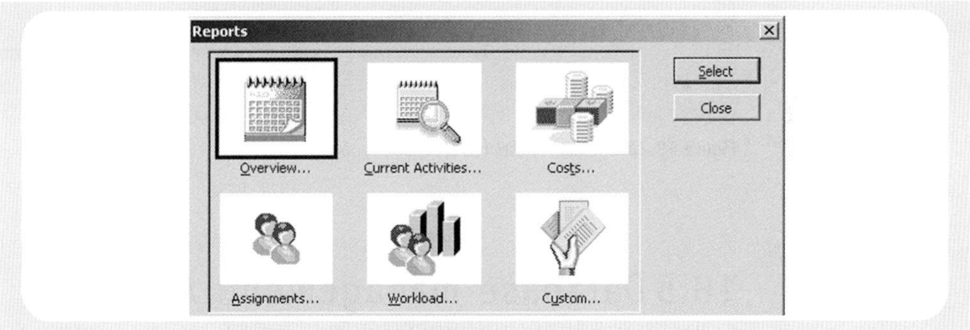

Figure 18.25: Project – Ways in which data can be displayed

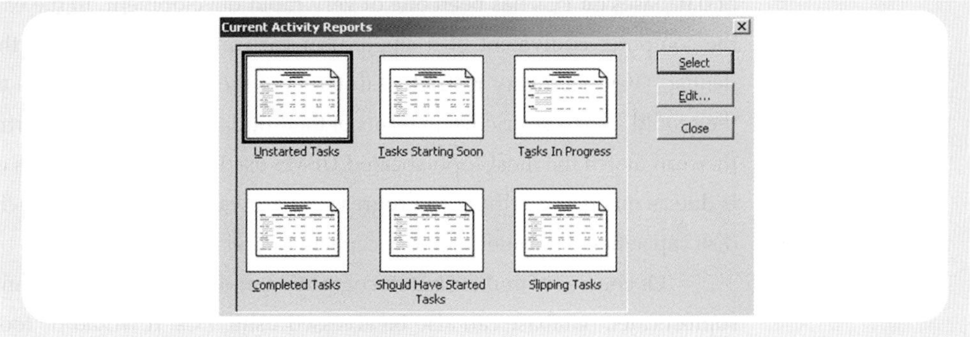

Figure 18.26: Project – Current activity reports

Completed Tasks as of Tue 20/12/05
Project1

ID	Task Name	Duration	Start	Finish	% Comp.	Cost	Work
January 2006							
1	A	10 days	Mon 02/01/06	Fri 13/01/06	100%	€0.00	0 hrs
2	a1	0.33 days	Mon 02/01/06	Mon 02/01/06	100%	€173.33	8 hrs
3	a2	1 day	Mon 02/01/06	Tue 03/01/06	100%	€200.00	8 hrs
4	a3	1 day	Tue 03/01/06	Wed 04/01/06	100%	€280.00	16 hrs
5	a4	3 days	Wed 04/01/06	Mon 09/01/06	100%	€0.00	0 hrs
6	a5	1 day	Mon 09/01/06	Tue 10/01/06	100%	€0.00	0 hrs

Figure 18.27: Project – Completed tasks

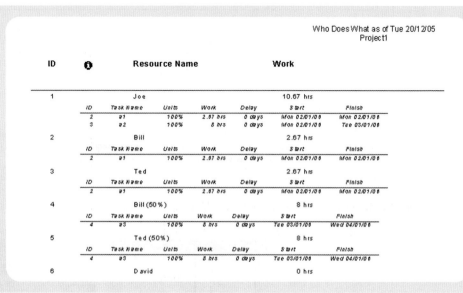

Figure 18.28: Project – Resource allocation report

18.5 Database management: Access

In this section we look at databases on personal computers (PCs). In particular, we look at Access, the database software that is part of the Microsoft Office suite of programs. The history of databases on PCs has been one of very rapid development. In the early days they were just the computer equivalent of a single user card index, for example, the patient records of a dentist. They were very restrictive in terms of the number of records that could be held, and they could also be restrictive in terms of their update, retrieval, and sorting abilities. Nowadays they can match the most sophisticated DBMS used on large computers of a few years ago. This is due as much as anything to progress in hardware development, such as processing speeds, disk capacities, computer memory capacities, and the like.

Of course, with distributed computing and client–server systems, the largest and most sophisticated database can also be accessed using PCs. It would be foolish to put numbers to speeds, maximum number of users, size of databases, and so on as the technology changes so

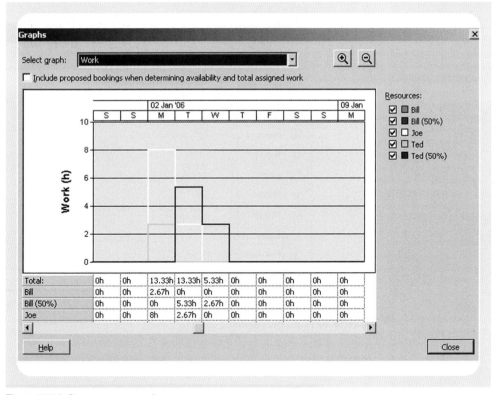

Figure 18.29: Resource usage graph

rapidly. All we can say for sure is that PC databases are likely to be smaller, with fewer users, using fewer tools, compared to corporate databases at any one time.

At the time of writing, there are three versions of Access in common use: Access2000, Access2002 and Access2003. Like all the tools discussed in this chapter, a full discussion of Access requires a series of books, so we will look only at aspects which give a flavour of the package.

On opening Access, we see the screen as shown in Figure 18.30. It shows seven basic object types. 'Tables' are the relations, then there are options 'queries', 'forms', and 'reports' to help query the data in the database, enter data conveniently, or provide reports from the data, respectively. The option 'pages' enables the creation of web pages. A 'macro' performs a set of commands in sequence and a 'module' is a larger program written by the user in VisualBasic. These are used for more complex or non-standard applications of the database. The database will consist of a series of related tables and associated queries, reports, forms, and so on.

In Figure 18.31, we show the design form for designing a table. This table has a composite key (see the key icon to the left) of order number and item code. We have specified most fields as text and two fields as numeric. There are many other options, such as date/time, currency, automatically generated number (often used as a primary key), a yes/no alternative, and hyperlink to a web address. You will notice that the Access interface has a 'look and feel' similar to other Microsoft products with a series of menus along the top (for manipulating files, editing, and so on) and icons held in toolbars (for saving, printing, spell-checking, creating key fields, accessing help facilities, and so on).

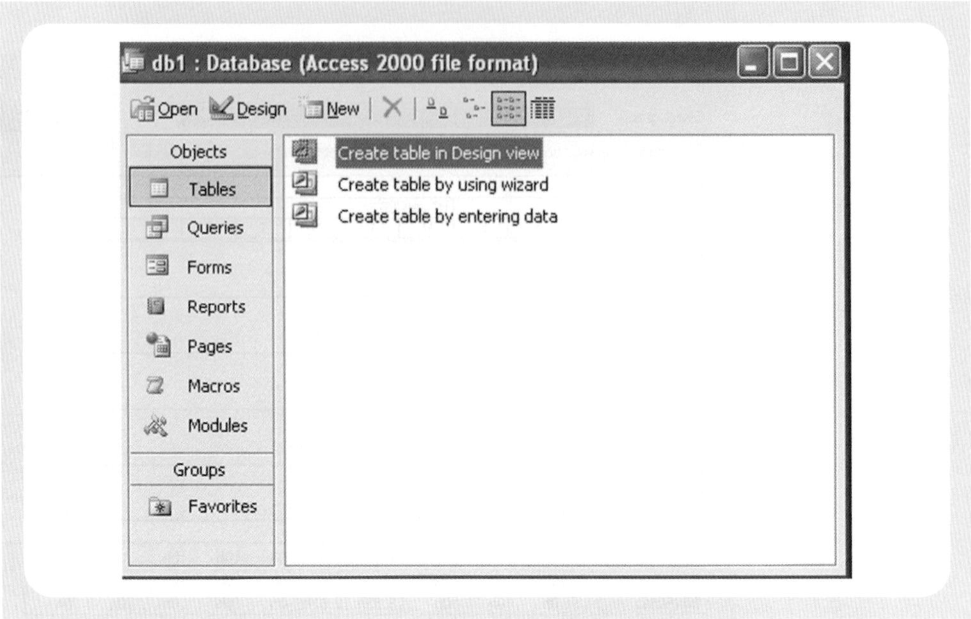

Figure 18.30: Access basic options

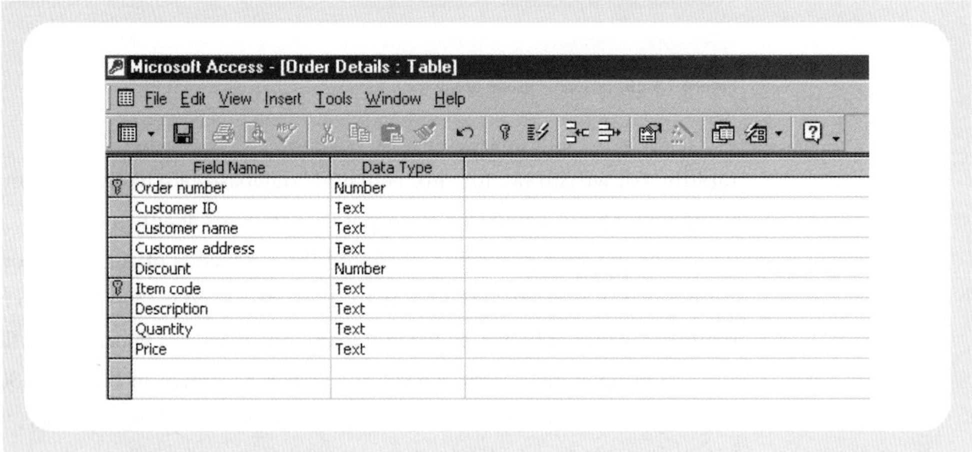

Figure 18.31: Designing an Access table

For each field in the table we can define a number of properties (Figure 18.32). The field size limits the number of characters for a text field, the format controls the display of the field in datasheet view (for example, currency, true/false type, or percentage), decimal places are relevant for a numeric or currency field, the input mask is useful when, for example, including a plus sign and hyphens for an international phone number, a caption provides a label for the field, and a default value is useful when a field will normally have one value more than others and hence save some keying in. A validation rule stops invalid data from entering the database.

As we see in Figure 18.33, once we begin to enter data in datasheet view, Access rightly prevents us from completing the data because rows of data would be formed with

General	Lookup	
Field Size		Long Integer
Format		
Decimal Places		Auto
Input Mask		
Caption		
Default Value		0
Validation Rule		
Validation Text		
Required		No
Indexed		No

Figure 18.32: Field properties

Order Details : Table

Order number	Customer ID	Customer name	Customer addre	Discount	Item code	Description	Quantity	Price
342	AS12	Brogan	1 Temple Close	10	Z02	Blue Spotted	2	89.99
342	AS12	Brogan	1 Temple Close	10	Z02	Green	7	49.99
0				0				

Microsoft Office Access ✕

⚠ The changes you requested to the table were not successful because they would create duplicate values in the index, primary key, or relationship. Change the data in the field or fields that contain duplicate data, remove the index, or redefine the index to permit duplicate entries and try again.

[OK]   [Help]

Figure 18.33: Access warning message – duplicate keys

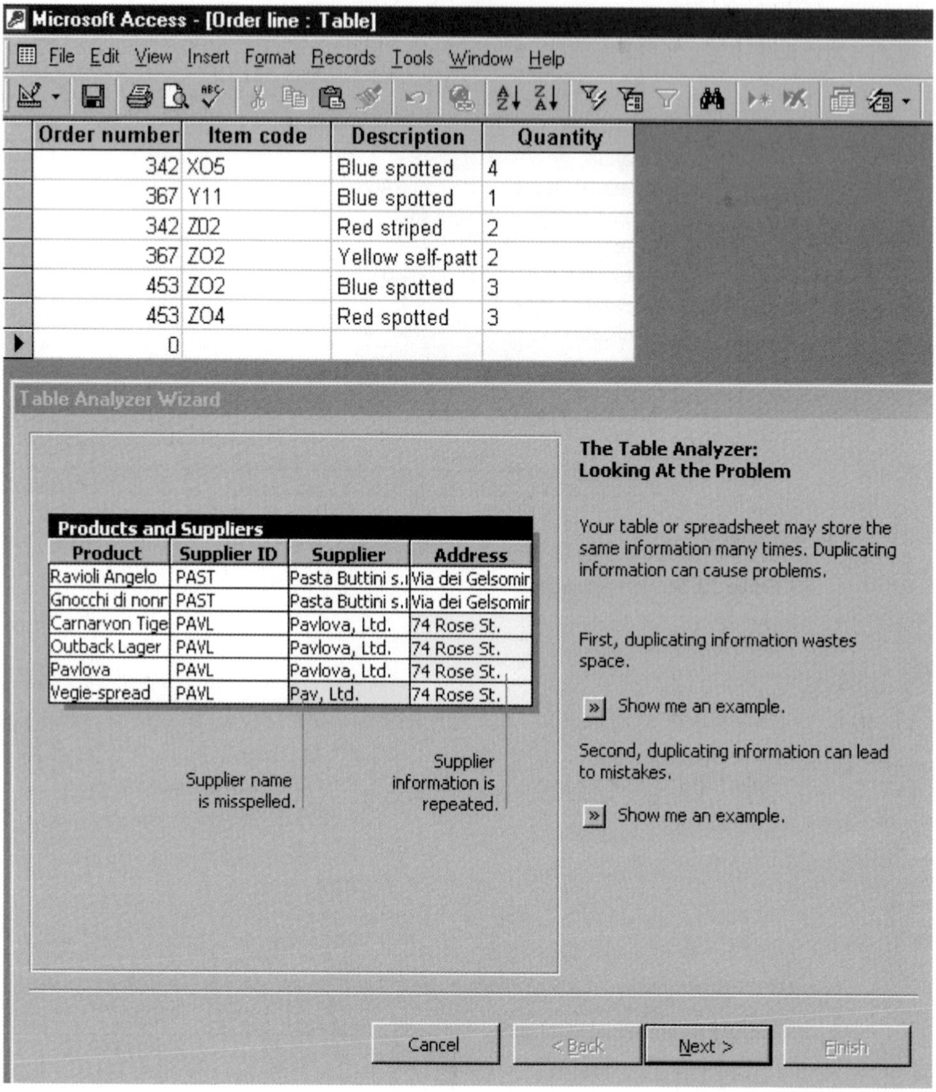

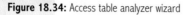

Figure 18.34: Access table analyzer wizard

duplicate keys. According to the relational model, there should not be rows with duplicate keys.

The wizard (tool) analyses the tables on the database (Figure 18.34). In our order line table, the wizard has identified (through duplicate item codes and descriptions) data not in third normal form (Section 11.2). The program will help us to create a new table of item code and description, linked to the order line table using the item code. Figure 18.35 shows the initial attempt at doing this with the relationship between the two tables being mapped.

In Figure 18.36 we begin to create a query, again using an Access wizard, which is formed from data held in two tables: order details and order line. However, as the warning message informs us, we have not yet linked these tables together by a relationship. In Figure 18.37 we show how the relationship is formed using the field order number to join the tables as it is in both relations, as our linking field. As shown in Figure 18.38, Access shows the relationship as one-to-many, with an infinity sign on the many side.

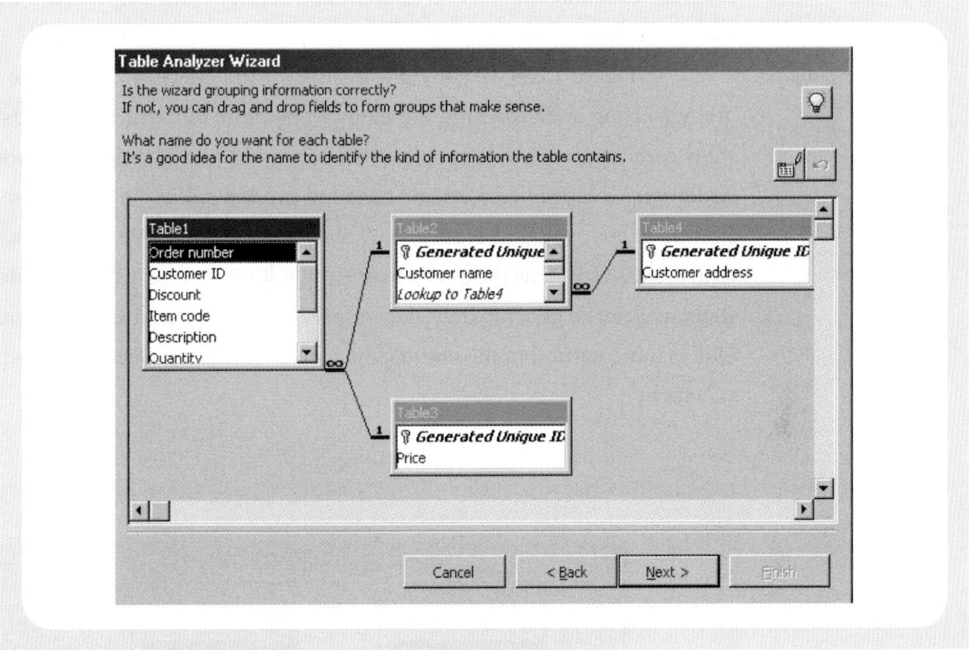

Figure 18.35: Access table analyzer wizard creating a new table

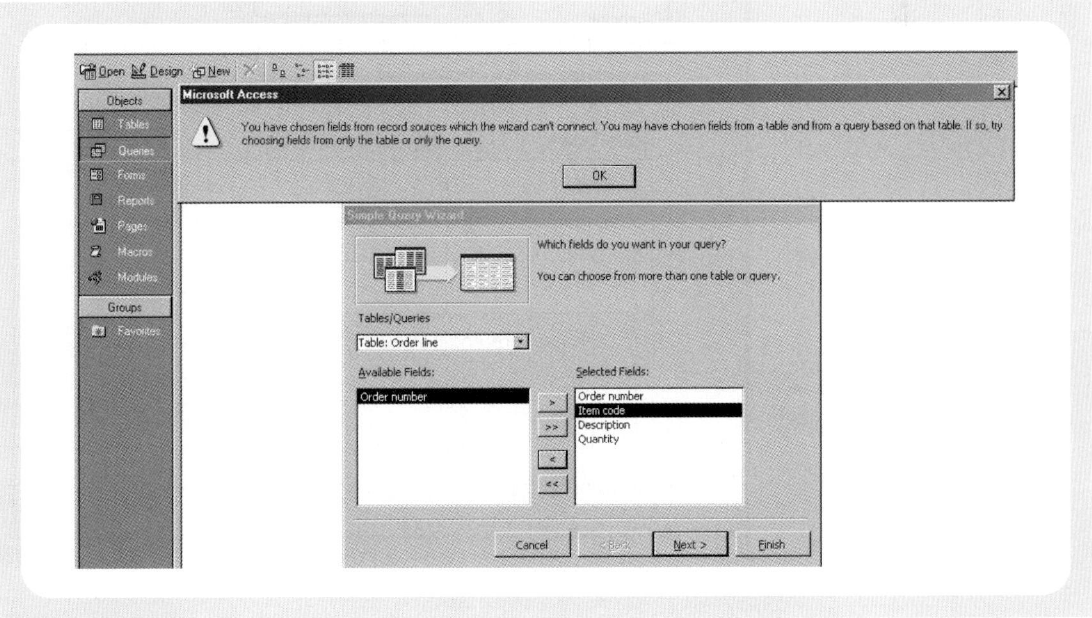

Figure 18.36: Creating a query

We notice in Figure 18.37 the possibility of enforcing referential integrity rules. The first of these states that you cannot enter data in the field that is used for the join in the related table, if the join field in the primary table does not have matching contents. The second rule prevents you deleting records from the primary table if there are matching records in the related table. Finally you cannot edit primary key values in the primary table if related records exist. Like the rules for normalization, these referential integrity rules help ensure that the database is well formed and easy to use.

In Figure 18.38 we show the design for the query, including the opportunity to sort the query (sorting and searching are likely to be more efficient if the fields are indexed) and only show certain records in the criteria list. For example, we could have chosen to query only those customers who ordered a certain range of products. We did not choose to limit the number of records in any way and Figure 18.39 shows the full list. We can use SQL language statements with Access as well as the query-by-example (QBE) method shown. Indeed, the equivalent SQL statement can be generated by Access, as shown in Figure 18.40. But constructing a query using QBE is more natural to the non-technical user and much easier than coding the equivalent SQL statement.

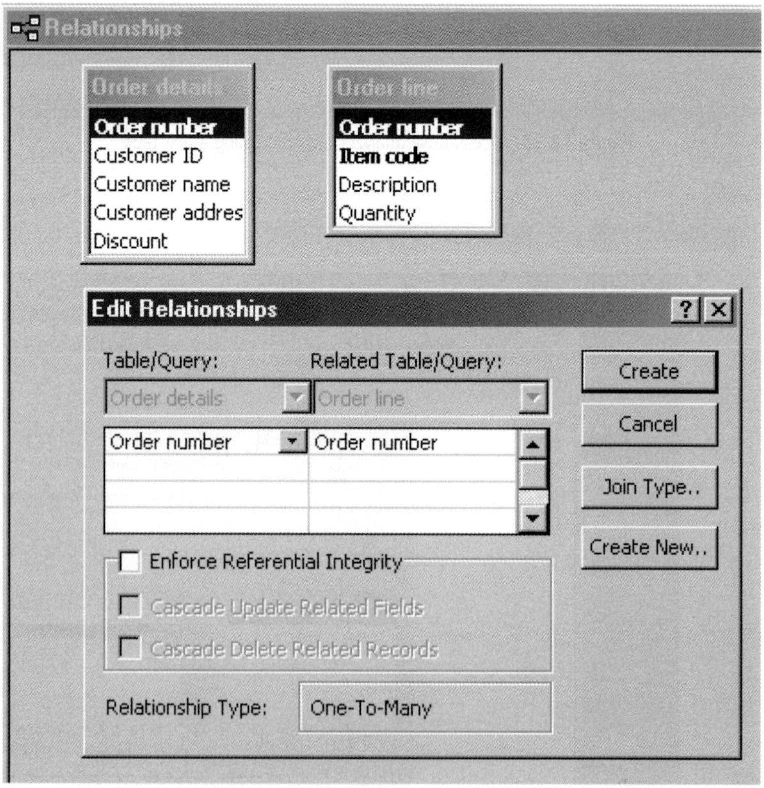

Figure 18.37: Establishing the relationship for the query

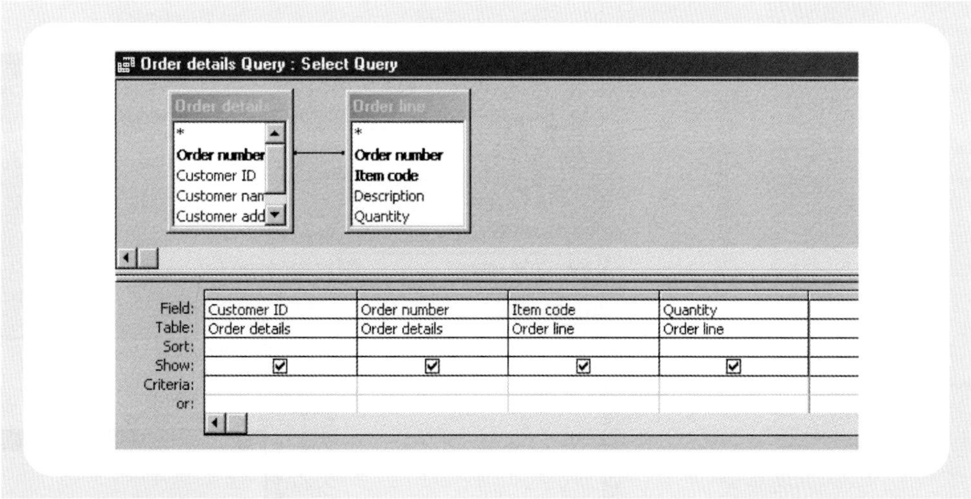

Order details Query : Select Query

Field:	Customer ID	Order number	Item code	Quantity
Table:	Order details	Order details	Order line	Order line
Sort:				
Show:	☑	☑	☑	☑
Criteria:				
or:				

Figure 18.38: Access query

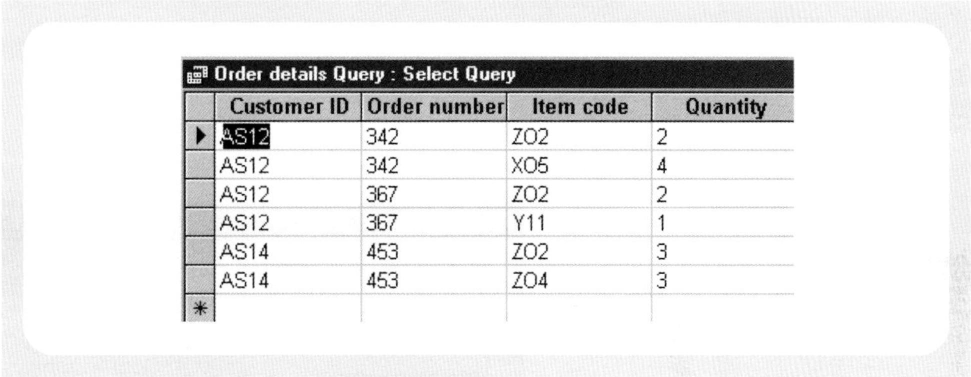

Order details Query : Select Query

Customer ID	Order number	Item code	Quantity
AS12	342	ZO2	2
AS12	342	XO5	4
AS12	367	ZO2	2
AS12	367	Y11	1
AS14	453	ZO2	3
AS14	453	ZO4	3

Figure 18.39: Displaying the query results

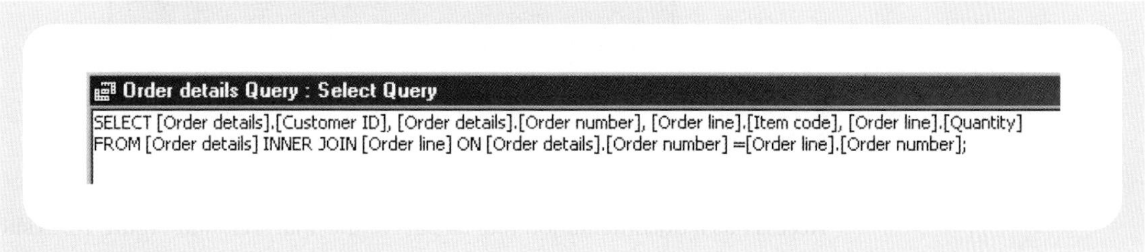

Order details Query : Select Query

```
SELECT [Order details].[Customer ID], [Order details].[Order number], [Order line].[Item code], [Order line].[Quantity]
FROM [Order details] INNER JOIN [Order line] ON [Order details].[Order number] =[Order line].[Order number];
```

Figure 18.40: The SQL statements for the query

Forms for entering data can easily be created by using the form wizard. Figure 18.41 shows one stage in the process and Figure 18.42 shows the form created using one of the default styles. There are similar wizards for creating reports in all manner of styles and levels of detail.

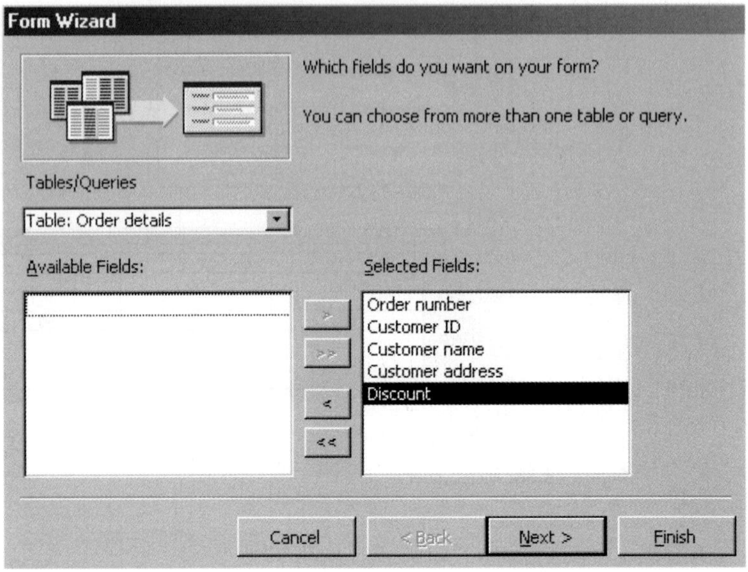

Figure 18.41: The form wizard

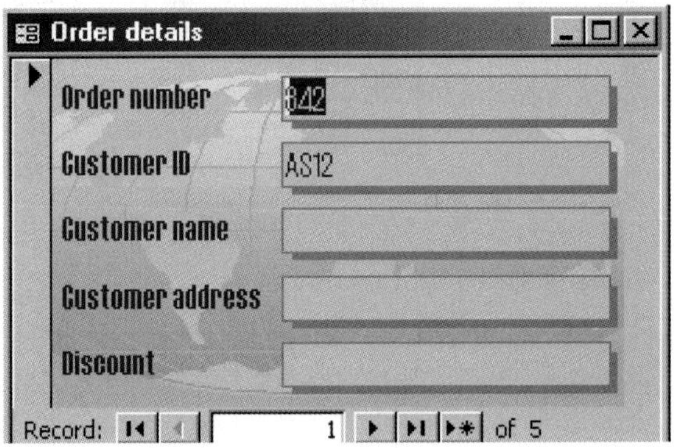

Figure 18.42: A wizard-generated form for other details

Access has a documentation tool as part of the analyzer, and Figure 18.43 shows part of the automated documentation for the item table and Figure 18.44 that for the relationship between the item table and the order line table.

Access has a number of security features. A database can be opened read-only or exclusively (via a password). Access even enables encryption and decryption of the data.

As well as enquiries and reports, you can display data using charts, such as pie charts, line graphs, bar charts, column charts, area charts, and so on. Figure 18.45 is a pie chart showing the different proportions of products ordered.

Macros are used to perform repeated actions such as opening tables and forms, regularly printing certain reports, answering particular queries, and finding particular records. Modules are programs written in the computer programming language VisualBasic for Applications. Compared to many computer programming languages, VBA is not that difficult

			28 May 2002
Table: Order details			Page: 1

Properties

Date Created:	28/05/2002 22:36:32	GUID:	Long binary data
Last Updated:	28/05/2002 22:37:59	NameMap:	Long binary data
OrderByOn:	False	Orientation:	0
RecordCount:	5	Updatable:	True

Columns

Name	Type	Size
Order number	Text	50

AllowZeroLength:	False
Attributes:	Variable Length
Collating Order:	General
ColumnHidden:	False
ColumnOrder:	Default
ColumnWidth:	Default
Data Updatable:	False
DisplayControl:	Text Box
GUID:	Long binary data
Ordinal Position:	1
Required:	False
Source Field:	Order number
Source Table:	Order details
UnicodeCompression:	True

Name	Type	Size
Customer ID	Text	50

AllowZeroLength:	False
Attributes:	Variable Length
Collating Order:	General
ColumnHidden:	False
ColumnOrder:	Default
ColumnWidth:	Default
Data Updatable:	False
DisplayControl:	Text Box
GUID:	Long binary data
Ordinal Position:	2
Required:	False
Source Field:	Customer ID
Source Table:	Order details
UnicodeCompression:	True

Figure 18.43: Access documentation tool – tables

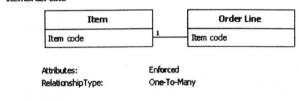

Relationships

Relationships

ItemOrder Line

Item		Order Line	
Item code		Item code	

Attributes: Enforced
RelationshipType: One-To-Many

Figure 18.44: Access documentation tool – relationships

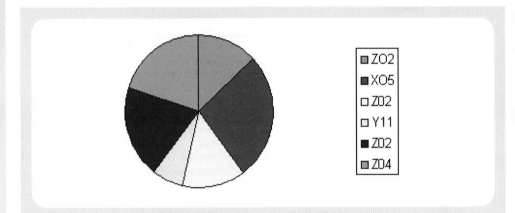

Figure 18.45: Pie chart

to use. It gives flexibility in using the database. When we add the much greater sophistications and individualized applications enabled by macros and modules, it is readily evident that Access is a very powerful database management system and yet is available on personal computers. Access is part of the Microsoft Office suite (Professional version) and information and a demo version can be accessed from the website (Microsoft, 2005).

Summary

- In this chapter we illustrate the use of various tools that are commonly used to develop information systems. These include groupware to help decision support, website development software, a drawing package to support the use of graphical techniques, software for management of time and other resources used in a project, and a database management system. All the application packages described are PC-based.

Questions

1. Use the various packages described in this chapter, or any equivalents available to you, to draw diagrams as follows:
    - the various diagrams used in Part IV (on techniques), using Visio;
    - plan an information systems development project, using Project;
    - set up your own personal website, using Dreamweaver;
    - develop a database application to organize and report on your CD collection, using Access.

Further reading

Bucki, L.A. (2000) *Managing with Microsoft Project 2000*, Prima, Rocklin, California.

Failla, A. (1996) Technologies for co-ordination in a software factory, in C.U. Ciborra (ed.) *Groupware and Teamwork: Invisible Aid or Technical Hindrance?*, John Wiley & Sons, Chichester, UK, pp. 61–88.

Microsoft (2001) *Developing Microsoft Visio Solutions (Pro-Documentation)*, Microsoft Press, Seattle.

Oliver, P.R.M. and Kantaris, N. (2001) *Microsoft Access 2002 Explained*, Babani, London.

Towers, J.T. (2001) *Dreamweaver 4 for Windows and Macintosh: Visual QuickStart Guide*, Peachpit, Berkeley.

19 Toolsets

19.1 Introduction

Over several years, software support tools have been available that help the information systems development process as a whole, not just in the drawing of some individual diagrams. These *were* generally known as **CASE** (Computer Aided Systems (or Software) Engineering) tools but we use the term information systems development toolset or, more simply, *toolset* for:

> Any integrated computer software system that is specifically designed to support a significant part of the information systems development process of an information system and the management of these tasks and processes.

Whereas the tools discussed in Chapter 18 in the main support one type of activity in the information systems development process, for example, project management, group systems work, or drawing diagrams, toolsets support several tasks; indeed, they provide a full set of tools for the analyst.

Originally CASE tools were divided into Upper and Lower CASE tools. The purpose of this distinction was to indicate which stages of the life cycle they addressed. Upper CASE includes tools that helped the strategy/planning, analysis, or logical design stages, whereas Lower CASE tools were concerned with aspects of physical design, programming, and implementation, including automatic code generation. Most toolsets now integrate these elements into a single, fully integrated development and support facility.

Integration in these toolsets is particularly important. They are integrated both horizontally and vertically. Horizontal integration is the integration of different tools at a particular stage of the development cycle. At the analysis stage, for instance, there are a number of different techniques which are supported by the toolset, for example, data flow diagrams, entity models, function decompositions, and so on. These are regarded as horizontally integrated if information is shared between the tools and if changes made in one diagram (using the diagram support tool) are reflected in the other diagrams, where appropriate.

Vertical integration, on the other hand, is the integration of tools between different stages of the life cycle, and this means that the results from one stage should be available in an automated form to the other stages. The results should be capable of being passed forward to subsequent stages of the life cycle (known as forward integration). Further, for highly integrated tools, the results or changes made at later stages of development should be capable of being passed backward, and be reflected in, earlier stages. This is known as reverse integration.

Another important aspect of integration is interpersonal integration. System development, in organizations of any size, is a matter of co-ordinating the work of many people and perhaps many development teams, frequently in different locations. Additionally, the work may be performed in parallel, at least within development stages, and possibly in parallel across stages. Toolsets should ensure that the work is co-ordinated and consistent. An important aspect of this is **version control**; that is, the organizing and handling of the large numbers of different versions of systems that exist.

Figure 19.1 presents a generic integrated toolset. In this chapter we look at Information Engineering Facility, Oracle and Select Enterprise. Following this, we provide an evaluation of toolsets. Before doing so, however, we discuss one of the key features of toolsets separately. This is the **repository**, seen to the middle left in Figure 19.1. This is so important because it enables the integration of all the models, definitions, and mapping of stages.

The precursors of systems repositories were **data dictionary systems** (DDSs), software tools for managing the data resource. They enabled the recording and processing of 'data about the data' (metadata) that an organization uses. They were originally designed as documentation tools, ensuring standard terminology and providing a cross-reference capability about

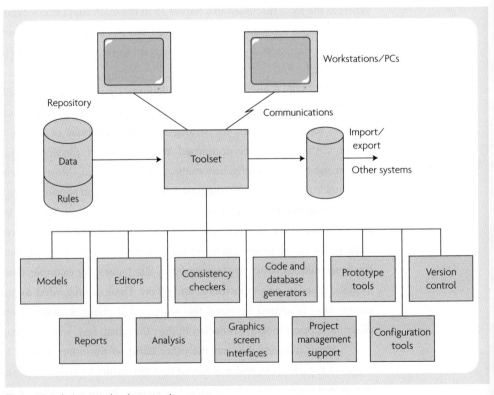

Figure 19.1: An integrated toolset: generic

data (and later processes) at the modelling and implementation levels. They have subsequently developed into systems repositories, which provide a central catalogue for all aspects of an information systems development project, containing all the information necessary to integrate the different stages of information systems development.

The repositories will normally contain information about the physical and operational elements of data and processes, for example, physical data items, processes, modules, code, and test data, and information about logical and functional levels as well, for example, data and process models and diagrams. But they do much more than that.

Active repositories contain information that enables the rules of a technique or an information systems development methodology to be applied. These may permit analysis, validation, and consistency and completeness checking. There may be a separate element of the repository (sometimes known as the repository manager), in which knowledge or rules are embedded. In some repositories these 'rules' (which could be the rules of the methodology) are locked (or hard-coded) into the repository. In other repositories they are more flexible and easily changed or defined in an expert systems language. In theory at least, therefore, a repository contains everything needed to support the creation and maintenance of information systems in organizations.

Repositories do not store diagrams as such, but a series of definitions about the objects in diagrams. This means that objects that appear in more than one diagram are only stored once and that diagrams are generated from the current information in the repository, as and when they are needed. When changes are made to one diagram, the effects of that change are automatically reflected when other diagrams in which the object appears are generated. This enables the basic repository information to be displayed in a number of ways according to needs. For example, an object may feature in an entity model, a data flow diagram, and an action diagram (see Chapter 12).

The repository can contain information beyond that which is needed to create software systems and include models of the organization and environment, that is, the framework of the systems. This is a view of the organization in terms of business areas, functions, hierarchies, departments, locations, strategic relationships, critical success factors, objectives, plans, and so on. Figure 19.2 shows another view of the repository reflecting the enterprise information and its mapping to data and processes in the information system, showing the mappings between them.

Many repositories are themselves relational databases, which reflects the dominant technology when they were developed. This is now somewhat inadequate, and one based on an object-oriented database management system is likely to deal better with the wide variety of data types such as image, voice, and graphics of multimedia systems.

19.2 Information Engineering Facility

Information Engineering Facility (IEF) is an example of an integrated toolset or CASE tool. It was originally developed by Texas Instruments to support the Information Engineering (IE) methodology (see Section 21.3, which might be read in conjunction with this section). IEF was

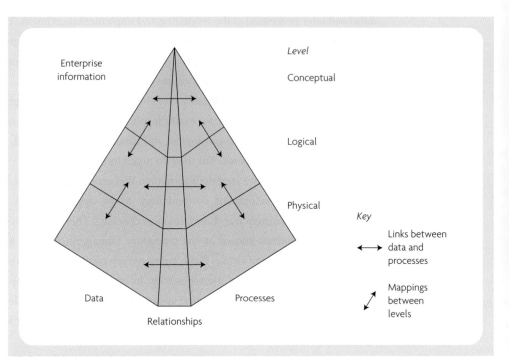

Figure 19.2: Repository with enterprise information

renamed Composer and bought by Sterling Software in 1997 who were themselves taken over by Computer Associates in 2000. It is now known as Advantage Gen and is marketed as a rapid application development tool which generates code from graphical business process models. The toolset has been enhanced and expanded and now supports component-based development (see Section 8.5), client–server and web-based development (Section 7.5) and the generation of C and Java code. In this section we examine IEF as a generic example of an integrated toolset rather than a specific product and concentrate on the concepts embedded in such a toolset.

IEF supports planning, analysis, design, construction and implementation. These three elements are integrated via an encyclopaedia (the IEF term for the repository). The main encyclopaedia was originally designed to be held on a centralized mainframe but now runs in a client–server environment. IEF is described as being 'integrated', and this means that across the toolset there is a common look and feel, the use of an encyclopaedia to enable consistency, and the automatic generation of objects and code from higher-level objects. We will consider each of the five aspects.

1 Planning

The planning software is designed to support a strategic approach to systems development following the identification of high-level business requirements. It helps in the production of three architectures: the information architecture, the business system architecture, and the technical architecture. First, planners build a subject area diagram, which is a high-level entity model of the subject areas of interest to the business. The data modelling tool enables the users (usually described as planners) to build the subject model and view the whole model or individual subject areas, which can then be 'exploded' to reveal the component entity types of the subject area. The tool ensures consistency across the diagrams and shows any relationship aggregation lines. A relationship aggregation line represents a relationship between entities in different subject areas.

Planners also construct high-level function hierarchy diagrams using the activity hierarchy diagramming tool, which enables the functions and their position in the hierarchy to be captured and displayed in various ways. The activity dependency diagramming tool is then used to document the sequence in which the functions must be performed, with arrows indicating the direction of the dependency showing what function must occur before another. There is also an organizational hierarchy diagramming tool which enables the construction and manipulation of an organization chart. Finally the matrix processor enables various interactions to be identified and depicted.

The matrix processor can be used to show the interactions between any two sets of things, but particular support is provided for 40 standard matrices, which include objectives, strategies, goals, critical success factors, entities, subjects, and functions. The tool is basically a two-dimensional matrix (Section 12.7) with, for example, functions as rows and objectives as columns, and an entry is made in the relevant cell when a particular function supports a particular objective. The matrix may reveal that some objectives are not supported by any functions, and therefore that objectives have been lost, or that only one or two functions support the majority of the objectives and therefore these are the key functions. Planners use such matrices in a variety of ways, and the software makes the comparison and analysis that much easier. One important use of the matrix processor is to plot the interactions between functions and entities, that is, which functions affect which entities. This may be create, read, update, or delete (CRUD) (see Section 12.7). The matrix processor is also able to cluster, and this might lead to the automatic production of a new matrix where all entities affected by a particular function are clustered together.

The diagramming software for planning is integrated. Once, for example, the entities are entered using the data modelling software, they will appear automatically in the matrices concerned with entities. This means that elements are only entered once, and when they appear in a matrix none will be 'forgotten'.

2 Analysis

Analysis refines a particular area of the business identified in the planning stage, and any information captured in the planning software is automatically available to the analysis software using the encyclopaedia. In this stage (business area analysis in IE), analysts enter further details, often using the same software as in the planning stage. For example, the subject area diagram is expanded into an entity-relationship diagram using the data modelling software. Relationships between entities are now defined, including their cardinality and optionality. This is done by the use of dialogue boxes within the tool, as is the definition of the entity attributes and properties. The user can specify entity subtypes, and the diagram can be viewed with these expanded or contracted. Functions are further refined into a series of lower-level processes using the activity hierarchy diagramming tool. For each process, the software enables details to be entered. These details will include name, description, type, and any entities that the process uses. Similarly, the function dependency diagrams from the planning stage are expanded, where necessary, into a series of process dependency diagrams using the activity dependency diagramming software, and details of events that trigger processes and relationships with external objects are added. Some further matrices are also produced using the matrix processor.

The elementary processes are then defined as action diagrams (Section 12.8) using the process action diagramming tool. The action diagram defines the steps required for each elementary process and the way that they interact with the entities. The software automatically begins constructing the action diagram using information derived from other diagrams, such as the entity model, the process hierarchy, and dependency diagrams, and the analyst can insert extra actions and manipulate the action diagram as required. The process action diagramming software applies action diagramming rules and only allows entries that are semantically and syntactically correct. The information for checking semantic correctness is derived from the earlier diagrams represented in the encyclopaedia. In addition, the tool allows processes to be 'synthesized'. This may be achieved by the software asking certain relevant questions to clear up areas that have not been completed or, in certain circumstances, by automatically generating process logic based on the entity model and information from the matrices. For example, if an entity REGISTRATION has a compulsory one-to-one relationship to entity STUDENT, then a certain action logic is implied by this that can be generated by the software. All this provides the detail of the process logic upon which any subsequent code generation will be based. In certain situations, IEF can be used in analysis without the preceding planning stage having been undertaken, in which case all information is entered from scratch in the analysis software.

3 Design

In IE terms, this includes support for both business systems design and technical design, and the design software enables the designer to take the results of analysis and transform them into designs. The first stage is for each process to be transformed into a set of procedures by the dialogue design software. Dialogue flow diagrams are produced, and the software allows the designers to specify control and sequences of screens. Next, the screens themselves are designed using the screen design tool. This will automatically produce an initial attempt at the screen design, based on the information in the encyclopaedia, which can then be modified. The software also attempts to provide previously defined screen elements to the designer and so encourage consistency in design. Indeed, designers can create templates that enforce standards across applications. The screens are then prototyped, showing the layout and flow of screens, and input validation can also be performed using the tool.

The process action diagrams from analysis can also be converted into procedure action diagrams using the action diagramming software. Designers can specify the detailed logic statements and associate processes with commands from the keyboard, specify run-time error routines, and so on. Again, as in the process action diagrams, the software identifies potential errors and prevents them occurring.

Up to this stage, the design has been at the logical level and not dependent on any target hardware or software environment. Technical design now requires the business systems design to be taken to the next stage, which is physical design where the target environment is specified. Using the software, the designers can specify the target environment and the constraints that this implies. Common specifications are already stored in the encyclopaedia. Physical design is then initiated, and the data model is transformed into a physical database design of records, fields, linkages, entry points, and so on, depending on the actual database specified in the physical environment. These would be tables, rows, columns, and foreign keys in a relational database environment. The physical design can be modified and tuned by the designers as necessary, usually for performance reasons. Such changes do not modify the conceptual designs and only changes as business requirements change the logical models.

4 Construction

The next stage is construction, and the automatic production of complete application systems by the generation of code. The software supports the generation of code in a number of languages and embedded SQL calls to the database. The tool also produces screen definitions, graphical user interfaces, database definitions, referential integrity triggers (to control database deletions), and a transaction control program. The code generated is based on the logic specified in the action diagrams and the entity models.

Developers can test and modify the code without tampering with the source code. Changes are made to the code by changing the action diagrams (or further back in the analysis stage) and regenerating the code. The software 'remembers' any changes made for tuning purposes and then reapplies them to the regenerated code. Further, for small changes, not all the code needs to be regenerated and, using various dialogues, the developers can specify which components require regeneration.

5 Implementation

Finally, the implementation software (which resides on the target hardware) enables the installation of code and database on the computer. This includes the compilation, linking and binding of the application, and the allocation and building of the database, plus a facility that enables the running and testing of the application on the target computer.

6 The encyclopaedia

The encyclopaedia is the key to IEF and enables the storing of models, the concurrent use of these models, the progression from one stage to the next, and the transformation of the models ultimately into complete developed applications. The main encyclopaedia resides on the server, and subsets of the models can be downloaded to workstation clients to enable individual work and to enable teams to work on developments concurrently. Any changes to models are then uploaded back to the main encyclopaedia in a controlled way from the local encyclopaedias to ensure consistency. The encyclopaedia also provides version control, which means that multiple copies of models for different purposes can be stored and used without confusion.

The encyclopaedia stores definitions of models rather than graphical representations of the model, and therefore the software uses these definitions to produce the diagrams that are required. Many of the diagrams are based on common information, and this enables a high degree of consistency between diagrams as they are constructed as and when necessary from the stored definitions. This also enables any changes or updates made by developers to one diagram to be reflected in other diagrams automatically. For example, in an entity diagram, a change to a relationship between two entities from optional to mandatory will automatically be reflected in all other diagrams that use this information when they are next displayed, because they will access the one definition of that relationship in the encyclopaedia. The storing of definitions in the encyclopaedia rather than the storing of graphical representations not only enables models to share information but allows the information to be passed forward to subsequent stages easily.

The encyclopaedia also enables consistency checking. This can be initiated at any stage and on the whole development or subsets, such as an entity model, functions, processes, action diagrams, and so on. Some checks are enforced when information and definitions are entered. For example, an attempt to use the same name for two functions or entities would be highlighted as they were entered. Other checks will only be made upon request or before proceeding to a subsequent stage. An entity model, for example, would be checked for completeness, in that it has attributes and relationships, and that these are consistent with the functions and activities defined, before proceeding from analysis to design.

In this section we have concentrated on the functions of IEF that support the development of applications following the Information Engineering methodology. As a product, Computer Associates have developed Advantage Gen to be appropriate to a wider range of application types, including application integration, web services and links with other tools (see CA, 2005).

19.3 Oracle

History

In 1979, Oracle was the first commercial relational database based on the relational language SQL. During the 1980s the company developed its database software. Through this tool in particular, Oracle became the second largest software company in the world, after Microsoft (Sections 18.3, 18.4 and 18.5 describe only three of its software packages) and before SAP (Section 9.3). Originally Oracle consisted entirely of the relational database system and a few trappings. It was designed to be used by professional programmers. Later in this period tools were provided to be used with the Oracle database which made it easier for use by users as well as IT professionals. CASE tools, similar to those provided by Information Engineering Facility (Section 19.2) were provided. The major difference, however, was that IEF was designed as a toolset supporting one IS development methodology, Information Engineering. Oracle's tools, on the other hand, could support many methodologies. Oracle users would purchase the database, but it was possible to purchase some tools that make up the Oracle product range but not others.

During the 1990s there were a number of toolsets, each with its suite of tools. These included the Oracle database suite, Designer/2000, Developer/2000 and Discoverer/2000. Other tools provided are outside the scope of this chapter. Many Oracle applications are developed in a PC-Windows environment, but Oracle has been designed to run on many hardware/software platforms. In this section, we will discuss this range of products and then discuss the present enterprise grid architecture, released in 2003. A full discussion of Oracle would require several books, so we provide here an overview of features particularly relevant to the IS development process. Even so, we omit many interesting features, such as the customer relationship management suite (released in 1998) and the web development suite (released in 2000). In particular, we also omit discussion of Oracle's ERP environment, including customer relationship management. With its Project Fusion, Oracle's former ERP application suite is being merged with features of its acquired companies JD Edwards and PeopleSoft. You are encouraged to explore the website at www.oracle.com. There are some interesting presentations (video and/or sound) included in the site.

Oracle 2000 suites

Designer/2000 was formally known as Oracle CASE, and is helpful in data and process modelling. Developer/2000 is used to build an application, once it has been designed using Oracle Designer/2000. Discoverer/2000 consists of a suite of user-friendly query tools designed for ad-hoc reporting. The Oracle Database Management System is central to these tools, though many end-users are hardly aware of its presence. We will first look briefly at the Oracle Database Management System and then look at Designer/2000 and Developer/2000.

Oracle Database Management System

Being a relational database system, we see data expressed in rows and columns. The main method of communicating directly with the Oracle database is by using the SQL language. Although this stands for Structured Query Language, it enables the experienced user to do more than just handle queries. It is certainly easier to use than conventional computer programming languages, but it is nevertheless not easy to use by untrained users. There are extensions to the basic SQL language provided in Oracle and together they form PL/SQL. The SQL optimizer attempts to make each SQL statement as efficient as possible when executed. SQL procedures can be triggered by certain events, for example, after updating or deleting a record (row).

The security features are now very sophisticated; you need system privileges to access the database and object-level privileges at different levels to query, insert, update and delete any object stored in the database. Security also implies the ability to roll back to the previous database state should there be a problem with a particular transaction. Data is validated through data constraints, that is, only allowing permitted values of data to be entered into the database.

Designer/2000

The aim of these tools is to construct applications based on the Oracle Database Management System as quickly as possible; indeed its use is often referred to as rapid application development. Most of its tools are either data modelling or application-generating tools. The latter are mainly used for form and report generation, but it can also be used to generate other programs in VisualBasic and for reverse engineering. They build up and require a central encyclopaedia of information about business processes and functions as well as the data resource.

The modelling tools provided include those supporting entity-relationship, function hierarchies, data flow and process flow diagrams. There is also help to define matrices, for example, to cross functions with entities and functions with attributes, and write reports. To give only one example of the power of the tools, the repository reporting tool has more than 100 standard reports and there are also various utilities provided to help the database administrator set up, maintain and use the repository.

There is a generator to define the subsequent database, along with its relations, constraints and so on. Help is provided to support normalization (Section 11.2) and resolving m:m relationships and other potential database design problems. Application module definitions are also supported that can specify screens, reports and PL/SQL subprograms. 'Wizards', similar to those provided in the Microsoft Office suite of programs, can be used to guide the designer to generate the database and applications. Again, diagrammer tools support designing the structure of each module and many other tasks. Relevant documentation can be generated at the same time, including user and training manuals. Along with supporting a 'brand-name' methodology such as Information Engineering, these tools can be used for rapid application development (Chapter 7) and other approaches. The process modeller is designed to support business process re-engineering and reverse engineering of existing systems by representing the essential aspects of each process.

The forms generator comes with a number of standard form templates that can be customized for each form. Constraints can be added to ensure data entry using the forms is validated, such as a range of numeric values or one of a list of possibilities provided in a combo box. Similarly, there is a reports generator which wherever possible has a similar look and feel to the forms generator. It is possible to see the SQL statements generated and modify this generated program directly. Oracle Designer/2000 can also be used to generate programs in VisualBasic and more recently web applications using the webserver generator and charts using the graphics generator.

Of course, such tools and diagrammers can only support information systems development, they do not do away with the necessity of having a good strategy for applications development, good planning to carry out the strategy, good fact-finding procedures and communications, and good analysis and design.

Developer/2000

The main features of this development suite are Developer/2000 Forms, Reports and Graphics that run with the Oracle DBMS and the application server. This latter tool carries out the processing to produce the reports, etc., and is integrated with the web server, for web-based applications, as well as the Oracle DBMS. It should be remembered that the toolset comprising Oracle Developer/2000 consists of very powerful, complex and sophisticated elements and is not really designed for the end-user, who is much better suited to the Discoverer/2000 toolset, with its data browser and data query tools. Further, to operate these applications efficiently requires a well-designed, modelled and normalized database. The Oracle toolset supports the production of such databases, but human skills are still crucial.

Oracle Developer/2000 Forms can be used to create data entry screens and menus, but it can also be used to start programs and other Developer tools and create database applications. PL/SQL libraries (of reusable SQL code) and object libraries can be used so that Forms becomes a very powerful applications generator. Oracle Developer/2000 Reports enables report generation. Some reports can be used interactively. The efficient use of Reports requires a good knowledge of SQL. Oracle Developer/2000 Graphics is used to add visual 'splendour' to a report or form. Again, though the results may be user friendly, their creation requires technical expertise, though over 50 predefined chart templates are provided.

Oracle Developer Suite 10*g*

The Oracle enterprise grid architecture was released in 2003 and is referred to as Oracle 10*g*. The **grid** concept is meant to represent the server and storage technology and tool support as one integrated platform. We will look at the Oracle Developer Suite 10*g*, which is part of that initiative and is the most relevant to the subject of this chapter: what toolsets are available to support the process of developing information systems, that is, tools which aim to make the developer more productive. It consists of many tools covering most aspects of IS development, but these are integrated so that the components are able to 'talk' to each other as well as to the developers in a consistent way.

The tools consist of:

- **Team support**, which facilitates communications between developers, for example, consistent software configuration and version control between development teams;
- **J2EE programming**, enabling Java design, programming and development, covering the full life cycle, including the generation of application code from tool-supported UML diagrams;
- **Reporting**, to enable the easy creation of reports based on different sources, such as relational files and XML files and, using Oracle Discoverer, end-user querying (the end-user need not, of course, know or understand the database structure and other technical aspects, though the database administrator needs to ensure end-users have appropriate security privileges);
- **Modelling**, to help the modelling and design of the applications, including the design of the supporting databases from diagrams and reverse engineering legacy database applications to the Oracle standard;
- **Business intelligence**, to support the effective use of organizational knowledge by querying the databases and the data warehouse, including end-user query and analysis

features providing visual display of tables, cross-tabulation and graphical analysis, and there is also an Oracle Warehouse builder available for the design, population and management of the data warehouse;

- **Rapid application development tools**, which along with reporting and modelling, are developments from earlier case and designer tools, for example generation of applications, including web applications, from forms for developers experienced in using SQL 2003; and finally

- **Web services development**, for example, using forms to develop web applications. It also includes Oracle Designer, based on Designer, discussed above.

Although the complete toolset is complex and expensive, it does provide a consistent and integrated package to try to merge fragmented heterogeneous and monolithic legacy applications as well as all aspects of IS development, including professional and end-user support so that developers, business analysts and others collaborate on rapid process development.

19.4 Select Enterprise

Select Software Tools was a software company that developed an early CASE tool that evolved over the years as a professional toolset which has also been widely used for teaching purposes, particularly in UK universities. Select is now owned by Aonix, and marketed by Select Business Solutions. Aonix is itself owned by the Gores Technology Group known for its Software through Pictures (StP) product. The product appears to have been subsumed into a number of other products, including Select Component Factory. Select Enterprise itself is still available in Europe and is described as 'allowing the modelling of complex IT applications before commencing development, uncovering costly problems earlier in the development life-cycle. It also acts as a store for intellectual property, which is invaluable during system maintenance, improvement and integration. Select Enterprise gives you the modelling techniques to design the right system, and design it right' (Aonix, 2005). It is this product that we will discuss here as being most relevant to the subject matter.

Select Enterprise is a much expanded version of the original Select tool, designed for object modelling and supporting UML. Amongst other features, Select Enterprise has the following:

- Business process modelling
- UML profile
- Graphical simulations of UML designs
- Database modelling and code generation
- Design patterns and optional component-based techniques
- Scaleable Enterprise Repository
- Intelligent document generation
- Traceability and impact analysis
- Java, VisualBasic, C++ code synchronization
- Integration with a range of other tools.

As well as Select Enterprise there are a number of related products. For example, there is a *Reviewer for Select Enterprise* which locates errors in syntax and UML (Section 13.2); and *Select SSADM* and *Select Yourdon* which support users of these information systems development methodologies (Sections 21.1 and 20.2 respectively).

The tools of the Select Enterprise product (Version 6.1) are illustrated in Figure 19.3. As can be seen, the repository is the centre of the system, which contains the various data stores and usually resides on the server. The repository management tool is the Repository Administrator, which manages the data stores. The Models Neighborhood is for managing the various models on the client workstation. The Select Enterprise tool itself enables UML modelling and more. Integrated with Select Enterprise are a number of other tools, for example, the Document Generator for creating presentations from models and the Model Copy tool which copies diagrams and other items between models.

The Select Enterprise user interface (Figure 19.3) has four separate areas or windows:

- The Explorer window (left-hand side) contains folders. There are four sets of folders available in this window accessed by various tabs (see bottom of window):
  - diagram types within the model (this tab displayed) (i.e. Process Hierarchy Diagrams, Process Thread Diagrams, Use Case Diagrams, etc.);
  - dictionary items within the model (a folder for each item);

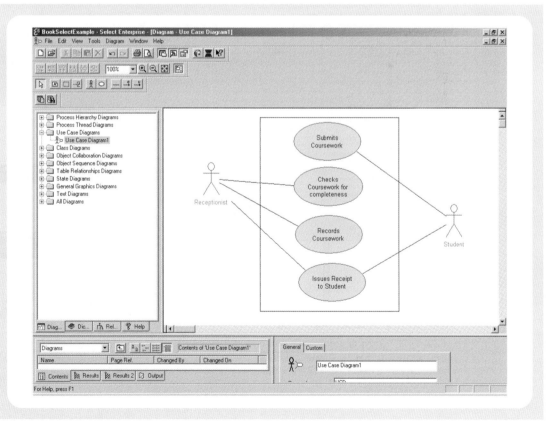

Figure 19.3: Select user interface

- relationships which show associations between dictionary items and their hierarchies; e-help contents.
- The Output window (lower left) which has the following tabs:
  - contents displays details of the currently selected diagram or item in the Explorer window; this tab can be used to navigate through the models;
  - results displays the model objects according to the previous action performed;
  - output displays textual information (e.g. results of a consistency check).
- The Diagram window (main) displays the selected diagram (a use case in this example) and is used to create, edit, and generally manage the diagrams.
- The Property page (lower right) displays properties for the currently selected diagram; the tabs available depend on the actual diagram selected.

Figure 19.4 shows the Tools menu, which gives an indication of the various tools and features available in Select. The Bridge tools interface with other products (e.g. Platinum's ERwin and Mercury's TestDirector).

The Repository Administrator window is illustrated in Figure 19.5. It is accessed from within Select via the Tools menu (Figure 19.4) and is used to create, start, and stop data stores (a data store has to be started before it can be used) and backing up and recovering data stores.

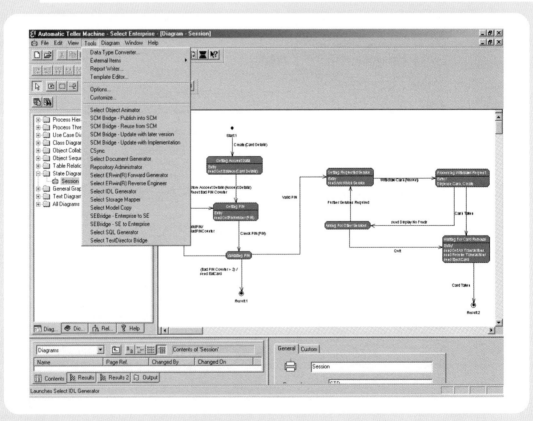

Figure 19.4: Tools menu

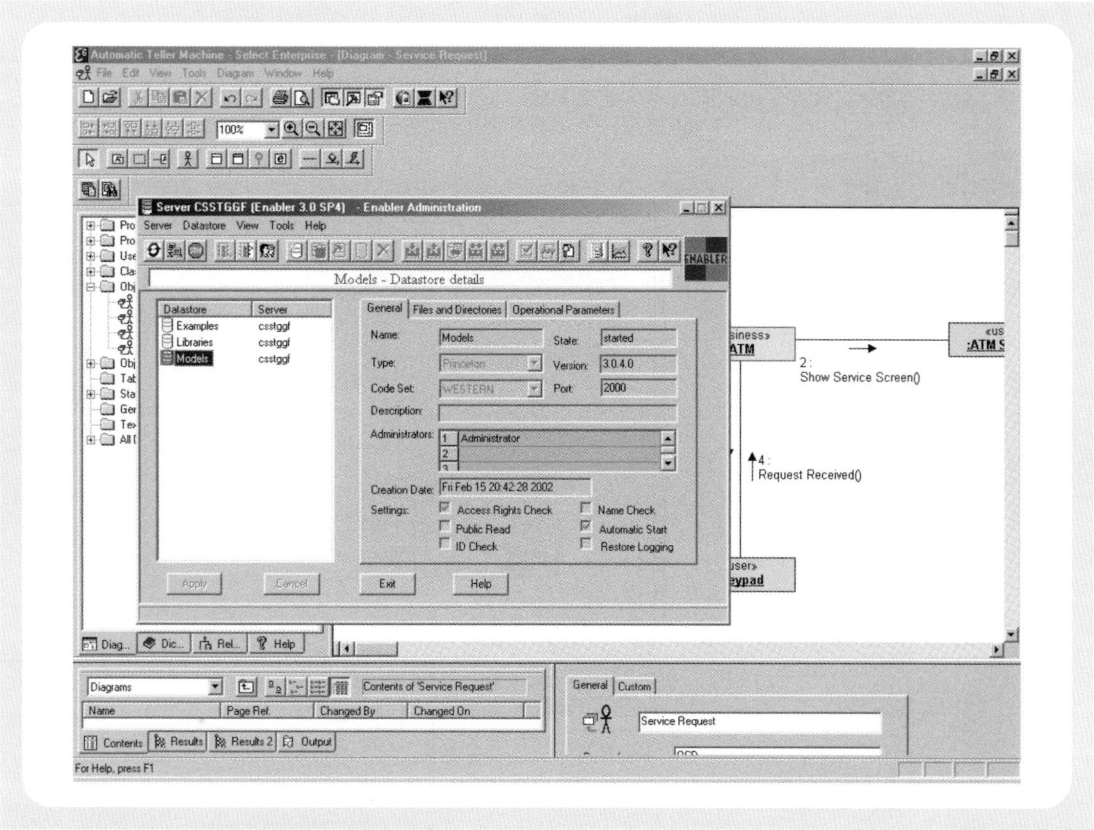

Figure 19.5: Repository Administrator Window

Figure 19.6 shows a Class Diagram displayed together with some menus accessed with a right click on the mouse button showing the various options for Association, Aggregation, etc. The functions available depend on the item selected.

Figure 19.7 displays an example Sequence Diagram (see Section 13.2 for a discussion of these UML diagrams). Figure 19.8 illustrates the Object Animator, one of the facilities of Select. The function is accessed via the Tools menu (Figure 19.4) and it animates the object interactions that are defined in a Sequence Diagram. Graphically representing objects and their interactions helps users and others to understand the sequences defined in a Sequence Diagram. Obviously the dynamic interaction cannot be shown in a static screenshot, but in Figure 19.8 the line made up of little squares has to be imagined moving from object to object, indicating the sequence of activities that occur.

As a further illustration of Select, the UML diagrams of Section 13.2 have been generated using Select Enterprise (except the Activity Diagram, Figure 13.9, which was developed using Rational Rose).

19.5 Discussion

We will now look at some of the potential benefits and problems that might accrue from using integrated toolsets. We begin with the potential benefits.

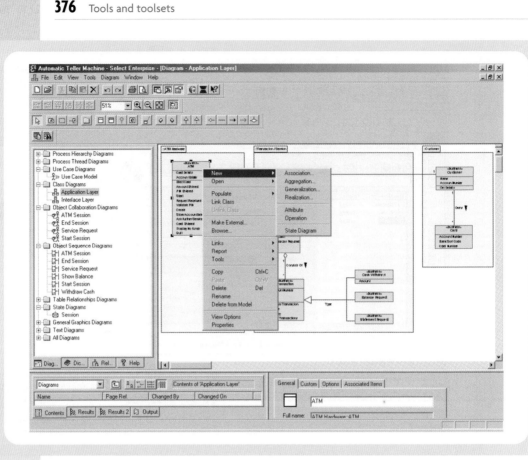

Figure 19.6: Select Class Diagram

1 Improvements in management and control

Applications development, particularly for large projects, is inherently difficult to manage and control. The process must therefore be tightly managed, and the IT profession has not historically been very good at this, particularly in the areas of estimation and keeping to budgets and schedules. Toolsets can help in this process by providing a central repository of information concerning the project, including rules and standards to be followed, and experience from other projects, such as the length of time certain activities actually take in the organization. They can also help with estimation, risk analysis, project planning, and the monitoring of project progress. Particular support for techniques may be included, such as function point analysis, CoCoMo, PERT, and critical path analysis. These were discussed in Chapter 14. Toolsets can also structure the work of developers; for example, by handling the devolvement of tasks to developers so that work can be completed in parallel and the subsequent recombination of the work put into a coherent whole. They can also support the change control process and ensure that new versions and releases are well organized and managed.

2 Improvements in system quality

The problems associated with specifications have already been discussed. It is argued that tools can help overcome these problems by providing better and more complete specifications

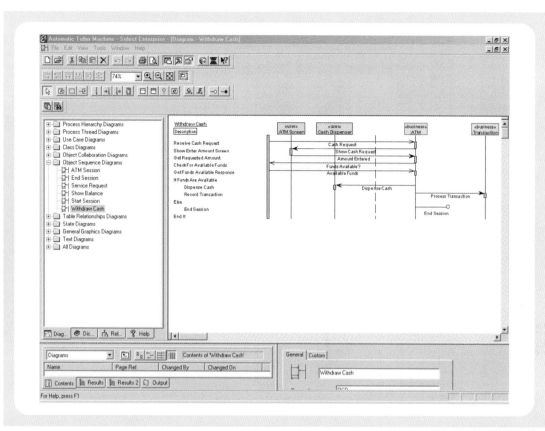

Figure 19.7: Select Sequence Diagram

through the use of diagrammatic representations that are easily modifiable by developers and users. These should also represent the real requirements better, partly because there is less resistance to changing diagrams because the tools make it easy.

3 Improved designs, better reflecting the specifications

Another problem of systems development is that the designs produced do not always accurately reflect the requirements specification. This can be the result of incomplete or conflicting information, and designers often make guesses or opt for whatever is the easy solution. Tools can help by providing the necessary information from the specification, having it available to designers from the repository, and by automating some of the process. This also helps to produce consistent designs, including that across applications.

4 Automated checking for consistency according to the rule base

Tools can automatically check the consistency of information input at the analysis and design stages, including information input using models and diagrams. They can highlight information that has been missed, areas that have been missed, interfaces that do not match, as well as incomplete information. This kind of automated consistency checking can be based on a set

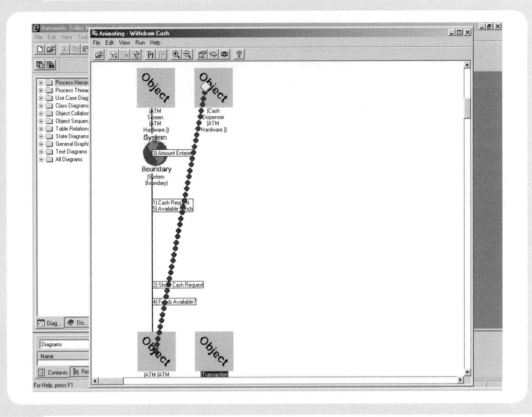

Figure 19.8: Select Object Animator

of rules concerning the methodology as a whole or those of the various diagramming techniques themselves. These rules would be included in the systems repository. Such checking should improve the consistency and quality of deliverables and thus the final information system, leading to less reworking and change at later stages.

5 Greater focus on analysis rather than implementation

It is often argued that the use of toolsets enables and encourages the focus of development to be changed from the later stages of the life cycle, such as design, coding, and implementation, to the earlier stages of analysis and requirements determination. Such a change of emphasis is likely to lead to better quality systems, as problems are detected and corrected at an earlier stage than with purely manual methods. The earlier in the development process that problems or errors are detected, the cheaper they are to correct.

6 Enforcement of standards and consistency

The use of tools can also help with the definition and enforcement of various standards in development. The tool itself often embodies certain conventions and standards that can help to ensure consistency in the development of individual projects and across different projects in an organization. The tool may enforce discipline by not allowing developers the freedom to ignore or contravene certain rules, procedures, and standards. It can also ensure consistency in the use of techniques, definitions, and terminology in the organization. As well as enforcing standards, they may ensure that the rules and objectives of a particular information systems development methodology are followed. Again, this may be achieved by the tool not allowing the developers the freedom to diverge from the requirements of the methodology and by the tool itself adhering to the methodology in what it does.

7 Improvements in productivity

Perhaps the most cited benefit of using tools is that of improved productivity due to a number of factors. First, it is argued that information systems are developed more quickly than with conventional methods. This is obviously a very attractive benefit, given the enduring problems of systems development in this respect. Faster systems development is achieved by improved management and control, the ability to create and change diagrams and specifications, and the automation and elimination of various manual stages, including the automatic generation of some aspects of design and the automatic generation of code. This latter benefit can potentially make a significant improvement to productivity, as the writing of code has always been a very labour-intensive part of the development process.

The second element is the ability to develop systems with fewer people. The automated support for much of the process and the automation of some tasks, it is argued, means that fewer developers are required. With fewer people, there is the added benefit that the number of interfaces and the communication required between developers are also reduced, which is also likely to enhance the speed and quality of development. The law of diminishing returns applies to systems development. This suggests that after a certain figure is reached, the more developers that are added to a project, the longer it will take to finish. This indicates that a kind of inverted economy of scale may apply to systems development, because the more interfaces there are between people, the more slowly things happen.

The third element contributing to improved productivity is the reduced costs of development. This is essentially an effect of the first two factors, that is, faster development and fewer people.

A fourth element of improved productivity is the ability to reuse existing development objects. The information captured by a toolset over a number of projects may eventually provide a repository of models or objects of various kinds that can be used again. These may include analysis and design models of all types and libraries of common code that can be utilized in future developments. Depending on circumstances, these models may be used in their entirety or are amended according to the requirement of the new project. This can save a significant amount of development effort and help achieve consistency between applications as well, as the standards in the original models will be incorporated into the new developments.

8 Reductions in maintenance

It is argued that the use of tools helps reduce the large degree of effort required for both maintenance and enhancement of existing systems. First, they can produce good and consistent documentation, which can lead to easier maintenance. Second, the better quality specification and analysis means there will be less change and thus less maintenance. Third, the improved design and implementation, including some automation, results in fewer errors at the programming, testing, and implementation stages, thus leading to less maintenance.

Accurate and effective testing is also an important element in reducing the maintenance load. Traditionally, this has been carried out by a separate group of people because the developers themselves were not trusted to test a system that they had developed effectively. Programmers were thought to be the worst people to test their own systems, as they assumed the system would work because they had written it. The consequence of having a different set of people performing the testing is that they have no knowledge of how the system was developed. Further, they could not take advantage of any verification and validation potential that the development methodology might have to offer. Toolsets can provide this enhanced testing by helping to administer and control the activity and by generating test data, applying (or even simulating) it, and analysing and comparing the results. The type of test data generated can be derived from, and reflect, the requirements of the analysis and design stages. For example, if there is a requirement that when a particular process is invoked then a subsequent process must also be performed, then this can be captured as knowledge in the repository and a particular set of tests and relevant data automatically produced. An example may be that a debit from one account must be accompanied by a credit to some other account. Such testing is more likely to test the requirement effectively than random testing, and the benefits apply both to the initial systems development and any subsequent maintenance or enhancement to the system.

Further, the traditional form of maintenance is made obsolete because changes are not made by directly reworking the code in response to errors and changing requirements, but by going right back to the analysis and design stages and amending the original diagrams and specifications and regenerating the code automatically. This helps eliminate the frequently encountered problem of introducing new errors as a result of correcting existing ones.

9 Re-engineering (or reverse engineering) of existing systems

The problem in many organizations is not so much that of developing new systems, but the maintenance and enhancement of their old systems, some of which are based on 1960s designs, third generation languages, and dated file and access methods. Re-engineering is the application of tools, techniques, and methods to extend the useful life of application systems cost-effectively. Re-engineering changes the underlying technology of a system without affecting the functioning of that system. Therefore, for example, the hardware platform and environment, including the programming language of the applications to reduce maintenance costs, may be changed. The rapid developments in technology have rendered many existing systems, even some that are relatively recent, obsolete, not in the functional sense, but in the programming language used and the hardware on which the system runs. Such systems can carry a high maintenance workload and be difficult to enhance. Further, many manufacturers refuse to maintain old hardware, and it is difficult to integrate legacy systems with more recently developed systems. For many organizations, the cost and resource implications of scrapping these old systems and redeveloping them from scratch on new hardware platforms are prohibitive. The normal use of tools provides support for 'forward development', that is, the top-down, linear approach to the development of new systems. Some tools are designed to support reverse engineering as well, providing the ability to capture the primary elements from current systems, such as their process logic and the data they use, including entities, attributes, names, locations, sources, edit criteria, and relationships. From this captured information, the tool can help to clean up the data definitions, produce entity models, restructure the process logic and build process hierarchies, and construct the repository for the old system. The tool can then be used in the normal forward development mode to produce new systems.

10 Strategic contribution

The potential for information systems to contribute to the achievement of the strategic objectives of the organization has been discussed in Section 4.2, and the use of these tools can help by improving the quality of systems and the speed at which those systems are developed and enhanced. Additionally, the planning elements of an integrated toolset may help to identify and prioritize those systems which are most likely to contribute to the business strategy.

11 Improved responsiveness

This is really a function of improved maintenance and enhancement of systems that have already been discussed, but the particular element emphasized here is that systems developed using a tool are likely to be more easily and quickly enhanced leading to improved responsiveness to changing and evolving business needs.

12 Portability

Many toolsets make it easier to move systems from one hardware platform or environment to another. This is really a function of the ability of some tools to generate code for a variety of different languages on different hardware platforms. For a particular application, the code can be regenerated for a different environment without affecting the functionality of the application. While this makes it easier, it does not in practice mean that there is no manual intervention required.

13 Keeping up with the state of the art

Some people argue that tools enhance the credibility of the information systems group. It indicates to the rest of the organization, and perhaps the world at large, that they are at the cutting edge of the latest technological developments. This is perhaps not a totally justifiable benefit, as it seems to be an argument for technology for technology's sake. A better justification is that it helps to attract and retain good information systems staff and increases satisfaction among developers.

Having identified the potential benefits of integrated toolsets, we will discuss the other side of the equation. The most obvious cost is that for the software and hardware which can be considerable, but probably the more important costs are those associated with the adoption of these tools:

- *Staff education and training costs.* These are costs that apply not only for the professional developers, but for users and user management as well. These costs are not just one-off, as is sometimes assumed, but are a long-term requirement, because of staff turnover and new versions of the tool.

- *Consultancy and training costs.* In some cases, the training may only be available from the vendor, and vendors may make most of their profits from consultancy and training. An associated problem relates to staff turnover, which tends to increase as experienced tool users and developers are currently much sought after.

- *Development of appropriate conditions.* The setting of standards, working practices, and the resolution of conflicts all need to be sorted out, and an appropriate environment and culture for the use of the toolset developed. Again this takes time and effort and is an initial cost of adopting a tool.

- *Integration of the new tool.* The toolset needs to be integrated into the existing development environment so as not to cause conflict. It is very unlikely that any organization will be able to change to such a development environment except in well-thought-out and managed stages. All this may require organizational change and will certainly take management time and effort.

- *Customization of the tool.* There can be quite a major effort required to tailor the tool for use in the particular organization. This will take time and other resources, and systems support and consultancy from the tool vendor, which may be expensive.

- *People's time.* Often the time of people using toolsets is not properly costed, as time is often assumed to be free and to have no opportunity cost. For example, if someone goes on a course, the cost of the course is usually included but not the time lost by that person. The time put in by users is also frequently ignored.

- *The cost of recruiting experienced staff.* Probably an organization will not train everyone from scratch but will seek some developers from outside who are experienced in using and managing the toolset. Recruiting costs can be particularly expensive as these staff are in demand.

- *Other hardware costs.* It is frequently the case that the hardware needs upgrading and more workstations are required as more projects are developed. Again, these costs may not have been included in the initial estimates.

Far too many organizations have ignored the softer costs in their cost/benefit analyses and concentrated solely on the direct hardware and software costs associated with integrated toolsets. The former can easily amount to two to five times the amount spent on the tools themselves.

The evaluation of costs vs potential benefits is difficult, and the whole area of toolset use is surrounded, as are many IT developments, by a degree of 'hype', much of it emanating from vendors trying to market their products, but also, perhaps, from overenthusiastic developers seeking quick solutions. The IT community is characterized by the greeting of new approaches and products with great enthusiasm and a belief that, contrary to previous experience, this latest innovation is going to solve all known problems. Inevitably, there is then a backlash against this overly optimistic view, and a certain pessimism sets in whereby people begin to condemn the innovation as either worse than useless or nothing new and simply 'old wine in new bottles'.

Organizations that jump on the bandwagon, expecting a panacea, experience difficulties and problems. They then turn against that particular innovation and vehemently condemn it before rushing on to the next one. The truth with any innovation is usually somewhere in the middle of the two extremes. We have already seen evidence of a backlash. This is perhaps not altogether surprising considering some of the hype; for example, one vendor suggests that productivity of 25 times that of traditional development can be achieved with their product.

There are further concerns related to the particular context of this book. The first is that the technology of such toolsets might distract people from the real issues of information systems development, that is, a concentration on the tool rather than the development approach that lies behind the tool, and as a result the tool being used indiscriminately and inappropriately. Second, the tool may force people to use some methods that are not relevant or well enough defined. It has been suggested that tools are sometimes purchased and used without enough thought being given to the processes that they enforce. In other words, the tool enforces a particular approach to systems development, and it is this approach that needs to be carefully considered rather than the look and feel of the software. Some companies have found themselves implementing a particular development methodology without quite realizing it, due to their use of a particular tool. A further issue is the degree to which the methodology is supported and the way the support is implemented.

The evaluation of any toolset is not a simple process, and the statement that a toolset supports a particular methodology is only the beginning of the story. The way in which the rules are enforced is also critical. A further problem is that the vendor may interpret the rules of the methodology somewhat differently to the author of the methodology or the organization adopting the tool.

Whatever the degree of improved productivity claimed, there appear to be a growing number of indications that achieving them is more difficult, and takes longer, than might be thought. The learning period for most toolsets is long. Both developers and users need time to learn, assimilate, and become effective in using the tools. In terms of productivity, it has been suggested that the learning curve (productivity benefits plotted against time) may fall in the early stages. The length of this early stage, before improved performance is reached, has been estimated to be between six months and two years. It perhaps makes more sense to measure the length of the learning curve, not in time, but in terms of numbers of projects, in which case it has been suggested that it is not until the third or fourth project when productivity benefits begin to accrue. Clearly toolset use is not an instant panacea and some organizations may not be prepared for the kind of long-term investment that is required. The introduction of a toolset must be handled with care and people's expectations relating to benefits, problems, and timescales need to be realistically managed.

It might be that the adoption of toolsets is most successful when it is seen as part of a process of changing the culture of an IT department, that is, when it is seen as a process of organizational development. It may also be seen in the context of changing the organization as a whole, perhaps empowering users or centralizing power. It may be analysed in terms of its effect on the hierarchical structure of an organization or as an element of organization learning. The tools are not implemented in an organizational vacuum, and there are many indications that success or otherwise is heavily influenced by a range of organizational and human factors. In essence, these are characteristics of the organizational fit of the tool, or, as it is sometimes termed, the compatibility of the innovation with its context:

- the management approach;
- power structures in the organization;
- the degree of organizational creativity;
- the organizational culture;
- work patterns;
- teamwork;
- the incentive and reward systems;
- perceptions of job security;
- satisfaction levels;
- the role of champions and sponsors;
- change agents;
- the history of innovation and experimentation in the organization.

Another important aspect of the organizational dimension of analysis is the maturity of the IT department and the software development process in the organization. As we saw in Section 4.5, there are a number of models of the 'stages of growth' of IT in organizations, and there are indications that certain stages of maturity need to have been reached to allow the effective introduction of toolsets. Further, the type of tool, and its objectives and justification, might be different depending on the stage of growth at which it is introduced.

A further organizational dimension concerns the way that innovations are adopted and diffused in organizations. One strategy might be *laissez-faire* in which the tool is adopted without any deliberate organizational encouragement. A second is 'cautious', which is a slow but deliberate approach. The third is 'active', which is fast and requires a high degree of organizational and managerial push.

They also identify three types of innovation in terms of the degree and nature of the change that occurs. The first is 'compatible' innovation if the tool fits in with and does not change current working practices, such as the methodology. The second type is 'incremental', where the tool involves only small changes to current working practices, and the third is 'radical', if it requires major change and differs significantly from current experiences.

A further dimension beyond that of the project is that of the individual affected by the tool. These individuals, or stakeholders, are, first, the developers (or the primary tool users), and their perceptions and feelings are important aspects in determining success. These perceptions can be analysed in relation to the degree of change to work practices, job satisfaction, reward, communication, teamwork, and so on. It has been argued that tools sometimes require the primary users to unlearn old practices and learn new ones, and that this may result in a perceived loss of status. Further, there are indications that such changes may be more difficult for older, more experienced developers to make. However, it is not always made clear whether these difficulties are the result of the introduction of a toolset or whether it is due to the introduction of an associated methodology. We suggest that the introduction of a tool together with a new methodology into an organization is a more difficult innovation than the introduction of a tool to support an existing and well-established methodology, simply because of the greater degree of learning (and unlearning) involved.

The reaction of individual developers is not always negative, as some perceive the use of toolsets as enriching their work. Others perceive it as de-skilling, reducing their creativity, and increasing the ability of management to exert control, in much the same way that supermarket checkout systems monitor and control their operators. The reaction of individuals appears to be difficult to predict, but it is likely to be a key element in the ultimate success or failure of the introduction of tools in an organization. The planning and management need to focus on addressing these personal perception and motivation issues.

A second set of individuals, who are potentially as important as the primary users, are the secondary users, that is, the people who use (or manage) the systems that are developed with the tool. They may be involved in the development process as well as being the users of the information system produced. Their perceptions of the tool, its impact, and effects are also important, although frequently forgotten. If the secondary users perceive that the tool results in better quality systems, or faster production of systems, or improved identification of their requirements, or enabling their greater participation, or whatever advantage, then this is likely to result in the organization as a whole regarding such tools in a favourable light. Of course, the reverse is also true. Unfortunately when tools are introduced there seems relatively little emphasis on involving the secondary users or recognizing them as an important component of success.

19.6 Framing influences

In Chapter 17 we looked at framing influences in relation to techniques. We argued, following Adams and Avison (2003), that the use of techniques may restrict people's understanding by framing the ways that they think about the problem situation because of the way that the problem is presented to them by the technique. In effect, techniques frame or blinker the way their users perceive the problem under review. We will not repeat the arguments here, but we suggested that the adoption of techniques should be based on appropriateness to the problem situation rather than because it just happens to be part of a methodology being used or the habits of the individual or organization. We also suggested that different types of techniques might be used together to provide different perspectives to a problem situation.

These arguments apply also to tools and toolsets, but sometimes even more so, because tools may well impose certain techniques on the user and limit choice to a small number of techniques offered in the package. Most techniques offered in such tools and toolsets are closed rather than open, and very few are non-prescriptive. We will consider some of the tools and toolsets discussed in Part V from this perspective.

Some tools, by definition, cover a limited area of application, and therefore the number of techniques offered will also be limited. For example, Microsoft Project supports project management, and the techniques offered are the conventional ones of resource allocation, networks, critical path analysis and so on. These are very structured techniques with input being factual data such as names, job details, times, inter-dependencies, cost figures and the like.

Again, using a package such as Access limits the user to using the relational database approach. Data is input by the user in tabular form and output to the user in tabular form. Data relationships are described in entity-relationship form. There are specific ways that the data can be represented and the database queried. All these may well be appropriate, but because of the impact of the framing effect, potentially limiting. Indeed it could have a worse impact because information may be lost if it is not possible to represent some of the information in the models and techniques used by the tools. There are no ways, for example, of representing the application area as a rich picture and the system interpreting and converting it into an Access database.

The drawing package Visio does offer a range of drawing possibilities, including the more open mind-mapping technique, for example. However, most of the techniques that the drawing tool supports are structured ones, along with the more structured methodologies (see Figures 18.13 and 18.14).

GroupSystems supports group discussion, and of course group discussions are potentially open and unstructured. But using the facilities of GroupSystems and similar groupware may impose potentially undesirable structure and reduce flexibility. Some of the spontaneity and potential changes of direction possible in a face-to-face meeting may be restricted in a computer-facilitated meeting.

Some toolsets are designed for use with a particular methodology. Information Engineering Facility provides such an example. This imposes much on the user, certainly restricting the range of techniques to be used, but perhaps also the way in which they are used and when they are used. Automating aspects of the information systems development task even further by using Oracle Designer and Developer or Select Enterprise, for example, can impose techniques on the users, whether they are aware of this restriction or not. They may be frustrated if they think alternative techniques would be more appropriate.

Users need to be aware of the restrictions that tools and techniques impose on them and are therefore recommended to use as wide a range of techniques as feasible if available with the tools or even to go outside this range where those provided are too limiting. A balance of techniques, covering both the open and closed paradigms and prescriptive and non-prescriptive paradigms, is recommended.

Summary

- A toolset is any integrated computer software system that is specifically designed to support a significant part of the information systems development process of an information system and the management of these tasks and processes.
- A repository contains information about the physical and operational elements of data and processes. It will also hold the rules of a technique or an information systems development methodology thus permitting analysis, validation, consistency, and completeness checking.
- In this chapter we looked at three commercial toolsets and discussed some of the potential benefits and problems that might accrue from using integrated toolsets.

Questions

1. If you have access to one of the toolsets discussed in this chapter (or an alternative), use it to draw some individual diagrams, for example those of Question 1 in Chapter 18.
2. Use the toolset to develop a small part of a system. Identify how the models integrate and how changes in one model are reflected in others.
3. Does the toolset follow the 'rules' of a particular methodology?
4. Can the toolset be customized to adapt to alternative:
   - methodologies;
   - models;
   - notations?
5. What are the benefits of using this particular toolset as compared to the theoretical benefits identified in the chapter?

Further reading

Allen, P. and Frost, S. (1998) *Component-based Development for Enterprise Systems: Applying the Select Perspective* (Managing Object Technology Series No. 12), Cambridge University Press, Cambridge.

Stone, J. (1993) *Inside ADW and IEF: The Promise and Reality of Case*, McGraw-Hill, New York.

Stowe, M.W. (1999) *Oracle Developer/2000 Handbook*, Prentice Hall, New Jersey.

Part VI: **Methodologies**

Contents

Chapter 20 – Process-oriented
       methodologies   p 395

Chapter 21 – Blended methodologies   p 419

Chapter 22 – Object-oriented methodologies p 451

Chapter 23 – Rapid development
       methodologies   p 469

Chapter 24 – People-oriented methodologies p 487

Chapter 25 – Organizational-oriented
       methodologies   p 507

Chapter 26 – Frameworks   p 537

We begin, as in Part V, with a 'road map' of the relationships between themes, techniques, tools, and methodologies. We adopt the perspective of methodology in this 'road map' (see Diagram 4) and use SSM and WISDM to show the links to themes, tools, and techniques.

In Part VI, we look at a number of information systems development methodologies that are well used, respected, or which typify the themes described in Part III. Again we had a problem devising categories wherein each methodology fits like a glove. We surely have not succeeded!

We first look at process-oriented methodologies in Chapter 20. The first methodology described reflects the process modelling theme and was proposed by Gane and Sarson (1979). The main techniques used are the process-oriented ones of functional decomposition, data flow diagrams, decision trees, decision tables, and structured English. Functional decomposition gives structure to the processes reflected in particular by the most important technique of data flow diagrams. This emphasis on structure gives the name of the methodology: Structured Analysis and Design of Information Systems (STRADIS). Yourdon Systems Method (YSM) was originally very similar to STRADIS; indeed, Gane and Sarson were at one time colleagues of Yourdon. However, more recent versions of YSM suggest that a 'middle-up' approach to analysing processes called event partitioning might be more appropriate than the top-down approach (functional decomposition). Although emphasis is placed on the analysis of processes, when compared to STRADIS, there is greater emphasis on the analysis of data as well.

Jackson's program design methodology, Jackson Structured Programming (JSP), has had a profound effect on the teaching and practice of commercial computer programming. Jackson Systems Development (JSD) is a development from JSP into systems development as a whole. An information system is seen, in effect, as a very large program. The approach is somewhat different from the methodologies described before as it concentrates on the design of efficient and well-tested software, which reflects the specifications. It is particularly relevant to applications where efficiency is paramount, for example, in process control.

In Chapter 21 we look at methodologies which exemplify blended approaches. Structured Systems Analysis and Design Method (SSADM) has been the standard in most UK government applications and can be said to be the more modern version of the traditional information systems development life cycle approach discussed in Chapter 3. It includes the techniques of data flow diagramming and entity life histories, and recommends the use of toolsets.

Merise is a widely used methodology for developing information systems in France. Unlike other methods described above which emphasize either process or data aspects of information systems analysis and design, Merise has been designed so that both are considered equally important, and these aspects are analysed and designed in parallel. It now incorporates an alternative object-oriented approach.

Whereas STRADIS and YSM emphasize processes, Information Engineering (IE) has more emphasis on data. Similarly, whereas the fundamental techniques of the process-oriented approaches are functional decomposition and data flow diagrams, the basic approach in IE is the data-oriented entity-relationship approach. However, like the development of methodologies described in Chapter 21, IE has process-oriented aspects embedded and has been extended to include a planning phase, which is the first phase of the methodology, reflecting some of the discussion in Section 4.4.

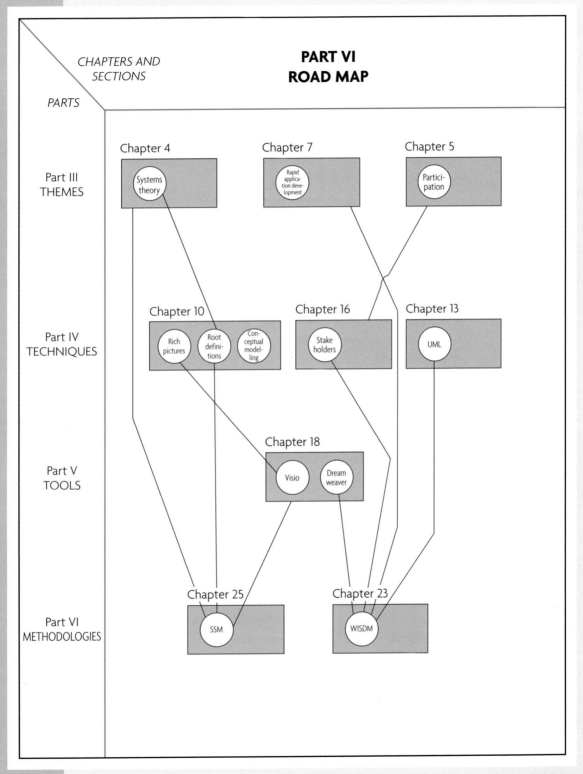

Diagram 4: SSM and WISDM methodologies 'Road Map' (shows relationships between methodologies (SSM and WISDM) *and* themes, techniques and tools)

Finally, in this chapter, we look at the methodology proposed by Welti (1999) for implementing enterprise resource planning systems, which can be regarded as a blend of applications rather than techniques and methods.

Chapter 22 looks at the object-oriented approaches, object-oriented analysis, and RUP. Coad and Yourdon's (1991) Object-oriented Analysis is significantly different from the approaches that have been discussed so far in Part VI. It is an approach that reflects the view that in defining objects (both data and processes encapsulated together) we capture the essential building blocks of information systems. It is also a unifying approach, as analysis and design can be undertaken following this approach, and applications developed using object programming languages and toolsets. The object-oriented theme leads to consistency throughout the development process.

The need to develop information systems more quickly has been driven by rapidly changing business requirements. Rapid Application Development (RAD) is a response to this need. It is based on the evolutionary and prototyping approaches discussed in Sections 7.1 and 7.2 (see also Section 7.3), and is usually enabled by toolsets. User requirements in RAD are often determined through joint applications development (JAD), which was introduced in Section 16.2. So Chapter 23 looks at methodologies aimed at rapid development of applications, James Martin's RAD, DSDM and extreme programming, along with WISDM, an approach designed to develop web applications quickly.

People-oriented methodologies are looked at in Chapter 24. Effective Technical and Human Implementation of Computer-based Systems (ETHICS) is a methodology proposed by Mumford (1995). It is a people-oriented approach (based on participation, Section 5.1) and, as the name implies, attempts to embody a sound ethical position. It encompasses the socio-technical view that, in order to be effective, the technology must fit closely with the social and organizational factors in the application domain. KADS is an approach to develop expert systems applications, whereas CommonKADS is broader and used to develop knowledge-based systems. They are people-oriented in the sense that they attempt to capture the expertise and knowledge of people in the organization.

Chapter 25 introduces what we have categorized as organizational methodologies. Soft Systems Methodology (SSM), a methodology proposed by Checkland and Scholes (1990), is influenced by systems theory (Section 4.1). Whereas many of the earlier approaches stress scientific analysis, breaking up a complex system into its constituent parts to enable analysis, systems thinking might suggest that properties of the whole are not entirely explicable in terms of the properties of the constituent elements. This is normally expressed as 'the whole is greater than the sum of the parts'. SSM addresses the 'fuzzy', ill-structured or soft problem situations, which are the true domain of information systems development methodologies, not simple, technological problems.

Information Systems Work and Analysis of Change (ISAC), a methodology developed in Scandinavia by Lundeberg et al. (1982), seeks to identify the fundamental causes of users' problems, and suggests ways in which the problems may be overcome (not necessarily through the use of computer information systems) by the analysis of activities and the initiation of change processes. It is therefore a people-oriented approach with emphasis on the analysis of change and the change process in organizations.

Davenport's (1993) Process Innovation does most to tie business process re-engineering (Section 4.3) with information technology and information systems, IT being seen as the primary enabler of process innovation as it gives an opportunity to change processes completely. We also look at PRINCE, a methodology designed for project management (Chapter 14), and Renaissance, an approach designed to ensure legacy systems (Section 8.1) are not neglected.

Chapter 26 looks at frameworks. These are approaches that describe themselves as frameworks rather than traditional methodologies. Multiview, for example, is a hybrid methodology, which brings in aspects of other methodologies and adopts techniques and tools as appropriate. In other words Multiview is a contingency approach: techniques and tools being used as the problem situation demands. It has been influenced particularly by aspects of SSM and ETHICS, but also by the proponents of process modelling and data modelling approaches. However, readers will see aspects of a number of approaches described earlier in Part VI.

SODA, described next, is an approach designed to provide consultants with a set of skills, tools, and techniques to help clients deal with messy work situations. Then we look at CMM (the Capability Maturity Model). This is a framework for evaluating processes used to develop software projects, it defines various levels of maturity and provides guidance relating to what organizations have to do in order to move from one level to another.

The last section discusses Euromethod, which results from a European initiative. It is also a framework but this time for the planning, procurement, and management of services for the investigation, development, or amendment of information systems. Other methodologies, such as SSADM, Merise, and Information Engineering, have influenced its design. This framework and associated standards will, it is hoped, help overcome the problems posed by the current diversity of approaches, methods, and techniques in information systems and help users and service providers to come to common understandings concerning requirements and solutions in information systems projects.

We have not described similar methodologies, even if they are well used, but reference this similarity where appropriate. The methodologies are described largely uncritically so that readers can follow their principles and practice, although we have commented on aspects of the methodologies where they reveal important features. However, the descriptions of the methodologies represent interpretations of the methodologies by the authors of this text, and these views may not correspond to those of the methodology suppliers! We return to the issue of interpretation in Part VII.

20 Process-oriented methodologies

20.1 Structured analysis, design, and implementation of information systems (STRADIS)

The major statement of Gane and Sarson's methodology of systems development called STRADIS comes in their book entitled *Structured Systems Analysis* (Gane and Sarson, 1979). The development of this structured systems approach to analysis came as a result of the earlier development of a structured approach to design. The structured design concepts were first propounded in 1974 by Stevens et al. (1974), and these ideas were later developed and refined by Yourdon and Constantine (1978), and Myers (1975 and 1978). The work of Jackson (1975) was also influential.

Structured design is concerned with the selection and organization of program modules and interfaces that would solve a predefined problem. However, it makes no contribution to the defining of that problem. This proves to be a practical limitation as the development of an information system requires both analysis and design aspects to be addressed, and, while structured design was acknowledged to provide significant benefits, these benefits were wasted if the definition of the original problem was not well stated or inaccurate.

A number of people have therefore attempted to take the concepts of structured design and apply them to systems analysis, in order to develop a method of specifying requirements and to provide an interface to structured design. In this way the techniques of structured analysis were developed. Apart from Gane and Sarson's work, DeMarco (1979), Weinberg (1978), and Yourdon (1989) are all texts on structured analysis covering some of the same ground and utilizing very similar techniques within each approach.

Gane and Sarson only relatively briefly outline a methodology of systems development in their book. The majority of the book is devoted to descriptions of the techniques which the methodology utilizes. This is in direct contrast to some other methodologies. SSADM (Section 21.1) and ISAC (Section 25.2), for example, lay out the steps of the methodology in great detail. Therefore the most important aspect of the STRADIS methodology is the bringing together of many of the techniques which were described separately in Part IV of this book. Nevertheless, we will continue to use the term 'methodology' in the context of STRADIS. These techniques are utilized, in some form or another, by many different methodologies, and therefore STRADIS is not unique but, along with the Yourdon Systems Method (Section 20.2), may be regarded as epitomizing those methodologies based on functional decomposition (Section 12.5) and the use of the data flow diagram (DFD), described in Section 12.1.

STRADIS is conceived as being applicable to the development of any information system, irrespective of size and whether or not it is going to be automated. In practice, however, it has mainly been used and refined in environments where at least part of the information system is automated. The methodology is envisaged to be relevant to a situation in which there is a backlog of systems waiting to be developed and insufficient resources to devote to all the potential new systems.

1 Initial study

The starting point of the methodology is an attempt to ensure that the systems chosen to be developed are those that most warrant development in a competing environment. The most important criterion in this selection process is argued to be the monetary costs and benefits of each proposal. Systems are viewed as contributing toward increasing revenues, avoiding costs or improving services. The initial study to discover this information is conducted by systems analysts gathering data from managers and users in the relevant areas. The analyst is to review existing documentation and assess the proposal in the light of any strategic plans relating to systems development that may exist within the organization. The initial study usually involves the construction of an overview data flow diagram of the existing system and its interfaces, and an estimate of the times and costs of proceeding to a detailed investigation. In addition, some broad range of final system development costs might be estimated. The initial study normally takes between two days and four weeks, depending on the size and importance of the application.

On completing the initial study, a report is reviewed by the relevant management, and they decide on whether to proceed to a more detailed study or not. If they approve of the proposal, they are committing themselves to the costs of the detailed study but not necessarily to implementing the proposed system.

The initial study might be thought to be quite close to the traditional notion of a feasibility study outlined in Chapter 3. However, there are some important differences. STRADIS does not include a review of alternative approaches to the proposal, and it is not, perhaps, as major or as resource-intensive a task as a traditional feasibility study. Furthermore, a traditional feasibility study, if approved by management, is usually in practice a commitment to the implementation of the complete proposal. Gane and Sarson do address all these aspects, but at later stages within their methodology.

2 Detailed study

This takes the work of the initial study further. In particular, the existing system is examined in detail. As part of this investigation, the potential users of the system are identified. These users will exist at three levels:

1. The senior managers with profit responsibilities, whom Gane and Sarson call the 'commissioners', whose areas will be affected. They initially commissioned the system proposal.

2. The middle managers of the departments affected.

3. The end-users; that is, the people who will actually work directly with the system.

Having identified these three sets of users, the analysts ascertain their interests and requirements by interviewing them. Next, the analyst prepares a draft logical DFD of the current system. This will usually involve constructing a DFD that extends well beyond the current system under consideration, in order to be clear exactly what and where the boundaries are in relation to other systems and to identify the interfaces between various systems.

Figure 20.1 depicts a data flow of part of a university admissions procedure. The system under consideration is that enclosed by the dotted line, but, in order to appreciate the context, a larger system is depicted which enables the interfaces to be clearly identified. Any data flow that crosses the dotted line must be addressed by both the external system and the system under consideration. In this case the diagram has highlighted the fact that those applications where the qualifications are not known require a decision to be made (see the data flow marked with the asterisk). This is a non-obvious interface which might otherwise have been neglected.

The **boundary** may be drawn in any place and could be moved. It may, for example, be more logical to include other processes within the boundary in order to minimize the number of interfaces to the external system. This is particularly important when the automation boundary is being chosen.

STRADIS describes in detail the drafting of DFDs at various levels, showing how each level is exploded into lower levels through to the level where the logic of each process box in the low-level DFD should be specified using the appropriate process logic representation, for example, decision trees, decision tables, or structured English (Sections 12.2–12.4). They suggest that DFDs and other outputs should be reviewed or 'walked through' with a number of users, so as to check their validity, and alterations made where necessary.

The detail of the DFDs and the process logic is entered into the data dictionary. The data dictionary can be either manual or computerized. On the DFD, data flows and data stores are defined using a single name which is meaningful. All the details that the name represents must be collected and stored in the data dictionary (Section 19.1).

The extent of detail that the analyst goes to at this stage in the methodology is not made clear, but it appears that not all low-level processes are specified in process logic and that not all data flows and data stores are specified in the data dictionary. It is usual to specify in detail only the most significant at this stage.

The initial study estimated the costs and benefits of the proposed system in outline. These estimates are further refined within the detailed study. The analysts need to investigate the assumptions on which the estimates were based, and ensure that all aspects have been considered. They also need to consider the effects and costs of the proposed system from the point of view of organizational impact. In other words, they need to have a better estimate on which a final decision can legitimately be made.

In summary, the detailed study contains:

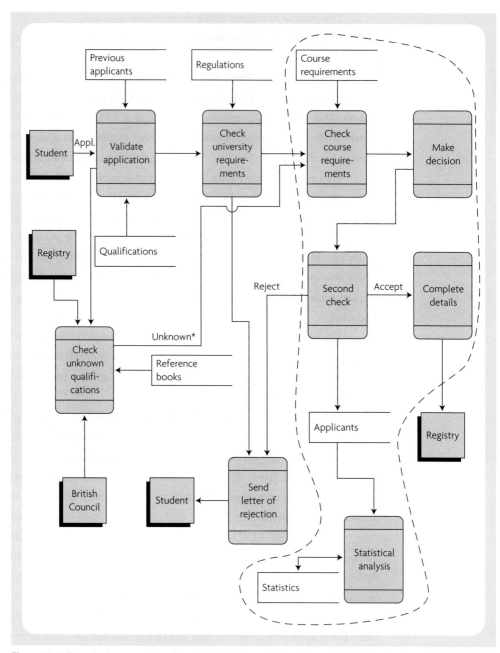

Figure 20.1: Example of a system boundary (dashed line) on a DFD

- a definition of the user community for a new system, that is, the names and responsibilities of senior executives, the functions of affected departments, the relationships among affected departments, the descriptions of clerical jobs that will be affected, and the number of people in each clerical job, hiring rates, and natural attrition rates;
- a logical model of the current system, that is, an overall data flow diagram, the interfacing systems (if relevant), a detailed data flow diagram for each important process, the

logic specification for each basic process at an appropriate level of detail, and the data definitions at an appropriate level of detail;

- a statement of increased revenue/avoidable cost/improved service that could be provided by an improved system, including the assumptions, the present and projected volumes of transactions and quantities of stored data, and the financial estimates of benefits where possible;
- an account of competitive/statutory pressures (if any) including the system cost and a firm cost/time budget for the next phase (defining a menu of possible alternatives).

The results of the detailed study are presented to management, and a decision will be made either to stop at this stage or proceed to the next phase.

3 Defining and designing alternative solutions

The next phase defines alternative solutions to the problems of the existing system. First, the organizational objectives, as defined in the initial study, are converted into a set of system objectives. An organizational objective is a relatively high-level objective having an effect on the organization. This could include increased revenue, lower cost, or improved service. A system objective is at a lower level and relates to what the system should do to help management achieve the organizational objectives.

The system objectives should be strongly stated. This means that they should be specific and measurable, rather than general. For example, 'improving the timeliness of information' would be a weakly stated objective, and it would be preferable to state this objective more strongly, for example, 'to produce the monthly sales analysis report by the fourth working day of the following month'.

The analyst uses these objectives to produce a logical DFD of the new or desired system. The existing system DFD would normally be used as the basis for this, and the desired system may involve the introduction of new or changed data flows, data stores, and processes. The new DFD should be constructed to a level of detail which shows that the most important system objectives are being met.

The methodology then enters a design phase. At this time, analysts and designers work together to produce various alternative implementation designs which meet a variable selection of the identified system objectives. The alternatives should cover three different categories of design. First, a low-budget, fairly quick implementation which may not initially meet all the objectives; second, a mid-budget, medium-term version, which achieves a majority of the objectives; and third, a higher-budget, more ambitious version achieving all the objectives. Each alternative should have rough estimates of costs and benefits, timescales, hardware, and software.

The report of this phase of the project should be presented to the relevant decision makers, and a commitment made to one of the alternatives. The report should contain the following:

- a DFD of the current system;
- the limitations of the current system, including the cost and benefit estimates;

- the logical DFD of the new system.

For each of the identified alternatives, the design will include statements covering:

- the parts of the DFD that would be implemented;
- the user interface (terminals, reports, query facilities, and so on);
- the estimated costs and benefits;
- the outline implementation schedule;
- the risks involved.

4 Physical design

The design team then refines the chosen alternative into a specific physical design which involves a number of parallel activities:

1. All the detail of the DFD must be produced, including all the error and exception handling, which has not been specified earlier, and all the process logic. The content of the data dictionary is completed and report and screen formats designed. This detail should be validated and agreed with the users.

2. The physical files or database will be designed. They will be based on the data-store contents previously specified at the logical level. Data stores are defined in the DFD as the temporary storage of data needed for the process under consideration. This has the effect of introducing many data stores scattered all over the DFD. Many of them will be very similar in content and have a significant degree of overlap.

3. The data stores need rationalizing, and the technique of normalization (described in Section 11.2) is utilized to consolidate and simplify the data stores into logical groupings. The actual process of mapping and the design of the physical files (or database) are not defined.

4. Derive a modular hierarchy of functions from the DFD. The designer seeks to identify either of two structures that any commercial data processing system is thought to exhibit. The first structure is the simplest. Here all transactions follow very similar processing paths (Figure 20.2). Such a system is termed a **'transform-centred'** system. The second structure is one in which the transactions require very different processing. This is termed a **'transaction-centred'** system and is illustrated in Figure 20.3.

The first step therefore is to identify which type of system is being described. It is recommended that the raw input data flow is traced through the DFD until a point is reached at which it can no longer be said to be input, but has been transformed into some other data flow. The output is traced backward in a similar fashion until it can no longer be considered to be output. Anything in-between is termed the **'transform'**. The transform is then analysed to see if it is a single transform or a number of different transforms on different transaction types. Once one or other of these high-level functional hierarchy types have been identified from the DFD, the detail of the modules in the hierarchy and the communication between them is constructed.

The final task in this phase is the definition of any clerical tasks that the new system will

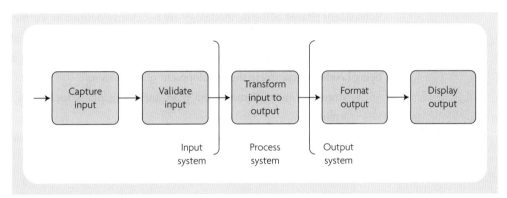

Figure 20.2: Transform-centred system

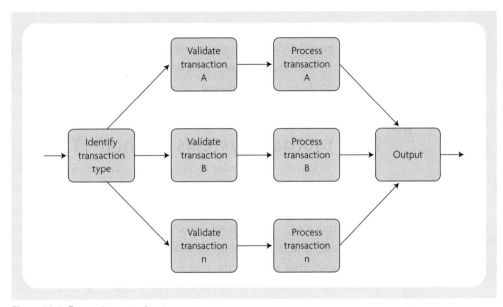

Figure 20.3: Transaction-centred system

require. The clerical tasks are identified according to where the automated system boundary is on the DFD and according to what physical choice of input and output media has been made.

The above activities are pursued to a level of detail at which it is possible to give a firm estimate of the cost of developing and operating the new system. The major components of these costs are identified as:

- the professional time and computer test time required to develop the identified modules;
- the computer system required;
- the peripherals and data communication costs;
- the professional time required to develop user documentation and train users;
- the time of the users who interact with the system;
- the professional time required to maintain and enhance the system during its lifetime.

Subsequent phases of the methodology are not clearly defined as the methodology is effectively concerned mainly with analysis, to a lesser extent design, and hardly at all with implementation. However, the following list indicates the remaining tasks that Gane and Sarson envisage as being needed to complete the development of the system:

- draw up an implementation plan, including plans for testing and acceptance of the system;
- develop concurrently the application programs and the database/data communications functions (where relevant);
- convert and load the database(s);
- test and ensure acceptance of each part of the system;
- ensure that the system meets the performance criteria defined in the system objectives, under realistic loads, in terms of response time and throughput;
- commit the system to live operation and tune it to deal with any bottlenecks;
- compare the overall system facilities and performance to original objectives, and amend to resolve any differences, where possible;
- analyse any requests for enhancement, prioritizing these enhancements, and placing the system in 'maintenance' state.

20.2 Yourdon Systems Method (YSM)

YSM was originally very similar to STRADIS. Functional decomposition or top-down design, in which a problem is successively decomposed into manageable units, was the basis of the approach. However, although based on the structured approach it uses an approach known as event partitioning. This approach is neither pure top-down nor bottom-up, but is described as 'middle-out'. The analyst begins by drawing a top-level context diagram which indicates the system boundaries and thus the sources and sinks. Then, following interviews with the users, a textual list of the events in the environment to which the system must respond is constructed. Following this, some of the techniques described in Chapter 12 are used to document the system further.

YSM covers both the activities of the organization (although this could be at department level as well as at the level of the organization as a whole) and the system itself. Enterprise requirements need to be modelled as well as system requirements. For example, analysts may create an entity-relationship diagram and other information about data for the department, but only some of this will be appropriate for the system. Modelling at a department rather than system level will ensure consistency as well as avoiding the duplication of time and effort. Emphasis is therefore placed on modelling both the organization and the system. Many of these modelling methods are appropriate to the use of support tools, particularly toolsets (Chapter 19).

There are three major phases in the YSM approach, as shown in Figure 20.4. The feasibility study looks at the present system and its environment. Phase 2, essential modelling, aims to describe the essence of a software system in terms of how the required system must behave and what data must be stored to enable this to happen. It assumes that there are no limitations

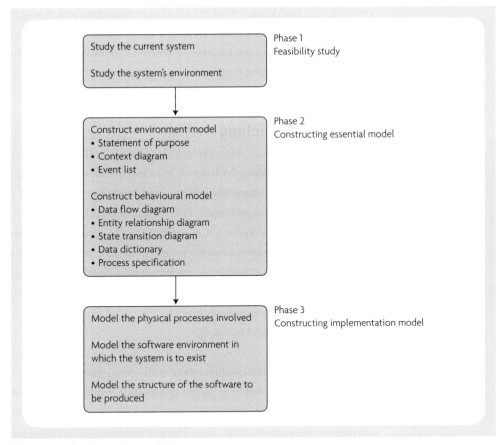

Figure 20.4: The Yourdon Systems Method

affecting implementation, that is, it assumes unlimited resources, unlimited power of technology, and so on. It is the major phase of the approach. The final phase, implementation modelling, aims to incorporate those features found in the customer's statement of requirements using the essential model and will be dependent on the appropriate use of available technology. We will look at these three phases in more detail.

Although the only enterprise model described in detail in Yourdon (1993) is the enterprise essential model, the creation of an organization as well as system-level model might suggest an enterprise or strategic planning phase for information systems development. Indeed, some followers of YSM include a strategic planning phase before the feasibility study of each proposed system in that plan. However, an enterprise implementation model is suggested and this will point to proposals for the hardware and software decisions for the organization as a whole, not a decision which should be dominated by the needs of one particular application.

1 Feasibility study

The feasibility study looks at the present system, its environment, and the problems associated with it. The objective here is to get a general understanding and an overview of the existing system. It is to understand what the existing system does (not how it works). The analyst will tend to draw an overview data flow diagram for the current system and its interfaces, and the analyst may also start to put together an entity-relationship diagram. The information required will normally be obtained from interviewing the users. This phase is much the shortest of the three, normally taking only a few weeks to complete.

2 Essential modelling

This stage gains the most emphasis in YSM. There is both an enterprise and a system essential model. We will emphasize the latter in our description as it is essentially the sum total of the systems models. The same considerations and models are reflected at the organizational level where, unusually, the 'organization' to be considered and the 'system' are the same. Having an overview of the present system, it is possible then to construct an essential model. The system essential model is a model of what the system must do in order to satisfy the users' requirements. It does not say anything about how this system is to be implemented. Therefore, it is the new logical model. In the 1993 version of YSM, essential modelling itself also has two major components:

- environmental model building;
- behavioural model building.

In some descriptions, the creation of the entity-relationship diagram is seen as part of a third parallel component, referred to as the information model. In the following description, these aspects form part of the environmental model and behavioural model building phases. The activities are the same in either case. The key difference is, perhaps, more subtle. It represents the change in the approach from one which emphasizes processes to one which emphasizes both process and data aspects. It also enables stress to be made on the importance of comparing the data and process aspects to ensure consistency of models and therefore the integrity of the overall specification. This also separates YSM from STRADIS.

The **environmental model** defines the boundary between the system and the environment in which the system exists. The data coming from and to the environment are identified. The model consists of a statement of purpose, context diagram, and an event list.

The statement of purpose is a brief, concise, textual statement about the purpose of the system. It is provided for top management, user management, and others who are not directly involved in the development of the system. It is only about a paragraph long.

A DFD type context diagram is used to depict the system and its environment (see Figure 12.6 in Section 12.1 for an example). The context diagram represents the system in a circle in the middle of the page, along with the main sources and sinks of the data entering to and from it. It identifies the people, organizations, and systems with which the system communicates. The data coming into the system that are processed in some way and then output in a different form are also identified along with any intermediary data stores. It also shows the boundary between the system and its environment.

The event list names the 'stimuli' that occur in the environment of the system to which the system must respond. An event may be flow-oriented, temporal, or a control event. A flow-oriented event is one associated with a data flow. A temporal event is triggered by reaching a particular point in time. Control events occur at an unpredictable point in time and are therefore a special case of a temporal event.

A first-cut data dictionary which describes the composition of each data element and a first-cut entity-relationship diagram highlighting the relationship between stores (the entities) may also be constructed at this time, but both are very early versions.

The **behavioural model** is a model of what the internal behaviour of the system must be in order to deal with the environment successfully. It includes a first-cut data flow diagram, entity-relationship diagram and state transition diagram and adds information to the data dictionary. A state transition diagram shows how the properties of an entity change over time and is therefore similar to an entity life cycle (Section 12.9). Note that behaviour refers to the behaviour of the system and does not imply any emphasis on people-oriented aspects. The processing behaviour of the system, that is, how the system uses its inputs to produce the required output, is shown using data flow diagrams. The structure and use of the data in the system are shown using a data dictionary and a set of entity-relationship diagrams. The dynamic behaviour of the system, describing how events in time affect behaviour, is modelled by extending the data flow diagrams (which represent control) and state transition diagrams (which represent control behaviour).

From the event list obtained in the environmental model, a data flow diagram is constructed with one process representing the system's response to each event in the event list. Stores are then drawn as needed to enable the processes to access the required data, and the input and output flows are connected to and from the processes. The data flow diagram or diagrams are then checked against the context diagram for consistency. In parallel, the control transformations are specified and the data relationships are modelled. By the end of this stage, the data flow diagrams are completed by a process of levelling out, and the data dictionary, process specifications, entity-relationship diagram, and state transition diagram are also completed.

The process of levelling data flow diagrams involves restructuring so that there is a set of data flow diagrams, some the result of levelling upward and some the result of levelling downward. This is the key to the claim that YSM is middle-out rather than top-down. If the first-cut data flow diagram is too complicated with many processes, then related processes are grouped together into meaningful aggregates, each of which will represent a process in the higher-level data flow diagram. A rule of thumb is suggested that each data flow diagram will have around seven processes and stores in total. Downward levelling may be necessary where it is found that a process at the middle level is not a primitive process but needs to be expressed in more detail in a lower-level data flow diagram. This means that the initial process, which was a response to an event, is too complex for that middle level. The levelling process is seen in Figure 20.5.

Note that, as shown in Figure 20.6, processes are illustrated by circles in the YSM standard. Other shapes are also used to represent sources and sinks (a simple rectangle with no shadowing) and data stores (two parallel lines). Further, Yourdon (1989) recognizes two types of data flow. Discrete data flows arrive at their destination at discrete points of time (arrow) whereas continuous data flows are always available at their destination (arrow with two

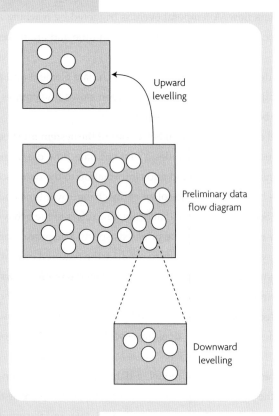

Figure 20.5: Upwards and downwards levelling of YSM data flow diagramming

heads). This indicates that, although many of the techniques are common to a number of methodologies, there are often variances in the way they are used, drawn, and so on.

Process specifications are then drawn up for every 'functional primitive', that is, every process in the bottom-level data flow diagram. These are referred to as minispecs, which are detailed specifications of each data process. Essentially, they state the rules that convert the inputs to the outputs. The process specification may take the form of structured English (Section 12.4), decision tables (Section 12.3), or any other method appropriate for the process which can be verified and communicated easily to the users. These will be cross-checked with the data dictionary and entity-relationship diagram, and it might be necessary to modify the data flow diagrams as a result of this further detail.

The entity-relationship diagram also needs to be completed. YSM advocates iteration, and this diagram will also be refined from its first-cut form in stages. The knowledge gained when refining the data flow diagram will be used to help refine the entity-relationship diagram.

If the system being modelled has any real-time characteristics, then a state transition diagram is developed in addition to the entity-relationship diagram and data flow diagram. A state transition diagram specifies how a control process is to take account of its input control flows and how it is to output control flows. The effect which input control flows have depends on the 'state' of the system, and they may change the state of the system and cause control flows to be output. Moreover, the output control flows have an effect on the data flow model of system behaviour.

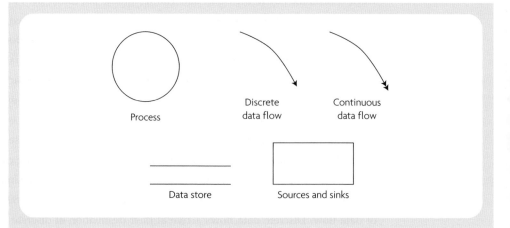

Figure 20.6: YSM data flow diagram symbols

Like other structured approaches, 'methods' are emphasized in YSM rather than a 'methodology', and many techniques are described in detail, including entity–event matrices and function–entity matrices (Section 12.7), the specification of entities, relationships, attributes, events, and operations (Section 11.1) and normalization (Section 11.2).

The models together should describe 'what will the system do?' (e.g. the data flow diagram), 'what happens when?' (e.g. the event list), and 'what data is used by the system?' (e.g. the entity-relationship diagram). Some models link these dimensions, for example, entity life cycles; entity–event matrices link time and information, and data flow diagrams link data and processes. Together, it is argued, the models provide a full description of the system.

3 Implementation modelling

This phase starts the systems design process. The limitations of such factors as the technology available, performance requirements, and feasibility modify the essential model. Data flow diagrams and state transition diagrams are examined so that, for example, boundaries of computerization are marked, and within them groups of processes are bounded for particular devices and processes. This also includes allocating software environments to groups of processes. The entity-relationship diagram is examined to look for pointers as to which database management system might be suitable and how the data might be stored. Implementation modelling, in short, bridges the gap between specification and systems design.

20.3 Jackson Systems Development (JSD)

Michael Jackson's program design methodology, Jackson Structured Programming (JSP), which is described in Jackson (1975), has had a profound effect on the teaching and practice of commercial computer programming. Jackson (1983), on Jackson Systems Development (JSD), argues that system design is an extension of the program design task and that the same techniques can be usefully applied to both. Aspects of JSP are diffused throughout JSD, so that the JSD methodology is a significant development on its precursor, and therefore should not be seen as a 'front end' to JSP but an extension of it, where JSP is the core. 'In principle', says Jackson, 'we may think of a system as a large program.' However, the primary purpose of JSD is to produce maintainable software, and its emphasis is on developing software systems. This leads to a potential criticism of JSD in that, in the context of this text, it is oriented toward software and not to organizational need.

Given this comment, therefore, it is not surprising that JSD does not address the topics of project selection, cost justification, requirements analysis, project management, user interface, procedure design, or user participation. Further, JSD does not deal in detail with database design or file design. At least as described in his book, Jackson's methodology is not comprehensive in the sense that it does not cover all aspects of the life cycle. However, the commercial version of JSD, because of practical necessity, was extended to include some of these aspects.

The emphasis in the methodology is solving what Jackson terms the hidden path problem, that is, the path between the presentation of a specification to the design/

programming group and the completed implemented system, which could be described as a 'bundle' of documentation, listings, and executable programs. Jackson asks, 'What reasons do we have to support the claim that we have delivered what is required in the specification?' The traditional response is that the answer is found in the processes of testing and checking. But there are two problems here. We cannot be sure that the tests are complete, and, in any case, when testing is possible, the system is already complete and it is usually rather too late and too costly to repair the damage.

JSD uses transformation through process scheduling as the answer to the hidden path problem, and a major contribution of JSD lies in the areas of process scheduling and real-world modelling. Further, JSD deals with the problem of time in systems modelling and systems design in a way that most other information systems design methodologies do not, as the latter tend to model static elements in the system.

There are three major phases in JSD: the modelling phase, the network phase, and the implementation phase. In the *modelling phase*, events and entities are identified and entity structures and entity life cycles formed. In other words, analysts ask what is happening in the real world and how might this be connected to the computer world. In the *network phase*, the inputs and outputs are added to the model so far derived so that the input and output subsystems can be analysed. In other words, the analysts ask what outputs are needed from the system and what processes and operations must be added to produce these outputs. The *implementation phase* is concerned with detailed design and coding, that is, how can the specification (model plus function) be transformed to run on the hardware:

Modelling phase
   1. Entity action step
   2. Entity structure step

Network phase
   3. Initial model step
   4. Function step
   5. System timing step

Implementation phase
   6. Physical system specification step.

In the entity action step the systems developer defines the real-world area of interest by listing the entities and actions with which the system will be concerned. In the entity structure step the actions performed or suffered by each entity are ordered by time. In the initial model step communications between entities are depicted in a process model linked to the real world. In the function step functions are specified to produce the outputs of the system, and this may give rise to new processes. In the system timing step some aspects of process scheduling are considered which might affect the correctness or timeliness of the system's functional outputs. In the physical system specification step the system developer applies techniques of transform-ation and scheduling that take account of the hardware and software available for running the system. JSD is applied iteratively, and, as increasing detail is revealed, data and functions will also be revealed. Each of these stages will be looked at in turn.

1 Entity action step

JSD aims to model the real world. In the entity action step, real-world entities are defined. These might include SUPPLIER, CUSTOMER, or PART, but, unlike the data analysis approaches, JSD is more concerned with the behaviour of the entity than its attributes or its relationships with other enti-ties. Conventional entity modelling presents a static view of the real world, whereas JSD is concerned with modelling system dynamics.

To be defined as an entity in JSD, an object must meet the following criteria:

1. It must perform or suffer actions in a significant time ordering.
2. It must exist in the real world outside the system that models the real world.
3. It must be capable of individual instantiation with a unique name.

Entities may also be collective (e.g. BOARD OF DIRECTORS) if the instantiation has objective reality without considering its component objects. Entities may be generic (e.g. SPAREPART) thus sup-porting the abstraction of classification, or specific (e.g. INNER-FAN-SHAFT). Entities that exist in the world may be ignored if it is impossible or unnecessary to model their behaviour. Therefore, only a relevant subset of the real world is modelled.

An action describes what an entity does within a system. Since the distinctive feature of JSD entities is that they perform or suffer actions, it is necessary to specify the criteria for some-thing to be an action. These are as follows:

1. An action must be regarded as taking place at a point in time, rather than extending over a period of time.
2. An action must take place in the real world outside the system and not be an action of the system itself.
3. An action is regarded as atomic and cannot be decomposed into subactions.

Since the original version of JSD, more emphasis has been placed on the process of eliciting attributes, both action attributes and entity attributes. Whereas action attributes come from outside the system and trigger the action, entity attributes add information about the entity and will be updated by its entity actions. The actions and changes to the entity attributes form the entity life history. It is important to analyse when these changes occur so that the entity life history will have the correct time ordering. But we are now discussing the beginnings of the next step, the entity structure step.

The end result of the entity action step is a list of entities and their attributes and a list of actions and their attributes. The list of entities is liable to be much shorter than that produced by an equivalent data analysis process, particularly if the latter normalizes the entities, because the functional components of the system are excluded at this stage.

2 Entity structure step

The actions of an entity are ordered in time and are expressed diagrammatically in JSD. This is similar to the technique of entity life cycles (Section 12.9), although there are differences. They show the structure of a process in terms of sequence, selection, and iteration. The diagram shown as Figure 20.7 is read from top to bottom as a hierarchical decomposition. Actions are shown as the leaves, whereas components higher up represent aggregations of actions. Each **structure diagram** is intended to span the whole lifetime of an entity, including therefore an action that causes the entity to come into existence and one that causes it to cease to exist. The model must illustrate time ordering of these elements. The lifetime of an entity may span many years in the real world.

JSD structure diagrams do not support concurrency. For example, the entity CUSTOMER in a banking system, an example discussed fully in Jackson (1983), might have been specified as a sequence (as in Figure 20.7) of OPEN-ACCOUNT, OPERATE-ACCOUNT, and CLOSE-ACCOUNT. OPERATE-ACCOUNT is an iteration of TRANSACTIONS (hence the asterisk which illustrates iteration), each of which is

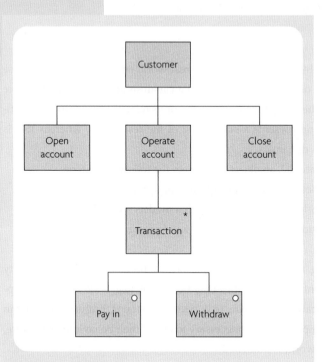

Figure 20.7: JSD structure diagram

a selection of either PAY-IN or WITHDRAW (hence the small circle which indicates selection). Such a structure would constrain a customer to having only one account. To relax this constraint, the systems developer may be tempted to redraw the structure diagram as in Figure 20.8. It now appears that the customer may have many accounts each being operated as in Figure 20.7. The diagram now specifies that the customer can have more than one account, but not more than one at the same time. A customer may only open a new account after an existing account has been closed. The JSD structure diagram cannot show the simultaneous operation of many accounts. The answer to this problem is to specify a new entity ACCOUNT whose life history proceeds in parallel to the life history of the CUSTOMER entity. CUSTOMER now appears as in Figure 20.9(a) and ACCOUNT as in Figure 20.9(b).

In JSD, discrimination between entity roles is necessary if an entity can play more than one role simultaneously. Jackson provides an example using the entity SOLDIER. A soldier enlists in the army and may be promoted to a higher rank at various points in his career. Soldiers are also given training and may attend training courses, which they may or may not complete successfully. If successful completion of a course always leads to promotion, then these facts can be accommodated in one structure diagram. If there is no necessary connection between training and a career, then two structure diagrams are required, one for the soldier's promotion career and one for his training career. The soldier in this example is playing two roles, one as a

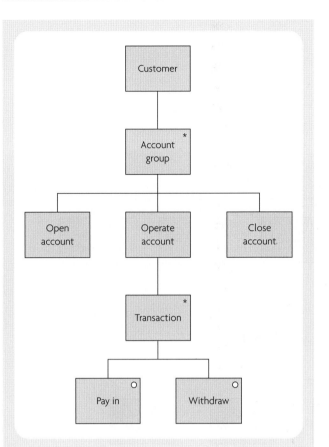

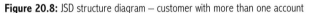

Figure 20.8: JSD structure diagram – customer with more than one account

person being trained and one as a person being promoted to a higher rank. Multiple roles may be synthesized into another structure diagram showing a selection of the possible activities in the possible roles that can be played.

Entity structure diagrams represent a sequence of activities ordered in time, without concurrency, from the 'birth' of an entity to its 'death'. One final problem addressed by the methodology in this step is that of the premature termination of the life cycle. In the real world, events may occur that prevent an entity making an orderly progression through its life cycle. For example, a soldier may be killed in battle without proceeding to retirement. It may not be feasible to draw a structure diagram for every possible variation on a prematurely terminated life cycle. JSD allows for a general specification of premature termination. This recognition of such a circumstance is an example of 'backtracking' in JSP.

The end result of the entity structure step in JSD is a set of structure diagrams. New entities and multiple roles for the same entity may have been generated during this phase.

3 Initial model step

In this third step the systems developer creates a model that is a simulation of the real world. For each entity defined in the preceding two phases a sequential process is defined in the model that simulates the activities of the entity in such a way that it could be implemented on a computer. This is not to say that the implementation necessarily has to be computerized, merely that it could be if this were required.

In the model there will be a sequential process for each instance of an entity type, not one process for all instances. Therefore, if there are a hundred instances of entity type CUSTOMER, there will be one hundred sequential processes in the model. Moreover, the processes notionally execute at exactly the same speed as the real-world processes. So, if a customer has a bank account for 50 years, the matching processes will also execute for 50 years.

The sequential processes specified in the initial model step are documented both by a diagram showing the interconnection of processes and by a **pseudo code** definition of each model process. Pseudo code is a language similar to structured English, which was described in Section 12.4, but nearer to a programming language in type. The pseudo code is known as structure text in JSD and resembles a high-level Algol-like programming language. Structure

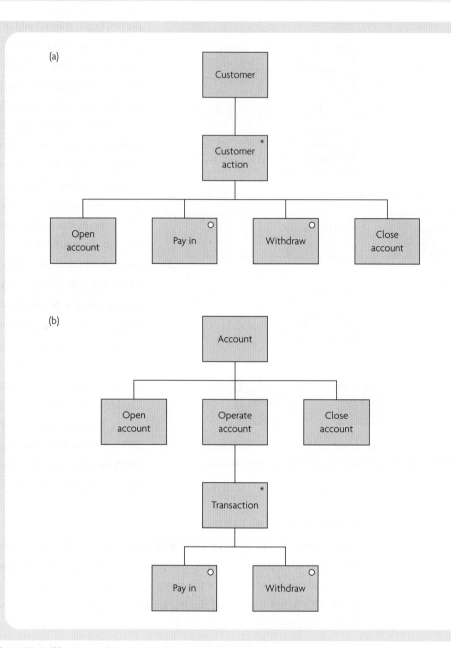

Figure 20.9: JSD structure diagram – simultaneous operation of many accounts

text exactly matches a corresponding entity structure diagram, and major constructs are sequence, selection, and iteration. The value of structure text is that it may be elaborated in later phases of the JSD methodology in a manner similar to the program design technique of stepwise refinement used in JSP. This process should be straightforward. An example of structure text for the ACCOUNT entity is provided in Figure 20.10.

Process connection in JSD is achieved by either data stream connection or state vector connection. In data stream connection one process writes a sequential data stream, consisting

```
ACCOUNT-1 seq
  read data-stream
  OPEN-ACCOUNT; read data-stream
  OPERATE ACCOUNT itr while (PAY-IN or WITHDRAW)
    TRANSACTION sel (PAY-IN)
      PAY-IN; read data-stream
    TRANSACTION att (WITHDRAW)
      WITHDRAW; read data-stream
    TRANSACTION end
  OPERATE ACCOUNT end
  CLOSE-ACCOUNT;
ACCOUNT-1 end
```

Figure 20.10: JSD structure text

of an ordered set of messages, and the other process reads this stream. This is similar to process connection in a data flow diagram.

The JSD **system specification diagram** (SSD) models the system as a network of interconnected processes showing how they communicate with each other. In the diagram (Figure 20.11) processes are shown in boxes, with the data streams that connect processes shown as a circle with its identification given (in the example, the identification is 'C'). Arrows show the direction of data stream movement.

In Figure 20.11, CUSTOMER-0 is intended to represent a real-world instance of a customer, sending messages about his or her actions to a process that simulates this behaviour (CUSTOMER-1). A circle in an SSD indicates data stream connection. CUSTOMER-1 is sending a stream of messages to ACCOUNT-1. Since a customer can have many accounts, a double bar is used on the diagram to represent this multiplicity. Data stream connection is appropriate in the banking example, as it is not practical to telephone the customer every ten minutes to find out if he or she has paid in or withdrawn money.

Jackson (1983) also gives the example of a lift system that finds out whether a button has been pressed in a lift by linking the button via a **state vector** to a process that models the button's behaviour. The alternative state vector connection is appropriate here because the button is essentially a switch, denoting an on or off state.

In the state vector connection, one process inspects the state vector of another process. A state vector is the internal local variable which describes and is owned by a particular process. State vector connection has no equivalent in data flow diagramming because the data flow technique permits process connection via logical files. There are no logical files in a JSD

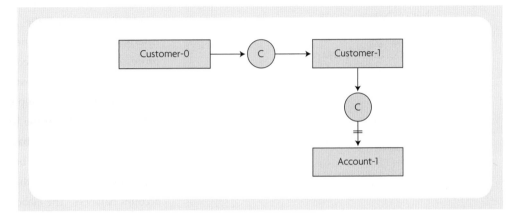

Figure 20.11: JSD system specification diagram

system specification diagram. State vectors are shown as a diamond on the SSD. In Figure 20.12, the data relating to the account is used to produce a report. Normally a data stream is used where a long-term view of events is required and a state vector used where a short-term snapshot is required.

Data stream connection is considered to be buffered (a data stream will be read on a first in, first out basis), so writing processes are never blocked; reading processes may lag behind writing processes. State vector connection is also unbuffered, and again no blocking occurs. State vector inspection therefore depends on the relative speeds of the processes involved. This is not true of data stream connection. If there is more than one input to a process, rules must be specified for determining which input is to be taken next. The determination may be made by fixed rules (fixed merge), or specified as part of the message stream (data merge), or determined simply by the relative availability of messages (rough merge). Such careful attention to synchronization details is absent from most other methodologies. JSD also allows for time grain markers (TGM) to indicate the arrival of particular points in real-world time (see Step 5).

The end result of the initial model step is a systems specification diagram depicting a set of communication processes each of which is specified by a pseudo-code structure text.

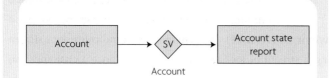

Account

Figure 20.12: State vector connection between processes

4 Function step

The model created in the first three phases of JSD has no outputs; it models the dynamic behaviour of the real world. In the function step, further elaboration takes place, and functions are added to the model to ensure that the required outputs are produced when certain combinations of events occur. The addition of functions may require no change to the SSD, in which case structure text is elaborated to specify the functions required. Alternatively, it may be necessary to create new processes, which are added to the SSD and specified with new structure text.

To give an example, we may wish to provide the facility in the banking application to interrogate customer balances on demand. Therefore functions must be added to the existing SSD and structure text that record and display account balances. The elaboration of the ACCOUNT text is shown in Figure 20.13. Clearly, the state

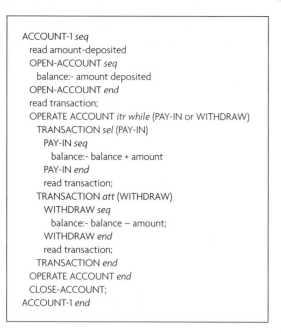

```
ACCOUNT-1 seq
  read amount-deposited
  OPEN-ACCOUNT seq
    balance:- amount deposited
  OPEN-ACCOUNT end
  read transaction;
  OPERATE ACCOUNT itr while (PAY-IN or WITHDRAW)
    TRANSACTION sel (PAY-IN)
      PAY-IN seq
        balance:- balance + amount
      PAY-IN end
      read transaction;
    TRANSACTION att (WITHDRAW)
      WITHDRAW seq
        balance:- balance – amount;
      WITHDRAW end
      read transaction;
    TRANSACTION end
  OPERATE ACCOUNT end
  CLOSE-ACCOUNT;
ACCOUNT-1 end
```

Figure 20.13: Addition of functions to structure text of Figure 19.10

vector of ACCOUNT now includes knowledge of the customer's balance, because the structure text has been elaborated to update that balance. The SSD can now be amended (as in Figure 20.14) to show the new interrogation process. INTERROGATE can inspect the state vector of any ACCOUNT-1 process (as indicated by a diamond symbol). It will do so when it receives a message specifying an account enquiry, and it will produce an output showing the balance of the customer's account. Therefore, the addition of functions to an initial model may cause the elaboration of existing structure texts and/or lead to the specification of new processes with their own structure texts.

Whereas the model of processes and their input and output data streams correspond to the basic system and are fairly stable, the functions represent ancillary processes and are likely to be less stable. They relate to reports, queries, and the user interface, which might change much more frequently, but the partitioning of parts of the system that change frequently into separate functions is relatively easy to carry out.

While the earlier versions of JSD did not provide many guidelines regarding data analysis, later versions do require the identification of attributes relating to entities, and these are referred to as fields of the state vector record. Update procedures and integrity constraints are also defined. However, data analysis is not as complete as in other methodologies in that, for example, it does not suggest normalization.

The end result of the function step is an amended system specification diagram with associated structure texts.

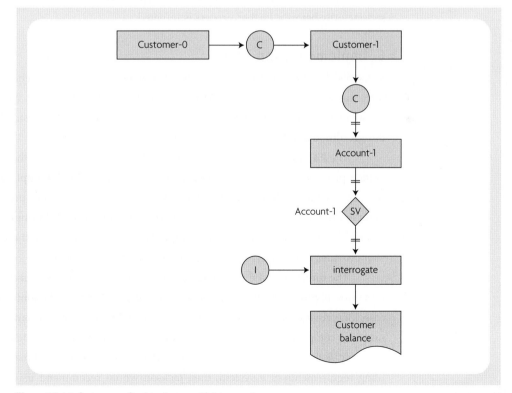

Figure 20.14: System specification diagram with interrogation process

5 System timing step

The JSD modelling process so far described has not yet explicitly raised the question of speed of execution of processes and their timing. Implicitly, the model must lag to some extent behind the real world because input must take some time to arrive. In the system timing step (sometimes included within the function step) explicit consideration is given to permissible delays between receipt of inputs and production of outputs. Different parts of the system may be subject to different time lags. Timing is important both within a process and between processes. JSD uses messages known as time grain markers which act like data streams but which contain timing information. They are rough-merged with other data streams to control the arrival of messages and the timing of the execution of processes. They are used to trigger actions within processes, start and stop processes, and generally aid the synchronization of processes.

Time constraints will derive either from user requirements (e.g. for a monthly report or for an immediate response to an enquiry) or from technical considerations. Examples of the latter are state vector retrievals that must be sufficiently frequent to capture changes of state (as in a process control application) but not so frequent that they capture too many instances of the same state. The system timing step will gather information usable in the next phase when decisions are made. These decisions may concern questions relating to online, real-time, or batch implementation of aspects of the model system.

The end result of the system timing step is a specification of timing constraints using the TGMs associated with processes. The step does allow for the addition to the SSD of synchronization processes whose sole function is to ensure that certain actions have been completed satisfactorily before a further process is initiated.

6 Implementation step

Jackson's (1983) account of systems implementation is not a comprehensive treatment of all implementation considerations. Moreover 'implementation' in JSD includes activities that would be regarded in other methodologies as 'systems design', for example, file and database design, although JSD does not describe these processes in any depth.

The JSD implementation step concentrates on one major issue, the sharing of processors among processes. A system specification diagram can be directly implemented by providing one processor for each sequential process. Since there is one sequential process for each instance of an entity type (e.g. one for each customer of a bank) this might imply many thousands of processors. If this is an unacceptable implementation of the model, then the implementation step provides techniques for sharing processors among processes.

The direct opposite of providing one processor for each process is to provide one processor for all processes, which could be provided by a centralized mainframe computer. In this case, JSD provides for a transformation of the model into a set of subroutines. JSD is not recommending computer users to write their own operating systems and teleprocessing monitors, however. If these items of software are available on a machine and match the process scheduling requirements of a system, then Jackson would recommend using them. In fact, most computer-based systems are scheduled by a mixture of administrative, clerical, and software action. The structure diagram for a scheduler can be drawn alternatively to represent an online system, a batch system, or a mixture of these, together with actions that may, in fact, be performed by human beings.

The JSD systems implementation diagram (or SID) is, in a way, an abstraction of all these real-world scheduling possibilities. Figure 20.15 shows the sequence of processes hierarchically: the scheduler is next to the top (drawn as a vertical bar in a box); the processes (contained in the SSD) are also shown in boxes; inversion sequences (i.e. the hierarchy of processes in terms of a main program calling subprograms) are shown as parallel lines representing the data stream as a pipe and use data contained in a state vector file (soft box) and data streams as buffers. Therefore, if we return to the banking example, the buffer could represent a data stream of credit amounts, with the processes arranged hierarchically as deposit, interest being added, and deposit account. The state vector file in this example provides data about the rate of interest. The scheduler handles the overall sequencing, and this can be represented as a process structure diagram and as structure text.

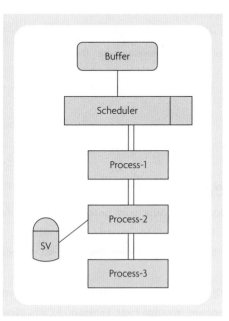

Figure 20.15: System implementation diagram

The system may be implemented by one or more real processors, thus giving rise to possible implementations that range from completely centralized to completely distributed. In general, multiple instances of common processes would share process texts (i.e. programs) as well as processors. To make process text sharing possible, it is necessary to separate the state vectors of processes from the shared process text. A concatenation of state vectors is then transformed into a file or a database. Therefore, the implementation step in JSD can give rise to perfectly conventional data processing solutions, and the information provided in the above steps gives the programmer all that is necessary to code the system using JSP procedures.

JSD's strength as a methodology lies in its determined and detailed attempt to model the dynamic aspects of real-world systems. Its treatment of concurrency, timing, and process scheduling is far more comprehensive than any other methodology discussed in this book. The model is, as far as possible, kept up to date so that it is always a fair reflection of the real world that this abstract model represents. In this way, it is possible to see what is happening in the real world and address it in the decision-making process. Entity modelling, it is argued, only represents a static view of the real world, whereas this process-oriented approach is dynamic.

It is, however, self-consciously incomplete as a methodology. Jackson wishes to say nothing about areas that are satisfactorily treated by other methodologies, only those that are not. Jackson is critical of methodologies that rely on structured decomposition, on the grounds that they confuse a method of documenting a design with the design process itself. JSD is not top-down design. Jackson is similarly critical of data modelling approaches: 'It is not much more sensible to set about designing a database before specifying the system processes than it would be to declare all the local variables of a program before specifying the executable text: the two are inextricably intertwined' (Jackson, 1983).

Summary

- In this chapter we have described three methodologies that we have identified as being process-oriented. These are also called structured systems analysis approaches.

Questions

1. Identify the key elements of each of the methodologies in this chapter.
2. Construct a table that highlights their differences and common aspects.
3. Pick one of the methodologies and identify important elements of systems development that might be missing.

Further reading

Gane, C. and Sarson, T. (1979) *Structured Systems Analysis: Tools and Techniques*, Prentice Hall, Englewood Cliffs, New Jersey.

Jackson, M. (1983) *Systems Development*, Prentice Hall, Hemel Hempstead, UK.

Yourdon Inc. (1993) *Yourdon Systems Method: Model-Driven Systems Development*, Yourdon Press, Englewood Cliffs, New Jersey.

21 Blended methodologies

21.1 Structured Systems Analysis and Design Method (SSADM)

SSADM is a methodology developed originally by UK consultants Learmonth and Burchett Management Systems (LBMS) and the Central Computing and Telecommunications Agency (CCTA), which is responsible for computer training and some procurement for the UK Civil Service. (CCTA is now part of the Office of Government Commerce.) SSADM has been used in a number of government applications since 1981, and its use was mandatory in many Civil Service applications. This description is based on SSADM, Version 4+ (Weaver et al., 1998), which is very similar to the later version of SSADM4.2

The methodology provides project development staff with very detailed rules and guidelines to work to. It is highly structured. Another reason for its success has been in the standards provided (often exercised by completing pre-printed documents or through supporting software tools). Documentation pervades all aspects of the information systems project.

SSADM has seven stages (numbered 0 to 6 in Figure 21.1) within a five 'module' framework (the bullet points) with its own set of plans, timescales, controls, and monitoring procedures. The activities of each stage are precisely defined as are their associated end products (or deliverables), and this facilitates the use of project management techniques (the project management method PRINCE is recommended, see Section 25.4).

These modules cover the life cycle from feasibility study to design, but not implementation and maintenance. It is assumed that planning has been completed, and the stages following design are presumably seen as installation-specific, and therefore not covered by the methodology. We will now look in outline at each of the seven stages of SSADM.

0 Feasibility

This stage is concerned with ensuring that the project which has been suggested in the planning phase is feasible; that is, it is technically possible and the benefits of the information system will outweigh the costs. PRINCE may be used at this stage (and following stages) to help plan the project.

This phase has four steps: prepare for the study, which assesses the scope of the project; define the problem, which compares the requirements with the current position; select feasibility option, which considers alternatives and selects one; and assemble feasibility report.

Systems investigation techniques, such as interviewing, questionnaires, and so on, discussed in Chapter 3 in the context of the SDLC, are used in this stage as are the techniques of data flow diagramming (referred to as data flow models). The latter have different symbols (see Figure 21.2), but essentially the technique is the same as described in Section 12.1. Entity models (referred to as logical data structures) similar to that described in Section 11.1 are drawn. As one would expect at the feasibility stage, these are all done in outline and in not too great a detail. This detail will come in later stages.

The requirements of the new system, in terms of what the system will do and constraints on the system, are partly defined by considering the weaknesses of the present system.

Once the problem has been defined in this way, it is possible to consider the various alternatives (there might be up to five business options and a similar number of technical solutions) and recommend the best option from both the business and technical points of view. All this information is then published in the feasibility report.

1 Investigation of current environment

The second module, requirements analysis, has two stages: investigation of current requirements and business system options. This module sets the scene for the later stages, because it enables a full understanding of the requirements of the new system to be gained and establishes the direction of the rest of the project.

The first of these stages repeats much of the work carried out at the feasibility study stage but in more detail. For example, at the feasibility stage, the data flow diagrams may not have included much of the processing which is not related to the major tasks nor decomposed to

■ **Feasibility study**
0 *Feasibility*
   – Prepare for the feasibility study
   – Define the problem
   – Select feasibility options
   – Create feasibility report

■ **Requirements analysis**
1 *Investigation of current environment*
   – Establish analysis framework
   – Investigate and define requirements
   – Investigate current processing
   – Investigate current data
   – Derive logical view of current services
   – Assemble investigation results

2 *Business systems options*
   – Define business system options
   – Select business system options
   – Define requirements

■ **Requirements specification**
3 *Definition of requirements*
   – Define required system processing
   – Develop required data model
   – Derive system functions
   – Enhance required data model
   – Develop specification prototypes
   – Develop processing specification
   – Confirm system objectives
   – Assemble requirements specification

■ **Logical system specification**
4 *Technical system options*
   – Define technical system options
   – Select technical system options
   – Define physical design module

5 *Logical design*
   – Define user dialogues
   – Define update processes
   – Define enquiry process
   – Assemble logical design

■ **Physical design**
6 *Physical design*
   – Prepare for physical design
   – Create physical data design
   – Create function component implementation map
   – Optimize physical data design
   – Complete function specification
   – Consolidate process data interface
   – Assemble physical design

Figure 21.1: SSADM stages

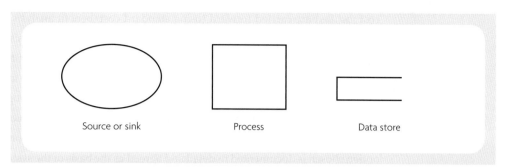

Figure 21.2: Data flow diagram symbols in SSADM

more than two levels of detail (Level 3 diagrams would be the norm at this stage). Further, the conflicts and ambiguities of the entity model need to be resolved. Indeed, in some projects the feasibility stage is carried out very much in outline and the investigation of current environment may have much less of a basis for the tasks of this stage.

The results of the feasibility study are examined, and the scope of the project reassessed and the overall plan agreed with management. The requirements of the new system are examined along with investigating the current processing methods and data of the current system, again in more detail than that carried out at the feasibility stage. The present physical data flow model is mapped onto a logical data flow model, and this helps to assess the present functionality required in the new system. Matrices (Section 12.7) might be constructed which, for example, show the relationship between processes and entities (i.e. which processes access the information in the various entities). Catalogues will be created, such as the user catalogue, which lists the activities carried out in each job, and the requirements catalogue, which lists the functional and non-functional requirements. Again, there is a complete description of the results of this stage assembled and reviewed as the deliverable.

SSADM suggests that decisions be made at this stage regarding customization of steps and techniques used for the particular problem situation. Users of the methodology are encouraged to think about the appropriateness of each of the steps. For example, some may be dropped or reduced in scope if a key objective is for rapid application development. Customization factors include:

- risk assessment;
- application type;
- situational factors;
- project objectives;
- available technology;
- control procedures;
- organizational constraints.

2 Business systems options

It is at this stage that the functionality of the new system is determined and agreed. The user requirements were set out in Stage 1, but it is at Stage 2 that only those requirements that are cost-justified are carried forward (using standard cost/benefit analysis techniques), and these requirements are specified in greater detail. Function point analysis (Section 14.1) is recommended for estimation. A number of business system options are outlined, all satisfying this minimum set of user requirements, and a few of these are presented to management so that one can be chosen (or a hybrid option chosen, taken from a number of the options presented). Each of these will have an outline of its cost, development timescale, technical constraints, physical organization, volumes, training requirements, benefits, and impacts on the organization. The option chosen is documented in detail and agreed as the basis of the system specification, which is the next stage of SSADM. Data flow diagrams and entity models are developed, but this stage is largely a specification in narrative.

3 Definition of requirements

This stage leads to the full requirements specification and provides clear guidance to the design stages, which follow. It is at the centre of SSADM where investigation and analysis are replaced by specification and design. For example, stress is placed on the required system design rather than the functionality of the current system. The requirements catalogue will be consulted and updated, and the logical entity model extended followed by normalization (Section 11.2) of the relations (to third normal form). The data flow model is also extended and used as a communication tool with users. User roles are defined for the new system. But it is the entity model that is emphasized at this stage and is the essential basis of the logical design of the new system. Documentation forms for all the entities and attributes (see Section 11.1) are completed.

Although the data model is emphasized at this stage, the components of each function (in terms of inputs, outputs, and events or enquiry triggers) are defined. Each function is documented in detail, and a form is used which includes space for function name, description, error handling, data flow diagram processes, events, and input and output descriptions. Jackson structure diagrams (Section 20.3) are used to show the input and output structures. Further documentation shows other details, such as the relationship between user roles and functions (via a user role/function matrix).

This stage in SSADM also has an optional prototyping phase. The methodology suggests demonstrating prototypes of critical dialogues and menu structures to users, and this will verify the analysts' understanding of the users' requirements and their preferences for interface design. As well as verifying the specification, this phase can have other benefits such as increased user commitment (Section 7.3). But these prototypes are not used as part of incremental development: they are used to form a clearer understanding of user requirements.

Entity life histories (called entity life cycles in Section 12.9) are also constructed during this phase. These document all the events that can affect an entity type and model the applicable business rules. Events affecting each entity may have been identified previously by constructing an event/entity matrix (similar to the CRUD matrix, Figure 12.20 in Section 12.7). Again, this is useful for verification purposes, as an entity should normally have at least one creation and one deletion event, and every event should lead to the update of at least one entity. Finally, at this stage, the system objectives are verified, the functions checked for completeness of definition, and the full requirements specification documented.

The diagramming conventions used in drawing entity life cycles (in SSADM called entity life histories) are very similar to the entity structure step conventions of Jackson System Development (JSD) (Section 20.3), and the technique is described in more detail there because it is of such crucial importance in JSD. In SSADM the diagrams look like hierarchies, but they are meant to be read from left to right, and, in so doing, progressively suggest the different states of the entity. Using an example from the academic world, Figure 21.3(a) shows how the entity 'student' changes over time, as an applicant, registered student, and graduate (there

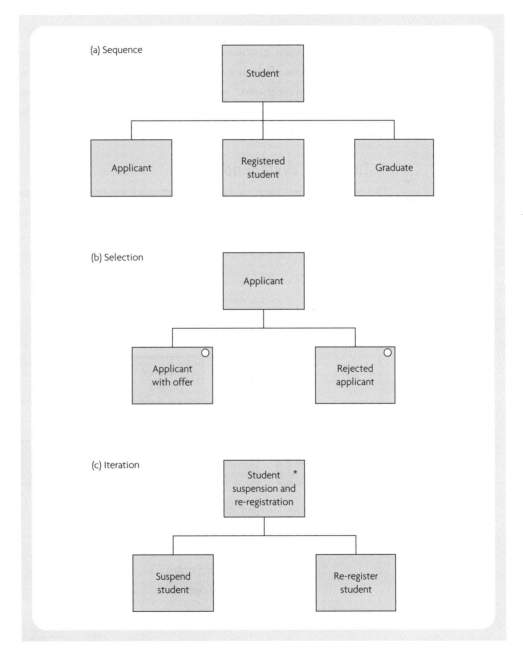

Figure 21.3: SSADM entity life history constructs

will be other intermediary states). Figure 21.3(b) shows the use of the selection construct, whereby the small circle in the 'applicant with offer' and 'rejected applicant' boxes denote alternative conditions (these are mutually exclusive). Figure 21.3(c) shows the iteration construct, marked with an asterisk, which shows an event that may repeat (in this example, the possible repeated suspension and re-registration of a student who might regularly pay fees late).

Figure 21.4 presents an SSADM entity life history. The first level contains the events that cause an entity to be initiated into the system and those events that terminate the entity from the system. There is an iteration construct relating to whether the student is accepted conditionally or not and to reflect suspended or registered states. There are four states for the end condition: withdrawn, graduated, non-graduated, or rejected. These are all mutually exclusive (the selection construct). Notice that in this model it is not possible to show that 'graduated' can only happen from 'registered' and that 'suspended' can only terminate with 'non-graduated'. So, some information is lost when compared to the earlier entity life cycle representation described in Section 12.9.

4 Technical system options

This stage and the following logical design stage are carried out in parallel. In the technical system options stage, the environment in which the system will operate, in terms of the hard-

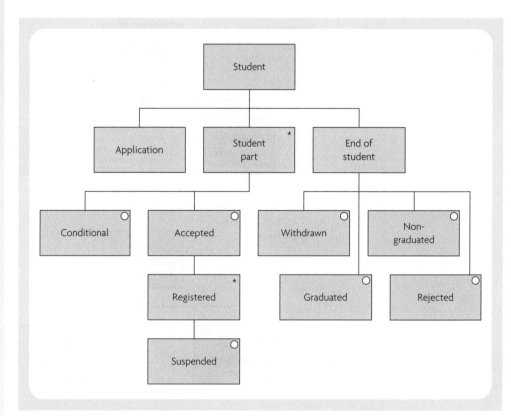

Figure 21.4: Student entity life history

ware and software configurement, development strategy, organizational impact and system functionality, is determined.

The definition of technical options will be implementation-specific, because there are so many alternative hardware, software, and implementation strategies. The analysts need to identify constraints; for example, the hardware platform may be 'given' along with time and cost maxima and minima. System constraints might include performance, security, and service level requirements that must be met, and these will limit choice. Technical system options need to meet all these constraints, and a chosen option has to be agreed with management.

The analysts may perform an impact analysis of the various technical system options, focusing on organizational, personnel and operating changes, training requirements, systems documentation, savings, and testing requirements.

5 Logical design

This is a statement of what the system is required to do rather than a statement about the procedures or program specifications to do it. The latter is the realm of the final stage (Stage 6), the physical design. In Stage 5 the dialogue structures, menu structures, and designs are defined for particular users or user roles. User involvement is recommended at this stage, and the prototypes developed in Stage 3 are referred to. Furthermore, following the entity life cycles designed in Stage 3 (which are developed further), the update processes and operations are defined along with the processing of enquiries, including the sequence of processing. In other words, it is at this stage that further detail about how the system will apply and control the operations following each event will be defined. Detail such as the rules of validating data entered into the system will be specified. All the requirements to start designing the physical solution are now in place.

6 Physical design

It is at this final stage that the logical design is mapped onto a particular physical environment. A **function component implementation map** (FCIM) documents this mapping. The phase provides guidelines regarding physical implementation, and these should be applicable to most hardware and software configurations. However, this stage will be carried out with the actual configuration in mind. The roles of the technologist, the programmer, and database designers, in particular, are stressed in this phase, although the analyst and user need to be available to verify that the final design satisfies user requirements.

The logical data model will be converted into a design appropriate for the database management system available. The database mapping will be a key aspect of final implementation and include not only the way data and data relationships are held on the database, but also key handling and access methods. Much will depend on performance measurement so that database access is efficient, and again this will depend on the actual hardware and software configuration (including the database management system).

The function component implementation map lists the components of each logical function and their mapping onto the physical components of the operational system. The principles of the FCIM are well specified in SSADM, although the form of the FCIM is somewhat ambiguously expressed. Presumably, this is seen as dependent on the standards of the particular organization. Designs are optimized according to storage and timing objectives. From this stage it should be possible to design and develop the programs necessary to provide the required functionality. It is at that point the SSADM stops and detailed software design and testing starts.

The well-defined structure of SSADM makes it teachable, and many UK university courses in information systems have used this methodology for in-depth treatment and discuss other methodologies in overview only for comparative purposes. Its three basic techniques, entity models, data flow diagrams, and entity life histories, are common to a number of methodologies and they ensure that there has been a detailed analysis of the target system. Along with the well-defined tasks, and guidance with the techniques, the methodology defines the outputs expected from the stage, and gives time and resource management guidelines.

SSADM is expected to be used with a toolset (see Part V). The proponents of the methodology also recommend '**Quality Assurance Reviews**' based on structured walkthroughs (see Section 12.6). They are meetings held to review identifiable end products of the various phases of the methodology, such as entity models, data flow diagrams, entity life histories, and process details. Post-implementation feedback is also encouraged, and there is an audit at this time.

The successful implementation of the methodology relies on the skills of key personnel being available, though the techniques and tools are widely known, and the project team method of working, along with systems walkthroughs, encourages good training procedures and participation.

SSADM has evolved to allow a more 'pick and mix' approach, and so a rapid application development (RAD) model or a reuse approach, or an SSM front end, may be integrated as appropriate. Nevertheless, SSADM is traditionally associated with well-defined, large-scale systems development projects requiring heavy documentation, for use particularly in large, bureaucratic organizations.

21.2 Merise

Merise is the most widely used methodology for developing information systems in France. It is used in both the public and private sectors. Its influence has spread outside France to Spain, Switzerland, and Canada. Like SSADM, Merise has become influential in the European approach, Euromethod (Section 26.4).

The essentials of the approach lie in its three cycles: the decision cycle, the life cycle, and the abstraction cycle, which cover data and process elements with equal emphasis. Although it is prescriptive to some extent, Merise permits the participation of end-users and senior management as well as data processing professionals in its decision cycle. Again, like SSADM, there are a number of software tools for use with Merise.

The project which led to Merise was launched by the French Ministry of Industry, and included research groups, consultancy and engineering firms, and academics, the inspiration coming from Hubert Tardieu. Merise has since developed into a very thorough and comprehensive methodology.

The core of the Merise approach lies in its three cycles: the decision cycle, which relates to the various decision mechanisms; the life cycle, which reflects the chronological process of a Merise project from start to finish; and the abstraction cycle, the key to Merise, which describes the various models for processes and data in each of three stages. Each of these three cycles will be considered in turn, with the major emphasis being placed on the abstraction cycle.

1 Decision cycle

The decision cycle, sometimes referred to as the approval cycle, consists of all the decision mechanisms, including those for choosing options, during the development of the information system. Decision making is a joint process concerning senior management, users, and systems developers. Decisions will include:

- technical choices regarding hardware and software;
- processing choices, such as real-time or batch;
- user-oriented choices relating to the user interface;
- identification decisions regarding the major actors of the information system and the organization;
- financial decisions relating to costs and benefits;
- management decisions concerning the functionality of the information systems.

Each decision point during the development of an information system is identified by Merise. It is essential to know who takes the decisions, particularly those relating to the validation of the various models used by the method, and when to complete one stage to start the next. The Merise authors suggest that the decision-making process will follow the scheme as shown in Figure 21.5. The groups of users and systems developers will together discuss various options (1), and it is the responsibility of the user team to produce a report reflecting these deliberations (2). This is then discussed at a joint meeting (3) of senior management, users, and application developers, and the decision made at this point.

It is necessary to specify how a compromise should be reached in the case of conflicting views. This will depend on the norms of the specific organization, but there is a strong user element suggested in the decision-making process and this will influence the acceptance of the final system, from the point of view of operational and technical criteria and usability.

Therefore, in Merise, there are opportunities for user influence and participation, but this is not spelt out in detail as it will depend largely on the norms of the organization.

2 Life cycle

The life cycle shows the chronological progress of the information system from its creation, through its development, until its final review and obsolescence. Each of these stages is well defined in Merise. The main phases of the life cycle are:

1. *Strategic planning (at the corporate level)* – which maps the goals of the organization to its information needs, and partitions the organization into 'domains' for further analysis (such as purchasing, manufacture, finance, and personnel). For each of these a schedule

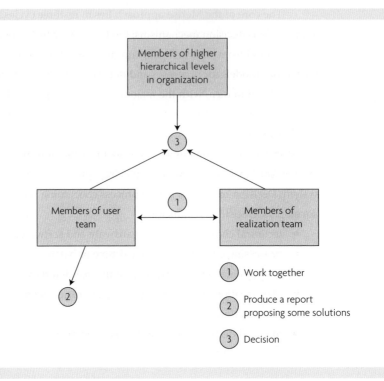

Figure 21.5: Schema of decision-making process at each step

of applications is devised to include a policy for human resources, software and hardware products, and system development methodology implementation. Within the frame of the strategic plan, the analysis that has just been carried out for one domain should be done for all the others, and then it will be possible to understand better and more coherently the connection between them.

2. *Preliminary study (for the domain of interest)* – which describes the proposed information systems, discusses their likely impact, and details the associated costs and benefits, which should be consistent with the strategic plans.

3. *Detailed study (for a particular project)* – of only those aspects which will be automated, including detailed specifications for the functional design (the requirements specification) and the technical design (the technical architecture of programs and files).

4. *Schedules and other documentation* – for development, implementation, and maintenance (all three at the application level).

Sometimes the last stage is defined as consisting of three separate stages: development, implementation, and maintenance.

The whole of this second cycle is similar to the conventional life cycle as found, for example, in SSADM and other methodologies, and for this reason will not be discussed further. It should be pointed out, however, that, unlike many alternative approaches, Merise includes a strategic planning phase, and in this respect Merise is similar to Information Engineering. The objective of this stage is to link the goals of the business with the information systems needs.

Nevertheless, like SSADM, the emphasis is on the analysis and design of the database and corresponding transactions. The reference to SSADM made earlier is apposite, for the nearest UK equivalent to Merise is SSADM, being the most used methodology in the UK and widely adopted by the UK Civil Service and other public and private sector organizations.

3 Abstraction cycle

The abstraction cycle is the key to Merise. Unlike many alternative approaches, the separate treatment of data and processes is equally thorough and both are taken into account from the start. The data view is modelled in three stages: the conceptual, the logical, through to the physical. Similarly, the process-oriented view is modelled through the equivalent three stages of conceptual, organizational, and operational. Each of these six abstraction levels in the abstraction cycle is a representation – albeit a partial one – of the information system, and they should be consistent.

The abstraction cycle is a gradually descending approach which goes from the knowledge of the problem area (conceptual), to making decisions relating to resources and tasks, through to the technical means on which to implement it. The conceptual stage looks at the organization as a whole; the logical stage addresses questions, such as who must do what, where, when, and how; whereas the physical stage looks at resources and technical constraints surrounding what will be the operational system. Merise is therefore independent of the technology until the later phases.

The modelling logic of Merise, outlined above, is shown in Figure 21.6. At the conceptual level the group of entities dealt with by the information system will be represented in a totally independent way from the organization and from the existing or future technical means for developing the project. At this level it is necessary to find out what the business does and the essence of the problem situation. At the logical level it is necessary to make choices (using methods developed at the conceptual level) in terms of the organization for the processing and with regard to the database models for the data, which will be part of the automated system.

LEVEL	CONCERN	DATA	PROCESSING
CONCEPTUAL	What do you want to do?	Conceptual data model	Conceptual processing model
LOGICAL OR ORGANIZATIONAL	Who does what, when, where, how?	Logical data model	Logical or organizational data model
PHYSICAL OR OPERATIONAL	By what means?	Physical data model	Operational processing model

Figure 21.6: Merise by levels of data processing

The physical level is the level at which constraints related to the operating system, database management system, and programming languages are going to be introduced.

An initial overall view of the system is given in the Merise flow diagram, and the construction of this precedes the conceptual models (both data and processing). The Merise flow diagram is not to be confused with the more conventional data flow diagrams. The Merise flow diagrams bring to light the information flows between the various actors in the domain studied, together with the environment. They serve as a base for developing the conceptual data model and the conceptual processing model. The actors are described in the ellipses, and arrows represent the information flows between them. So, the flow diagram showing the accounts, suppliers, and customers might be as shown in Figure 21.7, where the actors who are shaded are external to the information system. From it, we can see directions for the conceptual data model (concerning customers, accounts, and suppliers) and for the conceptual processing model (concerning the settlement of invoices) in this example.

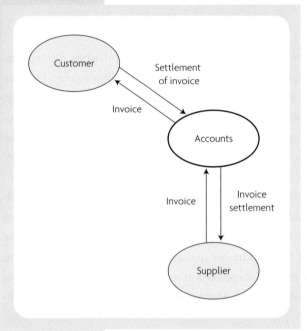

Figure 21.7: A Merise flow diagram

At the conceptual level, the information system is represented independently of its organization and of the physical and computing means that it could use. The objective of the conceptual level is to answer the question 'what?' and to understand the essence of the problem. The rules evidenced at the conceptual level are the 'management rules' of the domain under analysis.

The graphical representation of the conceptual data model (Figure 21.8) is the entity–attribute relationship model similar to that described in Section 11.1. An entity is represented by a box in Merise; a relationship is represented by an ellipse. Merise has a number of rules, which enable the verification of the model.

The conceptual processing model describes the activities of the organization. The concern of the conceptual processing model is with events, operations, and their synchronization.

Many other approaches which include the concepts of operations and events do not include that of the **synchronization** of an operation. This is the condition or conditions (events) which must have occurred to trigger the operation and a rule or set of rules regarding the necessary condition for the operation to be triggered. For example, payslips should be produced if it is the 28th of the month. The conceptual processing model related to the production of payslips might include a synchronization (28th of the month when payslips are produced) following the event that a new day has dawned (see Figure 21.9). This will trigger the process to produce payslips and routines which are dependent on whether the payslips are valid or

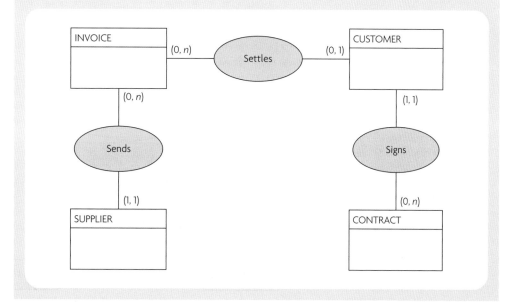

Figure 21.8: A conceptual data model (part)

invalid. In the figure, we also see an issuing rule, related to whether the payslips are OK or not OK, and, depending on this state, the processes that follow are different. Cheques are signed, however, whether the event is A or B, that is, a validated payslip is received or a corrected invalid payslip is received. These diagrams are types of **Petri nets**. Merise also provides a series of rules to enable the verification of the conceptual processing model.

Having established what to do at the conceptual level, at the logical level all the organizational alternatives are identified in order to discover who will do what, where, and when, and how the processing will be carried out. The information system is represented by taking into account the constraints imposed by these alternatives. The rules brought to light at the logical/organizational level are the 'organizational rules' of the domain under analysis. The organizational processing model is used again to clarify all the concepts described in the conceptual processing model. It is therefore a question of describing how the processing methods are executed within the organization, which could be manual (where the procedure is carried out without computing resources), conversational automatic (where the procedures are carried out by computer but with the intervention of people), or automatic (where the procedures will, once started, run without human intervention). The organizational processing model is used to define who carries out the processing, and when and how it is achieved. The organizational processing model will be based on the conceptual processing model with some changes, such as the names of departments where the processing will take place. Figure 21.9 might be amended so that processing is allocated to the personnel department (signing cheques), computing department (producing payslips), and accounts (records).

The logical data model is situated between the conceptual data model and the physical data model. It represents the world of data, described in the conceptual data model, but which

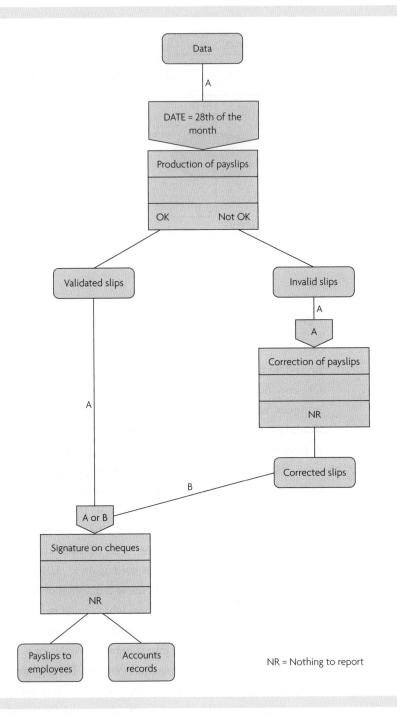

Figure 21.9: Conceptual processing model of the process 'production of payslips'

takes into account the type of database management system chosen. In other words, the logical data model transforms the conceptual data model into a form that is suitable for computerization. Merise also offers a full description of the normalization process (Section 11.2).

One of the most important aspects of Merise is the detailed rules for converting from one model to the next (as well as specifying rules for creating each model), for example, there are 10 verification rules for the conceptual processing model. Therefore, the rules for mapping the conceptual data model to the logical model in relational Boyce–Codd Normal Form (BCNF) are shown in Figure 21.10, and the rules for special cases, such as those of binary and *n*-ary relationships, are described in detail. Rules for mapping to other data models and optimizing the relational and other data models, taking into account volume and activity, are also provided.

At the physical level, all the technical alternatives are identified which make it possible to define the computing needs, and this definition represents the last stage before the development of the software. The objective of the physical level is to answer the question 'with what means?'. The rules brought to light at the physical level are the 'technical rules'. It therefore takes into account the physical resources (database management system, hardware, support tools, and so on). Typically, the physical data model might be represented as a series of SQL definitions and the operational processing model as structured English, along with equivalent SQL queries to the database. The diagrammatic representation of the database processes and queries might be in the form of data flow diagrams and mapped onto structure diagrams. Again rules of mapping are provided.

Another important aspect of Merise is that the methodology has been designed to reflect the world where change is common, and new needs and directions can be incorporated into the design as the information systems develop. For example, guidelines are given to show how the conceptual data model and the conceptual processing model can be modified to take account of new management rules for data and processing. Further, unlike a conventional database methodology which is directed toward the static, data-oriented aspects of information systems, Merise, as we have seen, includes a thorough analysis of events, operations, and synchronization, all dynamic aspects of an information system.

Each of the six models of the abstraction cycle has a graphic formalism, with the possible exception of the physical data model, hence the methodology lends itself to the use of support tools. Many tools support all three levels of the abstraction cycle, both for data and processes, and thus ease the task of drawing the models. Some will validate, or partially validate, each of the models, help generate the required documentation at each stage, and may

CONCEPTUAL DATA MODEL		LOGICAL DATA MODEL IN BCNF
Identifier of entity	becomes a	Key
Property of identifier	becomes an	Attribute
Entity	becomes a	Relation
Relationship not of cardinality (1, 1)	becomes a	Relation
Relationship of cardinality (1, 1)	disappears	

Figure 21.10: Rules for mapping the conceptual data model to the normalized relational model

incorporate an applications generator, query language, and data dictionary interface, which again will lighten the task of developing information systems.

Significantly, there are a number of support tools which help the user of Merise. These tools are varied: some are design tools (e.g. to develop the various conceptual and logical models and diagrams), others are modelling and prototyping tools (to give alternative views of the final system), and yet others are execution and code generation tools (to generate the future application).

One of the important reasons for adopting an information systems development methodology is that of common standards, and, as well as the graphical support tools for each of the six models which are part of the abstraction cycle, there are standards and tools supporting strategic planning, project planning, requirements specification and the file of options, and the documentation of each entity, relationship, attribute, event, and operation defined. The strategic plan, for example, is likely to contain a diagnosis of the present situation, perspectives on the evolution of the organization, a description of the conceptual solution (what we want to do), and plans for development regarding organizational and technical solutions. It will also include, for the adopted solution, a description and information about its impact (including advantages, risks, and means), as well as reasons for rejecting other solutions.

For the processing, the overall specification is likely to include the organizational processing model, a detailed description of processes, and the operational processing model, with a list of applications, transactions, and batch chains and their arrangement into computer programs. For the data, it will include the conceptual, logical, and physical data models. It will also include a study of constraints (security and control policy), details about interfacing with existing applications and responsibilities. Appendices are likely to include definitions of relations (depending on the eventual database approach chosen), a list of states and screens for each process and their sequence, and a physical description of records or relations.

The above description can be regarded as 'classic Merise'. However, Merise has several versions in use. One significant adaptation has been the adoption of a third area to model, that of **state transitions**, to be modelled alongside data and processes. Essentially, these are entity life histories. Therefore, Figure 21.6 might be amended to include another column, that of state transition. This version is sometimes known as Merise/2. Another version separates the logical and organizational level (Figure 21.6) into two levels, forming a four-level schema. Here, organizational questions include who?, when? and where?, with the separate logical question being how? It is only at this logical stage (and the subsequent physical stage) that decisions about potential IT solutions are raised. Yet another version includes the incorporation of object modelling as an alternative to relational modelling.

21.3 Information Engineering (IE)

The origins of Information Engineering (IE) differ according to which source is referenced. It appears that Clive Finkelstein first used the term to describe a data modelling methodology that he developed in Australia in the late 1970s. In early 1981 he renamed his consultancy company IE and wrote a series of articles on the methodology. In the same year he collaborated with James Martin on a two-volume book entitled *Information Engineering* (Martin and Finkelstein, 1981) and then Martin produced a later version (Martin, 1989).

Since these early days there has evolved, rather confusingly, a number of versions of IE which, while very similar in concept, have tended to develop along somewhat different lines. The reason for this is that James Martin, who is generally credited with evolving and popularizing the methodology, set up a number of independent companies based on the methodology of IE. One such initiative was an association with Texas Instruments (TI) to develop the IEF (Information Engineering Facility), a toolset to support the methodology. Martin believes that IE 'should not be regarded as one rigid methodology but, rather, like software engineering, as a generic class of methodologies' (Martin, 1991).

The version of IE described here is based upon a number of sources and is sometimes termed 'classical' IE. There also exist a number of variants of IE for different development environments. These include a package-based approach, a Rapid Application Development (RAD) approach (see Section 23.1) and an object-oriented-based version (Martin and Odell, 1992).

IE is claimed to be a comprehensive methodology covering all aspects of the life cycle. It is viewed as a framework within which a variety of techniques are used to develop good quality information systems in an efficient way. The framework is argued to be relatively static and includes the fundamental things which must be done in order to develop good information systems. The techniques currently used in IE are not part of those fundamentals, but are regarded as the best currently available to achieve the fundamentals. Therefore, the techniques can and do change as new and improved techniques emerge. The framework is also a project management mechanism, which reflects IE's philosophy of 'practicality and applicability'. It is not just a set of ideas, but is argued to be a proven and practical approach. It is also said to be applicable in a wide range of industries and environments.

There are a number of philosophical beliefs underpinning IE. One of the original was the belief that data are at the heart of an information system and that the data, or rather the types of data, are considerably more stable than the processes or procedures that act upon the data. Therefore, a methodology that successfully identifies the underlying nature and structure of the organization's data has a stable basis from which to build information systems. Methodologies which are based only upon processes are likely to fail due to the constantly shifting nature of this base as requirements change. This is the classic argument of the data modelling school of thought (Section 6.3). However, IE also clearly recognizes that processes have to be considered in detail in the development of an information system and balances the modelling of data and processes as appropriate.

A further aspect of the philosophy of IE is the belief that the most appropriate way of communication within the methodology is through the use of diagrams. Diagrams are very appealing to end-users and end-user management and, it is argued, enable them to understand, participate, and even construct for themselves the relevant IE diagrams. This helps to ensure that their requirements are truly understood and achieved. The diagrams are regarded as being rigorous enough on their own to ensure that all necessary information is captured and represented.

Each IE technique is oriented toward diagramming, and a diagram is a deliverable of each major stage in the methodology. One of the key elements in IE is the use of standard diagrams which initially are defined at high levels of abstraction and, as the methodology proceeds, they are gradually evolved, becoming more and more concrete and detailed until they ultimately form potentially executable applications. Standard symbols are used throughout, for example, boxes with square corners represent data and boxes with round corners represent activities.

The primary IE model consists of three components: data, activity, and the interaction of the data and activities (as shown in Figure 21.11). The interaction may be a matrix indicating at a high level which subject areas are used in which functions. At a lower level it may show which entity types are used by which processes. At an even lower level still, the interaction may be expressed as an action diagram (Section 12.8) and finally as actual program code.

Automated support, that is, the use of an appropriate toolset, is identified as a basic imperative for the IE methodology. A description of IEF is given in Section 19.2.

The methodology is top-down and begins with a top management overview of the enterprise as a whole. In this way separate systems are potentially related and coordinated and not just treated as individual projects which enables an overall strategic approach to be adopted. As the steps of the methodology are carried out, more and more detail is derived and decisions concerning which areas to concentrate upon are made. Based on the overall plan, the business areas to be analysed first are selected and then a subset may be chosen for detailed

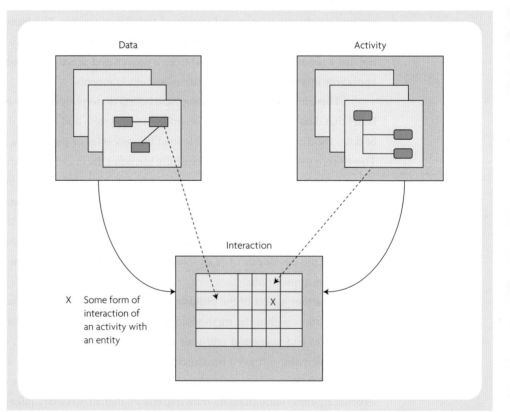

Figure 21.11: Data, activity and interaction

design and construction. This approach to the management of the complexity of the information system requirements of an organization is termed 'divide and conquer', and is illustrated in Figure 21.12. The objectives and focus change as the methodology progresses with each stage having different objectives, although the overall objectives remain consistent. Progress is controlled by measuring whether the objectives have been achieved at each stage, not by how much detail has been generated.

The methodology is divided into four levels or layers. These levels are represented in Figure 21.13. The four levels are:

- *Information strategy planning.* The objective here is to construct an information architecture and a strategy which supports the overall objectives and needs of the organization. This is conducted at the enterprise level. One part of this planning is the identification of relevant business areas.
- *Business area analysis.* The objective here is to understand the individual business areas and determine their system requirements.
- *System planning and design.* The objective here is to establish the behaviour of the systems in a way that the user wants and that is achievable using technology.
- *Construction and cutover.* The objective here is to build and implement the systems as required by the three previous levels.

The first two levels are technology independent, whereas levels three and four are dependent on the proposed technical environment.

1 Information strategy planning (ISP)

Much of this level is really concerned with the overall corporate objectives. It may not always be part of the IE methodology, as it would normally be performed by corporate management

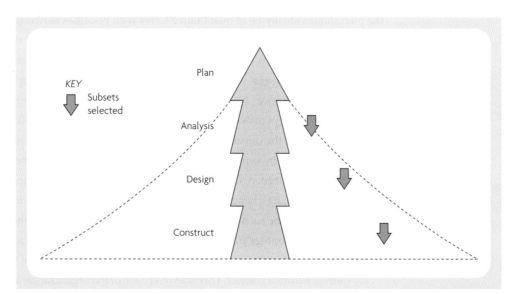

Figure 21.12: Divide and conquer approach of IE

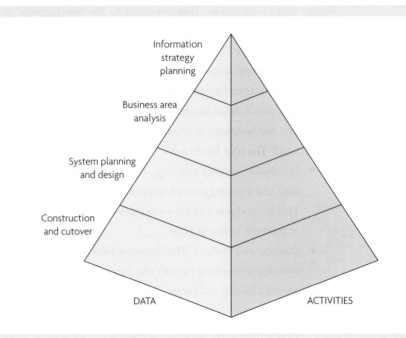

Figure 21.13: The four levels of IE

and planners. However, it is recognized as a fundamental starting point for the methodology. It implies that the organization's information system should be designed to help meet the requirements of the corporate plan and that information systems are of strategic importance to the organization. The corporate or business plan should indicate the business goals and strategies, outline the major business functions and their objectives, and identify the organizational structure. Such analysis includes any necessary re-engineering of the business processes (Section 4.3). The plan should ideally be in quantitative terms with priorities between objectives established.

ISP involves an overview analysis of the business objectives of the organization and its major business functions and information needs. The result of this analysis is what is termed 'information architectures' which form the basis for subsequent developments and ensure consistency and coherence between different systems in the organization. The resulting information strategy plan documents the business requirements and allocates priorities, which are the rationale for the development of the information systems. The plan enables these high-level requirements to be kept in view throughout the development of the project. In many other methodologies, it is argued, these needs get lost, if they are ever identified at all. It also provides a means of controlling changes to assumptions, priorities, and objectives, should it become necessary. Apart from such changes, the information strategy plan should remain relatively static. Information strategy planning is a joint activity of senior general management, user management, and information systems staff. It involves the performance of four tasks as follows:

1. *Current situation analysis.* This is an overview of the organization and its current position, including a view of the strengths and weaknesses of the current systems. This overview

will include an analysis of the business strategy, an analysis of the information systems organization, an analysis of the technical environment, and a definition of the preliminary information architecture (data subject areas, such as customer or product, and major business functions).

2. *Executive requirements analysis.* Here, managers are provided with an opportunity to state their objectives, needs, and perceptions. These factors will include information needs, priorities, responsibilities, and problems. This also involves the identification of goals of the business and how technology can be used to help achieve these goals and the way in which technology might affect them. Critical success factors (CSFs) for the overall organization are identified, and these are also decomposed into CSFs for the individual parts of the organization (see Section 15.2).

3. *Architecture definition.* This is an overview of the area in terms of information (the identification of global entity types and the decomposition of functions within the subject areas described in the preliminary information architecture in the current situation analysis above), an analysis of distribution (the geographic requirements for the functions and the data), a definition of business systems architecture (a statement of the ideal systems required in the organization), a definition of technical architecture (a statement of the technology direction required to support the systems including hardware, software, and communications facilities), and a definition of information system organization (a proposal for the organization of the information systems function to support the strategy).

4. *Information strategy plan.* This includes the determination of business areas (the division of the architectures into logical business groupings, each of which could form an analysis project in its own right), the preparation of business evaluations (strategies for achieving the architectures, including migration plans for moving from the current situation to the desired objective), and the preparation of the information strategy plan itself (a chosen strategy including priorities for development and work programs for high-priority projects).

2 Business area analysis

The business areas identified in the information strategy plan are now treated individually, and a detailed data and function analysis is performed. Maximum involvement of end-users is recommended at this stage. The tasks of business area analysis are as follows:

- *Entity and function analysis.* This is the major task of the stage. It involves the analysis of entity types and relationships, the analysis of processes and dependencies, the construction of diagrammatic representations of the above, such as entity models (Section 11.1), function hierarchy diagrams, similar to that shown as Figure 6.1, and process dependency diagrams (a kind of data flow diagram (Section 12.1) but without data stores), and the definition of attributes and information views.

- *Interaction analysis.* This examines the relationship and interactions between the data and the functions, that is, the business dynamics. Figure 21.14 is an example of a

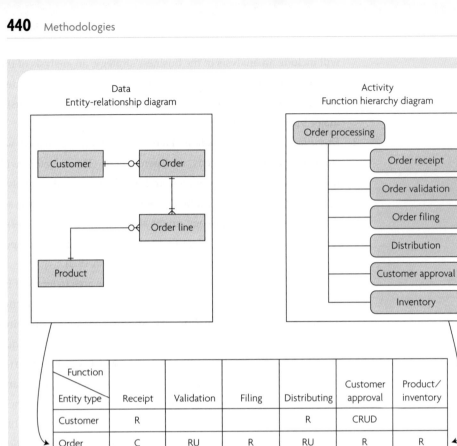

Figure 21.14: Example function/entity type matrix

function/entity-type interaction matrix. In this example, the order processing function of a business is shown in the form of an entity-relationship model, a function/process hierarchy diagram, and the matrix of interactions between the two. In this example, the matrix maps the interactions in terms of whether the function creates, reads, updates, and/or deletes the entities (CRUD matrix). The example is somewhat simplified, but it readily shows that orders and order lines are never deleted. This may be an error, or it may indicate that an order archiving function is required. Interaction analysis also involves an analysis of entity life cycles (Section 12.9), an analysis of process logic (Sections 12.2–12.4), and the preparation of process action diagrams (Section 12.8).

- *Current systems analysis.* This models the existing systems in the same way as for the entity and function analysis task in order that the models can be compared in the confirmation task (below), so that a smooth transition from one to the other can be achieved. The phase includes the construction of procedure data flow diagrams (Section 12.1) and the preparation of a data model by **canonical synthesis**. Because this is a technique not described previously, an example of its use will be provided. Canonical synthesis is a technique for pulling together all the data identified in separate parts of the organization, whether they be reports, screens, forms, diagrams and so on, in fact all sources, into a coherent structure, which is the entity model. The technique involves the drawing of bubble charts (user view analyses) and synthesizing all the data into an entity model. A bubble chart is a graph of directed links between data item-types (Figure 21.15). A double ellipse represents a key, an arrow represents a one-to-one dependency, and a double arrow represents a one-to-many dependency. In this case, the key completely determines (or identifies) the attributes, therefore the data are normalized. A separate bubble chart is constructed for each separate user view of the data. Figure 21.16 shows

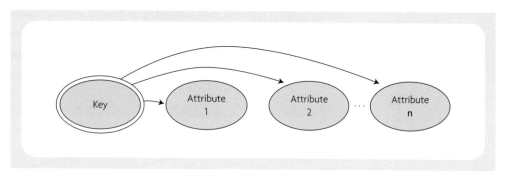

Figure 21.15: A bubble chart

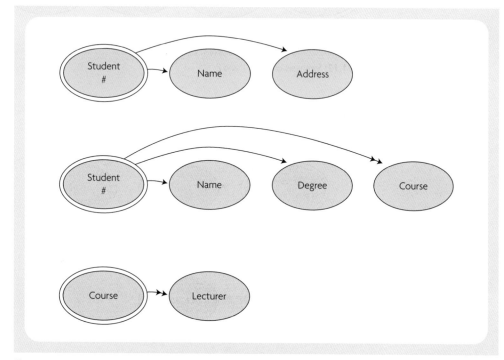

Figure 21.16: Three user views (courtesy of James Martin Associates)

an example of three user views in a university environment. View 1 might be a secretary's view, View 2 a registrar's view, and View 3 a course manager's view. The process of canonical synthesis combines the separate views into one data model. Each view is normalized (Section 11.2) and combined with another, and any duplications in the graph are eliminated. Figure 21.17 is the result of the combination of Views 1 and 2, and Figure 21.18 is the synthesis of all three views.

- *Confirmation*. This is the cross-checking of the results of the above, in terms of completeness, correctness, and stability. Hypotheses concerning business changes are also examined to see what effects these might have.
- *Planning for design*. This step includes the definition of design areas (which identify those parts of the model to be developed), the evaluation of implementation/transition sequences, and the planning of design objects. This includes the identification of areas where existing reusable objects (models and code) or components could be utilized. This may involve the reuse of objects generated internally or the purchase of objects externally. This is now an important area for IE, with the objective of speeding up the development process. In addition, areas where objects being designed for this particular

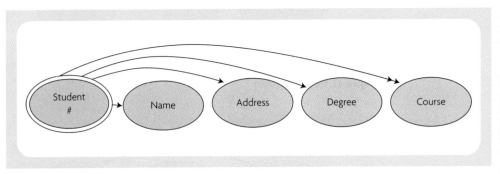

Figure 21.17: Synthesis of views 1 and 2

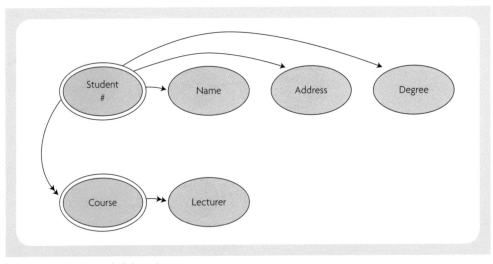

Figure 21.18: Synthesis of all three views

area might be required themselves in the future for reuse should be designed with this in mind. Design for reuse involves additional requirements such as flexibility and ease of use (Section 4.6). The effective identification of objects for reuse requires an appropriate repository (Section 19.1).

The output from business area analysis is the business area description, which contains the business functions, and each function is broken down into its lower-level processes and the process dependencies. On the data side, the entity types, relationships, and attributes are described, along with their properties and usage patterns. The level of detail here is much greater than that arrived at during the construction of the architectures performed during the information strategy planning stage. This information provides the basis for the broad identification of business processes requiring computer support.

3 System planning and design

This level is divided into business system design and technical design. In some versions of IE, these are termed external design and internal design.

In the area of business systems design, for each design area identified, the facts gathered are used to design a system to fulfil the identified business requirements. The design is taken up to the point at which technical factors become involved, therefore it is the logical design. The steps involved are as follows:

- *Preliminary data structure design*. In order to ensure integration and compatibility for all systems in the business area, this step is performed at the level of the whole business and not just the design area. It involves a first attempt at converting the entity model to the structure of the chosen database management system. This includes a summary of data model usage (basically an analysis of the way the data are used by the functions to produce a quantifiable view, sometimes referred to as volumetrics) and the preparation of the preliminary data structure.
- *System structure design*. This involves the mapping of business processes to procedures, and the interactions are highlighted by the use of data flow diagrams. This phase involves the definition of procedures and the preparation of data flow diagrams.
- *Procedure design*. This stage involves the development of data navigation diagrams (access path analysis, which examines the types and volumes of access required to particular entity types), the preparation of dialogue flows (i.e. the various hierarchies of control of user interaction), and the drawing of action diagrams (Section 12.8). Hierarchically structured dialogues and menus can be represented in action diagrams, but non-hierarchical menus require an alternative diagrammatic representation. For this purpose, IE uses dialogue flow diagrams (sometimes called dialogue structure diagrams), a simple example being illustrated in Figure 21.19. In overview, the horizontal lines represent screens and the vertical lines potential jumps or transfers between screens. This will depend on the choices made by the user. A horizontal line with bars at both ends represents a procedure, which could be a screen or a menu. Horizontal lines

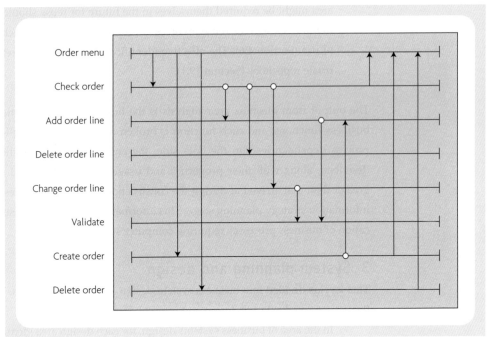

Figure 21.19: A dialogue flow example

without bars represent a procedure step, that is, a part of a procedure. A vertical line between procedures indicates that control is transferred from the procedure at the tail of the arrow to the procedure or step pointed to by the arrow. If the tail of the arrow has a loop, this represents a link transfer which means that control may return to the originating procedure with the context retained. The example shown in Figure 21.19 is the structure of a menu system for handling customer orders. The first procedure is the main order menu, and the user can select three options: check order, create order, or delete order. In check order, the customer order number is entered, and the options of adding an order line, deleting an order line, or changing an order line can be made. These are link transfers as the context, that is, the order details for the selected customer, is retained on return. Note that add order line and change order line have a link transfer to validate, which is a procedure step of some other procedure.

- *Confirmation.* Again, as part of business systems design there is a stage to confirm completeness, correctness, and usability. Matrices are used to analyse completeness. For correctness, the question 'does it follow the IE rules?' is asked. For usability, verification is normally achieved by the users commenting on a prototype.
- *Planning for technical design.* The final phase of this stage involves the definition of implementation areas and the preparation of technical design plans.

At the end of this stage a business systems specification is produced which details, for each business process, the information flows and user procedures, and for each computer procedure, the consolidated and confirmed results of business area analysis, plus the dialogue design, screens, reports, and other user interfaces. The scope of the proposed computer systems is defined along with the work programmes and resource estimates for the next stage.

The computerized aspects of the business systems identified above are designed at a technical level such that the final construction and operation of the systems can be planned and costed. The tasks of this technical design phase are as follows:

- *Data design* – which includes preparation of data load matrices, refinement of the database structure, design of data storage, and the design of other files.
- *Software design* – which includes the definition of programs, modules, reuse templates, integration groups, the design of programs/modules, and the definition of test conditions.
- *Cutover design* – which includes the design of software and procedures for bridging and conversion, the planning of system fan-out (the phases in which it should be implemented by location), and the definition of user training.
- *Operations design* – which includes the design of the security and contingency procedures, the design of operating and performance monitoring procedures, and the design of software for operations.
- *Verification of design* – which includes benchmark testing and performance assessment.
- *System test design* – which includes the definition of system tests and acceptance tests.
- *Implementation planning* – which includes a review of costs and the preparation of the implementation plan.

The output from this stage is the technical specification, including the hardware and software environment, its use, standards, and conventions. It also includes the plan and resources for subsequent construction and cutover.

4 Construction and cutover

This level includes the stages of construction, cutover, and production. Construction is the creation of each defined implementation unit and includes the following tasks:

- *System generation* – which includes the construction of the computing environment, preparation of development procedures, construction of database and files, generation of modules, generation of module test data, performance of integration tests, and generation of documentation.
- *System verification* – which includes the generation of system test data, performance of system tests, generation of acceptance test data, performance of acceptance tests, and obtaining approval. The use of test support tools is recommended.

Construction is completed once the acceptance criteria are satisfied.

Cutover is the controlled changeover from the existing systems and procedures to the new system. The tasks are:

- *Preparation* – which includes the preparation of the cutover schedule, training of users, and the installation of hardware.

- *Installation of new software* – which includes the conversion to the new software and execution of trial runs.
- *Final acceptance* – which includes agreement of the terms for acceptance and transferring fully to the new system.
- *Fan-out* – which means the installation at all locations.
- *System variant development* – which is to identify requirements, revise analysis and design, and perform construction and cutover where a particular location requires a variance from the norm.

Cutover is regarded as complete when the system operates for a period at defined tolerances and standards, and passes its post-implementation review.

Following cutover, production is the continued successful operation of the system over the period of its life. The tasks are to ensure that service is maintained and that changes in the business requirements are addressed:

- *Evaluate system* – which includes performance measurement, comparing benefits and costs, user acceptability, and making a comparison with the design objectives.
- *Tune* – which includes monitoring performance, tuning software, and reorganizing databases.
- *Maintenance* – which includes correcting bugs and modifying the system as required.

The levels, stages, and tasks of IE outlined above are described in a sequence that would suggest that a top-down classic waterfall model is in operation. This is not necessarily the case, and much of the development after the information strategy planning level, and particularly after the business area analysis level, can be performed in parallel. To support parallel development a co-ordinating model is constructed. This is essentially a high-level model of data and processes, which identifies and highlights dependencies and necessary interfaces between systems and subsystems. This enables complex activities to be broken down into manageable components that can be developed independently. IE is also claimed to be able to support a variety of paths through the development layers. For example, reverse engineering starts at the bottom of the framework, that is, with an existing implementation, and deduces business rules from that system. This might be useful when existing legacy systems are to be included in an IE framework. It can support re-engineering which is a combination of forward and reverse paths through the framework. Developers may reverse engineer an existing application back to an appropriate point and when the design rules have been identified, combine these with some new requirements, and then forward engineer it to implementation. As mentioned above, IE is dependent on a suitable toolset with a sophisticated repository to provide the capability to reverse and re-engineer.

21.4 Welti ERP development

In this section we consider Norbert Welti's approach to developing ERP projects, as described in Welti (1999). The description is based on his experience implementing SAP R/3 projects, which is the most common ERP solution base, but the approach can be used for other ERP projects. He has used the approach for ERP projects involving many countries as well as many sites; indeed, most ERP projects are large and complex. SAP have developed their own methodology called Accelerated SAP, and we look at this briefly at the end of this section.

The ERP system, along with the organizations in which it is implemented, is large and complex. The components of SAP R/3, for example, include separate modules for financial accounting, controlling, fixed assets management, project systems, workflow, industry solutions, human resources, plant maintenance, quality management, production planning, materials management, and sales and distribution. Of course, not all these need to be implemented in one time slot for any organization: they can be implemented in an evolutionary fashion or only few modules chosen. Even so, the ERP project is likely to be the largest and perhaps most expensive IT project for most organizations that choose this option. Welti argues, therefore, that such projects require a different approach than that for most other IT projects in organizations.

Figure 21.20 describes the project life cycle tailored for this approach. In the following, we highlight only those aspects that take on special importance in an ERP project. The four major phases are:

1. *Planning.* This involves defining the scope of the project (including locations and departments involved), allocating resources (e.g. human, both internal and consultancy, hardware and software), suggesting objectives and targets (e.g. response times, relia-

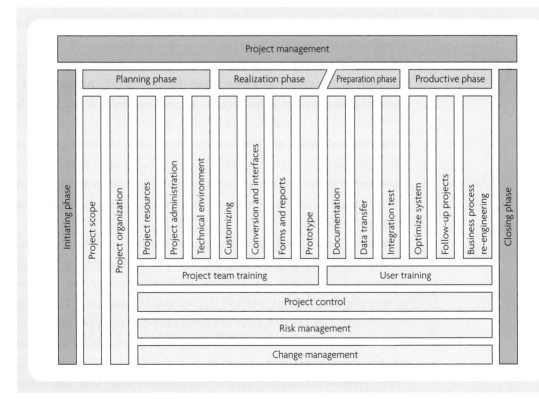

Figure 21.20: Project life cycle – ERP development (Welti, 1999)

bility targets, and other savings), planning the detailed activities, and setting up the technical environment. Obviously a major aspect of this phase relates to determining which modules of the ERP system will be adopted for which sites, whether these modules are likely to necessitate much adaptation, and whether much (expensive) ERP consultancy time might be necessary. Further, because of the size, scope, and risk of an ERP project, there needs to be decisions related to adaptation by the organization: 'The company must be made aware of the fact that adopting the philosophy of the standard software means adapting its own organization, processes and procedures to the software standard, and not vice versa' and later 'company strategy must follow the processes described in the standard software, and not the other way round' (Welti, 1999). The approach suggests that an external consultant needs to be on the project team to advise on customization scheduling and project control, and another on the steering committee to advise on project management issues in particular. Other external consultants may be hired, but this is expensive and experience suggests that little know-how is transferred from them to the company employees. Even worse, because ERP consultants are in such high demand, there is a risk of taking on poorly qualified people. They should be evaluated very carefully for a high level of experience, skills, and competence. New hardware, including networks, is likely to be required and this takes time to acquire and implement.

2. *Realization.* This involves developing a prototype, customizing the system, and creating reports and forms. The prototype needs to reflect the organization's processes and procedures in the ERP system's structures, with customization where necessary. Skilled consultants can be invaluable in this phase, as customization can be particularly difficult, although there are now tools to help in this work. Data needs to be converted to the standards of the appropriate ERP modules. Further, the ERP modules need to interface with the legacy systems in the organization, unless they will also be implemented as ERP modules. The conversion and interface aspects are likely to be particularly time consuming and critical, and often very problematical. Regarding reports and forms, the ERP system is likely to produce standard ones, which might need customization. Other issues to be decided in this stage are the authorization of access to sensitive data. The system will produce the information, and it is necessary for the organization to decide on access rights.

3. *Preparation.* There is likely to be some parallel operation of the realization and preparation stages. The preparation phase involves developing the final system, final customizing, integrating the modules, performing quality tests, documenting the system, migrating the data, and final preparations for changeover. This changeover may be immediate for all modules and/or sites or a step-by-step approach, involving one site and/or module at a time over a period of some years. On the other hand, it often involves some compromise between the two. Apart from the large and complex nature of ERP projects, the various aspects of this phase are likely to be similar to many other projects and methodologies.

4. *Productive*. This involves management acceptance of the system and fine-tuning. The latter will include optimization to ensure that all the modules are running together efficiently. It is particularly interesting that Welti emphasizes business process re-engineering at this time. It is not surprising, however, in view of the approaches insistence that the organization needs to change to fit in with the ERP system (improving and adapting its organization, processes, and procedures to the software). This is not ideal: in our view the organization might have to change, but for other reasons than fitting in with IT constraints. However, the implication is that the implementation of ERP systems is too complex for it to happen any other way. Even so, the approach strongly suggests that BPR should happen *after* the ERP system is implemented, since political and other problems of BPR would make implementation of the ERP unmanageable at the same time, and it is only after implementation that the functionality and the potential of the ERP system will be apparent.

Running throughout all four phases are the activities of training, project control, risk management and change management. These are similar to other development projects in nature, but are likely to take on greater significance due to the complexity, cost, risk, and political implications of such a large change impacting on all departments, sites, and people.

SAP itself has developed a methodology called Accelerated SAP or ASAP, which is designed for smaller and more straightforward ERP projects. The methodology has six phases:

1. *Project preparation*. This phase sets up the project team and planning for the project.
2. *Design business blueprint*. This is an outline of the expected SAP system to be implemented.
3. *Simulation*. In this phase the design and configuration of the ERP system are completed in detail and agreed.
4. *Validation*. In this phase the planned system is implemented and tested fully.
5. *Final preparation*. In this phase the interfaces between the ERP modules are written and the system becomes operational.
6. *Support*. This refers to the ongoing maintenance and upgrading, where necessary, of the ERP system.

ASAP is appropriate only for straightforward applications where the ERP modules 'fit' in well with the organization (or the organization's processes can be adapted easily to the ERP). In addition it would only be appropriate for the central ERP modules of finance, sales, distribution, and control. In these cases, however, it is expected that the ERP system can be implemented well within a one-year period.

Summary

- In this chapter we have described four methodologies which we classify as being blended, that is, being formed from parts of, or perhaps, the 'best of', other methodologies, techniques, and tools (and applications in the case of ERP).

Questions

1. In what ways are the first two methodologies (SSADM and Merise) discussed similar and different?
2. Why do you think that both these methodologies became recommended standards for use in government departments in the UK and France, respectively?
3. Do you think the first three methodologies conform fully to the method engineering theme that was discussed in Section 8.4?
4. Is the Welti ERP methodology distinct because it applies to a particular type of information system?

Further reading

Eva, M. (1994) *SSADM Version 4: A User's Guide*, second edition, McGraw-Hill, Maidenhead, UK.

Martin, J. (1989) *Information Engineering*, Prentice Hall, Englewood Cliffs, New Jersey.

Quang, P.T. and Chartier-Kastler, C. (1991) *Merise in Practice*, Macmillan, Basingstoke, UK (translated by D.E. and M.A. Avison from the French: *Merise Appliquée*, Eyrolles, Paris, 1989).

Weaver, P.L., Lambrou, N., and Walkley, N. (1998) *Practical SSADM+*, Pitman, London.

Welti, N. (1999) *Successful SAP R/3 Implementation*, Addison-Wesley, Harlow, UK.

22 Object-oriented methodologies

22.1 Object-oriented analysis (OOA)

There have been many different approaches to the analysis and design of object-oriented systems. Books have proliferated, and it appears they continue to do so. For example, Booch (1991, 1994), Coad and Yourdon (1991), Coad and Argila (1996), Jacobson et al. (1999), Kruchten (2000), Martin and Odell (1992), Mathiassen et al. (2000), and Rumbaugh et al. (1991).

Of these competing methodologies, particularly those focusing on analysis, possibly the most well known is the Coad and Yourdon Object-oriented Analysis (OOA) methodology. This approach was published before the advent of UML and was first described in 1990. It has been updated and enhanced since (e.g. Yourdon and Argila, 1996). As is sometimes the case with methodologies it is difficult to identify which is the definitive version of OOA, indeed there does not seem to be one and perhaps marketing considerations have had some influence here. For example, the version of OOA that Yourdon has written with Argila includes additional elements from other authors, such as Jacobson. It also includes some elements from the older structured analysis methods, which is interesting, for Yourdon was closely associated with structured methodologies (see Section 20.2) before deciding that object-oriented methods were the answer. So rather than describe these more hybrid versions we stick with the more pure 1991 description of Coad and Yourdon which for our purposes includes all the basics that are needed. Yourdon now uses the UML notation (Section 13.2) in his approach but this description sticks with the original notation.

OOA consists of five major activities:
- finding class-&-objects;
- identifying structures;
- identifying subjects;
- defining attributes;
- defining services.

Coad and Yourdon emphasize that these are activities that need to be performed. They should not necessarily be seen as stages or sequential steps. They point out that many analysts prefer to iterate around the various activities in a variety of sequences. Nevertheless, we shall describe them in this order which progresses from a high level to increasingly lower levels of abstraction.

1 Finding class-&-objects

This activity is about increasing the analyst's understanding of the problem domain and, as a result, identifying relevant and stable classes and objects that will form the core of the application. Coad and Yourdon describe this as the 'system's responsibilities'. The problem domain is the general area under consideration, and the system's responsibilities are an abstraction of those elements that are required for the system that is conceived. It is the system's responsibilities which are modelled. The analysis of the problem domain is not particularly original nor examined in great detail by Coad and Yourdon. The approach recommended is first-hand observation, talking (or rather listening) to 'domain experts', reading (or 'read, read, read' as they suggest), gathering experience from previous, and related, systems, and finally prototyping. Later versions of the approach from Yourdon and Argila also recommend the use of entity models, data flow diagrams, and linguistic analysis techniques to help in this activity. See Section 6.4 for an example of why finding the correct class-&-objects is so important.

The relevant classes and their associated objects are filtered out from the problem domain. The specific term class-&-objects (represented by a particular symbol as shown in Figure 22.1) includes a class (the bold inner box) and the objects in that class (the outer box). In the symbol, the class is divided into three parts. The top part is for the name of the class-&-object, the middle part for the attribute names, and the lower part for the services. An object in OOA is an abstraction from the problem domain, about which we wish to keep information (attributes of the object) and with which we can interact (the services). A class is a description of one or more objects with a common set of information and interactions.

An example in the domain of university student administration might be the classes of registration, student, course, registration-clerk, and so on. For the class registration, attributes might include date, number, and fee. Services might include create, renew, terminate, suspend, approve, and check-qualifications. An object might be an instance of the class Student, for example, the attributes and processing for student Smith. An object embodies the notion of encapsulation (see Section 6.4).

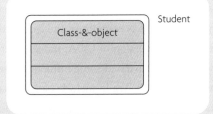

Figure 22.1: Objects in a class

Coad and Yourdon offer a set of helpful hints in order to find relevant class-&-objects:

1. Most importantly, look for structures which is the second activity of OOA as is discussed later.
2. Look at other systems with which the system under consideration interacts as a way of prompting potential class-&-objects.
3. Ask what physical devices the system interacts with. In our student administration system it is difficult to think of any example, although perhaps the photo booth might qualify. In a manufacturing system it is more obvious, for example, a weighing platform or a bottling machine. (It should be noted that these devices are not the technology with which the system might be implemented.)
4. Examine the events that must be remembered and recorded, for example, the date of

registration, then the roles that people play, for example, the owner, manager, and client.

5. Examine the physical or geographical locations of relevance and also the organizational units, for example, divisions and teams.

An examination of all these factors will help to reveal relevant class-&-objects. This is by way of a checklist, and may or may not lead to the identification of all the relevant ones, but it is argued that it is a useful starting point. However, even with object orientation, the traditional problems of systems analysis remain, including users and stakeholders not really knowing what they require. Yourdon suggests that a common problem in OOA is the identification of too many objects, and so a criterion for evaluating objects is provided. This Coad and Yourdon term 'what to consider and challenge' is somewhat similar to the criteria that are applied when building entity models:

- *Needed remembrance.* Is there anything, that is, any data, that must be kept by the system for this object? If there are no data, it probably means it is not an object.
- *Needed behaviour.* Is there any behaviour, that is, processing or functionality, that must be kept by the system for this object? If there is no behaviour, it probably means it is not an object, and it will certainly not be an object if there is no needed remembrance and no behaviour.
- *More than one attribute.* An object is likely to have more than a single attribute, and it should be reviewed if it has only one.
- *More than one object in a class.* If there is only one object in a class, then it should be seriously challenged, indeed Coad and Yourdon suggest that this is a 'suspect' object.
- *Always applicable attributes.* Are the attributes common, that is, applicable to each object in the class? If not, it is probable that the model should contain a class hierarchy. If the object is 'student' and the attribute 'employer', but this does not apply to full-time students, then it is likely that we have two subtypes of student, full-time and part-time. (This is examined in a subsequent activity.)
- *Always applicable behaviour.* Similarly, we apply the same test to the behaviour, or in Coad and Yourdon terminology, services. If certain services do not apply to all instances of the object, then we should consider subtypes or breaking down the structure.
- *Domain-based requirements.* Ensure that all the objects are derived from the domain and not from implementation considerations. For example, 'student' is clearly derived from the domain, whereas 'registration-card' or 'application-form' is about a particular design of implementation. It is recommended that the model is kept at the highest possible level of abstraction, because the concept of a registration-card may not exist in some possible implementations, that is, it is a design consideration. Application-form is a similar case. It might preclude a design which enables a direct application via the telephone. In this case, no application form is completed, and therefore the object should be less specific and focus on the logical requirement or event rather than the document or the implementation. In the examples we might prefer to consider a registration-

complete object or an application-event object rather than the registration-card or application-form, respectively.

- *Not merely derived results.* Derived results, that is, things that can be derived or calculated or implied from other attributes, should be avoided. For example, holding a student's examination grade (A, B, or C) as well as the percentage mark is not relevant, as the grade can always be derived from the mark. Such consideration avoids duplication and helps to simplify things.

The end result of the class-&-objects activity is a set of relevant classes and, for each class, the associated objects modelled using the appropriate conventions. These classes and objects should have been challenged and accepted or modified according to the guidelines outlined above. These class-&-objects will form the basic structure of the system under consideration.

2 Identifying structures

The next activity is to organize the basic **classes-&-objects** into hierarchies that will enable the benefits of inheritance to be realized (see Section 6.4). This involves the identification of those aspects or objects that are common or generalized, and separating them from those that are specific. (Yourdon points out that some analysis of structure will probably already have occurred in the class-&-objects activity.)

Coad and Yourdon use the terms 'generalization' and 'specialization' (which they shorten to 'gen-spec') for what is otherwise known as the identification of superclasses and sub-classes. First, each class-&-object is examined to identify the gen-spec structure for each class. In other words, the generalized form of the class is examined and any specific subclasses are identified. There may be many ways of breaking down the generalized elements into specific elements but what should ideally be identified are those that will lead to the greatest degree of inheritance.

(1) Gen-spec structure

The gen-spec structure is graphically modelled as in Figure 22.2 and usually reflects a hierarchy of classes. Therefore, for example, we may break our student-class into full-time-student and part-time-student, keeping as much as possible that is common to both the lower-level classes in the higher-level class. This will enable all the common aspects (data and behaviour) to be inherited from the higher level to the lower levels. The benefits of this were seen in Section 6.4. Any specific factors or requirements (known as specializations) for the lower levels can then be added to the general ones inherited from the higher level. Therefore, for example, all those aspects common to all types of student will be included in the high-level class of student, such as registration and qualifications. Those attributes and services specific to part-time-student, possibly employer details (full-time students would not have employers) and processing of student progress reports to employers, would be added at the lower level. This would mean that in implementation, the code for part-time-student could mostly be inherited from student and only a small addition need specifically be added for part-timers.

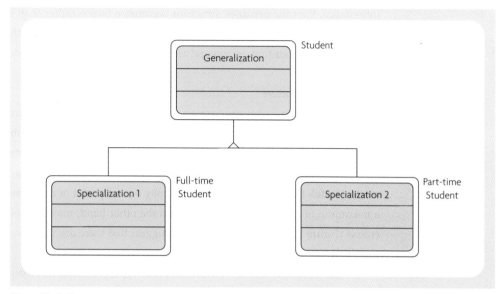

Figure 22.2: Gen-spec structure

As a way of testing the gen-spec structures, Coad and Yourdon suggest asking the following questions of each specialization, that is, the lower-level classes:

- *Is it in the problem domain?* Does it make sense in terms of the business or organization? In the example, do we really have a distinction between part-time and full-time students that is of relevance in this context? Objects should not be broken down just for the sake of it. Some students are male and some are female, but we must ask whether this is of importance and relevance in the problem domain. In most cases it probably is not and we would distinguish between male and female simply by an attribute of gender at the generalization level rather than identifying separate classes of male-student and female-student.

- *Is it within the system's responsibilities?* Again, if the system does not need to make a distinction then it should not be broken down.

- *Will there be inheritance?* Are there some attributes and/or services that are common (shared) and some that are specific? If there are not, then there is not much point in breaking it down.

- *Will the specializations meet the 'what to consider and challenge' criterion?* This will be detected from the class-&-objects activity above.

At this stage, the diagrams are drawn to indicate the class hierarchies. They are usually hierarchies, although multiple inheritance is allowed. This is where a specialization (lower level) inherits elements from more than one generalization (higher level). An example might be a specialization course-exam which inherits some aspects from the exam, such as common examination standards and procedures which apply to all courses in the university, and some aspects from the course, such as examination weightings and course-specific data and procedures, for example, that it is laboratory-based or has practical sessions. If there is multiple inheritance, then the gen-spec structure becomes a lattice rather than a hierarchy (see Figure 22.3). It should be noted that at this stage the model simply indicates that there are some attributes and behaviour that are general and some that are specific, they do not specify the detail.

(2) Whole-part structure

OOA also identifies what are termed whole-part structures. These are hierarchies of objects which indicate that one object is composed or made up of a series of subobjects. The notation is illustrated in Figure 22.4. The distinction between the gen-spec structure and the whole-part structures is indicated by the triangle. The cardinality of the relationship may also be indicated on the model, for example, that a course may be composed of a minimum of one module and a maximum of six modules. A module, on the other hand, may not necessarily be part of any course (Figure 22.4). Coad and Yourdon suggest that there are three types of whole-part structures that might be considered:

1. The 'assembly and its constituent parts'-type, for example, an organization is composed of various departments.
2. The 'container and its contents'-type, for example, a lecture hall and its seats.
3. The 'collection and its members'-type, for example, the football club and its players and helpers.

A set of criteria for considering and challenging the identification of whole-part structures are similar to those used for gen-spec structures as outlined above, with the exception of the inheritance test.

Whole-part structures often present people with difficulties, and this is usually to do with their purpose in relation to an object-oriented approach. It seems that they have been introduced to capture elements that have been found to be needed but not captured in the tra-

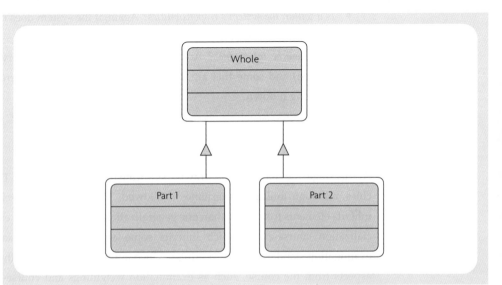

Figure 22.3: Gen-spec structure – multiple inheritance

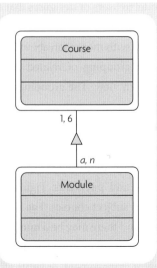

Figure 22.4: Whole-part structure

ditional object-oriented methods, for example, in gen-spec structures. Whole-part structures do not imply any notion of inheritance by the parts from the whole, they simply indicate that an object is composed of various other objects or parts. Coad and Yourdon suggest that whole-part structures are particularly useful for identifying class-&-objects at the edges of the problem domain, and these objects are dealt with by other systems.

Diagrams may be constructed that include both gen-spec structure and whole-part structure together.

3 Identifying subjects

The third activity of the OOA methodology is the identification of subjects. The purpose of this is to reduce the complexity of the model produced so far by dividing or grouping it into more manageable and understandable subject areas. This is somewhat analogous to the levelling of a DFD in other

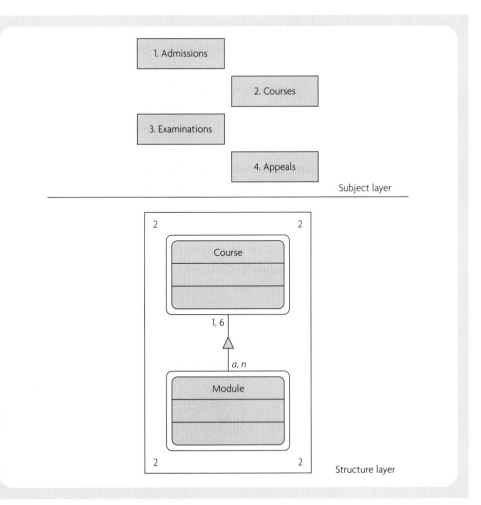

Figure 22.5: Subject and structure layers

approaches, and is about presenting relevant chunks of the model to users or designers that are understandable on their own but are also set in context as part of a larger whole. Obviously in a small system there may be no need for this, but in larger systems with more than about 20–30 classes it becomes important. Guidelines are provided for the grouping of related classes together and it is a bottom-up process which produces a top-down view. The groupings may be based on any criteria that are relevant to the area of concern, and might involve a traditional, functional decomposition but could also be based on problems or issues that emerge from the problem domain. For example, in a university problem domain, the subject layer might be admissions, courses, exams, appeals, and so on, where admissions might be composed of classes concerning applications, criteria, acceptance, references, and payments. Figure 22.5 illustrates the notation used and shows the structure layer for the subject courses. The subject identification provides a particular view or picture of the system and there may be a number of relevant, and overlapping, views. At any particular point, the most useful view is used depending on the objectives, which might be explanation to senior management, or verification by a user, or the creation of a work-package for an analyst or designer.

4 Defining attributes

In this activity, the attributes of the class-&-objects are defined. This is very similar to the identification of attributes for entity models (Section 6.3). It is the data elements of the object that are defined. The only difference is that attributes that define the state of the object are perhaps given more prominence; for example, things that might be defined using an entity life history diagram are emphasized (Section 12.9). Examples of attributes for an object student might be student-number, name, address, date-of-birth, suspended, or current. Attributes are normally listed by name in the middle part of the class-&-objects box (Figure 22.6). Attributes that 'point' to other objects are included. In relational database terms this means that foreign keys are included. For example, the attribute course above is in effect a pointer from the student object to another object called 'course'. In the model these objects would be connected with a line to indicate a relationship, and the degree and cardinality of the relationship expressed. This part of the identification of attributes is termed 'instance connection' by Coad and Yourdon and indicates that the connection is between instances

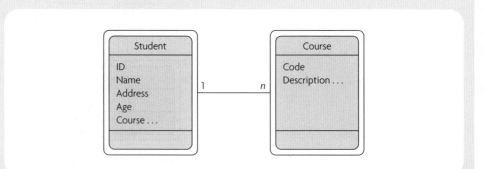

Figure 22.6: Defining attributes

of the individual objects rather than between classes. This is the same for entity models, where the relationships connect occurrences of the entity type rather than the entities themselves.

For each named attribute that has been identified, a short description is specified along with various constraints that apply to the attribute. These may include domain information, the allowable range, any default value, the various states that may apply, and any constraints implied by other attributes. An example would be: 'if in object student the attribute registration-suspended is set, then attribute fee must be zero'.

All implementation considerations associated with attributes are deferred to design. Therefore, the attributes should not be normalized nor performance considered, and specific identification keys are not defined.

5 Defining services

In this, the final activity of the OOA methodology, the services of the class-&-objects are defined. In the object-oriented approach, as we have seen, the objects are composed of data and processing. The previous activity defined the data, and this activity defines the processing, or in Coad and Yourdon's terminology, the services. The terms 'method' and 'behaviour' are also used to mean the same thing. A service is the operation or processes performed by the object in response to the receipt of a message.

In the previous activity, attributes that defined the state of an object were identified. In this activity, the services required to change or modify states are identified. In the student object, we might have identified a state of suspended-registration. This would imply that we need to define the services to suspend a student's registration and perhaps to unsuspend, or reinstate, the registration of that student. All the services needed to achieve the changes of state identified in the entity life history diagrams should be fully defined.

Next, what are termed the 'algorithmically simple' services are defined. Yourdon later changes the terminology and calls these 'implicit' services. These are the ones that are likely to appear in some form for each class-&-object in the model; they are create, release (i.e. delete), connect, and access. The services required for create, delete, and access are fairly obvious. For example, what is required to create an object student might be a check for a valid registration form, followed by the allocation of the next available registration number, the creation of a new object, and then the return of a 'successful creation' message. The connect is perhaps less obvious, but means a service that creates or terminates a mapping between objects, that is, the establishment of a relationship between objects. An example might be the allocation of a new student to a personal tutor as this would require the creation of a mapping from object student to object tutor. These algorithmically simple services are usually not included on the OOA diagrams.

As can be imagined, if the algorithmically simple services are identified, this will be followed by the identification of any algorithmically complex services. These are classified into two types:

1. Services concerned with calculations.
2. Services concerned with monitoring the external environment; that is, the services required to detect and respond to events.

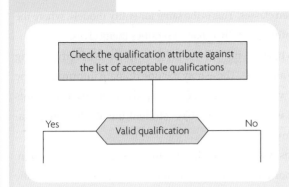

Figure 22.7: Service chart

Finally, the services required for processing a message received from another object, and any processing triggered by the message, are defined.

Once the required services have been identified they are specified in detail using either a form of structured English notation (Section 12.4) or via a service chart, which is a kind of program flowchart. Figure 22.7 provides an example of a service chart.

It should be emphasized again that the description of the methodology as a linear series of activities is not necessarily how it would actually be approached. Some analysts might identify a few key class-&-objects, then drive down through the activities, and then iterate the process with other objects. It does not really matter as long as the outcome is a complete set of OOA models and diagrams. A further aspect that needs highlighting is the importance of the activities of identifying reusable objects, classes, and services, and therefore, ultimately, reusable code.

The methodology of OOA, as its name implies, does not include design and implementation phases, although the authors address design in some detail in other sources. In this book we will not extend the description into detailed design because that is not our focus. However, an important aspect of the transition from analysis to design in object-oriented methods is that it is not a question of changing, or introducing, new concepts. The transition is simply a matter of extending the detail of the object-oriented models and specifications, and adding components concerning human interactions (such as dialogue design), task management (such as real-time tasks, communication, and hardware considerations) and data management (e.g. designing the database). The detailed design stage slowly becomes program language-dependent, that is, we need to know what the target program language is, and the actual implementation will normally take place in an object-oriented programming language in order to utilize most easily the object-oriented concepts. Coad and Yourdon point out that the results of the OOA can be implemented in a non-object-oriented language, although it would be much more difficult.

22.2 Rational Unified Process (RUP)

In 1998 Jacobson stated that the methods war was over (Jacobson, 2000). He believed that a standard had been achieved, and that standard was the Unified Modelling Language (UML) (Section 13.2). He states, 'that all the different methods found in the software industry are now moving to one modelling language: UML', and that this new standard is far better described and more complete than any previous modelling language. This may or may not be true, and, of course, Jacobson was talking from quite a narrow perspective concerning modelling methods; nevertheless, for object-oriented approaches the standardization of UML was quite an advance.

However, Jacobson recognized that having a standardized modelling language, such as UML, was not all that was needed. As he says, 'You also need to know how to use it', and this has led to the development of the associated software development process or, in our terms, a methodology. The process by which a system is developed, for example, the type of things that have to be done, how the requirements are discovered, the stages and tasks, etc. are not part of UML. UML is only a modelling language, discovering what it is that needs to be modelled is quite different. For this Jacobson and others (Rumbaugh, Booch, Kruchten, and Royce) developed and evolved a process that has become known as the Unified Process and which utilizes UML for modelling.

Jacobson's early work was in the telecommunications industry with Ericsson where some of the ideas behind the Unified Process were developed, but he left Ericsson in 1987 to establish a company called Objectory, based in Stockholm. Objectory is a concatenation and abbreviation of the term 'Object factory', and the product of the company was a process for developing systems. In 1995 the Rational Software Corporation acquired Objectory. Rational had also been developing a number of software development practices, particularly in relation to software architecture and iterative development, and these concepts were combined with those of Objectory to form what was first called the Rational Objectory Process (first released in 1997) but became the Rational Unified Process (RUP) in 1998. Jacobson et al. (1999) describe RUP as a 'full-fledged process able to support the entire software development life-cycle'.

Jacobson does not like the term 'method' (or methodology) used to describe the Unified Process. He says, 'a method is usually a set of interesting ideas and general step by step descriptions. However, it typically does not guide developers in how to use it in commercial product development.' For Jacobson methods are embodied in textbooks and are primarily used for educational purposes, but they do not guide developers as to how to use them to develop commercial software. To achieve this the method has to be 'processified' into a real software engineering process. Therefore, the Unified Process as described in textbooks is essentially the fundamental and conceptual ideas of its methodology whereas the Process (e.g. RUP) is the processified software engineering process, typically converted and delivered as a product. Jacobson says that he would never recommend his book (Jacobson et al., 1992) to be used for commercial development, but it would be useful for acquiring and learning the basic ideas before developing a process on the basis of his method. Jacobson (2000) goes on to identify a number of shortcomings of methods:

- they are paper products, typically frozen in a book;
- they are rarely tried or tested in real projects (before publication);
- they are simple introductions (they are insufficient for use in real, commercial development projects);
- they focus on developing new systems and have little to say concerning evolving development or maintenance;
- they are rich in notations but lack semantics.

In 2002 Rational Software was acquired by IBM, including RUP. According to IBM (Mills, 2002) they purchased Rational because they liked the open industry standards that it represented and because the tools enable integration of business applications across companies and their value chains. These aspects complement and reflect IBM's strategy. IBM have pledged to continue developing the Rational products, including RUP, and to integrate them more tightly with IBM's own products, and RUP has now become the IBM Rational Unified Process. In this book we continue to refer to the methodology simply as RUP. There are a number of variants of RUP including one specifically for e-business, and indeed customers can tailor variants for their specific needs using RUP Builder and RUP Plug-ins. A trial version of RUP is available for download from the IBM website. The description provided here is the standard RUP.

The Unified Process is described as 'use-case driven, architecture centric, iterative and incremental' (Jacobson et al., 1999), and this perspective is argued to make the process unique. Use-cases have been described in Section 13.2 as part of UML. In the Unified Process they are used to capture the user requirements. A use case describes an element of the functionality of a system that gives a user 'a result of value'. They thus focus on things of specific value to a particular user (or group of users) and in this way overcome the growth of 'wish lists' or vague general functionality that might be nice to have but is not essential to the system. The sum of these use-cases (called the use-case model) depicts the total functionality of the system. But they are argued to be more than this because they are used to drive the development process, through design, implementation, and testing. Therefore, the system is designed based on the use-cases, it is implemented to support the use-cases, and is tested based upon the content of the use-cases.

RUP is an 'architecture-centric' process. Software architecture is compared to the architecture of a building, in that it is a blueprint of the building design that allows people to 'see' the building before it is actually built. The architecture of a system is similar but provides different views of the system. It contains details of the hardware, the operating system, the database, the network, etc. (i.e. the platform for the system), plus the non-functional requirements (non-functional requirements might concern reliability, performance, conformance, user interface standards, etc.). The architecture is defined in outline at the beginning but evolves and develops in tandem with the needs of the software system as it develops. Many other methodologies ignore the development of the architecture and just concentrate on the development of the software part of the project.

RUP is also described as iterative and incremental because its authors believe that user requirements cannot be fully and accurately defined initially at one go. Requirements evolve with improved understanding and change over time. Therefore, the project is not seen as one large activity with a 'big bang' outcome at the end but as a series of controlled incremental iterations which helps to minimize risk and reduce the chance of the system not meeting its goals at the conclusion. It is suggested that this means it is not a waterfall or life cycle approach (although as seen in Chapter 3 the life cycle is also supposed to be iterative). It is in fact much more like the spiral model (see Section 7.1).

These three key concepts of RUP (use cases, architecture, and iteration) are described as being like a three-legged stool, without one of the legs the stool falls over (Jacobson et al., 1999). Other authors identify six (or more) core elements of RUP. For example, Kruchten (2000) also identifies tool support and use of components as critical to RUP.

The RUP has a number of 'cycles' which together make up the development of a project and run throughout its life. Each cycle consists of four phases: Inception, Elaboration, Construction, and Transition. Figure 22.8 indicates the phases and the workflows of RUP. The shapes in the figure indicate the relative emphasis of each phase in that particular workflow. Therefore, the major emphasis in construction is on implementation, as would be imagined but there are elements of other workflows as well. Each phase is composed of a number of iterations involving all the core workflows (see the blue box which depicts the second iteration of the elaboration phase). The number will be determined by the circumstances. So a development is not one pass through the workflows but a series of iterations round them for each of the four phases.

A workflow is a sequence of activities that 'produces a result of observable value' (Kruchten, 2000), and there are nine core process workflows in RUP. The workflows are shown in Figure 22.8, with the first six being termed engineering workflows and the bottom three support workflows (i.e. Configuration and change management, Project management, and Environment).

An interesting aspect of RUP is the concept of a worker (i.e. someone who performs a role in the development process). It is not actually an individual but someone performing that role, or as Kruchten (2000) looks at it, it is someone wearing a particular 'hat' at a particular time. RUP actually defines a list of all workers (of which there are around 30) potentially involved in the process, ranging from architect, systems analyst, and designer, through to stakeholder, project manager, and change control manager. A worker is associated with a set of 'cohesive' activities, meaning activities best performed by one person, in relation to the

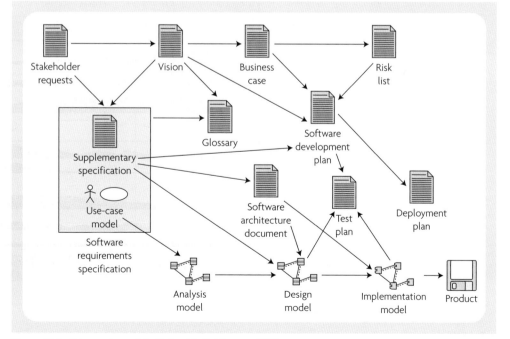

Figure 22.8: Major artefacts of the Rational Unified Process (RUP)

manipulation (e.g. creation, update) of an artefact. An artefact might be a use case, a model, a piece of code, etc. It is part of the philosophy that the artefacts are created and maintained in a CASE toolset. In RUP design artefacts might be stored in Rational Rose or Select (see Section 19.4) while the project plan might be in Project 2000 (see Section 18.4). The major artefacts of RUP are specified in Figure 22.9. RUP also has a set of guidelines which relate to artefacts, activities, or steps. They might specify, for example, what is good practice, what makes for a quality artefact, or a heuristic. So they might give advice on how to conduct a review, how to model a use-case, how to check something, good user interface design, etc.

The workflows themselves will now be examined, starting with the engineering ones followed by the support workflows. Different descriptions of RUP vary in the number of workflows that are identified. We look at the nine workflows described by Kruchten (2000), but in some descriptions the first and last of the engineering workflows are not included (i.e. business modelling and deployment).

1 The business modelling workflow

This starts with the development of the Business Model. This is essentially establishing the context for the system being developed and the shape of the organization in which the system is to be deployed. It might include the identification of current problems and areas for redesign or re-engineering, the identification of business rules, etc., depending on the nature of the proposed development. This workflow is not always necessary, for example, if the development

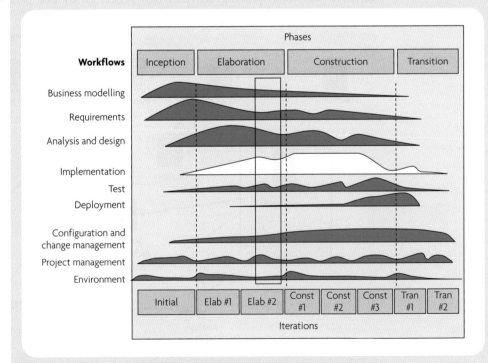

Figure 22.9: RUP process structure (reproduced with permission)

concerns simply adding a new feature to an existing system, but it would be essential for any major new business-oriented development. The objectives of this workflow are quite similar to that of other methodologies, such as the Business Planning phase of IE (Section 21.3), for example, but in RUP the same techniques are used as in the later stages of development (i.e. workers, artefacts, activities, and workflows), thus ensuring an end-to-end process and that everyone is talking the same language. The key artefacts of business modelling are:

- a business vision document, specifying the objectives of the development;
- a business use-case model, depicting the functions of the proposed development;
- a business object model, describing the realization of the business use-cases.

2 The requirements workflow

The objectives of this workflow are to establish with the stakeholders what the system should do and why, to define the boundaries of the system, and estimate the costs and timescales involved. A 'vision' of the system is developed which is then translated into a use-case model with some supplementary requirements specifications. Both functional and non-functional (e.g. minimum response times) requirements are collected and analysed. The key stakeholder and user needs and the high-level features are defined, and then these have to be converted into specific software requirements. Again the use-case model is used to express this.

3 The analysis and design workflow

The objective of this workflow is to convert the requirements from (2) above into an implementation specification. Analysis ensures that the functional requirements are met, typically ignoring the non-functional requirements and the run-time environment. So, analysis is a logical view of the system. Design takes the output of the analysis and adapts it to the constraints of the architecture and the non-functional requirements. This involves the activities of Defining and Refining the Architecture, Analysing the Behaviour (the functions), the Design of Components and the Design of the Database.

4 The implementation workflow

This workflow is to convert the designs into an implementation. This involves planning the process, converting the classes and objects from (3) above into components, testing the individual components, and then building an operational version of the system in parts, known as 'the builds'. The separate software components are then incrementally integrated into a complete system, usually over a period of time.

5 The test workflow

This workflow tests and verifies the interaction of components, that all requirements have been implemented, and that a quality product has been developed in terms of the absence of defects and fitness for purpose. The system is tested for reliability, functionality, and performance. Testing is not a single phase in the project, it occurs throughout the life cycle and at all stages. Clearly it differs in the early stages to the later stages. The type of testing may include

Benchmark tests, Configuration tests, Functionality, Installation, Integrity (resistance to failure), Load tests, Performance, Stress tests, etc. These might involve the creation of test cases, procedures, scripts, etc. to help to test comprehensively.

6 The deployment workflow

This workflow deploys the finished software to the users and involves:

- testing the software in its operational setting (beta-testing);
- training the end-users;
- migrating from existing software (including database conversion);
- packaging the software (if it is a shrink-wrap software package);
- installing the software.

The exact nature of the deployment workflow will depend on whether the system is a traditional custom-made, in-house developed piece of software or a package, and whether, for example, it is to be delivered via an Internet download. The deployment workflow is important in RUP, and it is suggested that deployment activities are often ignored by other methodologies and approaches.

7 The configuration and change management workflow

This workflow tracks and maintains the integrity of the project. The artefacts developed in the project represent a significant investment and their use should be maximized. Artefacts must be identified and stored, and the various history and versions controlled. The workflow involves monitoring and managing change requests, change costs, and keeping control of the various versions of products and artefacts. It also involves the management of the configuration of hardware and software. Tool support is advised for what is described as the 'tedious aspects' of the workflow.

8 The project management workflow

This workflow provides a framework for managing software projects and managing risk. It also provides guidelines for planning, staffing, monitoring, and generally performing project management. Again the use of tools such as Project 2000 and techniques such as Gantt charts and PERT are emphasized (see Sections 14.2 and 14.3). RUP recognizes that project management has been a particularly difficult aspect of software development and so devotes a workflow to helping ensure that a project is successful in this respect and in particular addresses risks.

In planning there are two levels of plans, a 'coarse-grained' plan (the phase plan) and a set of 'fine-grained' plans (the iteration plans). The phase plan relates to the major milestones of the project, from project approval through to product release, while the iteration plans relate to the detail of the current iteration and possibly the next iteration. Part of this workflow examines potential risks to the project, for example, requirement creep or difficulties in recruiting key specialisms. These risks are assessed as to their probability and their potential impact on the project. If a risk is assessed as serious then attempts are made actively to mitigate the risk together with the development of contingency plans to be enacted if the risk actually occurs. As part of project management RUP recommends that detailed project metrics (quantitative measures) be established and kept. These might relate to progress, productivity, levels of reuse achieved, customer satisfaction, etc. and are important not only for effectively managing the current project but for learning and improving the management of future projects.

9 The environment workflow

This workflow is about supporting the project with relevant processes, methods, and tools in an organization. As we have seen, the use of tools is a key element of RUP so the activities concern the selection, procurement, implementation, and management of appropriate tools and support processes.

RUP, according to IBM Rational, is based on experience and best practices and is suitable for a wide range of projects and organizations. It is use-case-driven and focuses on developing software iteratively, and provides phases, workflows, guidelines, and frameworks for software development. It is designed for and utilizes UML for its modelling elements and is based upon the integral use of tools to support its processes.

Summary

- In this chapter we have described two methodologies that are object-oriented: object-oriented analysis and the rational unified process.

Questions

1. What do you see as the common features and differences between object-oriented analysis (OOA) and rational unified process (RUP)?
2. In what ways is RUP different to other, non-object-oriented methodologies, for example, Information Engineering (IE), discussed in Chapter 21?
3. Why do the RUP authors not like to call it a methodology? Do you agree with them and why?
4. Do both these approaches (OOA and RUP) conform fully to the object modelling theme in Section 6.4, and do they use all the object techniques discussed in Chapter 13?

Further reading

Coad, P. and Yourdon, E. (1991) *Object Oriented Analysis*, 2nd edn, Prentice Hall, Englewood Cliffs, New Jersey.

Jacobson, I., Booch, G., and Rumbaugh, J. (1999) *The Unified Software Development Process*, Addison-Wesley, Boston.

23 Rapid development methodologies

23.1 James Martin's RAD

The goal of rapid development of applications has been around for some time and with good reason, as the objective of speeding up the development process is something that has been on the agenda of both general management and information systems management for a long time. The need to develop information systems more quickly has been driven by rapidly changing business needs. The general environment of business is seen as increasingly competitive, more customer-focused, and operating in a more international context. Such a business environment is characterized by continuous change, and the information systems in an organization need to be created and amended speedily to support this change. Unfortunately, information systems development in most organizations is unable to react quickly enough, and the business and systems development cycles are substantially out of step. In such a situation, the notion of rapid application development (RAD) is obviously attractive.

RAD (Rapid Application Development) has been discussed in general terms in Section 7.3, and this should be reviewed along with this description of James Martin's RAD methodology, which we term JMRAD in order to distinguish it from the general form of RAD and other specific RAD methodologies, for example, Dynamic Systems Development Method (DSDM), described in Section 23.2.

Martin is, of course, also known for the development of the Information Engineering (IE) methodology (Section 21.3), and it comes as no surprise to find that his version of RAD is set firmly in the context of IE, as illustrated in Figure 23.1.

JMRAD is actually a combination of techniques and tools that are, for the most part, already well known and dealt with elsewhere in this book. We identify the following as the most important JMRAD characteristics:

- it is not based upon the traditional life cycle (Chapter 3), but adopts an evolutionary/prototyping approach (Sections 7.1 and 7.2);
- it focuses upon identifying the important users and involving them via workshops at early stages of development;
- it focuses on obtaining commitment from the business users;
- it requires a toolset with a sophisticated repository (Chapter 19).

JMRAD has four phases which we shall describe in turn (Figure 23.2):

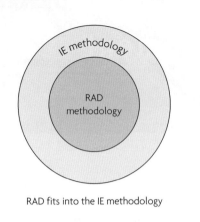

RAD fits into the IE methodology

Figure 23.1: Rapid application development and Information Engineering

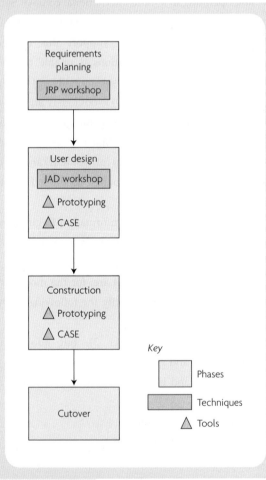

Figure 23.2: The phases of the RAD approach

1. Requirements planning.
2. User design.
3. Construction.
4. Cutover.

1 Requirements planning

JMRAD devotes a lot of effort to the early stages of systems development. This concerns the definition of requirements. There are two techniques used in this phase, both of which are really workshops or structured meetings. The first is joint requirements planning (JRP), and the second is joint application design (JAD). In some other RAD methods, these are not separated; indeed, Martin points out that they may in practice be combined with some of the functions of JRP being subsumed into JAD.

The role of JRP workshop is to identify the high-level management requirements of the system at a strategic level. The participants in JRP are senior managers who have a vision and understanding of the overall objectives of the system and how it can contribute to the goals and strategy of the organization. If this understanding does not already exist, the workshop may be used to help determine such an understanding or vision. The JRP is a creative workshop that helps to identify and create commitment to the goals of the system, to identify priorities, to eliminate unnecessary functions, and so on. Martin separates JRP from JAD because there are different people involved. In JRP, the participants need to have a combination of overall business knowledge and specific knowledge about the area that the proposed system is addressing along with its requirements. They also need to have the necessary authority and seniority to be able to make decisions and commitments. Applications often cross traditional functional boundaries, and ensuring the right people are involved is difficult but absolutely critical. Martin suggests that if the right people are not available the workshop should not take place. The participation of substitute personnel, without the authority to make decisions, who have frequently to refer back to their superiors, could negate one of the main workshop objectives which is to get the requirements identified and agreed in the shortest possible time.

The detail of the workshops will be discussed in the next section as they are the same as JAD workshops.

2 User design

JAD (Section 16.2) is the main technique of the user design phase; indeed, it contains little else. In fact, user design is in reality both analysis and design. As mentioned above, the JRP workshop may be combined with JAD in situations where the overall requirements are already well established. Normally, however, JAD would follow on from JRP. Prototyping (Section 7.2) is advocated to enable the quick exploration of processes, interfaces, reports, screens, dialogues, and so on. Prototyping may be used for the overall system or be used to explore particular parts of the system that are contentious or present particular problems. The user design is developed and expressed using the four diagramming techniques of entity modelling (Section 11.1), functional decomposition (Section 12.5), data flow diagramming (Section 12.1), and action diagrams (Section 12.8). The participants in the JAD workshop need to be familiar with these techniques, but the emphasis is on getting the requirements as correct as possible and to reflect the business needs. Therefore, the language used in the workshop and expressed in the diagrams is that of the business and the users, rather than the more technical language of information systems. The results of the user design are captured in a toolset (Chapter 19) which checks both internal consistency and that with other applications and corporate models. Where necessary, the terms used should be discussed, defined, and entered into the repository of the toolset. The use of a toolset enables the speedy, accurate, and effective transfer of the results into the next phase, the construction phase.

3 Construction phase

The construction phase in RAD consists of taking the user designs through to detailed design and code generation. This phase is undertaken by information systems professionals using a toolset, for example, IEF (Section 19.2). Construction in Martin's RAD methodology is highly dependent on the presence of an IE-based toolset and is performed by creating a series of prototypes which are then reviewed by the key users. In this way the screens and designs of each transaction are prototyped, and the users then approve them. If they do not approve them, they will request changes, and the process goes on through a series of iterations. By prototyping and the use of the toolset, these iterations are achieved quickly, and testing is enabled. Some of these key users will already have been involved in the earlier phase of user design. Construction is performed by small teams of three or four experts in the use of the toolset. These experts are known as **SWAT teams**. SWAT stands for 'skilled with advanced tools', and the approach requires them to work quickly, making maximum possible use of reusable designs and code that already exist. Teams are kept small so as to reduce the number of interfaces and interactions between people in the teams. One of the problems of traditional development is low productivity which, it is argued, results from the large teams of developers involved, the consequent large communications network, and the number of communications. Normally there is a SWAT team member allocated to developing each transaction in a system. In practice, there is often only one developer for a particular part of the system, and this reduces the number of potential interactions with other developers for the area to zero. Using this approach, it is argued that the core of a system can be built relatively quickly, typically in four to six weeks, and then it is progressively refined and integrated with other aspects developed by other team members.

Once the detailed designs have been agreed, the code can be generated using the toolset and the system tested and approved. Because of the way that the construction has occurred, there should not be any surprises to the users when they see the finished version. All associated documentation is then produced and database optimization is performed.

4 Cutover

The final phase is cutover, and this involves further comprehensive testing using realistic data in operational situations. The users are trained on the system, organizational changes (implied by the system) are implemented, and finally the cutover is effected by running the old and the new systems in parallel, until the new system has proved itself and the old system is phased out.

JMRAD adopts an evolutionary or timebox approach to development and implementation (see Section 7.3). Typically, it recommends implementation of systems in a 90-day life cycle. The objective is to have the easiest and most important 80 per cent of system functionality produced in the first 90-day timebox and the rest in subsequent timeboxes. This forces users and developers to focus on only those aspects of the system that are necessary and probably most well defined for development in the first timebox. Everything else is left until later. The knock-on benefit of this is that with experience and use of the basic system, developed in the first timebox, users often find that their requirements evolve in different directions from those originally envisaged. In other words, the benefits of an evolutionary approach accrue. The other advantage of the timebox is that it creates a focus on achieving an implementation in the specified period. In order to achieve this, the functionality must be trimmed accordingly. The timebox approach contrasts with the traditional approach where every conceivable requirement is implemented together, and the resulting complexity often causes long delays in implementation.

23.2 Dynamic Systems Development Method (DSDM)
DSDM Consortium

In 1994 a group of systems developers from companies interested in rapid development came together to form an independent and 'not for profit' Consortium to discuss and attempt to define a standard RAD (Rapid Application Development) method. There was some concern that RAD was becoming associated with a 'quick but dirty' image whereas the Consortium believed that RAD should not only be rapid but also disciplined and of high quality. The method they defined became known as DSDM (Dynamic Systems Development Method). Although DSDM acknowledges a debt to James Martin's RAD (Stapleton, 1997), the latter was not thought to be comprehensive nor was it felt to be independent enough of proprietary processes and tools. The Consortium aimed to define a method that was in the public domain, developed from practical experiences, and tool independent.

The original Consortium contained 17 members representing a mix of organizations, in terms of size, vendors and user organizations. The Consortium rapidly grew and now has 'hundreds of members' (Stapleton, 2002), including ICL, IBM, the British Ministry of Defence, British Telecom and British Airways. The Consortium does have various product and service vendors amongst its membership but has striven not to let particular proprietary products influence or define the method and they describe the method as 'vendor-independent'. The Consortium operates with a Management Committee, a Directorate and has various Technical Committees and Working Groups, one of which is a Training and Accreditation Group to help disseminate the approach and accredit practitioners in the use of the method. Certification is undertaken in association with the BCS (British Computer Society).

Rapid Application Development (RAD) is defined by the Consortium as 'a project delivery framework which actually works. It aids the development and delivery of business solutions to tight timescales and fixed budgets' (DSDM, 2005) and the first detailed definition of DSDM was published in 1995. After a period of monitoring and reviewing experience in practice, a second version was published later the same year and Version 3 in 1997. DSDM has become well known and influential, especially in the UK but also increasingly in Europe as a whole and to a lesser extent in the rest of the world. In 2000, eDSDM was produced which is a version of DSDM tailored to organizations and projects undertaking e(lectronic)-business initiatives, including organizational and architectural change. The framework continues to evolve and Version 4 was produced in 2001 which is described as a Framework for Business Centred Development. The most recent version is 4.2, which contains guidance for those wishing to use eXtreme Programming (XP) (Section 23.3) in conjunction with DSDM. DSDM now aligns itself more closely to the agile movement (Section 7.4) rather than to RAD.

DSDM, despite its name, is more of a framework than a methodology, which sometimes disappoints people looking for a tightly defined methodology. Much of the detail on how things should actually be done and what the various products will contain is left to the organization or individual to decide (Stapleton, 2002). To try and define every detail and possibility would be contrary to the RAD ideals. As Stapleton (1997, p.4) states '… it would be disastrous for a RAD method to become known as bureaucratic or over-complex'. Further, DSDM is defined at a higher level than just a set of techniques and tools because, it is argued, an organization needing a RAD approach really requires a fundamental change to their development process. Thus DSDM provides 'a framework of controls for building and maintaining systems which meet tight time constraints and provide a recipe for repeatable RAD success. The method not only addresses the developer's view of RAD but also that of all the other parties who are interested in effective system development, including the users, project managers and quality assurance personnel' (DSDM, 2005).

Whilst some of the detail may be missing, DSDM clearly identifies its underlying principles. These principles are critical to DSDM project success and if even one is ignored or not applied then the whole project is endangered. The general advice is to reconsider the use of DSDM if they cannot all be guaranteed. There are nine principles, as follows:

1. Active user involvement is imperative.
2. Teams must be empowered to make decisions. The four key variables of empowerment are: authority, resources, information and accountability.
3. Frequent delivery of products is essential.
4. Fitness for business purpose is the essential criterion for acceptance of deliverables.
5. Iterative and incremental development is necessary to converge on an accurate business solution.
6. All changes during development are reversible, i.e. you do not proceed further down a particular path if problems are encountered, you backtrack to the last safe or agreed point, and then start down a new path.
7. Requirements are baselined at a high level, i.e. the high-level business requirements, once agreed, are frozen. This is essentially the scope of the project.
8. Testing is integrated throughout the life cycle, i.e. 'test as you go' rather than testing just at the end where it frequently gets squeezed.
9. A collaborative and co-operative approach between all stakeholders is essential.

Although speed of delivery is not specifically mentioned in these principles, only alluded to, it is clearly a key principle of DSDM. Historically of course there have been problems in this area and DSDM recognizes that the business often needs solutions faster than they can be delivered. It is recognized that deadlines are frequently set with no reference to the work involved, i.e. the deadline is outside of the control of those tasked with the delivery of the project. In situations of tight deadlines it is tempting to introduce extra resources and people to a project. However, as Brooks (1995) has observed, this frequently makes things worse as there is a considerable learning curve for new people joining a project and existing people are diverted to help bring the new people up to speed. Thus if the deadline of a late-running project cannot be altered, the only thing left is to reduce functionality. This is the RAD solution and the one that DSDM adopts.

DSDM Process

This section should be read in conjunction with the more general description of RAD in Section 7.3, for DSDM embraces the general concepts of timeboxing, Moscow rules, JAD Workshops, and prototyping (Section 7.2). The DSDM framework defines a set of phases that any new or enhancement development project should undertake. This includes the initial identification of a problem or opportunity to be addressed through the development of the system to keeping the system operating successfully.

Figure 23.3, known in DSDM as the 'three pizzas and a cheese diagram' depicts the phases and the main products that need to be produced in each phase together with the various pathways through the process. As can be seen, the feasibility and business studies are performed sequentially and before the rest of the phases because they define the scope and

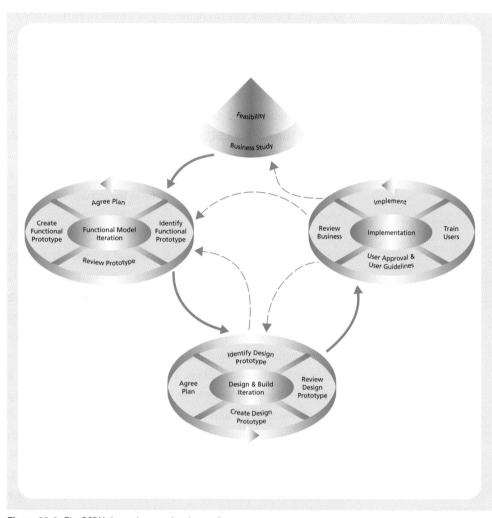

Figure 23.3: The DSDM three pizzas and a cheese diagram

justification for the subsequent development activities. The arrows indicate the normal forward path through the phases, including iteration within each phase, whilst the dashed arrows indicate the possible routes back for evolving and iterating the phases. In fact, the sequence that the last three phases are undertaken in or how they are overlapped is not defined but left to the needs of the project and the developers.

As can be seen, there are five main phases in the DSDM development lifecycle:

1. *Feasibility study*

   This includes the usual feasibility elements, for example, the cost/benefit case for undertaking a particular project, but also, and particularly important, it is concerned with determining whether DSDM is the correct approach for this project. DSDM recognizes that not all projects are suitable for RAD and DSDM. Some guidelines are provided but it is not simply that projects of type X are suitable and projects of type Y are unsuitable. It is more concerned with the maturity and experience of the organization with DSDM concepts. Further, where engineering, scientific or particularly computationally complex applications apply, DSDM is not usually advised. Howard (1997) says that DSDM is not recommended for real-time applications. Projects where all the requirements must be delivered at once may also not be suitable for DSDM. This occurs where there are no 'should haves or could haves' in Moscow Rules terms. General business applications, especially where the details of the requirements are not clear but time is critical, are particularly suitable for DSDM. The feasibility study is 'a short, sharp exercise' taking no more than a few weeks, in dramatic contrast to some traditional feasibility studies that take much longer. Thus it is not particularly detailed but highly focused on the risks and how to manage them. A key outcome is the agreement that the project is suitable and should proceed.

2. *Business study*

   This is also supposed to be quick and is at a relatively high level. It is about gaining understanding of the business processes involved, their rationales and their information needs. It also identifies the stakeholders and those that need to be involved. It is argued that traditional requirements-gathering techniques, such as interviewing, take too long and facilitated JAD-type workshops (see Section 16.2) are recommended. The high-level major functions are identified and prioritized as is the overall systems architecture definition and outline work plans. These plans include the Outline Prototyping Plan which defines all the prototypes to be included in the subsequent phases (encompassing the prototyping strategy and the configuration management plan). These plans get refined at each phase as more information becomes available. The other major output is the Business Area Definition, usually containing an overview DFD and Entity Model or business Object Model, if the environment is object oriented. DSDM is applicable to both structured and object-oriented development (and indeed to any other approach). DSDM advocates using 'what you know' and is not prescriptive concerning analysis and design techniques.

3. *Functional model iteration*

   Here the high-level functions and information requirements from the Business Study are refined. Standard analysis models are produced followed by the development of prototypes and then the software. This is described as a symbiotic process with feedback from prototypes serving to refine the models and then the prototypes moving towards first-cut software, which is then tested as much as is possible given its evolving nature.

4. *System design and build iteration*

   This is where the system is built ready for delivery to the users. It should include at least the 'minimum usable subset' of requirements. Thus the 'must haves' and some of the 'should haves' will be delivered, but this depends on how the project has evolved during its development. As indicated above, testing is not a major activity at this stage because of the ongoing testing principle. However, some degree of testing will probably be needed as in some cases this will be the first time the whole system has been available together.

5. *Implementation*

This is the cutover from the existing system or environment to the new. It includes training, development and completion of the user manuals and documentation. The term completion is used because, like testing, these should have been ongoing activities throughout the process. Ideally, user documentation is produced by the users rather than the specialist developers. Finally, a Project Review Document is produced which assesses whether all the requirements have been met or whether further iterations or timeboxes are required.

As mentioned above, DSDM adopts an incremental approach and uses the RAD concept of time-boxing. The normal concept of a timebox is the overall period from the start of the project to the scheduled end, when something tangible and usable is delivered to the users. This may be fol-lowed by a second timebox when something additional is delivered. However, rather confusingly, although using the term in this way DSDM also applies it to describe a sub-phase within a project. So in DSDM there are investigation timeboxes, analysis timeboxes, prototype timeboxes, development timeboxes, etc. Thus they apply timebox principles to any sub-phase or stage of the project, rather than just to the overall project. The principles relate to the fact that these are not activity based but a series of fixed deadline timeboxes in which as much is under-taken as is needed to make something or produce a particular output in the shortest possible time.

Figure 23.4 illustrates an example DSDM project broken down into timeboxes. Each timebox is recommended to be between two and six weeks long although, as is the case with many of the recommendations in DSDM, it depends on the nature of the particular situation. Every effort is made to stick to the timescales of these internal timeboxes but they are not com-pletely unchangeable and may have to be modified as the project evolves or as more information becomes available. DSDM addresses the management of these timeboxes in some detail but devotes relatively little attention to how to manage the overall timebox except to indi-cate that the Moscow rules should be applied and that not everything can, or should, be

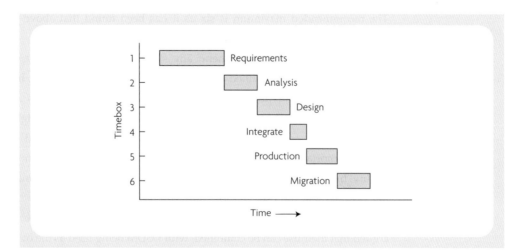

Figure 23.4: Examples of timeboxes for a DSDM project

delivered in a single increment. Indeed, Stapleton (1997) says that the reason for this is that 'incremental delivery has no distinct DSDM flavour'.

DSDM emphasizes the key role of people in the process and is described as a 'user-centred' approach. Overall there is a Project Manager, requiring all the skills of traditional project managers and more, as the focus is on speed! The project manager is responsible for project planning, monitoring, prioritization, human resources, budgets, timeboxes, re-scoping, etc. The use of software project management and control tools is recommended, for example PRINCE2 (see Section 25.4). Some people see the use of such project control tools to be in conflict with the dynamic nature of DSDM, but most DSDM users argue that this is not the case.

On the user side there are two key roles. The first is that of Ambassador User. This is someone (or more than one person) from the user community who understands and represents the needs of that community. The second is the Visionary User. This is the person who had the original idea or vision as to how the project might help in the business or organization. As well as defining the original vision, they have a responsibility to make sure that the vision stays in focus and does not become diluted. In other contexts this person might be described as the project champion.

On the IT side, although they are crucial, there are in general no particular specialist roles, i.e. no distinction is made between different IT roles, such as analysts, designers, pro-grammers, etc. Everyone has to have flexible skills and be capable of turning their hand to whatever is required at any particular time. Of course, in practice particular skills may have to be imported at times, but the key IT team members are generalists and do not change. One exception to this is the specific role of Technical Coordinator, responsible for the technical archi-tecture, technical quality and configuration management. A particular requirement for all is good communication skills.

DSDM recommends small development teams composed of users and IT developers. A large project may have a number of teams working in parallel, but the minimum team size is two, as at least one person has to be from the IT side and one from the business or user side. The recommended maximum is six as it has been found that above this number the RAD process can prove difficult to sustain.

DSDM does not recommend any specific software tools but does recommend that appropriate tools are identified and used appropriately in the process. Indeed, the use of some tools, such as prototyping tools, is implicit in the method. Others are recommended but optional. We have already seen that project management tools might be appropriate but analysis, design, construction and documentation tools, or integrated toolsets, are all poten-tially appropriate. The key factor is whether they will speed up the process and increase quality, but what they must not do is get in the way of the DSDM principles. Thus it is not so much what tools you use but more how you use and apply them.

DSDM has some similarities with Extreme Programming (see Section 23.3) and it is argued that their use can be combined, with DSDM being used as the overall methodology and Extreme Programming being used for the code development.

23.3 Extreme Programming (XP)

Extreme Programming or eXtreme Programming (XP) also attempts to support quicker development of software, particularly for small and medium-sized applications and organizations, and it is said to work best where the whole project requires three to ten programmers. Although larger groups are not uncommon, rapidly changing requirements are difficult to communicate and action in larger groups. XP is more a series of principles for developing software rapidly than a step-by-step methodology. It is said to be a 'lightweight methodology', in that it does not have many complex rules and documents which slow down progress, and it is not so much a software engineering approach (Section 8.2) but an **agile approach** (Section 7.4), though some of the principles of software engineering are taken to an 'extreme level'. Jeffries (2001) in his book *Extreme Programming Installed* defines Extreme Programming as 'a discipline of software development with values of simplicity, communication, feedback and courage'. A useful website is XP (2005), along with Beck (2000).

On communication, the approach stresses the role of teamwork and in most XP shops an informal meeting is held daily, usually called a 'stand-up meeting' to avoid long discussions but encourage communication about the basic progress. 'Open and honest communication' between managers, customers and developers is encouraged and the importance of communications with customers, in particular through continuous testing from day one, referred to as 'concrete and rapid feedback', is stressed. This emphasis on customer satisfaction should ensure quality and customer changes are 'acceptable' even late into the life cycle. Courage is necessary to develop the programs at high speed and to throw code away and start again if necessary. Programmers should look for simple solutions, not complex ones, and develop applications through a series of simple steps. The focus is therefore on incremental development, with each release being put into operation quickly, and functionality expected to change regularly.

A number of organizations are utilizing XP in their software development to reap its potential efficiency gains. This is a less structured, lightweight methodology, which places emphasis on teamwork. This team consists of customer, management and developers. Extreme Programming is a process that convenes these people and assists them in succeeding together.

The customer must define their requirements in **user stories**; these are the things that the system needs to do for its users, and therefore replaces the requirements document (Section 5.6). They may only consist of a few sentences, so they are much less detailed, but do give information relating to how long the requirements implied by each story will take to implement, normally around one to three weeks. A crucial component of defining a user story is the definition of test scenarios, so that testing is always up-front in XP, but the focus is on user needs and not how they will be implemented. Face-to-face discussions between developers and customers will eke out the detailed requirements later. Several acceptance tests will determine whether the user stories have been correctly interpreted as it is the customer who verifies whether the software is acceptable as part of quality assurance.

An **architectural spike** is an aid to figuring out answers to tough technical or design problems. This is usually a very simple program to explore the potential solutions; it builds a system, which only addresses the problem under examination and ignores all other concerns.

A spike is not a high-quality piece of code, therefore, it is expected that it will be thrown away. The purpose behind the use of a spike is to reduce the risk of a technical problem or increase the reliability of a user story's estimate.

Paired programming – two programmers per workstation – is particularly put to good use and reduces the potential risk when a technical difficulty threatens to hold up the system's development. Whilst one programmer is keying in the best way to perform a task, the other is 'thinking more strategically' about whether the whole approach will work, tests that may not work yet and ways of simplifying. The roles should reverse frequently. Further, another principle involves 'moving people around', so that knowledge is spread between all the developers (spreading their own individual skills as well as knowledge of the system).

A release planning meeting is where a release plan is created. The release plan is then used to create iteration plans for each iteration. Each iteration is one to three weeks of the project. In iteration planning acceptance tests are created from user stories and scheduled. The customer specifies scenarios to test when a user story has been correctly implemented. A story can have one or many acceptance tests, whatever it takes to ensure the functionality works. Indeed, tests are said to drive development in XP.

Although we have described XP as a 'lightweight methodology', we see phases as follows:

1. *Planning* relates to the scope of the project (what and how much is to be done), the priority of the functions (implying when each aspect is done), the members of the team (part of the resources aspect), the contents of each increment (in more detail), estimating financial costs and arriving at a schedule for release of the increments, and finally an agreed quality level (how well tested software should be). An attempt is made to balance business and technical aspects. Story writing by users and clients can help by suggesting scope, content, costings and timings for each programming task. The plan may suggest some user stories are not to be implemented or implemented later. The planning meetings are therefore crucial for all the stakeholders. Planning is sometimes divided between the release plan for the overall project and iteration plans, each iteration being a one- to three-week programming task. Making a number of small releases at frequent intervals is seen as desirable as customers see that they are getting results to the agreed quality and enables feedback to be given to developers. (The planning phase may be preceded by an *exploration* phase where the customer suggests change through storyboards and is assured that the technical people understand the requirements in outline.)

2. *Designing* is done on the principles of simplicity, feedback and courage, and enabling incremental change as described above. As we have seen, there will normally be a daily meeting of all the participants at this phase onwards. Simplicity suggests that the easiest approach to achieving the functionality required for that iteration should be adopted. It is thought that discussions involving all the stakeholders are likely to derive the best solutions, and this also has the result of what is called 'collective code ownership', which can be an outcome of participative approaches generally (Section 5.1). The approach is

therefore much less dependent on 'key programmers and analysts' when compared with a more conventional approach to software development. Creating a spike solution, as described above, can reduce the risk of missing the requirements laid out in a user story. (This phase may be engulfed by the urge to develop and be subsumed in that next phase.)

3. *Developing the code* includes the principles of paired programming, testing using programmer and user data, gaining rapid feedback, ensuring that the test works fully, and continuously integrating with already implemented code. Each release of the developing project may be rolled out to users every two to five iterations. Under this scenario, it is evident that the customers need to be always available to help the development team, indeed to be part of the development team, from answering questions about the requirements to accepting the software developed. This may imply full-time commitment to the project at times as well as a general agreement 'to be there when required'. Testing here refers to that of each iteration, with linked tests being part of the next phase. But each iteration may consist of coding a small part, then adding another feature and testing that as well, and then another and so on. Interestingly, many XP developers argue that overtime does not help and projects in difficulty should be replanned at a planning meeting. This supports Brooks' argument about the mythical man-month (Brooks, 1995).

4. *Productionalizing* may also be seen as part of developing, but the tests at this time ensure fitness for production of the 'whole system', perhaps parallel running, ensuring good performances for running the software, so that this phase naturally runs into *maintenance*. Unlike many methodologies, documentation is not a prime concern – delivering software is seen as the primary goal. This is seen as part of XP being a 'light' methodology.

23.4 Web Information Systems Development Methodology (WISDM)

There are a number of approaches that have been proposed for developing web applications. They are available in books and on the web. However, none have been taken up widely. It appears that many web developers simply develop their sites and associated applications by trial and error or by using software such as Dreamweaver to guide development (Section 18.2). Others use a more conventional approach and modify it as appropriate for web development. Indeed, it could be that web applications are like any other, but with different emphases, such as those on interface design and security, and also the usual necessity for rapid development, hence its inclusion in this chapter.

Here we offer an approach known as WISDM, which is a modification of Multiview (Section 26.1) for web development (see Vidgen et al., 2002):

Many of the approaches to Web development have focused on the user interface and in particular the look and feel of a Web site, but have failed to address the wider aspects of Web-based information systems. At the same time, traditional IS methodologies – from the waterfall lifecycle to rapid application development (RAD) – have struggled to accommodate Web-specific aspects into their methods and work practices. Although Web sites are characterized historically as graphically intense hypermedia systems, they have now evolved from cyber-brochures into database-driven information systems that must integrate with existing systems, such as back office applications. Web-based IS therefore require a mix of Website development techniques together with traditional IS development competencies in database and program design.

Essentially, the authors argue that Multiview can be drawn upon in a specific situation by particular people to create WISDM, a modification of Multiview used locally and uniquely for web development in practice. The book contains a complete case study related to a theatre booking system. As seen in Figure 23.5, emphasis is placed on design, and human–computer interaction and the user interface, in particular.

We will concentrate here on differences in emphasis when compared to the Multiview framework expressed in Section 26.1. Unlike conventional methodologies, there is no a priori ordering of the five aspects of the methods matrix, each one being emphasized alone (or with others) as appropriate during the lifetime of the project.

Organizational analysis represents *value creation* and stresses strategy as relationship building and maintaining with a broad range of stakeholders that includes customers, employees, government, suppliers, labour organizations, and so on. As the users of websites can be broader and different than conventional applications, a broad view of stakeholders is essential (Section 16.1). In an age where dotcom failure has been noticeable, the right strategy may also be emphasized more than normal. For some organizations, the website is the most prominent aspect of the company and its main source of income, so it is essential to get it right.

Information analysis represents the *requirements specification*. This is a formalized specification of the information and process requirements of the organization. The specification might be in the form of a document with graphical notations, but it might also be in the form of a software prototype (an executable specification). The indicative approach in WISDM is to use UML

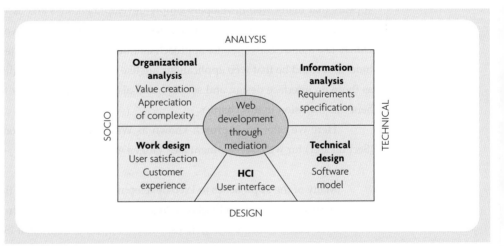

Figure 23.5: WISDM

(Section 13.2). It is used to create structural (class model), behavioural (use-case and interaction diagrams), and process-oriented (activity diagrams) models of the problem situation. In general, web content analysis and management is a major concern including such factors as consistency, navigability, and tracking, again concerns that will have more prominence in web applications development.

Work design represents *user satisfaction*. The traditional concern of the socio-technical approach to IS development has been with job satisfaction and genuine user participation in the development process (see ETHICS, Section 24.1). WISDM extends this view to incorporate the interests of external users, such as customers, who may be using the information system as part of their social activities. Customer satisfaction with a web IS is assessed using the method, and the Internet enables the designers to reach outside the organization to find potential external users during the development process as well as when the site is operational. WISDM uses an instrument known as WebQual to assess e-commerce offerings against a range of qualities using a 7-point Likert scale. As seen in Figure 23.6, there are 23 questions related to usability, information, interaction, and convergence (the overall view of the website).

Usability
1. I find the site easy to learn to operate
2. My interaction with the site is clear and understandable
3. I find the site easy to navigate
4. I find the site easy to use
5. The site has an attractive appearance
6. The design is appropriate to the type of site
7. The site conveys a sense of competency
8. The site creates a positive experience for me

Information
9. Provides accurate information
10. Provides believable information
11. Provides timely information
12. Provides relevant information
13. Provides easy to understand information
14. Provides information at the right level of detail
15. Presents the information in an appropriate format

Interaction
16. Has a good reputation
17. It feels safe to complete transactions
18. My personal information feels secure
19. Creates a sense of personalization
20. Conveys a sense of community
21. Makes it easy to communicate with the organization
22. I feel confident that goods/services will be delivered as promised

Convergence
23. What is your overall view of this website?

Figure 23.6: WebQual questions

Technical design represents the *software model*. A formalized model of the software in terms of data structures and program design is needed to support software construction. There are some technical restraints in developing web applications; for example, a variant of HTML will be used for the pages and an Internet server will host the pages. JavaScript is likely to be needed for some processing. The extensible mark-up language (XML) might be used to provide a notation for defining the content and presentation of data. The data are likely to use a relational database management system, such as Access or Oracle (Sections 18.5 and 19.3, respectively). ColdFusion might be used to convert files to HTML format, effectively automating much of the functionality of the implementation. Vidgen et al. (2002) recommends that in some instances functions may be supplied by a third party, for example, the handling of credit card payments.

Human–Computer Interface (HCI) represents the *user interface* and is located as an overlapping space in technical design and work design. The shape of this space reflects its role in pointing toward analysis but its foundations are solidly in design. The interface design draws on website design principles for page layout, navigation schemes, and usability in the context of work design. Templates and style sheets, such as those provided by Dreamweaver (Section 18.2), may be used for pages so that there is a common 'look and feel' to the site. Because the reach of websites can be very different when compared to conventional applications, this phase might include promotion plans and the monitoring of the use of the site. Emphasis is also placed on security issues, such as authentication.

Summary

- In this chapter we have described four methodologies that are oriented toward speed of development. RAD and DSDM concentrate on developing information systems rapidly, with XP mainly concerned with programming software quickly, and WISDM on web development.

Questions

1. What do you see as the common features and differences between the four approaches discussed in this chapter?
2. In what ways, if any, do they potentially sacrifice quality for speed of development?
3. Why is cultural change important in DSDM?
4. Would you use WISDM to develop your personal website? If not, why not? Will the use of WISDM guarantee a 'quality' website?

Further reading

Beck, K. (2004) *Extreme Programming Explained*, 2nd edn, Addison-Wesley, Boston.

DSDM Manual Version 4.2 (2005) DSDM Consortium, Tesseract Publishing, Surrey, UK (see also www.dsdm.org/version4).

Jeffries, R. (2001) *Extreme Programming Installed*, Pearson Education, London.

Martin, J. (1991) *Rapid Application Development*, Prentice Hall, Englewood Cliffs, New Jersey.

Stapleton, J. (2002) *DSDM: A Framework for Business Centred Development*, Addison-Wesley, Harlow, UK.

Vidgen, R., Avison, D.E., Wood, R. and Wood-Harper, A.T. (2002) *Developing Internet Applications*, Butterworth-Heinemann, Oxford.

24 People-oriented methodologies

24.1 Effective technical and human implementation of computer-based systems (ETHICS)

ETHICS is an acronym, but the name of this approach is meant to imply that it is a methodology that embodies an ethical position. ETHICS, devised by Enid Mumford (Mumford, 1995), is a methodology based on the participative approach to information systems development (discussed in Section 5.1). In addition, it encompasses the socio-technical view that for a system to be effective the technology must fit closely with the social and organizational factors.

The philosophy of ETHICS is thus different from most information system development methodologies and is also explicitly stated, which is also not common in most methodologies. The philosophy is one which has evolved from organizational behaviour and perceives the development of computer systems not as a technical issue but as an organizational issue which is fundamentally concerned with the process of change. It is based on the socio-technical approach of the social sciences as developed by a number of authors, one of the most influential being Davis (1972). Mumford (1983a) defines the **socio-technical approach** as:

> one which recognizes the interaction of technology and people and produces work systems which are both technically efficient and have social characteristics which lead to high job satisfaction.

Elsewhere, in Mumford and Weir (1979), **job satisfaction** is defined as:

> the attainment of a good 'fit' between what the employee is seeking from his work – his job needs, expectations and aspirations – and what he is required to do in his job – the organizational job requirements which mould his experience.

In order to ascertain how good this fit is, a theory for measuring job satisfaction has been developed based on the various views of what is important in job satisfaction and these have been integrated into a framework derived from Parsons and Shils (1951). Five areas of measurement are identified as follows:

1. *The knowledge fit* – a good fit exists when the employees believe that their skills are being adequately used and that their knowledge is being developed to make them

increasingly competent. It is recognized that different people have widely different expectations in this area, some wanting their skills developed, others wanting to remain static and opt for an 'easy life'.

2. *The psychological fit* – a job must fit the employee's status, advancement, and work interest (some of the Herzberg, 1966, motivators). These needs are recognized to vary according to age, background, education, and class.

3. *The efficiency fit* – this comprises three areas. First, the effort–reward bargain, which is the amount an employer is prepared to pay (as against the view of the employee about how much he is worth). Although this is probably the prime area of importance to management, it is in practice sometimes way down the list of employee needs. Second, work controls, which may be tight or loose but need to fit the employee's expectations. Third, supervisory controls, such as the necessary backup facilities, for example, information, materials, specialist knowledge, and supervisory help.

4. *The task-structure fit* – this measures the degree to which the employee's tasks are regarded as being demanding and fulfilling. Particularly important are the number of skills required, the number and nature of targets, plus the feedback mechanism, the identity, distinctiveness, and importance of tasks, and the degree of autonomy and control over the tasks that the employee has. This measure is seen to be strongly related to technology and its method of employment. Technology can affect the task-structure fit substantially and, it is argued, has reduced the fit by simplification and repetitiveness. However, it is also seen as a variable which can be improved dramatically by designing the technical system to meet the requirements of the task-structure fit.

5. *The ethical fit* – this is also described as the social value fit and measures whether the values of the employee match those of the employer organization. In some organizations, performance is everything, while others value other factors, for example, service. Some firms are paternal or welfare-oriented, others aim to achieve the characteristics of 'success', and so on. The better the match of an organization's values with those of the employee, the higher the level of job satisfaction.

A second philosophical strand of the ETHICS methodology is participation. This is the involvement of those affected by a system being part of the decision-making process concerning the design and operation of that system. Those affected by a system include the direct users and also the indirect users, such as management, customers, suppliers, and so on. Of course, there are limits to this. For example, competitors will be affected, but it is unlikely that they will be asked to participate. Participation is important in many methodologies, but has been described as vital in ETHICS (Hirschheim, 1985). In some other methodologies no more than lip service is paid to participation, sometimes being regarded simply as 'allowing the users to choose the colour of the workstations that they use'. In ETHICS, users are involved in the decisions concerning the work process and how the use of technology might improve their job satisfaction.

In ETHICS the development of computer-based systems is seen as a change process and therefore it is likely to involve conflicts of interest between all the participants or actors in that process. These conflicts are not simply between management and worker but often between worker and worker and manager and manager. The successful implementation of new systems is therefore a process of negotiation between the affected and interested parties. Obviously major affected and interested parties include the users themselves, and if these people are left out of the decision-making process, the process of change is unlikely to be a success. This is not just because of resulting disaffection among the user group but, more positively, because they have so much to contribute in making the implementation a success. They are probably the most knowledgeable about the current workplace situation and the future requirements. Mumford (1983a) summarizes:

> All change involves some conflicts of interest. To be resolved, these conflicts need to be recognized, brought out into the open, negotiated and a solution arrived at which largely meets the interests of all the parties in the situation . . . successful change strategies require institutional mechanisms which enable all these interests to be represented, and participation provides these.

It is recognized in practice that participation means different things to different people and that the parties involved may have quite different reasons for wanting participation and quite different expectations concerning the benefits. Management may see it as a way of achieving changes that would otherwise be rejected. This is perhaps not the ideal view for management to take but if the resulting participation is real then so be it, although the end result may not exactly turn out as they expect. The point being that it is not a prerequisite for everybody to hold the view that participation is a moral or ideological necessity, enlightened self-interest will do just as well.

The philosophical commitment to participation outlined above begs the question of exactly how it is to be achieved. There appears to be quite a degree of freedom involved, and, although there exist 'ideal' types of participation, in practice a variety of forms are acceptable for it still to be 'ETHICS'. In fact, it can be used by an expert group to design a system for another, non-participating group. However, this is not recommended. Nevertheless, it shows that ETHICS is a methodology which has quite a level of flexibility. It is better, Mumford argues, to use it in some form, rather than not at all. The implication being that its use, even stripped of some of its most important participatory trappings, is better than other more traditional methods which concentrate purely on technical and economic objectives.

Mumford distinguishes between structure, content, and process. Structure is the mechanism of participation which, as discussed in Section 5.1, can be consultative, representative, or consensus. Consultative participation involves the participants giving evidence to the decision makers which, possibly, will influence the decision makers but does not bind them in any way. This is the weakest form of participation and not recommended for detailed design. Representative participation is a structure where selected or, preferably, elected representatives of the various interests are involved in the decision-making process. This is most appropriate for the tactical or middle management type of decision making. In computing terms, this might be at the system definition stage where the system outline and boundaries are discussed and a fairly wide spectrum of interests are involved. The third form of participation is consensus, where all the constituents are involved in the decision making. This is most suitable at the detailed design stage where the decisions probably affect the day-to-day work practices of the people involved. Clearly it is difficult to involve everybody in everything, and what usually happens is that design groups are formed to do the work and present alternatives to the whole constituency, which takes the final decisions.

The content of participation concerns the issues and the boundaries of activities that are within the remit of participation. Generally, prior to any participation, management will want to keep certain things as their own prerogative. One objective of the process of participation is the gaining of relevant knowledge and information by the participants. In general, the users involved in participation will not have previously had the necessary knowledge, information, and, perhaps, confidence to discuss issues and make decisions. Without this, participation is only of a very limited kind. The users must have as much information and knowledge as is necessary to make informed decisions, or at least as much as anyone else. Without the acquisition of this information and knowledge they will be at a disadvantage and subject to undue influence from more powerful groups. True participation means equal knowledge and thus, it might be argued, equal power for all groups. Training and education of users is therefore a very important aspect of ETHICS.

Participation usually involves the setting up of a steering committee and a design group or groups. The steering committee sets the guidelines for the design group and consists of senior managers from the affected areas of the organization, senior managers from management services and personnel, and trade union officials (if the organization is unionized). It is recommended that the steering committee and the design group meet once a month during the course of the project. The design group designs the new system including:

- choice of hardware and software;
- human–computer interaction;
- workplace reorganization;
- allocation of responsibilities.

All major interests should be represented, including each section and function, grade, age group, and so on. The design group includes systems analysts, although their role is not the normal one of analyst and designer, but one of educator and adviser. This often involves the analysts in a learning process themselves. If the area of the design is large, involving many departments or sections, then a design group may first design in outline and then hand over to detailed design groups. A participative design requires the appointment of a facilitator to help the design group manage the project and educate the group in the use of ETHICS. The role is multifaceted and concerns motivation and confidence-building of the design group; it is not one of decision making or persuading. For this reason, the facilitator must be neutral and preferably external, if not to the organization, then to the department. The role is very important. In one situation that Mumford quotes, the facilitator withdrew, and the confidence of the design group declined and the importance of the group in the eyes of the management also declined. ETHICS has 15 steps (Mumford, 1986) as follows (unless stated otherwise, the work in the steps that follow are performed by the design group).

1 Why change?

The first meeting of the design group considers this rather fundamental question and addresses the current problems and opportunities. The result should be a convincing statement of the need for change. Presumably, if no convincing statement for change is arrived at, the process stops there, although this is not made explicit.

2 System boundaries

The design group identifies the boundaries of the system it is designing and where it interfaces with other systems. Four areas are considered: business activities affected (e.g. sales, finance, and personnel); existing technology affected; parts of the organization affected (e.g. departments and sections); and parts of the organization's environment affected (e.g. suppliers and customers).

3 Description of existing system

This is to educate the design group as to how the existing system works. In practice, it is found that people will know the detail of their own jobs and those that they interact with directly, but will probably have little knowledge of the whole system. In this step, two activities are undertaken. First, a horizontal input/output analysis is described with inputs on the left, activities in the middle, and outputs on the right. Second, a vertical analysis of the design area activities is made at five different levels. The lowest level is of the operating activities, that is, the necessary activities of a day-to-day nature. These should have appeared in the horizontal analysis. The problem prevention/solution activities are also identified. These are the key problems or variances that occur and how they are corrected. Third, the co-ordination activities are identified. These are activities that have to be performed together or in a particular sequence or at a particular time. These are both interdepartmental and intra-departmental co-ordinations. Fourth, the development activities are recorded. These are the things or areas that need improving. Fifth, the control activities are identified, indicating how the system is controlled, how it is judged to be meeting targets or objectives, and how it is monitored.

4, 5 and 6 Definition of key objectives and tasks

Three questions are asked in order to help define the key objectives. First, why do particular areas exist, what is their role, and what is their purpose? Second, given this, what should be their responsibilities and functions? Third, how far do their present activities match what they should be doing? From this, the key objectives can be listed and these form the design objectives of the new system. In addition, the key tasks that need to be performed to achieve the key objectives are defined in outline, along with their key information needs.

7 Diagnosis of efficiency needs

Weak links in the existing system are identified and documented. Mumford talks about them as variances, which is a 'tendency for a system or subsystem to deviate from some desired or expected norm or standard' (Mumford, 1983a). Mumford identifies two types of variance. First, systemic or key types, which are inherent in the system and cannot be completely overcome. They can only be eased. An example is provided by the variances connected with the *financial desire* to keep stocks small and the *service desire* to be able to supply customers with what they want. The second type of variance is operational. These are variances due to poor design or lack of attention to changing circumstances and can usually be completely eliminated in the new system. Examples could include bottlenecks, insufficient information, and inadequate equipment.

8 Diagnosis of job satisfaction needs

This step measures the job satisfaction needs. This is achieved by use of a standard questionnaire provided in the ETHICS methodology. The design group may alter the questionnaire to fit their organization and requirements. The results are discussed democratically and the underlying reasons established for any areas where there are poor job satisfaction fits. In addition, formulations for improving the situation in the new design are made, and everybody is encouraged to play a major part in this design work. Where there have been knowledge or task-structure problems of fit, these are susceptible to improvement by a redesign of the system. Other areas of poor fit, such as effort–reward or ethical, may be improved somewhat in this way, but will probably require changes in personnel policies, or more radically, organizational ethos.

9 Future analysis

The new system design needs to be both a better version of the existing system and able to cope with future changes that may occur in the environment, technology, organization, or fashion. Therefore, an attempt is made to try to identify these changes and to build a certain amount of flexibility into the new system. This may involve the design group in interactions with people outside the organization in order to identify and assess some of the potential changes.

10 Specifying and weighting efficiency and job satisfaction needs and objectives

Mumford identifies this as the key step in the whole methodology. Objectives are set according to the diagnosis activities of the three previous steps. The achievement of an agreed and ranked set of objectives can be a very difficult task and must involve everyone, not just the design group itself. Often objectives conflict and the priorities of the various constituencies may be very different. These differences may not all be resolved, but one of the stated benefits of ETHICS is that at least these differences are aired. Ultimately, a list of priority and secondary objectives is produced. The criterion for the systems design is that all priority objectives must be met along with as many of the secondary ones as possible. At this stage a certain amount of iteration is recommended, to review the key objectives and tasks from Steps 4 and 5.

11 The organizational design of the new system

If possible, this should be performed in parallel with the technical design of Step 12, because they inevitably intertwine. The organizational changes which are needed to meet the efficiency and job satisfaction objectives are specified. There are likely to be a variety of ways of achieving the objectives, and between three and six organizational options should be elaborated. The design group specifies in more detail the key tasks of Step 5 and addresses the following questions, the answers forming the basic data for the organizational design process:

- What are the operating activities that are required?
- What are the problem prevention/solution activities that are required?
- What are the coordination activities that are required?
- What are the development activities that are required?
- What are the control activities that are required?
- What special skills are required, if any, of the staff?
- Are there any key roles or relationships that exist that must be addressed in the new design?

Each organizational option is rated for its ability to meet the primary and secondary objectives of Step 10, and should identify the sections, subsections, work groups, individuals, their responsibilities, and tasks. In order to meet the job satisfaction objectives, it is almost inevitable that the design group will have to consider the socio-technical principles of organizational design and be provided with information and experience in relation to design. The socio-technical approach is the antithesis of Taylorism (Taylor, 1947) which is to break each job down into its elemental parts and rearrange it into an efficient combination. The traditional car assembly line, which requires its operators to perform small, routine, repetitive jobs, is regarded as the ultimate example of Taylorism in action. The requirements of the machine are given priority over the requirements of the human being. This has, it is argued, inevitably led to a bored, disaffected, and ultimately inefficient workforce.

Although ETHICS uses aspects of socio-technical design, the socio-technical school assumes a given technology, whereas, in ETHICS, the technology is part of the design. Further, they assume shop-floor situations, rather than the office and high-level organizational situations which concern ETHICS.

Mumford recommends the consideration of three types of work organization pattern. The first is task variety, and involves giving an individual more variety in work by providing more than one task to be performed or by rotating people around a number of different tasks. This is the more traditional approach, but is limited, especially where the expectations of job satisfaction are more sophisticated. In this case, the principles of job enrichment might be appropriate. This is where the work is organized in such a way that a number of different skills, including judgemental ones, are introduced. In particular, it involves the handling of problems and the organization of the work by the individual without supervision. This may require an increased skill level on behalf of the individual, but leads to enhanced job satisfaction. A further stage in job enrichment is the incorporation of development aspects into a job. This means that the individual has the freedom to change the way the job is performed. This leads to constant review and the implementation of new ideas and methods. Obviously this cannot be introduced into every job, but there are probably more opportunities than at first imagined.

As important as individual jobs is the concept of what Mumford calls self-managing groups. Here, groups are formed that have responsibility for a relatively wide spectrum of the tasks to be performed. These groups are preferably multi-skilled, so that each member is competent to carry out all the tasks required of the group. They are encouraged to organize themselves, their work, and their own control and monitoring, which may include their own target setting. This can provide a very stimulating and satisfying work environment for the group members. Again, self-managing groups are not always possible and require a good deal of management goodwill at first, but nevertheless can prove very effective.

12 Technical options

The various technical options that might be appropriate, including hardware, software, and the design of the human–computer interface, are specified. Each option is evaluated in the same way as the organizational options, that is, against efficiency, job satisfaction, and future change objectives. As mentioned in Step 11, the organizational and technical options should be considered simultaneously, as often one option implies certain necessary factors in the other. It is advised that one option should exist which specifies no change in technology, so as to be able to see how much could be achieved simply with organizational changes.

The organizational and technical options are now merged to ensure compatibility and are evaluated against the primary objectives and the one that best meets the objectives is selected. This selection is performed by the design group with input from the steering committee and other interested constituencies.

13 The preparation of a detailed work design

The selected system is now designed in detail. The data flows, tasks, groups, individuals, responsibilities, and relationships are defined. There is also a review to ensure that the detail of the design still meets the specified objectives. Obviously, the design detail includes the organizational aspects as well as the technical.

14 Implementation

The design group now applies itself to ensuring the successful implementation of the design. This involves planning the implementation process in detail. This will include the strategy, the education and training, the coordination of parts, and everything needed to ensure a smooth changeover.

15 Evaluation

The implemented system is checked to ensure that it is meeting its objectives, particularly in relation to efficiency and job satisfaction, using the techniques of variance analysis and measures of job satisfaction. If it is not meeting the objectives, then corrective action is taken.

Indeed, as time progresses, changes will become necessary and design becomes a cyclical process.

Quite a common reaction to ETHICS is for people to say that it is impractical. First, it is argued that unskilled users cannot do the design properly, and, second, that management would never accept it, or that it removes the right to manage from managers. In answer to the first problem Mumford argues that users can, and do, design properly. They need some training and help along the way, but this can be relatively easily provided. More importantly, they have the skills of knowing about their own work and system, and have a stake in the design. This is much more than many traditional analysts and designers have. To answer the second point, managers have often welcomed participation and can be convinced of its benefits. There are many success stories. It is not always the management that needs to be convinced, sometimes it is the users who are sceptical about participation, seeing it as some sort of management trick. The job of a manager is to meet the corporate objectives, not simply oversee people and make every last decision. This is often counterproductive to achieving those objectives, often resulting in very high staff turnover rates, which is not productive.

Mumford admits that it is not easy, quite the reverse, but the benefits are, she claims, worth it. Mumford shows how users can design their own systems and how they come to terms with their design roles, illustrated by experiences. For example, a group of secretaries at Imperial Chemical Industries (ICI) in the UK designed new work systems for themselves in the wake of the introduction of word processing systems, and a group of purchase invoice clerks helped design a major online finance system. One of the most interesting aspects, and most telling concerning the power of ETHICS, is the fact that the clerks designed three different ways of working with the computer system to do essentially the same thing. The one used depends on the clerk. Few professional systems design teams would design a number of alternative ways to achieve the same task.

ETHICS has also been used by a number of large companies to assist the building of very large systems. One of the first major uses of ETHICS in the development of a large system was Digital Equipment Corporation's XSEL, an expert system for their sales offices which helped to configure DEC hardware systems for particular customers.

In some situations it has become a method of requirements analysis in particular, and a version has been defined which is referred to as **QUICKETHICS** (QUality Information from Considered Knowledge). In order to create and maintain managers' interest, it can be organized as a drama having four 'acts':

- self-reflection;
- self-identification;
- group discussion;
- group decision.

In this approach, each manager describes his or her work role and relationships, along with information needs ranked as 'essential', 'desirable', and 'useful' on an individual basis. This provides an opportunity for self-reflection. Meeting then as a group, each manager gives a short description of his or her mission, key tasks, critical success factors, and major problems. This provides an opportunity for self-identification and encourages questions and discussion. Then managers may write the essential information needs on cards placed on a magnetic board explaining the reasons for their importance. This provokes group discussion because it soon becomes apparent that managers have many overlapping needs. These common needs can be agreed as forming a 'core' module of the proposed information system, delivering essential information needs in the group decision process. Once this is implemented, 'desirable' needs can guide future development. The requirements analysis phase may only involve two days of management's time. ETHICS has changed over time, as for all information systems development methodologies, but the importance of user involvement and participation in systems design has endured the process of change.

24.2 KADS

In this section we examine a formalized approach to expert systems development known as KADS (Wielinga et al., 1993; De Greef and Breuker, 1992) which had its origins as a European Union ESPRIT research project, with partners from commercial companies and universities. There is also a development of KADS known as CommonKADS. We will first look at KADS and then CommonKADS.

KADS adopts the view that developing an expert system is a modelling activity and that it is not the case that the system has to be filled only with knowledge extracted from a human expert. It is rather a computational model of desired behaviour, which may also reflect aspects of the behaviour of an expert. It is not the functional and behavioural equivalent of an expert; the system may actually do things in different ways and utilize different approaches to human experts.

Figure 24.1 illustrates the various models that are constructed in KADS. It also shows the relationships between the models in the sense that information in a higher-level model is used in the construction of the lower-level model. The model does not necessarily imply that the sequence has to be followed in the linear top-down fashion that the diagram suggests; in fact, KADS advocates a spiral approach rather than a waterfall life cycle. The spiral model (Section 7.1) has been adapted for KADS and is illustrated in Figure 24.2. It begins in the centre with the process circling around with each pass adding a degree of functionality or progress, and only when a number of circles have been made is the process complete. The breakdown or decomposition of the process into the development of these different models is the way that KADS attempts to address the complexity of expert systems. A model is an abstraction that eliminates certain detail but concentrates on certain key features of the area being modelled. Wielinga et al. (1993) call this a 'divide and conquer' strategy which is a term used by James Martin in describing the same concept in the context of Information Engineering (Section 21.3).

1 Organizational model

The first step of KADS is concerned with defining the problem that the expert systems (or knowledge-based system in KADS terms) is addressing in the organization. This step consists

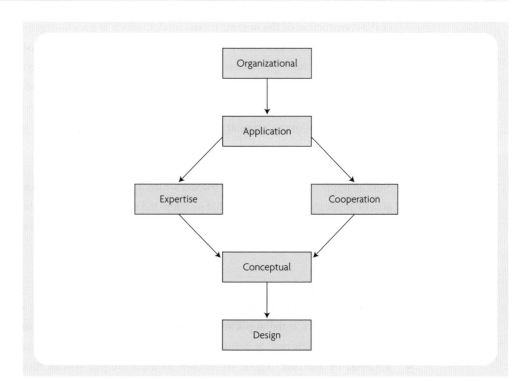

Figure 24.1: Models and processes of KADS

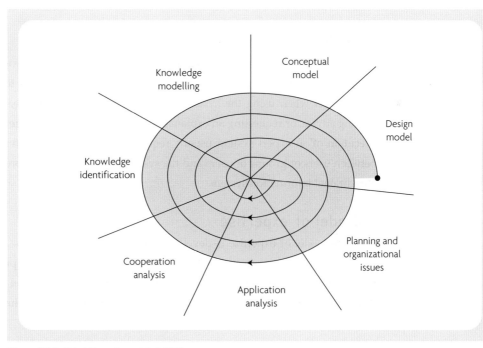

Figure 24.2: Spiral life cycle model of KADS

of the construction of the organizational model, which is a socio-organizational analysis of the environment. This is essentially a high-level model of the functions, tasks, and problems in the organization, including an assessment of the effects of the introduction of the expert system on the organization and the people. Unfortunately, the form that this model takes is not specified in detail, but it is a recognition by KADS that expert systems do not exist in an organizational vacuum and that organizational issues are important to the success of expert systems.

2 Application model

Next an application model is constructed which narrows the focus from the organization to the area of application and defines what problem the system is going to solve, and the overall function and purpose of the system. Again, we are not told the detail of this stage nor the exact form of the model.

3 Task model

The next stage is to develop the task model which decomposes the overall function of the system described in the application model into a series of tasks and subtasks that the system will perform. Each task is defined by the information input to the task and the goal that is achieved by the task. These are known as input/output specifications. The tasks are then assigned to agents which might be the expert system, the user, or some other system. This will depend on what or who could best perform each task. Finally, any constraints or specific requirements for the tasks are defined; for example, that the output of the task must be in the terminology of a particular type of user.

4 Model of co-operation

The next step is to define the model of expertise and the model of co-operation. The model of co-operation is the definition of how the system and the user will work together at the task level and interact when using the system in various modes, for example, in solving a problem, seeking advice, or requesting an explanation. This is an important stage because in practice the execution of an expert system usually requires a complex set of interfaces and interactions between users or groups of users and the system. This is typically more complex than that for a traditional information system.

5 Model of expertise

Next the model of expertise is developed. This is the key task in the methodology, and the model of expertise is effectively the functional specification of the problem-solving part of the expert system. Unlike many expert systems approaches, this model is a specification of the desired behaviour and the types of knowledge involved, rather than the specification of detailed rules.

In KADS, a process of knowledge identification is undertaken as a preparation phase before construction of the model of expertise. This is a kind of systems investigation of the area of concern prior to the building of the model. Data is collected and tasks identified using a variety of techniques, for example, structured interviews, rational task analysis, workflow analysis, and repertory grid analysis. A glossary and lexicon of terms are also developed to help understand and document the area of concern prior to any formal conceptualization of knowledge. After this, a knowledge-modelling phase is undertaken, which involves the identification and definition of four types or layers of knowledge. Some typical techniques and representations that might be used in this process are tree diagrams, decision trees, laddering, data modelling diagrams, think-aloud protocols, and scenario simulation. KADS suggests that a prototype of the problem-solving aspect of the system can be implemented as a way of validating the requirements and the knowledge models.

The first type of knowledge identified is the **domain knowledge** which involves the identification of concepts, properties, relationships between concepts, and relationships between properties and structures. Concepts are things or objects in the knowledge domain. For example, in the domain of credit rating for loans in a banking application, a concept might be customer, account, or application. In practice, these concepts should be as low level or elementary as possible. A property is an attribute of a concept; for example, a customer might be active or passive, or an application significant or not. Relationships between concepts are identified. For example, an applicant is a customer. Causal relationships between property expressions are also identified; for example, a customer with a transaction in the last year causes customer to be active or a value over a certain amount causes the application to be significant. Structures are also identified, and this is a way of breaking complex objects down into more manageable components. In practice, an account might be broken down into current or deposit, and deposit into short-term and long-term, open or closed, and so on. The examples used are illustrative and would in practice be at more elementary levels.

The second type of knowledge defined is **inference knowledge**. Inference knowledge is that which uses the domain knowledge to infer or produce additional information. KADS advocates the separation of domain knowledge from inference knowledge as a matter of philosophy, on the basis that it allows multiple use of the same domain knowledge. Inference knowledge is identified in the form of 'knowledge sources'. These are the processes that generate elementary pieces of information using domain information. For example, within a specific domain, the process by which a particular piece of information is compared to another piece of information, in order to see if they are similar, might be a knowledge source, or, in terms of our earlier example, the action or reasoning in the rejection of a loan application might be an inference knowledge source. Knowledge sources use domain knowledge, but the process is defined as independently as possible of the information in the knowledge domain and in principle could be applied to a variety of different situations. An extension of this idea is the generation of 'interpretation models' of typical inference knowledge for a particular task, which might be reused in a different domain. For example, a credit assessment interpretation model might form a template in the alternative domain of assessment for tenancy agreements. Such interpretation models would guide the knowledge engineer by providing a template for knowledge acquisition in the new domain and perhaps also save significant development time.

The third type of knowledge is **task knowledge** which is information about how the elementary knowledge sources can be combined to achieve a particular higher-level objective. These higher-level processes are called tasks; for example, the combination of various knowledge elements might be the definition of the task of verification of a hypothesis. The tasks are decomposed into a number of subtasks which may be inference tasks, problem-solving tasks, or transfer tasks. A transfer task is one that requires interaction with an external agent. The structure of the subtasks and the control dependencies are described using structured English (Section 12.4). Due to the independence of the knowledge sources from the domain knowledge, the task knowledge is also independent and may be thought of as representing fixed strategies for achieving problem-solving goals.

The fourth category of knowledge is **strategic knowledge** which identifies the strategies, goals, high-level rules, and tasks that are relevant to the solution of a particular problem. Although the need for strategic knowledge is identified in KADS, little work has been done on the definition of such knowledge, and Wielinga et al. (1993) concede that most systems developed with KADS have identified little or no knowledge of this type.

The description of the categories of knowledge relevant to KADS is described in what information systems people might feel is a bottom-up way and that it might be better if a top-down view were adopted. Figure 24.3 provides this architecture, showing the four levels and their interactions.

1 Conceptual model

The conceptual model is the combination of the model of co-operation and the model of

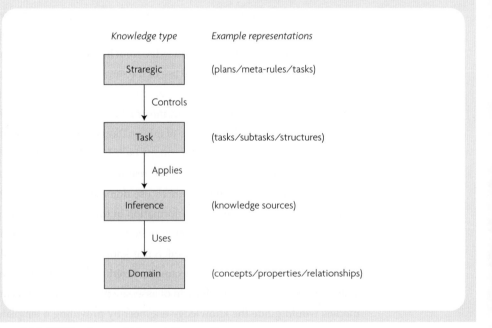

Figure 24.3: KADS four layer model for modelling expertise

expertise, and together this provides an application-independent specification of the system to be built.

2 Design model

The design model introduces the computational and representational techniques into the conceptual model. For example, the constraints of the hardware, software, and the external world are now defined for the purposes of implementation. This is similar to the separation of the logical and physical models familiar in much information systems development but has not usually been present in approaches to expert systems development. However, the transformation from the conceptual model to the design model is by no means automatic or indeed well defined. In principle, the designer in KADS is free to proceed in whatever way is thought best providing the conceptual specification is achieved. A completely independent approach can cause problems because it is then difficult to provide explanations of the systems reasoning in domain and conceptual language terms, unless there exist some clear linkages. KADS makes a few recommendations, but the process is essentially still in development and the subject of research. For inference knowledge, a computational technique that can realize the inferences needs to be selected. This requires elements that support (a) algorithms, (b) input–output data structures, and (c) a representation of domain knowledge (usually via production rules, as described in Section 5.3). For domain knowledge, a database is required that supports all the domain knowledge elements, for example, concepts, properties, and relationships. Most conventional databases are not adequate in this respect. For task knowledge, a control technique for executing tasks is required, which might be a blackboard (Section 5.3) or even a simple procedure hierarchy. For strategic knowledge, a production system that handles meta-rules is suggested.

As can be seen, KADS is a little different to other methodologies as it focuses on principles and models rather than the processes, phases, steps, and tasks that need to be followed and undertaken.

24.3 CommonKADS

KADS (above) relates essentially to the development of expert systems whereas CommonKADS relates to the wider domain of knowledge management. As we saw in Section 5.4 knowledge management is about levering knowledge as an important organizational resource. We have referred to information systems development greatly in this book, yet knowledge, that is, the ability to use information in action for a particular purpose, is even more important than information, which is much more commonplace. In this description of CommonKADS, we use the text of Schreiber et al. (2000) and the website http://www.commonkads.uva.nl as our major source. Along with KADS itself, its design has been influenced by soft systems methodology (Section 25.1) and the organizational learning literature, for example, Argyris (1993).

In this description, we will discuss in particular areas of divergence with KADS (Section 24.2). Figure 24.4 shows the general approach of CommonKADS from context, through concept, to artefact. Within this structure, there are six models:

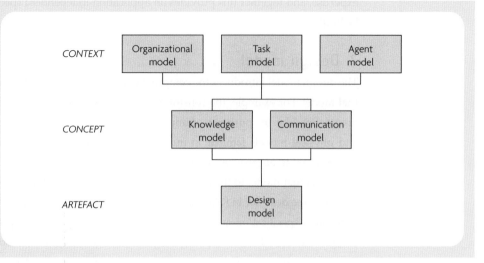

Figure 24.4: CommonKADS

1 Context

- *Organization model*. This concerns a review of potential knowledge-based solutions and comes from organizational analysis, which discovers problems, opportunities, and potential solutions for knowledge systems, establishes their feasibility, and assesses the organizational impacts.
- *Task model*. A task forms part of a business process. The task model analyses the global task layout, its inputs and outputs, needed resources (including agents) and competences, goals, preconditions, and performance criteria.
- *Agent model*. Agents carry out a task, be they human, information systems, knowledge, or culture/power related, or a combination of these. The agent model describes the characteristics of agents – competences, authority to act, and constraints.

2 Concept

- *Knowledge model*. The knowledge model details the types, roles, and structures of the knowledge used in a task, and the reasoning requirements of the prospective system.
- *Communication model*. This model defines the various needs and desires of the agents and the dialogue exchanges with other agents involved. The key aspect concerns the knowledge transfer between agents.

3 Artefact

- *Design model*. The above models together constitute the requirements specification for the knowledge system, and these are translated into a technical system solution. Therefore, a specification for the information system, including the software components (software architecture, algorithm design, data structure design, and hardware), is made here based on the knowledge and communication models.

Knowledge is modelled using a variant of the UML class diagram (Section 13.2). Modifications are necessary, as UML is not designed to model inferences and tasks, a requirement in CommonKADS. Modelling in this approach has also been influenced by Yourdon, from the structured analysis school, as well as object modelling. The authors argue that they 'take a middle line between data and process modeling'.

There are three stages in the process of knowledge-model construction: knowledge identification, knowledge specification, and knowledge refinement:

- *Knowledge identification.* At this stage, information sources that are useful for knowledge modelling are identified. Existing model components such as task templates and domain schemas are identified, and reusable components are also identified. Knowledge identification may start with a description of knowledge items in the organization model and the characterization of the application task in the task.

- *Knowledge specification.* At this stage the specification of the knowledge model is carried out. A task template is chosen and an initial domain model is constructed to which further detail is added. The inference knowledge, domain knowledge, and task knowledge are identified.

- *Knowledge refinement.* In the final stage, the knowledge model is validated and the knowledge bases completed. A technique used at this stage is the simulation of the scenarios gathered during knowledge identification. This should ensure that the knowledge model can generate the problem-solving behaviour required.

As for the CommonKADS approach as a whole, Schreiber et al. (2000) provide a detailed explanation of this aspect of the methodology. For example, for each of these stages, it lists a number of activities to be carried out at the stage and guidelines to help carry out these activities. Support is also provided on how to ascertain the knowledge of the experts in knowledge elicitation. These techniques include interviewing, protocol analysis (analysing how the expert solves problems), and repertory grids.

We have stated that the key aspect of the communication concerns the knowledge transfer between agents. This is illustrated in the dialogue diagram shown as Figure 24.5, which shows the overall information flow.

The design modelling stage includes design of the system architecture, choice of implementation (hardware and software) platform, specification of the architectural components, and the applications. In effect, it converts the requirements into the design of an information system. Quality criteria for the design are seen as code reuse, maintainability, adaptability, explanation, and support for knowledge elicitation. The last two are stressed in knowledge systems (and expert systems – see KADS in the previous section): the ability of the system to explain the rationale behind a reasoning process and support for obtaining knowledge from the knowledge workers. The approach authors see object orientation as the prevailing modelling approach and particularly suitable to these applications. Prototyping is suggested to form the basis for the reasoning system as well as develop the user interface aspects.

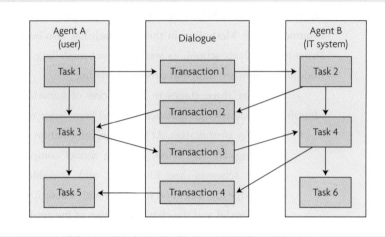

Figure 24.5: Knowledge transfer in CommonKADS

These are perhaps early days in the establishment of an accepted methodology for the development of knowledge management systems. However, CommonKADS, being based on KADS and the related knowledge-based expert systems applications, offers potential in supporting knowledge management systems development. The influence of its expert systems/knowledge engineering inheritance is, however, marked.

An alternative approach, perhaps more managerial in philosophy which puts knowledge management more in its organizational and environmental perspective, is given by Tiwana (2000).

Summary

- In this chapter we have described three methodologies that we have classified as people-oriented. ETHICS uses the tenets of participation and the socio-technical approach for information systems development. KADS and CommonKADS attempt to incorporate the expertise of the human expert in the system and manage knowledge for the organization.

Questions

1. In what ways does ETHICS correspond to the tenets of the participative approach (Section 5.1)?
2. Why is the name of ETHICS significant?
3. What is meant by a socio-technical approach, and why does it not feature in most methodologies other than ETHICS?
4. In what ways are KADS and CommonKADS people-oriented methodologies? How might they have been otherwise classified?

Further reading

CommonKADS (2005) www.commonkads.uva.nl

Mumford, E. (1995) *Effective Requirements Analysis and Systems Design: The ETHICS Method*, Macmillan, Basingstoke, UK.

Schreiber, G., Akkermans, H., Anjewierden, A., de Hoog, R., Shadbolt, N., Van de Velde, W., and Wielinga, B.J. (2000) *Knowledge Engineering and Management: The CommonKADS Methodology*, MIT Press, Cambridge, Massachusetts.

Wielinga, B.J., Sterner, Th. A., and Breuker, J.A. (1993) KADS: A modelling approach to knowledge engineering, in B.G. Buchanan and D.C. Wilkins (eds) *Readings in Knowledge Acquisition and Learning, Automating the Construction and Improvement of Expert Systems*, Morgan Kaufmann, San Mateo, California.

25 Organizational-oriented methodologies

25.1 Soft Systems Methodology (SSM)

As discussed in Section 4.1, general systems theory attempts to understand the nature of systems. Scientific analysis breaks up a complex situation into its constituent parts for analysis. Although this works in the physical sciences, it is less successful in the social sciences and in management science. One tenet of systems thinking is that the whole is greater than the sum of the parts: properties of the whole are not explicable entirely in terms of the properties of its constituent elements. Human activity systems are complex and the human components, in particular, may react differently when examined singly than when they play a role in the whole system. Something is lost when the whole is broken up in the 'reductionist' approach of scientific analysis.

The systems principle also implies that we must try to develop application systems for the organization as a whole rather than for functions in isolation. It might have taken only a few hours by Concorde to cross the Atlantic, but this progress was partly lost if it took as many hours to get from home to Heathrow Airport and from JFK Airport to the hotel in New York along with the requirement to be at the airport two hours before the flight. It is the transport system we should be looking at, not the airline system in isolation. Further, organizations are **'open systems'** and therefore the relationship between the organization and its environment is important. We should always be looking at 'the system' in terms of the wider system of which it is part.

Systems theory would also suggest that a multidisciplinary team of analysts is much more likely to understand the organization and suggest better solutions to problems. After all, specialisms are a result of artificial and arbitrary divisions. In the information systems context, a systems approach should prevent an automatic assumption that computer solutions are always appropriate. It will also help in problem situations which have been studied from only one narrow point of view. Such an approach is not appropriate in the study of large and complex problem situations.

Checkland (1981) has attempted to adapt systems theory into a practical methodology. By methodology he means the study of methods to achieve certain purposes. 'For any particular problem situation, the study will lead to a subset of principles which can be applied for that particular situation.' He argues that systems analysts apply their craft to problems which are not well defined. These 'fuzzy', ill-structured or soft problem situations, usually also complex, are common in organizations. The description of one category of systems, **human activity systems**, also acknowledges the importance of people in organizations. It is relatively easy, it is argued, to model data and processes (the emphasis placed in many of the preceding methodologies discussed), but to understand the real world it is essential to include people in the model, people who may have different and conflicting objectives, perceptions, and attitudes. This is difficult because of the unpredictable nature of human activity systems. There is no such thing as a repeatable experiment in this context. The claims for the soft systems approach are that a true understanding of complex problem situations is more likely using this approach than if the more simplistic structured or data-oriented approaches are used, which address mainly the formal or 'hard' aspects of systems.

Wilson (1990) gives an analogy. He considers two examples of problems. The first is a flat tyre. The solution is clear. The second is 'What should the UK government do about Northern Ireland?' (a more recent example might be 'What should the United Nations do about the Middle East?'). The solutions to these problems are not clear, and it would be difficult to find any solutions that satisfied all the interested parties. Wilson suggests that hard methodologies, that may be suitable for solving 'burst tyre-type problems', are inappropriate for organizational problem situations. It is not only a question of techniques and tools, but also culture, concepts and language.

Another difference between hard and soft systems thinking is that in hard systems thinking a goal is assumed. The purpose of the methods used by the analyst (or engineer) is to modify the system in some way so that this goal is achieved in the most efficient manner. Hard systems thinking is concerned with the 'how' of the problem. In soft systems thinking, the objectives of the system are assumed to be more complex than a simple goal that can be achieved and measured. Systems are argued to have purposes or missions rather than goals. Understanding is achieved in soft systems methods through debate with the actors in the system. Emphasis is placed on the 'what' as well as the 'how' of the system. The term 'problem' in this context is also inappropriate. There will be lots of problems, hence the term '**problem situation**' – 'a situation in which there are perceived to be problems' (Wilson, 1990).

The methodology of Checkland has been developed at Lancaster University. It was developed through '**action research**' whereby the systems ideas are tested out in client organizations. The analysts neither dictate the way the action develops nor step outside as observers: they are participants in the action and results are unpredictable. The practical work provides experience which can be used to draw conclusions and to modify these ideas. This proves to be a successful approach, because, as we have said, it is not possible to develop a good 'laboratory model' of human activity systems and set up repeatable experiments.

Each action research project therefore furthers knowledge which can be used in future soft systems work, and provides some benefits in a particular problem situation. Change is therefore achieved through the learning process as theory and practice meet and affect each other. Checkland's book is aptly titled *Systems Thinking, Systems Practice*!

Checkland has carried out extensive studies, both theoretical and practical, on the analysis of organizations in his action research projects. The central focus of the methodology is the search for a particular view (or views). This *Weltanschauung* (assumptions or world view) will form the basis for describing the systems requirements and will be carried forward to further stages in the methodology. The world view is extracted from the problem situation through debate on the main purpose of the organization concerned – its *raison d'être,* its attitudes, its 'personality', and so on. Examples of world view might be: 'This is a business aimed at maximizing long-term profits' or 'This is a hospital dedicated to maintaining the highest standards of patient care'.

The original version of soft systems methodology is given in Checkland (1981). We will outline this version (Mode 1) below. This will be followed by a brief outline of the changes made to the approach (Mode 2) as found in Checkland and Scholes (1990).

Figure 25.1 shows the seven stages of SSM. Stages above the dotted line are 'real-world' activities involving people in the problem situation, whereas stages below the dotted line are activities concerned with thinking about the problem situation. Checkland argues that this is a logical sequence for description purposes, but that it is possible to start a project at a place other than at Point 1. Further, the analyst is likely to be working on a number of stages simultaneously, and backtracking and iteration are essential. Therefore, this should be taken as more of a framework than as a detailed set of prescriptions which must be followed.

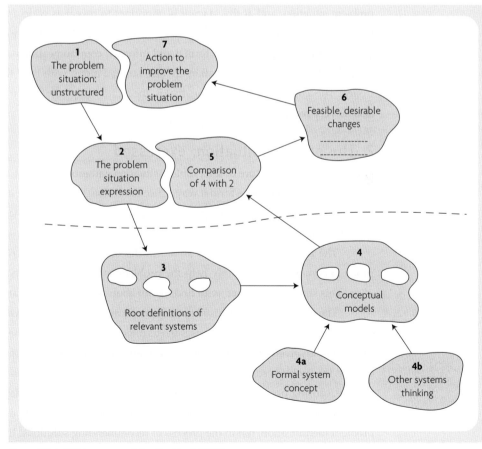

Figure 25.1: SSM in summary (after Checkland, 1981)

Stages 1 and 2 are about finding out about the problem situation. This unstructured view gives some basic information from the views of the individuals involved. The application of the CATWOE criteria (Section 10.2) gives some structure to the expressions of the problem situations, and in Stage 3 the analyst selects from these those views which he or she considers give insight to the problem situation. Stage 4 is to do with model building, that is, what the systems analyst might *do* (as against what the system *is* – the root definition). There must be one conceptual model for each root definition. Stage 5 compares the conceptual models from Stage 4 and the root definitions formed at Stage 2. This comparison process leads to a set of recommendations regarding change, and Stage 6 analyses these recommendations in terms of what is feasible and desirable. Stage 7, the final phase, suggests actions to improve the problem situation, following the recommendations of Stage 5.

1 The problem situation: unstructured

The first two stages are concerned with finding out about the problem situation from as many people in that situation as possible. There will be many different views as it is unlikely that the views of the problem owner, that is, the person or group on whose behalf the analysis has been commissioned, the other people taking part as 'actors' in the problem situation and other stakeholders, will coincide. There will be different views that the analyst can take regarding the problem situation, and at this stage it is important to reveal as wide a range of them as possible. The analyst will also look at the structure of the problem situation in terms of physical layout, reporting structure, and formal and informal communication patterns. These activities carried out in the problem situation are also studied along with how these activities are controlled. The 'climate' of the situation, that is, the relationship between structure and process can also be very revealing.

2 The problem situation: expressed

Given the informal picture of the problem situation gathered in Stage 1, it is feasible to attempt to express it in some more formal way. Checkland does not prescribe a method of doing this, but he and many users of the approach tend to draw rich picture diagrams of the situation (Section 10.1).

The rich picture is used as an aid in discussion between the problem solver and the problem owner, or may simply help the problem solver better understand the problem situation. This stage is concerned with finding out about the problem situation.

The rich picture can be used as a communication technique between the analysts and the users of the system, and therefore uses the terminology of the environment. It will usually show the people involved, problem areas, controlling bodies, and sources of conflict. From the rich picture, the problem solver extracts problem themes – things noticed in the picture that are, or may be, causing problems. The picture may show conflicts between departments, absences of communication lines, shortages of supply, and so on. Rich pictures are intended to help in problem identification, not in the process of recommending solutions. In general, SSM concentrates on understanding problem situations, rather than developing solutions.

Rich pictures prove to be a very useful way of getting the user to talk about the problem situation. They may stimulate the drawing out of some of the parts of the 'iceberg' which normally lie hidden when using traditional 'methods of investigation. Figure 25.2, an example rich picture chart (from Avison and Catchpole, 1987), highlights areas of conflict and concerns in the problem situation, a branch of a community health service in the UK.

3 Root definitions of relevant systems

Taking these problem themes, the next stage of the methodology involves the problem solver imagining and naming relevant systems. By relevant is meant a way of looking at the problem which provides a useful insight, for example:

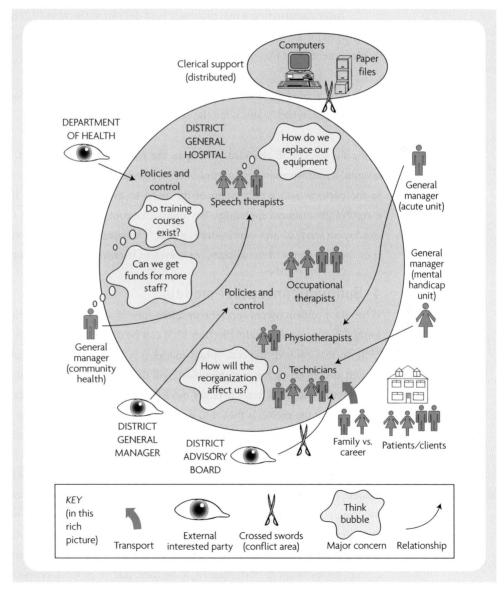

Figure 25.2: Rich picture diagram (early draft) for part of the paramedical services of health district

> Problem = Conflicts between two departments
>
> Relevant systems = System for redefining departmental boundaries

Several different relevant systems should be explored to see which is the most useful, but, in any case, this may be changed afterwards. It is at this stage that debate is most important. The problem solver and the problem owner decide which view to focus on, that is, how to describe their relevant system. For example, will the conflict resolution in our health example be 'a system to impose rigid rules of behaviour and decision making in order to integrate decisions and minimize conflict' or will it be 'a system to integrate decisions of actors through increased communication and understanding between departments' or even 'an arbitration system to minimize conflict between departments by focusing disagreements through a central body'?

After constructing a rich picture, a root definition (Section 10.2) is developed for the relevant system. A root definition is a kind of hypothesis about the relevant system and improvements to it that might help the problem situation:

> The root definition is a concise, tightly constructed description of a human activity system which states what the system is (Checkland, 1981).

Using the CATWOE checklist technique, the root definition is created. A root definition for a hospital administration system could be: 'to provide a service which gives the best possible care to the patients and which balances the need to avoid long waiting lists with that to avoid excessive government spending'. But alternative root definitions could have been 'a system for employing medical and administrative staff ', 'a system to generate long waiting lists to illustrate the high status of consultants', or 'a system to encourage the use of private health facilities'.

4 Building conceptual models

When the problem owners and the problem solvers are satisfied that the root definition is well formed, a conceptual model (Section 10.3) can be developed using this root definition (Figure 25.3). In this context, the conceptual model is a diagram of the activities showing what the system, described by the root definition, will do. (The term 'conceptual model' is used in some other methodologies to refer to entity modelling.) This stage in the methodology is about model building, but the model is meant to describe something relevant to the problem situation; it is not meant to be a model of the situation itself (otherwise it stifles radical thought about the problem situation).

This interpretation is different to that implied in a mathematical model, or an architect's model. The conceptual model, such as that shown in Figure 25.3, can again be used as a debating point so that the actors can relate the model to the real-world situation. Analysts need also to check the conceptual model against a general model of any human activity system (the 'formal system' model, Stage 4a in Figure 25.1). Checkland includes among these prerequisites a purpose, measure of performance, a decision-taking process, connectivity, and an environment. Checkland also suggests evaluating the conceptual model through looking at the work of

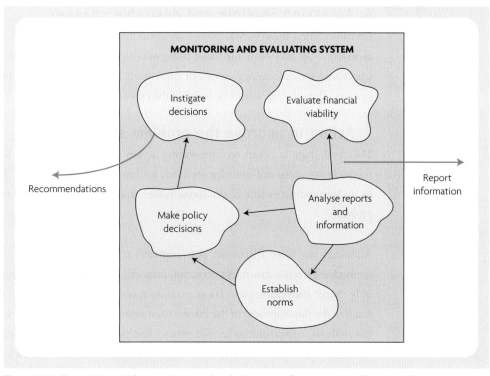

Figure 25.3: Conceptual model for a monitoring and evaluation system for community health (early draft)

other systems thinkers, such as Beer, Churchman, and Vickers, and thereby improve the model (Figure 25.1, Stage 4b).

Usually there is a conceptual model drawn for each root definition, and the drawing up of several root definitions and conceptual models becomes an iterative process of debate and modification toward an agreed root definition and conceptual model. An agreed statement is never easy to achieve as the process described is meant to draw out the different ideologies and conflicts, so that the final version represents an understanding of the problem situation. Similar processes were described in the ETHICS methodology. There is a danger that these final versions represent a conservative compromise. This is not what is intended. In SSM, if the ideological conflict is central to the problem situation, it has to be represented. The approach is meant to represent a holistic view to consider a complex problem, not a simplistic view representing a 'political' compromise.

5 Comparing conceptual models with reality

The next stage concerns the comparison of the problem situation analysed in Stage 2 through rich pictures alongside the conceptual models created in Stage 4. It is also about a comparison of views, and since these views are those of human activity systems, made by people, we may not be comparing similar things. The debate about possible changes should lead to a set of recommendations regarding change in order to help the problem situation.

6 Assessing feasible and desirable changes

On assimilating these views, Stage 6 concerns an analysis of proposed changes from Stage 5 so as to draw up proposals for those changes which are considered both feasible and desirable. Checkland's methodology does not limit itself to changing or developing new information systems, though this would be valid in the context of this book.

7 Action to improve the problem situation

This final stage is about recommending action to help the problem situation. Note that the methodology does not describe methods for implementing solutions. The methodology helps in understanding problem situations rather than provide a scheme for solving a particular problem.

Although we have discussed rich pictures, root definitions, and conceptual models, the methodology relies much less on techniques and tools than most other methodologies, particularly 'hard' methodologies. These provide tools for use in particular situations at particular times in the development of the information systems project. SSM provides all actors, including the analysts, opportunities to understand and to deal with the problem situation, that is, the human activity system. The analysts are perceived as actors involved in the problem situation, as much as those of the client and problem owner – they are not perceived as outside onlookers providing objectivity.

The process is iterative and the analysts learn about the system and are not expected to follow a laid-down set of procedures. This does present problems: it is difficult to teach and to train others, and it is difficult to know when a stage in the project has been satisfactorily completed. However, these features are also its strengths, because it does not have any preconceived notions of a 'solution' and use of the methodology gives a better understanding of the problem situation.

One possible way that SSM can be fitted into the information systems development process is by using it as a 'front end' before proceeding to the 'hard' aspects of systems development. This would seem to be appropriate because SSM concerns analysis whereas the harder methodologies tend to emphasize design, development, and implementation. The Multiview methodology (Section 26.1) also draws on SSM in the early parts of the systems definition process.

SSM Mode 1, described in Checkland (1981), is still the version most commonly referred to and possibly the most useful in an information systems context. However, there is an alternative version proposed in Checkland and Scholes (1990) based on lessons learnt from action research practice. This sees the 7-stage methodology given above as just one option in a more general approach.

In Mode 2, the problem situation is seen as a product of history and can be looked at in many different ways. People using SSM will follow two strands of enquiry, which together should lead to changes being implemented which improve the problem situation. These two strands are described as a 'logic-driven stream of enquiry' and a 'cultural stream'.

The logic-driven stream considers models of human activity systems, and a comparison of these models is made with an examination of our views of the real world and the ensuing debate concerns change. The cultural stream examines three aspects of the problem situation: the intervention itself, the situation as a social system, and the situation as a political system. The two streams are seen to interact and the process is viewed as a continual one.

A useful diagrammatic representation of Mode 2 SSM is shown as Figure 25.4. The four activities are:

1. Finding out about a problem situation, including culturally/politically.
2. Formulating some relevant purposeful activity models.
3. Debating the situation, using the models, seeking from that debate both (a) changes which would improve the situation and are regarded as both desirable and (culturally) feasible, and (b) the accommodations between conflicting interests which will enable action-to-improve to be taken.
4. Taking action in the situation to bring about improvement.

As Checkland points out, (a) and (b) are intimately connected and will gradually create each other.

Mode 2 can be considered as more of a framework of ideas for exploration rather than a methodology, although Mode 1 was never proposed to be used prescriptively nor without frequent iterative steps. It is perhaps this which is a key to interpreting SSM: Mode 2 is really

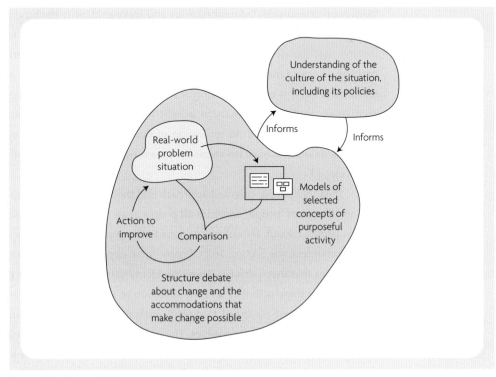

Figure 25.4: Mode 2 SSM

about suggesting that Mode 1 should not be used as a step-by-step prescription and that prac-titioners will tend to use some sort of compromise between Mode 1 and Mode 2. Essentially this represents a loose interpretation of Mode 1. SSM, more than any other methodology discussed so far, is very dependent on the particular interpretation followed by those who use the approach. This concept of framework rather than methodology and the role of the interpreter is taken further in an information systems context in Multiview. A 30-year retrospective of soft systems methodology is found in Checkland (1999). An alternative interpretation of soft systems methodology is found in Wilson (2001).

25.2 Information systems work and analysis of change (ISAC)

The ISAC methodology was developed by a research group at the Swedish Royal Institute of Technology and at the University of Stockholm. The methodology has been developed by use and experience in a number of commercial organizations and Swedish government agencies. Most users of the methodology are Scandinavian, although a number of users are claimed in other parts of Europe and North America. The methodology is closely associated with Mats Lundeberg (described in Lundeberg et al., 1982). It is interesting in that it was an early approach used which emphasized organizational issues rather than technological ones, but unlike SSM (Section 25.1), covers most phases of the life cycle.

Although the methodology covers information systems development as a whole, some users only apply the analysis and design parts of it, which are probably its best known aspects. ISAC is a problem-oriented methodology and seeks to identify the fundamental causes of users' problems. The methodology begins at an earlier stage than most methodologies and does not assume that the development of an information system is necessarily the solution to the problem. If a need for an information system is not identified, then the role of the methodology terminates. Need is established only if it is seen that an information system benefits people in their work, so that pure financial benefit to the organization, or some other benefit, is not thought to be enough of an indication of need for an information system. An information system is thought to have no value in its own right and without benefiting people should not be developed.

ISAC is a people-oriented approach and they are seen as the important factors in organ-izations. The term 'people' includes all people in an organization: users, managers, workers, as well as people usually thought of as outside the organization, such as customers and funders (i.e. stakeholders, see Section 16.1). People in an organization may have problems concerning the activities that they perform. These problems may be overcome, or the situation improved, by analysis of these activities and the initiation of various changes. The ISAC authors believe that the people best equipped to do this analysis, in terms of their knowledge, interest, and motivation, are the users themselves. The methodology attempts to facilitate this by providing a series of work or method steps and a series of rules and techniques which, it is claimed, can be performed by these users. An important part of this process is the education of the users to understand the organization better and to improve communication between people in the organization.

If the need for an information system is established, then the methodology emphasizes the development of a number of specific information subsystems rather than one 'total' system. The subsystems are local systems tailored to groups of individual needs, and these subsystems may well overlap in content and function. However, it is argued that the benefit that accrues is the specific relationship to the local needs. The assumption is therefore that solutions to sub-problems will give solutions to the organization's problems as a whole (which perhaps conflicts with the holistic, systems view described in Section 4.1).

In ISAC terms, an information system is an organized co-operation between human beings in order to process and convey information to each other; it does not necessarily involve any form of computerization.

The major phases of ISAC are:

1. Change analysis.
2. Activity studies.
3. Information analysis.
4. Data system design.
5. Equipment adaptation.

The first three phases are classified as problem-oriented work and focus on users and their problems, whereas the latter two phases focus on information systems-oriented work. Within each phase a number of work steps are identified and within these work steps various techniques are employed.

1 Change analysis

Change analysis seeks to specify the changes that need to be made in order to overcome the identified problems. Change analysis begins with the analysis of problems, the current situation, and needs. The following method steps are used:

- *Problem listing.* This is a relatively unstructured, first-attempt look at current problems and any anticipated future problems.
- *Analysis of interest groups.* ISAC acknowledges that problems are not necessarily objective. They are relative to the viewpoint of the participants in the organization. At this stage, the different interest groups are identified. These interest groups may be end-users, public users, departmental managers, and so on.
- *Problem grouping.* Here, the identified problems are grouped into related sets or categories.
- *Description of current activities.* The activities that the identified problems relate to, plus the activities undertaken by the concerned interest groups, are now modelled. This activity model is a functional view and shows processes performed on inputs to produce outputs. These aspects are not just concerned with information, but include physical activities, inputs, and outputs, such as the loading of a lorry. The activity model is documented in the form of an A-graph (an activity-graph).

An A-graph depicts three things:

1. Sets which can be real or physical sets, concerning, for example, people or services or stock, or they can be message sets, containing only information or they can be a combination of both.
2. Activities.
3. Flows, which can be shown in detail or in overview.

Activities are transformations of sets into new sets. Flows represent the movement of sets to and from activities. They are very similar to data flow diagrams (Section 12.1), except that they

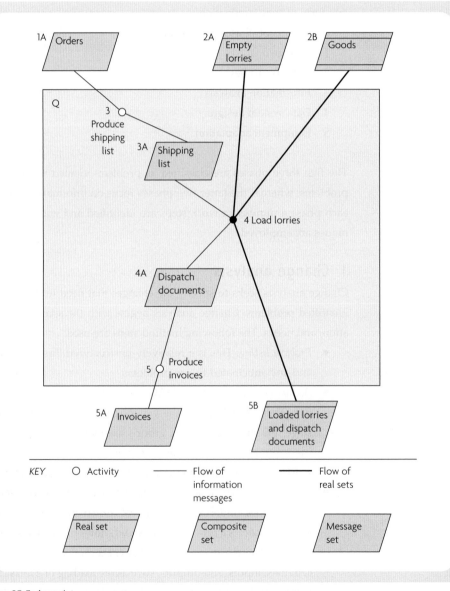

Figure 25.5: A-graph

also represent physical objects as well as data flows. Figure 25.5 shows an example of an overview A-graph concerned with the dispatch of goods. The message set 'orders' flows into the activity 'produce shipping list' which results in a message set 'shipping list'. This itself flows to an activity called 'load lorries' which has input flows of real sets 'empty lorries' and 'goods'. A-graphs exhibit a hierarchical structure capable of showing an overview picture which can then be broken down to show the detail at lower levels. The A-graph is supplemented by narrative or descriptive text and property tables. The property tables show quantitative information such as volumes:

- *Description of objectives.* The previously identified interest groups are perceived to have a variety of different general objectives and desires. Here, firmer and more specific objectives are defined and these are unified into a single set of objectives, via a process of negotiation and compromise. An attempt is made to reach a situation where the achievement of a set of agreed objectives solves the problems that have been identified.
- *Evaluation of current situation and analysis of needs for changes.* This is where all the previous work comes together and enables the methodology to progress. What is wanted (the objectives) is compared to what is available. What is available is described by the activity model, but this is tempered by the problems that have been identified. The differences between what is wanted and what is available are defined as the needs for changes. These needs are then evaluated and prioritized according to the values of the different interest groups involved. This evaluation of the importance of the various needs for changes leads directly into the next stage which is the generation and study of different change alternatives.

ISAC gives no guidance on how to generate ideas for changes, since this requires creativity in the context of the situation rather than the use of techniques, except to say that an analysis of flows and activities might be helpful. ISAC does not presuppose that the solution to the problems lies in automation or indeed in the construction of any form of information system. The type of solution is not constrained and may involve purely organizational and physical changes that do not result in the generation or modification of information systems. Once possible changes have been generated they are described through a new A-graph. The changes and the models are then analysed and evaluated from human, social, and economic feasibility viewpoints.

The final part of change analysis is to choose the most appropriate change approach based on the previous evaluations, and to document the reasons for the choices made. If the recommended changes do not involve information systems then the role of ISAC ceases. More likely, however, is that the recommendations involve a combination of types of change, and a plan is made concerning the necessary development measures for each type. An analysis of the effect and consequences of these parallel development measures is also made.

2 Activity studies

The starting point for activity studies is a proposal for a new system modelled and described in a number of ways, in particular, as an A-graph. The activity models that were produced in

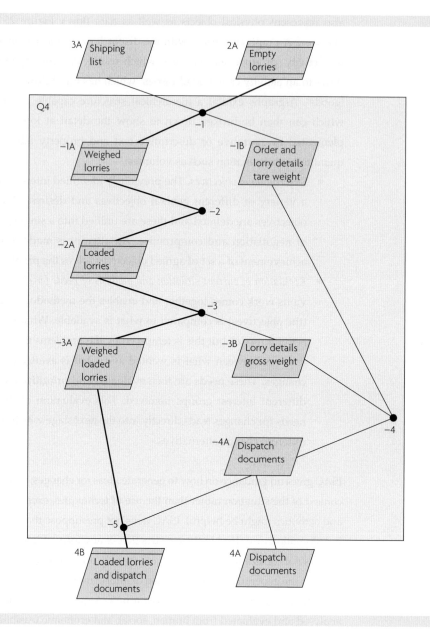

Figure 25.6: More detailed A-graph

change analysis for the purpose of identifying needs for changes were at a relatively high overview level, and these need to be expanded and investigated at a more detailed level. This is shown as Figure 25.6, which is a decomposition of Process 4 on Figure 25.5. The object is to get to the level at which the information system is separated from the human activities which it supports, such that each process on the graph has inputs and outputs that are unequivocally either information or some other flow (for example, materials).

The next step is to identify potential information subsystems relating to groups of users. The subsystems are not supposed to be identified in relation to artificial criteria, such as some common technical aspect, but only in relation to commonality in use. ISAC does not identify any overall information system totality. It is more relevant to identify a number of subsystems. These information subsystems are then classified according to whether they are formalizable or not. An unformalizable information system might be one concerning qualified decisions, informal contacts, know-how, and so on. The formalizable subsystems are divided into those that seem sensible to automate, in terms of cost, social desirability, and so on, and those that are not. The automated ones are further classified according to whether they involve calculations or simply involve storing and retrieving information. These classifications are the basis for subsequent steps of the methodology.

Each information subsystem is now studied separately in terms of its costs and benefits. ISAC attempts to do this cost/benefit analysis without making assumptions about technical implementation. To do this, ISAC refers to ambition levels for an information system, rather than particular technologies for implementation. For example, two different ambition levels for the same subsystem might be a one-second response time compared with a three-hour one. Each ambition level will have a different cost/benefit associated with it.

The steps in this phase are as follows:

- *Analysis of contributions.* This is a study of the benefits (not quantified) expected to accrue from the change. It is a refinement of the earlier work done in change analysis and the results are documented in a property table. It is emphasized that this analysis must be performed in the context of the environment and the way in which the environment uses the information. This may require a more detailed analysis of the environment than has been done up to this point.
- *Generation of alternative levels of ambition.* A number of alternatives are generated for each subsystem and documented. The alternatives must be realistic. There is no point in generating ambition levels that do not fulfil minimum requirements in terms of, for example, frequencies or volumes or cost.
- *Test of ambition levels.* Here the ambition levels are tested to see if they are practical. ISAC envisages a number of ways that they might be tested; for example, if similar information systems exist elsewhere, then it is likely that such a system can be created. Prototyping is also suggested, but not a prototype of the technology, rather one of the provision of information to the user.
- *Cost/benefit analysis.* This is a conventional cost/benefit study of each identified level of ambition.
- *Choice of ambition level.* The results of the cost/benefit analyses are evaluated in conjunction with the human and social analyses performed at an earlier stage, and a choice of ambition level made. One result may be that the development of an information system is discontinued.
- *Co-ordination of information subsystems.* This concludes the activity studies and is, in effect, the project plan. Special emphasis is given to the interrelationships between the different subsystems in order that they are sensibly co-ordinated. Priorities, resources, and schedules are allocated for the developments, and the plans are documented.

3 Information analysis

The transition from change analysis to activity studies would not be made unless the agreed proposal for change included the development of an information system. Similarly, the transition from activity studies to information analysis is made only if one or more information systems have been identified as formalizable. The techniques used in information analysis assume a formalizable and automatable information system, although it is indicated that a limited degree of information analysis might be appropriate for non-automatable systems.

For each information system, the input and output information sets are extracted from the A-graphs for the system. At this time, an iterative process of function and data analysis is performed.

The ISAC term for functional analysis is precedent analysis, because ISAC recommends that it be performed by reasoning about the precedents for each information set. If the output information set from an information system is clearly derivable from its input set, then precedence analysis stops. If, however, the derivation is not clear, then the information set that immediately precedes the output information set must be deduced. If the derivation of this set from the inputs to the system is not clear, then precedence analysis continues. The precedents from each information set are analysed until the input sets are reached. At each stage of precedence analysis the information system (considered as a set of processes) has been refined to a lower level of detail. Precedence analysis is in this way similar to functional decomposition in other methodologies. The reasoning process, however, is different in that, instead of enquiring about the logical structure of a process, ISAC concentrates on the transformation that must have been necessary to produce the output information set currently being studied. If this transformation is not clear, then a simpler problem must be solved. The question is asked: 'What was the nature of the information set that immediately preceded the transformation of the current output set?' In this way the definition of processes is implicit. At any stage a process is always a black box (or using ISAC notation a black dot).

The result of precedence analysis is a set of information precedence graphs (I-graphs). I-graphs describe information sets and precedence relations between information sets. They are more precise than A-graphs in that they not only show input and output sets but also show relationships between sets. Figure 25.7 shows an example of an I-graph derived from the A-graph of Figure 25.6. It shows the input and output information sets and the precedence relations. For example, it shows that, in order to derive the 'accepted dispatch' information set, we first need the 'order details', 'customer details', and 'product details' information sets.

Reasoning about the transformations that need to be performed on information sets requires knowledge of the structure of information, and that is why component analysis is performed at the same time as precedence analysis.

In component analysis the structure of the information sets is studied. An information set is either a data flow or a data store. An information set will either have been specified as a basic input to, or output from, the information system being studied; or will have been from a preceding process or a set of permanent information. An information set may be compound, that is, it may itself contain information sets. An elementary information set consists of one or more

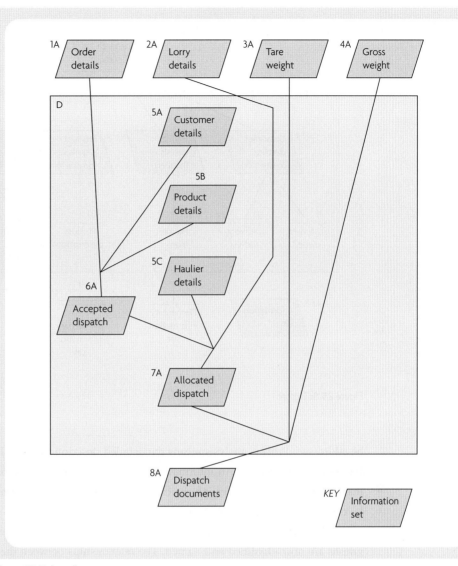

Figure 25.7: I-graph

messages, where each message consists of an identification term and one property term. An 'almost elementary' information set consists of a number of elementary information sets with common identification terms. Therefore, an almost elementary information set corresponds to a logical record with a key (identification term) and a number of data-item types (property terms).

ISAC documents information sets by means of a C-graph (component graph). This graph is a hierarchy showing the decomposition of an information set into subsets. Figure 25.8 is an example of a C-graph for the 'dispatch document' information set. The lowest level on the graph shows either elementary or almost elementary sets.

A further step in information analysis concerns process logic analysis. This has been ignored by ISAC during precedence analysis in order not to complicate things, but must now

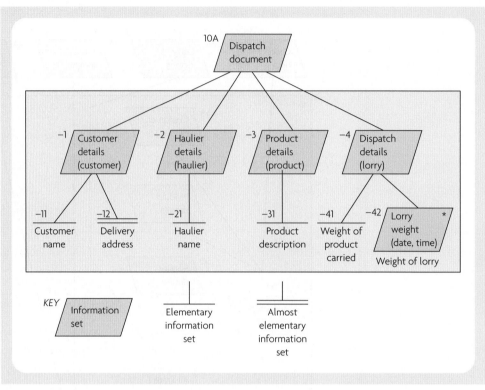

Figure 25.8: C-graph

be addressed. Process analysis means the detailed description of the information processes in the information system. An information process is the transformation of one or more information sets into some other information set or sets. These transformations are at the logical level and do not depict implementation aspects.

The processes which have previously appeared simply as nodes or black dots on the I-graphs are now identified and named. The relationships between the processes have already been described in the C-graphs, so that all that needs doing is for the content of the process to be specified individually. This is done using a process table that defines the prerequisites for the process, the conditions, permitted states for conditions, and required actions. A process table is effectively a decision table (Section 12.3) constrained by prerequisites.

The final stage of information analysis is property analysis. Precedence and component analyses are structural descriptions of an information system. Requirements that are specific to the environment in which the system is used must also be specified. Examples of such properties are volumes, response times, frequency, and security requirements.

4 Data system design

Transition to data system design implies a fundamental change in the application of the ISAC methodology. Up to this point the ISAC activities of change analysis, activity studies, and information analysis have concentrated on producing a specification of requirements for information systems. Data system design is the first part of data-oriented systems work, the purpose of which is to design a technical solution to meet the requirements specification. A data system is a means of implementing the requirements of one or more information systems. A data system will usually contain both manual and computerized parts, both of which must be designed. Data system design is followed in ISAC by an activity known as equipment adaptation. Data system design, therefore, assumes types of equipment but not specific equipment.

The design activity starts with a study of processing philosophy. Processes that are to be performed on a computer are differentiated from those that are not. A decision is made on the mix of batch and online processing and centralized vs distributed processing. The results of activity studies are particularly relevant to the determination of processing philosophy, and the identification of processes performed in information analysis enables preliminary 'process collections' to be made grouped by some common requirement (e.g. response time).

The next stage of data system design is data structure design and program delimitation and design. Information analysis has typically decomposed data and functions below the level of files and programs, and appropriate groupings are now made.

The design of a permanent data set is performed, first, by consolidating or aggregating more primitive objects (e.g. elementary information sets) into higher-level groupings, on the basis of functional dependency, and, second, by considering access requirements and search paths for efficient retrieval and storage.

Program delimitation consists of putting a boundary around a group of processes defined on an I-graph. The number of processes so grouped will clearly partially determine the size and complexity of the program, and these two factors are a constraint on the delimitation. The other important constraint is the nature of the decisions that have been made about file and database design.

Once programs have been delimited and files or a database has been designed, the overall structure of the system can be recorded in a D-graph (a form of program runchart). The next step is to specify each program, which is completed in some detail in the ISAC approach, and Jackson systems programming is recommended (see Section 20.3).

The final stage in data system design is the design of manual routines. Some information processing activities are naturally manual, and all computer-based systems will have associated manual routines. ISAC recommends that users design their own work routines.

5 Equipment adaptation

The preceding phase of data system design has produced an 'equipment-independent' solution. This is now adapted to fit particular equipment. Equipment adaptation consists of equipment study, adaptation of computer-based routines, and the creation of side-routines.

The equipment study involves collecting and evaluating technical specifications and cost and performance characteristics, and formulating an equipment strategy. Possible options might be, for example, the use of existing computer facilities, the purchase or rent of new equipment, or the use of bureau facilities. Accurate sizing of the configuration required is performed at this stage. The final choice of equipment is documented in an E-graph, which is a mapping of the D-graph onto a particular hardware configuration.

Adaptation of computer-based routines consists of two tasks. First, physical data structures must be designed, and, second, the program specifications must be adapted. Files and databases are mapped to specific storage devices, and retrieval and linkage mechanisms are specified. Outputs are mapped to specific output peripherals, such as workstations or printers. Input formats are mapped to specific methods of data capture. Any of these mappings may alter the data structures of computer programs, and thus the structure of the programs must be adapted. Finally, side-routines are specified. These are work-routines that are a necessary consequence of the choice of a particular set of equipment. For example, side-routines might be specified for data preparation or output handling.

The emphasis of the methodology is placed on analysis and design aspects of information systems development where appropriate. The methodology seeks to identify the fundamental causes of users' problems in the present system. These problems may be overcome or the situation improved by analysis of these activities and the initiation of various changes. The authors of this methodology believe that the people best equipped to do this analysis, in terms of their knowledge, interest, and motivation, are the users themselves. The methodology attempts to facilitate this by providing a series of work or method steps and a series of rules and techniques which, it is claimed, can be performed by these users. It is accepted that this might lead to a series of self-contained application systems which might be regarded as 'inefficient' from some points of view.

The methodology does not assume that the development of a computer information system is necessarily the solution to the problem. Need is established only if it is seen that an information system benefits people in their work. It has been traditional in Scandinavian countries, sometimes backed up by legislation, that technology is only implemented with the approval of the workers in that workplace. This people-oriented methodology has a wide view of the stakeholders of an information system, including users, managers, workers, and also those usually thought of as outside the organization, such as customers and funders. Development of ISAC has related to broadening the view toward business process analysis and BPR (Sections 4.3 and 25.3).

25.3 Process Innovation (PI)

Business process re-engineering was discussed as a theme in Section 4.3, and an approach to implementing classic BPR called Process Innovation (PI) has been devised by Tom Davenport (Davenport and Short, 1990; Davenport, 1993) which suggests five stages as follows:

- develop the business vision and process objectives;
- identify the processes to be redesigned;
- understand and measure the existing process;
- identify the IT levers;
- design and build a prototype of the new process.

1 Develop the business vision and process objectives

Davenport and Short argue that it is not enough to attempt to eliminate obvious bottlenecks and inefficiencies from processes; it is necessary to redesign the entire process according to a business vision. In this phase, the organizational strengths and weaknesses need to be identified, along with an analysis of the market and the opportunities it provides. A knowledge of the innovative activities of competitors will also be useful. But a business vision will only come as a result of the creative thinking of executives and others.

There needs to be an effective linkage between business strategy and business processes. Where strategy implies radical business change, this suggests radical changes to business processes and process innovation rather than the more usual incremental change. Examples given of such a vision include developing systems with a customer perspective, improving product quality, and taking best practice from the industry. Process innovation may lead to more complex processes. Davenport argues that process simplification or rationalization only leads to marginal change and therefore implies a lack of vision. A more radical vision, it is argued, will imply objectives which might include cost reduction, time reduction, increasing the quality of products, and empowering staff. Key activities in developing process visions include:

- assess existing strategy with respect to processes;
- consult with process customers for performance objectives;
- set up performance objectives and functionality targets.

2 Identify the processes to be re-engineered

At this stage the major processes are identified, along with their boundaries. The critical processes of the organization are considered for IT-enabled re-engineering. Processes which are of high impact, of great strategic relevance, or presently conflict with the business vision in some way are selected for consideration and a priority attached to them. It is unlikely that they can all be re-engineered in parallel. There may be somewhere between 10 and 20 processes identified for innovation. For businesses, these might include:

- customer contact;
- inventory management;
- product design;
- personnel support;
- product marketing;
- production;
- supplier management;
- customer feedback;
- human resource management;
- financial management.

The processes are classified according to beginning and end points, interfaces, owners, functions, users, and departments involved. It is important that the process owner, usually a senior manager, is motivated toward making the change. If there are difficulties in identifying these processes, senior manager workshops may help, as will interviewing senior managers. An alternative approach is to consider re-engineering all processes, but this may neither be feasible nor efficient. For example, there may be constraints preventing re-engineering of some processes due to the necessity of supporting some legacy systems. Some of these may need to be kept because of the degree to which they are embedded in the organization. It is rare that a 'clean slate' may be assumed.

3 Understand and measure the existing process

Processes cannot be redesigned before they are understood. The present processes need to be documented. This will help communications within the group studying the process. It will also help to understand the magnitude of the change and the associated tasks. Understanding existing problems should help to ensure that they are not repeated. It also provides measures which can be used as a base for future improvements. For example, measuring the time and cost consumed by process areas that are to be redesigned can suggest initial areas for redesign in a process. However, although designers should be informed by past process problems and errors, they should work as if in virgin territory, otherwise processes will be tampered with rather than redesigned. Key activities in this stage are:

- document the current process flow;
- measure the process in terms of new process objectives;
- measure the process in terms of new process attributes;
- identify problems and weaknesses of the existing process;
- identify short-term improvements in the process.

4 Identify the IT levers

The accepted view in information systems is that business requirements should be determined first before considering IT solutions. However, the benefits of simply automating existing processes are likely to be minimal. IT can be used to change processes completely. Davenport and Short argue that an awareness of IT capabilities can influence process redesign and should be considered in the early stages.

IT capabilities can enable better information access and co-ordination of processes. IT can make new process design options feasible, rather than simply support them. One distinguishing aspect of this approach, compared to most well-used information systems development methodologies, is that the latter concentrate on the development and implementation of information systems, whereas business innovation sees IT as the most powerful design tool providing opportunities for process re-engineering which is fundamental and broader.

The following list of IT capabilities, along with organizational impacts, is suggested:

- *Transactional.* IT can transform unstructured processes into routine transactions.
- *Geographical.* IT can transfer information rapidly and across long distances.
- *Automating.* IT can reduce the need for human intervention in processes.
- *Analytical.* IT can enable complex analytical methods to be incorporated into processes.
- *Integrating.* IT can support the co-ordination of tasks and processes.

- *Informational.* IT can bring in vast amounts of information to be included into a process.
- *Sequential.* IT can reorder the operation of tasks and allow them to be processed in parallel where appropriate.
- *Knowledge orientating.* IT can be used to capture and disseminate knowledge to improve the process.
- *Tracking.* IT enables monitoring the status, inputs, and outputs of tasks.
- *Simplifying.* IT can be used to simplify communication so that, for example, intermediary stages are not required.

5 Design and build a prototype of the new process

In this final stage, the process is designed and the prototype built through successive iterations. Design comes from a review of the information collected in the first four stages. It is suggested that the design team consist of key process stakeholders as well as those from the IT side who debate possible design alternatives.

Key activities at this stage are:
- 'brainstorm' design alternatives;
- assess feasibility, risk, and benefit of design alternatives and select the preferred process design;
- prototype the new process design;
- develop a strategy for changing to the new process;
- implement the new organizational structures and systems.

Davenport suggests process design at three levels: a process level, subprocess level, and activity level. At the process level, the inputs, outputs, interfaces, flow, and measures are specified. At the subprocess level, the objective, performance metrics, the performers, IT enablers, information needs, and activities in the process are defined. At the activity level, the information needed, decision points, the performers, and value-added are defined.

Process models may be rapidly generated and even automatically coded via the use of a toolset (Chapter 19). Such a design needs to satisfy general design criteria, such as satisfying the objectives set, simplicity, control mechanisms, and the generalization of tasks which can be executed by more than one person. It is suggested that prototypes are more likely to lead to systems which are accepted by the users as well as being produced faster than the conventional approach.

Davenport discusses the potential role of IT in process innovation. As well as in the design and build stage where the impact of IT is most obvious both in the design and the prototyping stages, all the phases can be supported by information systems and information technology. Executive information systems should provide managers with information about current business performance and market factors. Computer-supported conferencing may help in the brainstorming activities. When identifying the processes to be redesigned, information about the performance of present processes can be provided and simulation packages may help to identify alternative approaches which are potentially more successful.

Although the above concentrates on the role of IT in process innovation it is not to the exclusion of other factors such as the need for empowerment and participation in decision making and process planning, and the need for teamwork.

25.4 Projects in controlled environments (PRINCE)

PRINCE is a structured and standard approach for project management and like SSADM (Section 21.1) it was first developed for UK government applications although it is now used elsewhere and not just for IT projects. It was developed through the CCTA (the Central Computer and Telecommunications Agency), now part of the Office of Government Commerce (www.ogc.gov.uk) and the latest version is called PRINCE2.

A project is seen as having the following aspects:

- defined and unique set of products (we have tended to use the term deliverables in this book);
- set of activities and their sequence to construct the products;
- appropriate resources to undertake the activities;
- finite lifespan; and an
- organizational structure with defined responsibilities.

The PRINCE2 approach aims to deliver the end products at a specified quality within budget and on time. The emphasis is placed on the delivery of these products, not the activities to achieve their production.

All the stakeholders should be involved during the project as appropriate, but the management structure of a PRINCE project is expected to consist of:

- Project board, with a senior executive as member along with a senior user
- Project manager who fulfills the day-to-day management
- Team leaders who report to the project manager
- Management champion for the project, who agrees the business case and outlines the justification, commitment and rationale for the project.

The project is driven by the business case, and this is reviewed frequently. This suggests why the project is being done, the likely benefits and who is going to pay for it.

The formation of detailed plans is a cornerstone of the approach. The highest level of plan is the project plan, that is, an overall plan for the project, and this is broken down into stages. A detailed plan gives a further breakdown of activities within each stage. Although the project plan is important in showing the overall scope of the project, major deliverables, and resources required, it is the stage plans that are used for day-to-day control.

Bentley (2002) describes in detail the components of PRINCE2, including business case, organization, plans, controls, risk, quality, configuration management and change control, and these are as shown in Figure 25.9.

The technical aspects include the product breakdown structure as well as PERT charts (Section 14.2), which link the activities, showing their inter-dependencies, to create the end date. Resource plans identify the resource type, amount and cost of each resource at each stage. Gantt charts (Section 14.3) are used to help resource allocation and smoothing. However, the use of techniques and tools is seen as optional.

Meeting quality expectations, like times and costs, is seen as an important aspect of the approach. There are quality controls, which are defined in the technical and management procedures, and the descriptions of the deliverables in terms of fitness for purpose also represent a statement about quality. The acceptance criteria need to be explicitly stated.

Within a PRINCE project there are eight processes, as shown in Figure 25.10:

- Directing a project: As we saw above, there needs to be overall direction to authorize the project, approve the go ahead, monitor progress and closure.
- Starting up a project: This short stage includes the appointment of the project team, agreeing the aims of the project, deciding on the project approach, defining quality expectations and planning and drawing up the contract (whatever form that takes) between customer and supplier.
- Initiating a project: This includes agreeing that the project is justified, establishing management procedures, and creating the detailed plan and the project initiation document.
- Controlling each stage: This process describes the monitoring and control activities,

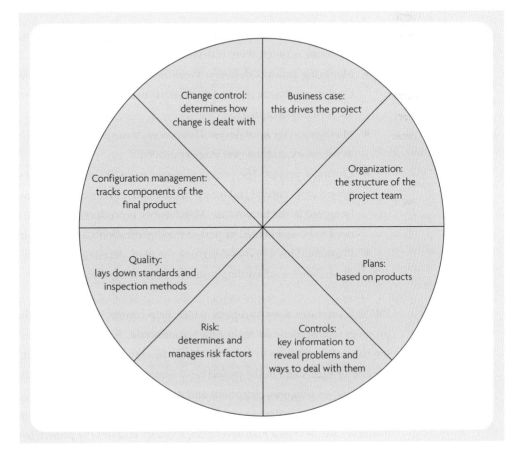

Figure 25.9: Components of PRINCE2 (modified from Bentley, 2002)

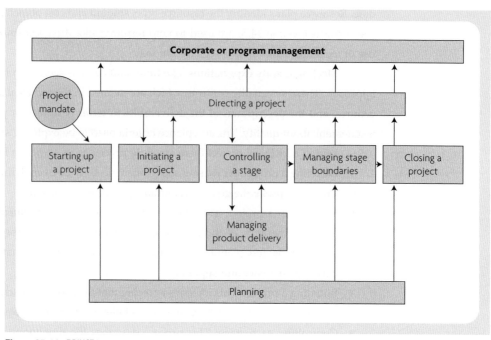

Figure 25.10: PRINCE2 processes

authorizing work to be done, gathering progress information, reporting and taking necessary action if there is deviance from the plans.

- Managing product delivery: Work is allocated to the team, planned and carried out. Quality criteria are checked and reports made to the project manager. Finished products are formally accepted.
- Managing stage boundaries: This process ensures that the plan is adhered to or updated as necessary and the next stage approved.
- Closing a project: The project manager gains the project board's approval to close the project at its natural end (or premature end if necessary). Customer satisfaction needs to be agreed in the former case. Maintenance procedures need to be in place. Lessons learnt need to be assimilated as part of the organization's knowledge management process.
- Planning: This involves designing the plan, defining the project's products, identifying dependencies, scheduling resources, allocating risks.

PRINCE produces a set of reports which help control, in particular the monitoring of actual progress against plan and against the business case. Some of these reports are expected to form the basis of discussions at meetings. These meetings are normally held at the initiation and at the end of each stage in the project (and sometimes mid-stage as well). The reports highlight such things as progress, exceptions and requests for change.

However, the first document produced is the project initiation report, which outlines the business case, defines a high-level plan, formally initiates the project, lists overall objectives for the project and defines personnel responsibilities. It may be more detailed, for example, identifying job descriptions for each person, providing a detailed cost/benefit analysis for the business case (similar to that in a feasibility report), examining the risks associated with a project, and dividing the project into well-defined stages.

End-stage assessment is a control point, which, if successful, signifies acceptance of the deliverables promised for that stage and provides authority to go on to the next stage. Mid-stage assessment may occur if the stage is of long duration, problems have been identified such as deviation from the plan, or there have been requests for change.

At the end of the project there is a project closure report, which lists the project's achievements in terms of deliverables achieved, performance in terms of comparisons of actual against forecast duration, cost and resource usage, and quality in terms of errors or exceptions. It also provides information to help organizational learning, for example, experience with the use of tools and development methodologies. The report also includes formal acceptance letters from senior technical staff, users and operations staff. There is also expected to be a post-implementation review.

Along with Bentley (2002), the website Prince2 (2005) proves a useful information resource on PRINCE2.

25.5 Renaissance

Renaissance is another European ESPRIT initiative involving a collaboration of company and university partners. The project addresses system evolution and maintenance, rather than new development, so it has particular links with the theme of legacy systems (Section 8.1). Its focus is generally on software engineering aspects rather than wider information systems development. As we will see, it also has links to object orientation (Section 6.4), toolsets, and other themes.

It provides an approach to ensure that legacy systems can be adapted to reflect changes in the environment. The main objectives of Renaissance are to propose 'a more methodological approach to evolution and re-engineering which is consistent with current development and maintenance practices used in industry' (www.comp.lancs.ac.uk/projects/renaissance/). Both this website and Warren (1999), which includes two interesting legacy system case studies, are good sources of material on this approach.

Renaissance first presents a framework for the evolution of legacy systems. The use of the term 'framework' suggests that it is expected to be adapted according to the particular organization. It has three viewpoints:

1. *Technical.* This includes an understanding of the technologies used, along with documentation and maintenance processes.
2. *Economic.* This includes an understanding of the business value of the system and cost and risk estimates for different evolution strategies.
3. *Managerial.* Decisions based on the previous two views.

The framework also identifies three role categories:

1. *Strategic.* This role is concerned with defining market strategies and identifying future

needs, aiming at cost reduction and quality improvement. There will be a focus on managerial and economic views.

2. *Operational.* This role is concerned with providing control over the evolution of legacy systems and negotiations with customers.

3. *Service.* This role is concerned with ensuring that the level of service required is achieved in the technical view.

A process model is defined which suggests various activities:

1. *Trade-off analysis.* This is a thorough investigation of open technical issues, technical market trends, and business goals. This view of the legacy system as a business asset along with an understanding of present-day technologies will prove a good basis for further work.

2. *Issue assessment.* This looks at the scope and direction of the project and in particular ensuring that the legacy system is a good basis for evolution from a technical and business point of view.

3. *Decision analysis.* This involves choosing the best evolutionary strategy from a number of possibilities from the point of view of costs, benefits, and risks, and developing a project plan to carry out this strategy. One strategy will be to do little apart from required maintenance, but many will involve much greater change.

4. *Solution implementation.* This involves the design of a solution that satisfies the requirements set out at the earlier phases, having understood the extent and effects of the required change. It also includes creating a programme for validating the evolved system.

5. *Solution deployment.* This concerns the validation and acceptance of the evolved system ready for running it operationally. There may be a period of parallel running before the organization is fully dependent on the evolved system.

6. *Kaizen improvement.* This phase is based on the Japanese approach to business improvement and suggests continuous evaluation, incremental change, and improvement through tuning. Therefore, the legacy system is continually being refined.

Renaissance suggests six possible evolution strategies:

1. *Continued maintenance.* This relates to the fixing of small problems that would be expected in any system maintenance.

2. *Revamp.* This might involve modernizing the user interface, such as changing from a command to a graphics user interface, but not involving change to the basic hardware and software. Software products are available that will enable the user interface to be changed without changing the application software otherwise.

3. *Restructure.* This will involve more fundamental change to the software, but not the hardware. It might concern changing old-fashioned 'spaghetti code' into more structured code resulting from good software engineering practice.

4. *Rearchitecture.* This involves change to both the hardware and software. Migration of the

application to the Internet may involve this more radical rearchitecture, though it might also be a much lesser revamp. The hardware platform may be changed from a mainframe environment to a client–server approach, and this would be classified as rearchitecture. Relational or object DBMS might replace conventional files (or hierarchical or network DBMS). This is perhaps a more risky and costly approach; for example, data migration may present a problem. However, the new design should be more flexible as well as more efficient.

5. *Redesign with reuse.* This involves more new software modules and is the most radical change, but it will reuse some parts of the present system.

6. *Replacement.* This involves replacing the legacy system with a new system.

The four strategies 2 to 5 are appropriate to the Renaissance approach, although legacy systems may undergo most of these during their lifetime (and sometimes a combination strategy at one time). Strategy 1 falls under standard maintenance of systems, while Strategy 6 requires developing a new system to replace the legacy system. Each approach requires planning, development, delivery, acceptance, and deployment, but in the spirit of evolutionary change this will be a continual process.

Summary

- In this chapter we have looked at five methodologies that we have classified as organizational oriented.

Questions

1. What do you think links these five methodologies? What are the differences between them? Can you suggest an alternative, better classification?

2. SSM has been very influential in terms of systems development ideas. Why do you think this is?

3. What makes ISAC different to other methodologies, for example, SSADM, described in Chapter 21?

4. How does PI relate to the 'classic' and 'new' forms of business process reengineering (BPR) described in Section 4.3?

5. Do you think that the use of Renaissance will 'solve' the problem of legacy systems?

Further reading

Bentley, C. (2002) *PRINCE2: A Practical Handbook,* Butterworth-Heinemann, Oxford.

Checkland, P. (1981) *Systems Theory Systems Practice,* John Wiley & Sons, Chichester, UK.

Davenport, T. (1993) *Process Innovation,* Harvard Business School Press, Cambridge, Massachusetts.

Lundeberg, M., Goldkuhl, G., and Nilsson, A. (1982) *Information Systems Development – A Systematic Approach,* Prentice Hall, Englewood Cliffs, New Jersey (for ISAC).

Warren, I. (1999) *The Renaissance of Legacy Systems: Method Support for Software-System Evaluation,* Springer-Verlag, London.

26 Frameworks

26.1 Multiview

Multiview 1

Multiview perceives information systems development as a hybrid process involving computer specialists who will build the system, and users for whom the system is being built. The methodology therefore looks at both the human and technical aspects of information systems development. In this aspect and others, it has been greatly influenced by the work of Checkland and Mumford, but has fused these ideas with those found in 'hard' methodologies, such as STRADIS and IE.

Multiview is a **contingent** methodology rather than highly prescriptive, because the skills of different analysts and the situations in which they are constrained to work always have to be taken into account in any project. Avison and Wood-Harper (1986) describe Multiview as an **exploration** in information systems development. It therefore sets out to be flexible: a particular technique or aspect of the methodology will work in certain situations but is not advised for others.

The methodology will be seen by readers of this text as truly 'multi-view', because it includes many of the techniques used by the other methodologies and its stages parallel those of other methodologies. The authors of Multiview claim, however, that it is not simply a hotch-potch of available techniques and tools, but a methodology which has been tested and works in practice. It is also 'multi-view' in the sense that it takes account of the fact that as an information systems project develops, it takes on different perspectives or views: organizational, technical, human-oriented, economic, and so on.

The five stages of Multiview are as follows:

- analysis of human activity;
- analysis of information (sometimes called information modelling);
- analysis and design of socio-technical aspects;
- design of the human–computer interface;
- design of technical aspects.

They incorporate five different views which are appropriate to the progressive development of an analysis and design project, covering all aspects required to answer the vital questions of users. These five views are necessary to form a system which is complete in both technical and human terms. The five stages move from the general to the specific, from the conceptual to hard fact, and from issue to task. Outputs of each stage either become inputs to following stages or are major outputs of the methodology. The Multiview methodology is shown in outline in Figure 26.1. The two analysis-oriented stages are shown in boxes and the three design-oriented stages in circles. The arrows represent information passing between stages, and the dotted arrows represent outputs of the methodology. The outputs of the methodology, shown as dotted arrows in Figure 26.1, are listed in Figure 26.2, together with the information that they provide and the questions that they answer.

The authors argue that to be complete in human as well as in technical terms, the methodology must provide help in answering the following questions:

1. How is the computer system supposed to further the aims of the organization installing it?
2. How can it be fitted into the working lives of the people in the organization that are going to use it?
3. How can the individuals concerned best relate to the machine in terms of operating it and using the output from it?

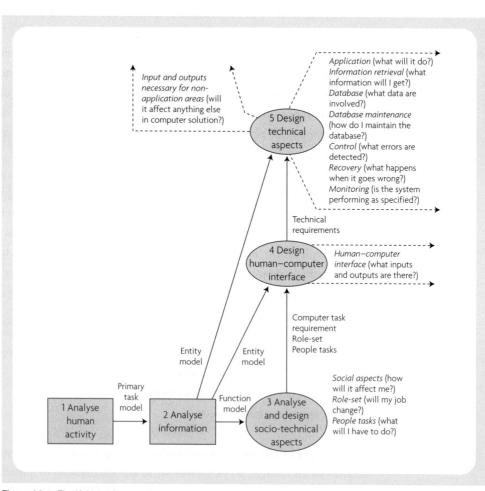

Figure 26.1: The Multiview framework

OUTPUTS	INFORMATION
Social aspects	How will it affect me?
Role-set	Will my job change? In what way?
People tasks	What will I have to do?
Human–computer interface	How will I work with the computer?
	What inputs and outputs are there?
Database	What data are involved?
Database maintenance	How will I maintain the integrity of data?
Recovery	What happens when it goes wrong?
Monitoring	Is the system performing to specification?
Control	How are security and privacy handled?
Information retrieval	What information will I get?
Application	What will the system do?
Input and outputs for	Will it affect anything else on the computer system?
non-application areas	

Figure 26.2: Multiview methodology outputs

4. What information system processing function is the system to perform?
5. What is the technical specification of a system that will come close enough to doing the things that have been written down in the answers to the other four questions?

Too often, the authors argue, methodologies and role players have only addressed themselves to a limited subset of these questions: for example, computer scientists to Question 5, systems analysts to Question 4, users to Question 3, trade unions to Question 2, and top management to Question 1. Multiview attempts to address all these questions and to involve all the role players or stakeholders in answering all these questions. The emphasis in information systems, it is argued, must move away from 'technical systems which have behavioural and social problems' to 'social systems which rely to an increasing extent on information technology'.

Because it is a multi-view approach, it covers computer-related questions and also matters relating to people and business functions. It is part issue-related and part task-related. An issue-related question is: 'What do we hope to achieve for the company as a result of installing a computer?' A task-related question is: 'What job is the computer going to have to do?'

The distinction between issue and task is important because it is too easy to concentrate on tasks when computerizing and to overlook important issues which need to be resolved. Too often, issues are ignored in the rush to 'computerize'. But you cannot solve a problem until you know what the problem is! Issue-related aspects, in particular those occurring at Stage 1 of Multiview, are concerned with debate on the definition of system requirements in the broadest sense, that is, 'what real-world problems is the system to solve?' On the other hand, task-related aspects, in particular Stages 2–5, work toward forming the system that has been defined with appropriate emphasis on complete technical and human views. The system, once created, is not just a computer system, it is also composed of people performing jobs.

Another representation of the methodology, rather more simplistic, but useful in providing an overview for discussion, is shown in Figure 26.3. Working from the middle outward we see a widening of focus and an increase in understanding the problem situation and its related technical and human characteristics and needs. Working from the outside in we see an increasing concentration of focus, an increase in structure, and the progressive development of an information system. This diagram also shows how the five questions outlined above have been incorporated into the five stages of Multiview.

The first stage looks at the organization – its main purpose, problem themes, and the creation of a statement about what the information system will be and what it will do. The second stage is to analyse the entities and functions of the problem situation described in Stage 1. This is carried out independently of how the system will be developed.

The philosophy behind the third stage is that people have a basic right to control their own destinies and that, if they are allowed to participate in the analysis and design of the systems that they will be using, then implementation, acceptance, and operation of the system will be enhanced. This stage emphasizes the choice between alternative systems, according to important social and technical considerations. The fourth stage is concerned with the technical requirements of the user interface. The design of specific conversations will depend on the background and experience of the people who are going to use the system, as well as their information needs.

Finally, the design of the technical subsystem concerns the specific technical requirements of the system to be designed, and therefore to such aspects as computers, databases,

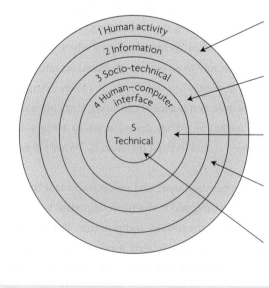

Figure 26.3: The Multiview framework

control, and maintenance. Although the methodology is concerned with the computer only in the latter stages, it is assumed that a computer system will form at least part of the information system. However, the authors do not argue that the final system will necessarily run on a large mainframe computer. This is just one solution, many cases show applications being implemented on a microcomputer.

1 Analysis of human activity

This stage is based on SSM (Mode 1). The very general term 'human activity' is used to cover any sort of organization. This could be, for example, an individual, a company, a department within a larger organization, a club, or a voluntary body. They may all consider using a computer for some of their information systems. The central focus of this stage of the analysis is to search for a particular view (or views). This *Weltanschauung* will form the basis for describing the systems requirements and will be carried forward to further stages in the methodology. This world view is extracted from the problem situation through debate on the main purpose of the organization concerned.

First, the problem solver, perhaps with extensive help from the problem owner, forms a rich picture of the problem situation. The problem solver is normally the analyst or the project team. The problem owner is the person or group on whose behalf the analysis has been commissioned. This picture can be used to help the problem solver better understand the problem situation. The rich picture diagram is also very useful to stimulate debate, and it can be used as an aid to discussion between the problem solver and the problem owner. There are usually a number of iterations made during this process until the 'final' form of the rich picture is decided. The process here consists of gathering, sifting, and interpreting data which is sometimes called 'appreciating the situation'. Drawing the rich picture diagram, examples of which are given in Figures 10.1 and 25.2, is a subjective process. There is no such thing as a 'correct' rich picture. The main purpose of the diagram is to capture a holistic summary of the situation.

From the rich picture the problem solver extracts problem themes, that is, things noticed in the picture that are, or may be, causing problems and/or it is felt worth looking at in more detail. The picture may show conflicts between two departments, absences of communication lines, shortages of supply, and so on. Taking these problem themes, the problem solver imagines and names relevant systems that may help to relieve the problem theme. Several different relevant systems should be explored to see which is the most useful. Once a particular view or root definition (Section 10.2) has been decided upon, it can be developed and refined. Thus, by using the CATWOE checklist, the root definition can be analysed by checking that all necessary elements have been included. For example, have we identified the owner of the system, all the actors involved, the victims/beneficiaries of the system, and so on?

When the problem owner and the problem solver are satisfied that the root definition is well formed, a conceptual model (or activity model) of the system is constructed by listing the 'minimum list of verbs covering the activities which are necessary in a system defined in the root definition . . .'. Examples of conceptual models are seen in Figures 10.3, 10.4, and 23.3. At this stage, therefore, we have a description in words of what the system will be (the root definition) and an inference diagram of the activities that the system will do (the conceptual model).

The completed conceptual model is then compared to the representation of the 'real world' in the rich picture. Differences between the actual world and the model are noted and discussed with the problem owner. Possible changes are debated, agendas are drawn up, and changes are implemented to improve the problem situation.

In some cases the output of this stage is an improved human activity system, and the problem owner and the problem solver may feel that the further stages in the Multiview methodology are unnecessary. In many cases, however, this is not enough. In order to go on to a more formal systems design exercise, the output of this stage should be a well-formulated and refined root definition to map out the universe of discourse, that is, the area of interest or concern. It could be a conceptual model which can be carried on to Stage 2, the analysis of entities, functions, and events.

2 Analysis of information

The purpose of this stage is to analyse the entities and functions of the application. Its input will be the root definition/conceptual model of the proposed system which was established in Stage 1 of the process. Two phases are involved: (a) the development of the functional model and (b) the development of an entity model.

A *Development of a functional model.* The first step in developing the functional model is to identify the main function. This is always clear in a well-formed root definition. This main function is then broken down progressively into subfunctions (functional decomposition), until a sufficiently comprehensive level is achieved. This occurs when the analyst feels that the functions cannot be usefully broken down further. This is normally achieved after about four or five subfunction levels, depending on the complexity of the situation. A series of data flow diagrams, each showing the sequence of events, is developed from this hierarchical model. This stage is therefore greatly influenced by the process modelling theme (Section 6.2). The hierarchical model and data flow diagrams are the major inputs into Stage 3 of the methodology, the next stage, which is the analysis and design of the socio-technical system.

B *Development of an entity model.* In developing an entity model, the problem solver extracts and names entities from the area of concern. Relationships between entities are also established. Again, this stage is greatly influenced by the data modelling them (Section 6.3). The preceding stage in the methodology, the analysis of the human activity systems, should have already given this necessary understanding and have laid a good foundation for this second stage. The entity model can then be constructed. The entity model, following further refinement, becomes a useful input into Stages 4 and 5 of the Multiview methodology.

3 Analysis and design of the socio-technical aspects

The philosophy behind this stage (influenced by ETHICS, Section 24.1) is that people have a basic right to control their own destinies and that, if they are allowed to participate in the analysis and design of the systems that they will be using, then the implementation, acceptance, and operation of the system will be enhanced. It takes the view therefore that human consider-ations, such as job satisfaction, task definition, morale, and so on, are just as important as technical considerations. The task for the problem solver is to produce a 'good fit' design, taking into account people and their needs and the working environment on the one hand, and the organizational structure, computer systems, and the necessary work tasks on the other.

An outline of this stage is shown in Figure 26.4. The central concern at this stage is the identification of alternatives: alternative social arrangements to meet social objectives and alternative technical arrangements to meet technical objectives. All the social and technical

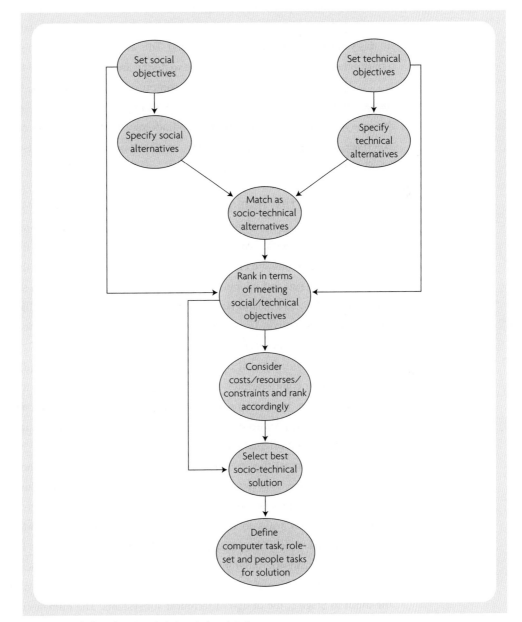

Figure 26.4: Outline of socio-technical analysis and design

alternatives are brought together to produce socio-technical alternatives. These are ranked, first, in terms of their fulfilment of the above objectives and, second, in terms of costs, resources, and constraints – again both social and technical – associated with each objective. In this way, the 'best' socio-technical solution can be selected and the corresponding computer tasks, role-sets, and people tasks can be defined.

The emphasis of this stage is therefore not on development, but on a statement of alternative systems and choice between the alternatives, according to important social and technical considerations.

It is also clear that, in order to be successful in defining alternatives, the groundwork in the earlier stages of the methodology is necessary and, in order to develop and implement the chosen system, we must continue to the subsequent stages. An important technique applicable to this stage is future analysis (Section 15.4). This aids the analyst and user to predict the future environment so that they are better able to define and rank their socio-technical alternatives.

The outputs of this stage are the computer task requirements, the role-set, the people tasks, and the social aspects. The computer task requirements, the role-set, and the people tasks become inputs to the next stage of the methodology, that is, the design of the human–computer interface. The role-set, the people tasks, and the social aspects are also major outputs of the methodology.

4 Design of the human–computer interface

Up to now, we have been concerned with what the system will do. Stage 4 relates to how, in general terms, we might achieve an implementation which matches these requirements. The inputs to this stage are the entity model derived in Stage 2 of the methodology and the computer tasks, role-set, and people tasks derived in Stage 3. This fourth stage is concerned with the technical design of the human–computer interface and makes specific decisions on the technical system alternatives. The ways in which users will interact with the computer will have an important influence on whether the user accepts the system.

A broad decision will relate to whether to adopt batch or online facilities. In online systems, the user communicates directly with the computer through a terminal or workstation. In a batch system, transactions are collected, input to the computer, and processed together when the output is produced. This is then passed to the appropriate user. Considerable time may elapse between original input and response.

Decisions must then be taken on the specific conversations and interactions that particular types of user will have with the computer system and on the necessary inputs and outputs and related issues, such as error checking and minimizing the number of keystrokes. There are different ways to display the information and to generate user responses. The decisions are taken according to the information gained during Stages 1 and 2 of Multiview.

Once human–computer interfaces have been defined, the technical requirements to fulfil them can also be designed. These technical requirements become the output of this stage and the input to Stage 5, the design of technical subsystems. The human–computer interface definition becomes a major output of the methodology.

5 Design of the technical aspects

The inputs to this stage are the entity model from Stage 2 and the technical requirements from Stage 4. The former describes the entities and relationships for the whole area of concern, whereas the latter describes the specific technical requirements of the system to be designed.

After working through the first stages of Multiview, the technical requirements have been formulated with both social and technical objectives in mind and also after consideration of an appropriate human–computer interface. Therefore, necessary human considerations are already both integrated and interfaced with the forthcoming technical subsystems.

At this stage, therefore, a largely technical view can be taken so that the analyst can concentrate on efficient design and the production of a full systems specification. Many technical criteria are analysed and technical decisions made which will take into account all the previous analysis and design stages. The final major outputs of the methodology might include:

- *the application subsystem* which is concerned with performing the functions which have been computerized from the function chart;
- *the information retrieval subsystem* which is for responding to enquiries about data stored in the system;
- *the database* in which all the data are organized;
- *the database maintenance subsystem* which permits updates to the data and provides the information necessary to check for data errors;
- *the control subsystem* which checks for user, program, operator, and machine errors and alerts the system to their presence;
- *the recovery subsystem* which allows the system to be repaired after an error has been detected;
- *the monitoring subsystem* which keeps track of all system activities for management purposes.

Figure 26.5 shows a schematic of these requirements for the technical specification. These subsystems cover all the things that have to be done by the computer system and the people operating it. These parts, or subsystems, may be implemented in different ways and in different combinations. For example, the information retrieval subsystem may be just another aspect of the database management system, and this may also include many of the necessary functions for control and recovery. The Multiview authors have separated them out because it is necessary to be sure that each one of them is catered for in the system.

Following full testing of all aspects of the system, there is a recognition that there will still be changes required, and this should be regarded as 'the norm'. Information systems will develop, and this requires an ongoing relationship between users, analysts, system creators, or owners. The authors recommend that the Multiview framework be applied for these changes (at least with a 'token run') so as to ensure that the system is still meeting its real objectives.

Multiview2

A new version of Multiview was published as Multiview2 (Avison et al., 1998). The original conception of Multiview posited a three-way relationship between the analyst, the methodology,

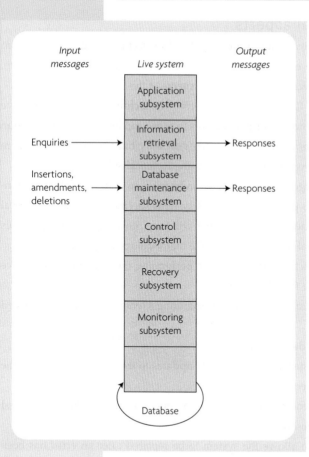

Input messages — Live system — Output messages

Application subsystem

Enquiries ⟶ Information retrieval subsystem ⟶ Responses

Insertions, amendments, deletions ⟶ Database maintenance subsystem ⟶ Responses

Control subsystem

Recovery subsystem

Monitoring subsystem

Database

Figure 26.5: Outline of the requirements for the technical specification

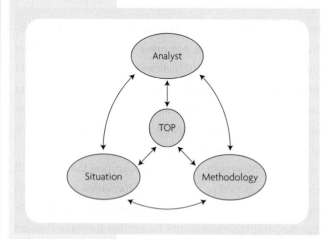

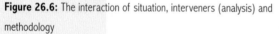

Figure 26.6: The interaction of situation, interveners (analysis) and methodology

and the situation. Avison and Wood-Harper (1990) suggested that parts of this relationship were missing in many descriptions of IS development (ISD), and that methodologies often contained unstated and unquestioning assumptions about the unitary nature of both the problem situation and the analysts involved in investigating it. Despite this criticism of other methodologies, Multiview1 itself offered no further guidance on how any given instantiation of the triad (analyst–methodology–situation) might come about in actual practice. Mitroff and Linstone (1993) is used to inform the particular occurrence of Multiview2 under any given set of circumstances (Figure 26.6).

Multiview2 offers a rich implementation of the multiple perspective approach as far as ISD is concerned. As we have seen, Multiview1 implemented such an approach through a five-stage methodology. These five stages were then typically presented as a waterfall structure. In Multiview2 the outcomes of ISD are posited as consisting of three elements: organizational behaviours, work systems, and technical artefacts, which are reflected in the parts: organizational analysis, socio-technical analysis and design, and technical design and construction, respectively (Vidgen, 1996). The fourth part of Multiview2, information modelling, acts as a bridge between the other three, communicating and enacting the outcomes in terms of each other (Figure 26.7). The proposed new framework for Multiview shows the four parts of the methodology mediated through the actual process of ISD.

Together with the change in the Multiview2 framework go changes to the content of the four parts, reflecting the experiences of applying Multiview through action research and developments in IS theory and practice. The major amendments made in the content of Multiview2 are summarized in Figure 26.8.

The Multiview2 parts of technical design and construction (T), socio-technical analysis and design (P), and organizational analysis (O) align well with the multiple perspective

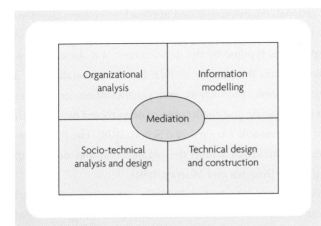

Figure 26.7: The Multiview2 framework

approach put forward by Mitroff and Linstone (1993). However, there are important differences of emphasis and definition. For example, the T perspective is equated by Mitroff and Linstone with rationality and functionalism, whereas in Multiview2 the technical design stage is concerned with the detail of computer system design and construction. In unbounded systems thinking, therefore, it would be perfectly feasible to have a T perspective of the analysis of human activity. The authors consider it more insightful to apply the TOP Multiple Perspective as a basis for informing the *approach* taken to ISD.

Part	Change	Rationale
Organizational analysis	Inclusion of strategic assumption surfacing and testing (SAST, Mason and Mitroff, 1989)	To strengthen the conceptual analysis of SSM with real-world stakeholder analysis (Vidgen, 1994)
	Radical change and business process redesign	IT as business enabler, rapid change in business environments (Wood et al., 1995)
	Introduction of ethical analysis	Stakeholders can have different moral ideals (Wood-Harper et al., 1996)
	Consideration of non-human stakeholders	To support a symmetrical treatment of social and technological factors (Vidgen and McMaster, 1996)
	Inclusion of technology foresight and future analysis	Consider the impact of the intervention on stakeholders (Avison et al., 1994, 1995) and the potential role of technology
Information modelling	Migration to Object-Oriented Analysis (from structured methods)	The principles of OO are more compatible with systems thinking than are the process/data separation and data flow metaphor of structured methods (e.g. the notion of systemic transformation and state change)
Socio-technical analysis and design	Ethnographic approaches to supplement ETHICS	Ethnographic approaches to socio-technical design (Randall et al., 1994; Avison and Myers, 1995) aid the analyst in understanding how work is accomplished (Sachs, 1995)
Technical design and construction	Construction of technical artefacts is incorporated within the scope of the methodology (Multiview 1 stopped at technical design)	Prototyping, evolutionary, and rapid development approaches to system development require that analysis, design, and construction be more tightly integrated (Budde et al., 1992)

Figure 26.8: Changes in the content from Multiview1 to version Multiview2 (Avison et al., 1998)

The T perspective reflects a rational, engineering-based approach to systems development in which the aim is to produce technical artefacts that will support purposeful human and organizational activity. The O perspective is typified by the development of a shared understanding and organizational learning, within the process of ISD. It can be visualized as a learning cycle including discovery, invention, production, and generalization, as well as double-loop learning to bring about the surfacing and challenging of deep-rooted assumptions which were previously unknown or undiscussable (Argyris and Schön, 1978). The P perspective represents the fears and hopes of individuals within the organization and deals with situations of power, influence, and prestige (Knights and Murray, 1994).

The TOP multiple perspective approach described by Mitroff and Linstone can be used to inform the different views that can be taken of the three sets of outcomes – organizational behaviours, work, and technical artefacts – within any given problem context. As an example of adopting a singular perspective of the system development process, Figure 26.9 offers a predominantly T perspective of Multiview2. This can be seen to be very much in line with what is often taken to be conventional wisdom as far as ISD is concerned, but here is taken to be the embodiment of the logical, rational view. It incorporates the agreement–analysis modes of thinking described by Churchman (1971) and Mitroff and Linstone (1993), and describes the technical interest of prediction and control identified by Habermas earlier. We also point out that alternate life-cycle models, such as iterative and evolutionary development, although generally more sympathetic to the O and P perspectives, may still be reduced in practice to a T-dominant view of ISD in which it is believed that the 'real' requirements are 'captured' more effectively than with a waterfall life-cycle model.

Multiview2 offers a systematic guide to any ISD intervention, together with a reflexive, learning methodological process, which brings together the analyst, the situation, and the methodology.

However, although the authors recommend a contingent approach to ISD, Multiview2 should not be used to justify random or uncontrolled development. An IS methodology, such as Multiview2, provides a basis for constructing a situation-specific method (Figure 26.10), which arises from a genuine engagement of the analyst with the problem situation (Wastell, 1996).

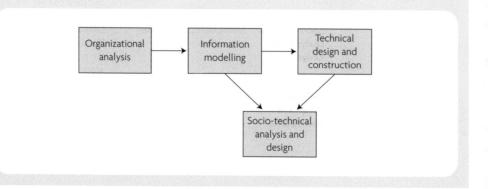

Figure 26.9: The T-dominant perspective of Multiview2

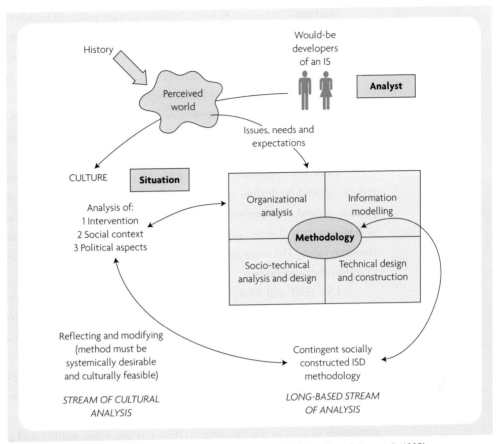

Figure 26.10: Constructing the information systems development methodology (from Avison et al., 1995)

Although Multiview has been in a continual state of development since 1985, the reflections on Multiview in action have suggested this radical redefinition of Multiview into Multiview2 which takes these experiences into account, along with the more recent literature and recognizing the new 'era' of the domain of information systems (see Chapter 27).

26.2 Strategic Options Development and Analysis (SODA)

We look briefly at SODA as an alternative framework approach. It is a general approach, in that it does not make assumptions about the problem situation being investigated, indeed, there is no assumption that a computer information system will be an outcome. It is a problem-structuring approach with its roots in operational research and the systems movement in the UK (it has some connections with soft systems methodology (Section 25.1) in this and some other respects). A very interesting introduction to the background to these approaches can be found in Rosenhead and Mingers (2001).

According to Eden and Ackerman (2001), SODA is: 'an approach which is designed to provide consultants with a set of skills, a framework for designing problem solving situations and a set of techniques and tools to help their clients work with messy problems.'

The consultant needs to have the skills to work with and facilitate the processes that lead to efficient and effective problem solving through helping a problem-solving team to work together efficiently and effectively. (The authors suggest that problem-alleviating might be a more appropriate term than problem-solving, as it reflects better the reality of such complex situations.) The consultant also needs the skills to construct a model of the content of the problem situation. The latter is seen as a complex mix of interconnected issues, problems, strategies and options. The approach may also support senior management teams on strategy making.

Thus the consultant works with the problem-alleviating and strategy-developing groups, should be flexible and needs to be aware of the important role of individuals in the process. Indeed, the clients are seen as small groups of individuals rather than an organization or business. It is also recognized explicitly that individuals in an organization may not share common views. The assumption that they do is implied, though of course not explicitly stated, in some other approaches.

Like SSM and Multiview, SODA is not seen as a prescriptive approach to problem solving, but more a framework to guide those interested in the problem area to investigate complex situations and suggest positive ways forward. Similarly, it is a cyclic and contingent approach, with all uses of SODA seen as being unique.

Indeed this guide through a complex situation has been developed more recently to form an approach known as **journey making**, suggesting that the journey towards strategy formulation is as important as the outcome. This stresses the impact of the process towards consensus as well as the outcome on personalities, roles, politics and power dimensions. The term journey is also an acronym representing the process of JOintly Understanding Reflecting and NEgotiating strategY.

The SODA framework is shown in Figure 26.11. It has four perspectives relating to the individual, nature of organizations, consulting practice, and technology and technique. The focus on the individual stresses the attempt to understand or 'make sense' of that part of the organization in order to 'manage and control' that world.

Cognitive mapping is seen as a very important technique (Section 10.4) to help this process of understanding following the SODA approach. The cognitive map is a formal way of expressing people's thinking about an issue. The perspective on the nature of organizations also focuses on the individuals within them as it stresses the political and power aspects as important explanations for decision making. Participants are seen as continuously negotiating their roles in organizations and having their own subjective views of the problem situation.

The third perspective on consulting practice brings together the individual and organizational perspectives but centres on the role of negotiation in effective problem solving or alleviating, the consultant facilitating this negotiation, and managing consensus and commitment. These three building blocks come together through appropriate technology and technique, the main technique being cognitive mapping.

Cognitive maps are usually created individually with clients and these are then merged to represent the views of all clients. This is not an easy process as individuals may use different

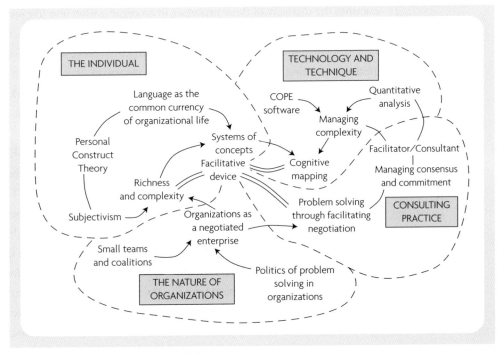

Figure 26.11: Theory and concepts guiding SODA

language constructs as well as having different views. There are different ways to deal with this difficult task.

Somehow, the maps may be 'averaged' in such a way that the result is a map having only shared beliefs and constructs. Alternatively, powerful individuals may have their views emphasized, a result of taking into account political realities. A more democratic approach would be to attempt to develop a consensus through debate in a workshop. One version of the merged map may help to set the agenda for the workshop that follows and be tuned during the workshop.

A workshop, typically lasting from two hours to two days, of the client team as a whole can help to ensure full client commitment to all the emerging issues (not just their own part of the map), understanding all the detail, yet moving to the specified actions. This requires negotiation, the merged map is not a *fait accompli*, and the workshop will probably include a number of cycles before agreement and commitment is reached. Typically, it will last one or two days and lead to agreements about the next courses of action related to the problem situation represented in the map.

A full case study concerning the UK National Audit Office is provided in Ackerman and Eden (2001) and following this case will provide greater insight into the approach.

26.3 Capability Maturity Model (CMM)

The Capability Maturity Model is strictly speaking not a methodology for systems development; it is a framework for evaluating processes used to develop software projects. The CMM classifies the maturity of these processes in an organization into five levels, with Level 5 being the most mature. The CMM framework specifies the characteristics that the various levels should have rather than prescribing any particular processes. It also provides advice and guidance relating to the improvements necessary to move from a lower maturity level upward. However, the CMM, although being a maturity framework and not prescriptive, does embody a certain philosophy concerning the way information systems and software should be developed.

CMM was created by the Software Engineering Institute (SEI) at Carnegie Mellon University for the US Department of Defense (DoD) to help assess the software engineering capability of their vendors and subcontractors (McGrew et al., 1999). The SEI was founded in 1984 and is a federally funded research and development centre sponsored by the DoD. Its original goal was to 'advance software engineering practice' in the light of the increasing dependence of the military on software and the increasing recognition that software was problematic in terms of its delivery, escalating costs, and customer dissatisfaction. A not unfamiliar story, and one that has also been found in the commercial sector. Today the mission of the SEI is to advance 'software engineering and related disciplines to ensure the development and operation of systems with predictable and improved cost, schedule, and quality' (SEI, 2005).

In 1987 the first version of a maturity framework was defined (Humphrey, 1989) and this evolved, with experience of use, into what became known as the CMM for Software (Paulk et al., 1993a, 1995; Weber et al., 1991). According to Paulk et al. (1993a) CMM 'provides software organizations with guidance on how to gain control of their processes for developing and maintaining software and how to evolve towards a culture of software engineering and management excellence. The CMM was designed to guide software organizations in selecting process improvement strategies by determining current process maturity and identifying the few issues most critical to software quality and process improvement.'

Since these early days, SEI has defined a number of other capability maturity models (CMMs), based on the success of the original CMM for Software. These relate to wider areas and other issues than just software engineering, for example, they defined a People model (P-CMM), a Software Acquisition model (SA-CMM) and a Systems Engineering model (SE-CMM), among others. More recently they have focused their attentions on defining a new model (or models) that integrates these previously separate and individual models. This model is known as the CMMI (Capability Maturity Model Integration). This work attempts to integrate the existing models into a common meta model with common terminology, processes and activities. Thus, CMMI replaces, or rather has subsumed, the previous separate models including the CMM for Software. The new integrated models are as follows:

- CMMI-SW is now the name of the software engineering model within the integrated model.
- CMMI-SE/SW is the name of the systems engineering and software engineering model.
- CMMI-SE/SW/IPPD is the name of the systems engineering, software engineering, and integrated product and process development model.

- CMMI-SE/SW/IPPD/SS is the name of the systems engineering, software engineering, integrated product and process development, and supplier sourcing model.

This makes things a little confusing and as in this book we are really only interested in the CMMI-SW that is what we will describe here, and we continue to describe it with the original term of just CMM. It should be remembered that there are also other maturity and quality process models not produced by SEI, so strictly speaking it should really be given its full name: the SEI CMMI-SW model. Other models or frameworks for software processes include ISO 9000 (now ISO/IEC 90003:2004), ISO 9001:2000, Trillium, SPICE, BOOTSTRAP and TickIT.

The CMM is designed to help organizations improve their software processes by providing a path for them to move from *ad hoc* development through to more disciplined software processes in a staged approach. The CMM framework provides a context in which policies, procedures, and practices are defined and established that enable good practices to be repeated, transferred across groups, and standardized. CMM has five maturity levels shown in Figure 26.12 and characterized as follows.

1 Initial level

Development is characterized as *ad hoc* or possibly chaotic. Processes are generally not defined, and success or failure depends on the capabilities of the individuals involved. Success can sometimes be achieved because individuals perform 'heroically', but this is generally not sus-

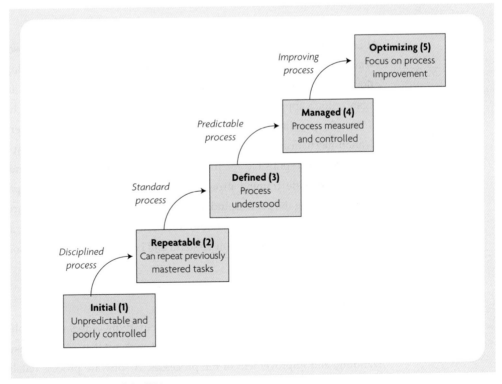

Figure 26.12: Five levels of the CMM

tainable over time as there are few repeatable processes and often those heroic individuals move on. Typically development is *ad hoc* and lurches from crisis to crisis, software is typically delivered late and over budget, and there is little effective management and control. This has been the norm in the past for software development and is still frequently seen in organizations. This is the initial stage from which organizations are trying to escape and move to a more mature process where development is not so dependent on the individuals but more on the characteristics of the management process. Fine (2001) says that this level might be better called the Heroic Level as any success is so dependent on a few star performers. He also reveals that CMM implementers often joke that many companies would be at Level 0 rather than 1 if such a level existed (i.e. the situation is quite poor out there).

2 Repeatable level

At Level 2 policies for managing software development are identified and established based on experience. Basic software development processes are in place; SEI characterize an effective process as one which is 'practiced, documented, enforced, trained, measured and able to improve' (Paulk et al., 1993a). Management process and controls are also established for planning, estimating, and tracking costs, schedules, functionality, etc. Project standards are defined and followed. Specific processes are not defined and may differ between projects. For an organization to be in Level 2, software projects and processes are essentially managed and under control with realistic plans based on performance of previous projects (i.e. previously mastered tasks are repeatable).

3 Defined level

At Level 3 the standard software engineering and management processes are documented and form a coherent, integrated, and standard approach to software development for the organization as a whole. The process is well defined, relatively stable, and recognized as the organization's approach to software development. Paulk et al. (1993a) characterize well defined as 'including readiness criteria, inputs, standards and procedures for performing the work, verification mechanisms (such as peer reviews), outputs, and completion criteria'. There should exist a group responsible for maintaining and improving these standard processes and an organization-wide training programme for communicating and imparting knowledge and skills concerning the process. Individual projects may tailor the standard process for the needs of the particular project but it should remain a well-defined, coherent, and integrated process. Overall, management should understand and be in control of quality and technical progress on each project.

4 Managed level

At Level 4 quantitative quality and productivity measures are established for key software development activities across all projects and goals set that will help ensure consistency, understanding, and improvement of the processes. A software process database should be used to collect and help analyse the data resulting in processes that are now predictable and operate within specified limits with any exceptions able to be quickly identified and remedied, ensuring high-quality software and predictable processes. The level is characterized as measured and predictable.

5 Optimizing level

At Level 5 the whole organization is focused on continuous process improvement on a proactive basis. The ability to identify strengths and weaknesses, to assess new technologies and process innovations, and take action to improve things on this basis is in place. The level is one of continuous process improvement on a planned and managed basis as a standard activity.

The levels provide a set of stages and criteria which organizations can measure themselves against and attempt to move up the levels as a way of achieving a more mature software process.

Having identified Maturity Levels, CMM defines a number of further elements to its structure. These are represented in Figure 26.13. Each Maturity Level (except Level 1) contains a number of Key Process Areas that need to be focused upon in order to achieve a particular Maturity Level. They are shown in Figure 26.14.

The goals specify the scope, boundaries, and intent of each of these Key Process Areas and help determine if they have been satisfied for the level. According to Paulk et al. (1993b) an example of a goal from the Software Project Planning key process area might be 'Software estimates are documented for use in planning and tracking the software project.'

Further, each Key Process Area has a number of attributes or Common Features that are used to test whether the implementation and institutionalization of a Key Process Area is effective, repeatable, and lasting. These Common Features are Commitment to Perform, Ability to

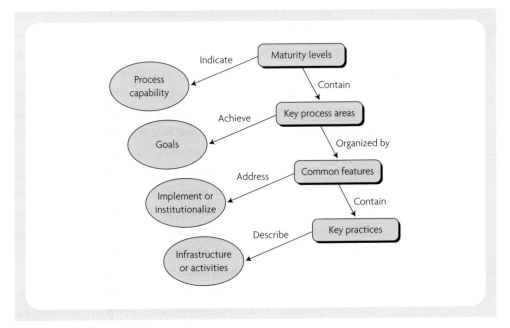

Figure 26.13: CMM structure

Process categories / Levels	Management — Software project planning, management, etc.	Organizational — Senior management review, etc.	Engineering — Requirements analysis, design, code, test, etc.
5 Optimizing		Technology change management Process change management	Defect prevention
4 Managed	Quantitative process management		Software quality management
3 Defined	Integrated software management Intergroup co-ordination	Organizational process definition Training programme	Software product engineering Peer reviews
2 Repeatable	Requirements management Software project planning Software project tracking and oversight Software subcontract management Software quality assurance Software configuration management		
1 Initial	*Ad hoc* processes		

Figure 26.14: Key process areas grouped according to level and process category

Perform, Activities Performed, Measurement and Analysis, and Verifying Implementation. Activities Performed relates to implementation activities whereas the rest relate to the institutionalization factors (i.e. the elements that ingrain the process into the culture of the organization). So, for example, Ability to Perform describes the conditions which must be present, such as adequate resources, necessary training, etc. Commitment to Perform specifies the organizational policies and senior management sponsorship which must be present to ensure that the process is created and that it endures over time. Finally the Key Practices describe the infrastructure and activities that ensure the effective Implementation and Institutionalization of the Key Process Area. An example of a Key Practice might be 'that adequate resources and funding are provided for planning the software project'; this might then be broken down into detail which specifies that developers with experience and skills are available and that relevant tools, such as spreadsheets, estimation software, project scheduling, etc. are also available.

It can be seen that the specification of all these Key Process Areas, Goals, Common Features, and Key Practices and their interrelationships is an enormous task on its own. The Report documenting the Key Practices is itself nearly 400 pages which has to be got to grips with before the processes can be implemented and institutionalized. Thus, CMM provides a very detailed set of elements and definitions for each level which defines what is necessary, and must be in place, for organizations to be considered as having reached a particular level and what they have to achieve in order to move to the next level. For example, according to McGrew et al. (1999), there are at least 121 Key Practices that must be implemented in order to move from Level 1 to Level 2 and that on average it takes 23 months (Johnson and Broadman, 1996). Organizations must move from level to level in sequence; that is, levels cannot be skipped, otherwise some of the essential foundations or building blocks of effective software processes will be missed.

By specifying the key practices in detail CMM emphasizes that it is not requiring or espousing 'a specific model of the software life cycle, a specific organizational structure, a specific separation of responsibilities, or a specific management and technical approach to development. The intention, rather, is to provide a description of the essential elements of an effective software process. The key practices are intended to communicate principles that apply to a wide variety of projects and organizations, that are valid across a range of typical software applications, and that will remain valid over time. Therefore, the approach is to describe the principles and leave their implementation up to each organization, according to its culture and the experiences of its managers and technical staff ' (Paulk et al., 1993b).

Although originally developed for the military, CMM has over the years become increasingly popular in commercial business (Diaz and Sligo, 1997; Jalote, 2000) as a basis for improving software processes and has become very influential in relation to software quality. For example, the processes of the two largest producers of software, IBM and Microsoft, have been compared against the standards of the CMM, with the basic assumption that the CMM is 'still the best way to produce quality software' (Phan, 2001).

It is difficult to know exactly how many organizations have formally adopted the CMM but certainly some of the major producers of software have done so. However, most organizations are still operating at Levels 1 or 2 (34.9 and 38.2 respectively), with relatively few at the highest levels, India having 50 companies at Level 5 and the US 20 (Bardoloi, 2002). However, according to McGuire (2000) the implementation of CMM leads to impressive improvements and he states that there are 'compelling statistics on reduced cycle time, increased productivity, fewer defects, and decreased risk throughout the software life cycle'. More specifically Lawlis et al. (1995), in a study of 52 defence projects, found lower costs and better scheduling with higher levels of CMM. Broadman and Johnson (1996) also found improvements in productivity, schedules, and quality, as did Herbsleb et al. (1994) suggesting that companies implementing CMM achieved averages of 35 per cent improvements in software productivity and 39 per cent reductions in post-delivery defects. More recently, General Motors using CMMI increased the percentage of milestones met during software development from around 50 per cent to 95 per cent with the average number of delays reduced from 50 to around 10 days. Lockheed Martin Management & Data Systems are reported to have achieved a 20 per cent reduction in software costs having moved from Level 2 to Level 5 in 2002 (Braue, 2004). Improvements are also reported in individual companies, for example, Hughes Aircraft (Willis et al., 1998), who were the first organization to achieve 'SEI-assisted' Level 3, Boeing (Fowler, 1997), Motorola (Diaz and Sligo, 1997), and Infosys (Jalote, 2000).

CMM is essentially about introducing discipline into software development by formalizing, standardizing, and institutionalizing a set of processes. These processes relate to a particular view of technical software development and embody the assumption that an engineering approach is the best way to develop such systems. It is based on a manufacturing and product-building view of systems development. This is not accepted by all, and they have argued that software development has conceptual differences, for example, it is not usually designed and built and then mass-produced, and it changes and evolves over time incrementally. These differences, it is sometimes argued, mean that software is not like a traditional product mass manufacturing process but is perhaps more like a creative art than a science, and needs to be treated as such. Also it is pointed out that CMM is really only concerned with the narrower software development process rather than the wider process of information systems development and thus misses many of the real, complex issues of systems development. Further, CMM, although including some processes relating to requirements identification, adopts the view that requirements are inherently definable and that it is just a question of working hard enough, and having the right processes in place, to identify the requirements successfully. Clearly this is just one view of the world and it contrasts sharply with other views (e.g. SSM, see Section 25.1), where requirements are more problematic, are more perceptual than absolute, and are not 'discoverable' in this way.

These criticisms relate to some fundamental differences in philosophies and assumptions relating to systems development; nevertheless it is clear that the CMM view is widely accepted in some quarters and is an important 'approach' to systems development practice, if not a methodology in its own right. Some have suggested that although there are these broad differences in approaches, of which CMM is one end of the spectrum, it does not necessarily mean that one is always right and the other wrong. For example, Fitzgerald and Fitzgerald (1999) have argued that in certain circumstances CMM may be an appropriate approach to systems development, such circumstances being where requirements are narrowly scoped and definable, where the circumstances are akin to manufacturing, where the environment is relatively stable, where the processes are predictable, and where the human element in the problem domain is not the primary issue. Fitzgerald and Fitzgerald (1999) use the example of the software elements of a telecommunication system, that is, the software controlling large switching and communications infrastructure systems, and indeed these have been successfully developed in a company having achieved a high CMM maturity level.

26.4 Euromethod

With the introduction of the single European market and the removal of various barriers to trade, it was predictable that at some point the European Commission would turn its attention to service and procurement standards in the information systems arena. The lack of standards in the area and the fragmentation of the information systems services and tools marketplace are perceived as a barrier to open competition and the principles of the single or open market. In 1989, the European Commission (EC), through its IT standardization policy unit, established Euromethod as an initiative to facilitate cross-border trading within the European Union (EU) in the acquisition of information systems. Clearly it was also intended to enhance and promote the overall competitiveness of the European information systems industry in a global context and thereby European industry in general. The description below is based on CCTA (1990, 1994a, b), Jenkins (1994), Stewart (1994), and Turner and Jenkins (1996).

The objective of Euromethod is to provide a public domain framework for the planning, procurement, and management of services for the investigation, development, or amendment of information systems. This framework and associated standards would, it was hoped, help overcome the problems posed by the current diversity of approaches, methods, and techniques in information systems used in Europe and help users and service providers to come to common understandings concerning requirements and solutions in information systems projects.

The focus of Euromethod is on the marketplace, and it seeks to smooth the path for those requiring and procuring information systems services and those potentially providing such services. It seeks to enable all suppliers to compete on an equal footing, no matter which European country they are from, by providing a common terminology that can bridge the different cultures and methods employed across the EU member states. Euromethod only addresses those arrangements where there is likely to be a contractual relationship between a customer and supplier. It does not address the situation where information systems services are provided to users by an in-house IT facility or department.

Euromethod was based on experiences with existing methods as follows:

- SSADM from the UK;
- Merise from France;
- DAFNE (DAta and Function NEtworking) from Italy;
- SDM (System Development Methodology) from the Netherlands;
- MEIN (MEtodológica INformática) from Spain;
- Vorgehensmodell from Germany;
- IE (Information Engineering) from the UK/US.

Some of these methodologies are described in Chapter 21 (i.e. SSADM, Merise, and IE).

The scope of Euromethod is supposed to cover all stages of procurement through to completion of an information systems adaptation and the associated planning and management elements. In practice some elements, such as the requirements specification stage, are addressed in more detail than others, and some things are missing altogether, such as technical architecture.

An information systems adaptation, to which Euromethod might apply, consists of any development or modification of an information system, including organizational, human, and technical elements, providing that the initial (or current) state and the required final state can be defined. For example, an information system adaptation might be a feasibility study, a system design, an enhancement, or a reverse engineering project. So, Euromethod can be applied at any stage of a project and applied many times in a development.

Euromethod focuses on the understanding, planning, and management of the contractual relationships between customers and suppliers of information systems adaptations. The

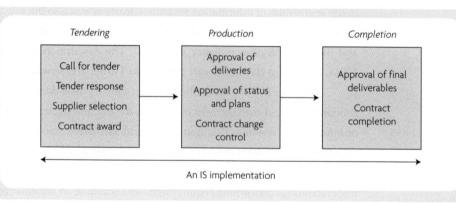

Figure 26.15: Types of transaction in an IS adaptation

key transactions required between customers and suppliers are identified. Figure 26.15, for example, summarizes the high-level transactions required in the process of tendering, production, and completion of an information systems adaptation. This focus results in Euromethod identifying and defining the deliverables required between customers and suppliers. Each transaction implies one or more decision points, each of which needs supporting by a deliverable. Euromethod identifies a hierarchy of deliverables as in Figure 26.16. The target domain is the part of the organization affected by the information systems adaptation, and the information systems descriptions are the documentation designs and specifications, whereas the operational items are deliverables that can be installed as part of the information system itself. This could be a screen or a prototype. The project domain is the temporary organization established to adapt an information system and manage the process. The deliverables on this side are divided into plans for the production process and reports to manage the process. The delivery plan is the key definition of the process of defining the customer/supplier transactions in the overall context of the information systems adaptation, that is, the problem situation. Figure 26.17 represents the planning process to produce the delivery plan.

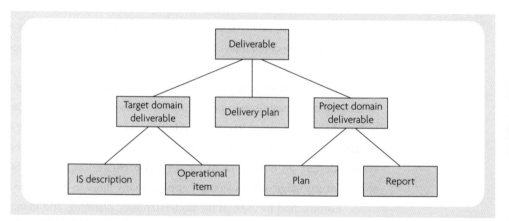

Figure 26.16: A hierarchy of deliverables

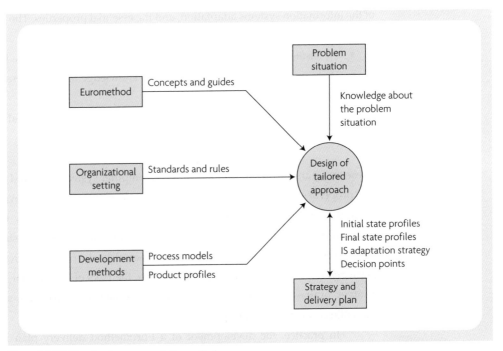

Figure 26.17: The planning process of Euromethod

Euromethod recognizes that different types of information systems adaptation require different approaches depending on the situation. Differences may be due to the complexity or uncertainty of the environment. Euromethod provides guidelines and support for a variety of different situations. Guidelines exist for the identification of important situational factors, the assessment of complexity, uncertainty and risks, and the definition of a strategy for the IS adaptation. The delivery planning process of Figure 26.17 illustrates the context of this situation-driven approach.

Euromethod is not another information systems development methodology but is a framework. It seeks to provide guidance as to how to map or bridge a particular method into the basic Euromethod framework. In this sense a method might include a European systems development methodology (e.g. SSADM), a project management approach (e.g. PRINCE), or a procurement standard (e.g. TAP, Total Acquisition Process). (TAP is the UK public sector pre-ferred procurement method for IS/IT services.) Guidelines are provided that help bridge between Euromethod concepts and those of the specific method.

These principles result in three Euromethod models. These are the transaction model, the deliverable model, and the strategy model. The transaction model helps understand, manage, and facilitate customer/supplier interactions across organizational boundaries during an information systems adaptation. It identifies a set of generic customer/supplier transactions in information systems adaptations, including, for each, the goals, the key roles, and responsibilities in both the customer and supplier organizations. The deliverable model defines the target domain for an information systems adaptation, what an information system is, the scope of the information systems adaptation, and the essential properties. The essential properties are expressed as a set of views as follows:

- *the business information view* – the relevant knowledge about the information resource;
- *the business process view* – the relevant knowledge about processes and their use of the information resource;
- *the work practice view* – the relevant knowledge about individual actors and their use of the information resource.

These views are further developed, and the computer systems required to support these essential properties are identified. The business information view is developed to identify the computer systems' data view, the business process view to identify the computer systems' function view, and the work practice view to identify the computer systems' architecture view.

The strategy model reflects the situation-driven principle above and helps define the approach required which reduces, or contains, the risks of a particular problem situation. The strategy options relate to systems installation, systems construction, systems description (or modelling), and project control.

In practice, the development of a Euromethod delivery plan involves the following tasks:

- assessment of the problem situation;
- definition of a strategy;
- definition of decision points.

The first stage, assessment of the problem situation, helps to clarify the requirements of the information systems adaptation for both customer and supplier. First, the initial state of the adaptation, that is, the current state of the existing information system, is defined. A good definition of the initial state is crucial to enable the supplier to understand the starting point of the adaptation. Second, the desired final state of the adaptation is defined. This includes the elements of the initial state that are to be changed, the additional items to be produced, the documentation, the data, the interfaces, and so on. The initial and final states are defined in terms of Euromethod essential properties and the views that they represent (see above). As various existing methods and methodologies will have already been defined in these terms, this should enable each final state deliverable to be assessed as to which approach might be most appropriate for its production.

This first stage also involves the determination of the situational factors and an assessment of their complexity and uncertainty. Such factors might include the complexity of the business processes, the heterogeneity of the users, the attitude and ability of the actors involved, the stability of the environment, and the requirements and the quality of the specifications.

These factors are now used for the next task, to define an appropriate strategy for the production of the final state deliverables based upon the strategy model. For example, the strategy option chosen for system construction might be evolutionary and participatory in situations where the complexity of requirements is judged to be high and the understanding of the business processes is low. In practice, the analysis is more sophisticated than this, as Euromethod provides heuristics for the suitability of the approach and examples of how to reduce risk.

Finally, once the strategy has been defined, the decision points are identified. These are a by-product of the deliverables exchanged between customers and suppliers and their identification is based on the transaction model. Euromethod identifies three categories of decision: investment, design, and systems acceptance. For each decision point, the decision to be made is identified plus the roles of those involved, the transactions they represent, and the deliverables that are exchanged. The full documentation of each deliverable is made according to Euromethod product profiles. Both project and target domain deliverables are identified. The profile for target domain deliverables consists of:

- the deliverable type;
- the information system view; that is, the essential properties of the transaction model as described above;
- the computer system view;
- the IS properties within each view;
- non-functional properties;
- scope (of the deliverable);
- state (of the deliverable);
- degree of formality (of the deliverable).

Some of this information is similar in type to that required for initial and final states, and this reflects Euromethod's approach to standard terms and definitions. These concepts are again used in the method's bridging process and enable relevant external methods for the decision points and deliverables to be identified and used.

The sequence and schedules of the decision points are also identified, and the total documentation provides the basis of the contractual agreement for the information systems adaptation between customer and supplier.

However, the high ideals of Euromethod are difficult to achieve in reality. There are different versions of Merise and other country standard approaches that are used in practice, so agreeing one for the whole of Europe is not easy. In practice, this means that Euromethod has influenced SSADM, Merise, and so on by providing agreed standards, rather than led to an imposition of one approach throughout Europe. However, it is often seen as adding yet another layer to already bureaucratic methodologies, such as SSADM and Merise.

Euromethod is a Europe-wide initiative that the EC have decided is important and in which they have invested. The concept of method bridging is clearly attractive, and it supports a common understanding about IS development across Europe. Whatever its originators set out to do, it has proved to be more of a contingency framework (like Multiview), to be adapted and to adapt local methodologies, such as SSADM, rather than a 'super-methodology'.

Summary

- In this chapter we have described four approaches that we classify as frameworks rather than methodologies, mainly because they provide guidance to the developer in choosing methods, techniques, and tools rather than a prescriptive step-by-step approach.

Questions

1. Do you agree with the distinction made between frameworks and methodologies?

2. Multiview might also be described as a 'blended' methodology (see Chapter 21). Why is this? What elements have been blended? Where have they come from?

3. What are the five levels of CMM? How does an organization move from one level to another?

4. Why does Euromethod include elements relating to 'procurement' rather than development of information systems? Do you think a European standard for ISD is feasible and/or desirable?

Further reading

Ackerman, F. and Eden, C. (2001) SODA – journey making and mapping in practice, in Rosenhead, J. and Mingers, J. (2001) *Rational Analysis for a Problematic World Revisited – Problem Structuring Methods for Complexity, Uncertainty and Conflict*, 2nd edn, John Wiley & Sons, Chichester, UK.

Avison, D.E. and Wood-Harper, A.T. (1990) *Multiview: An Exploration in Information Systems Development*, McGraw-Hill, Maidenhead, UK.

Paulk, M.C., Weber, C.V., Curtis, B., and Chrissis, M.B. (1995) *The Capability Maturity Model, Guidelines for Improving the Software Process*, Addison-Wesley, Reading, Massachusetts.

Turner, P. and Jenkins, T. (1996) *Euromethod and Beyond*, Thomson, London.

Part VII: **Methodology issues and comparisons**

Contents

Chapter 27 – Issues   p 567

Chapter 28 – Methodology comparisons   p 591

In the first edition of this book, published in 1988, the area of information systems development methodologies was described as a jungle, and we concluded that this was unlikely to change much in the future due to the continuing developments in information technology, information systems, and organizations as a whole. In the second edition published in 1995 we thought that not much had changed in this respect, and that it was still 'a methodology jungle'. Six years later in our third edition we felt that there was not quite the diversity of methodological approaches that there were.

Although some newer methodologies have been adopted, most notably the OO methods based on UML and the agile and rapid approaches such as DSDM and XP, many have disappeared, and perhaps the picture continues to be a little clearer in this our fourth edition. In the 1995 edition, we made reference to an estimate that there were over 1,000 brand name methodologies worldwide (Jayaratna, 1994), but we suggested that this figure should not be taken too literally because we felt that many of them were probably very similar and differentiated only for marketing purposes. Today we feel the figure would be considerably smaller, possibly less than a hundred – with the number of fundamentally different methodologies even smaller. Therefore, we conclude that there is some improvement in the situation and that the area is less of a 'jungle' than it was. Nevertheless it is still quite confusing, especially for newcomers to the field.

Chapter 27 attempts to review developments and discuss the various issues as a way of providing some help in understanding the 'jungle'. The treatment is both theoretical and philosophical, as well as practical. The nature of such discussions is that there are few hard and fast facts, and much of the chapter involves our interpretations and opinions. We hope that they are based on sound judgement and experience, but they also recognize that there will be those who make different interpretations. Nevertheless we hope that this chapter contributes to informed debate concerning methodologies.

In Chapter 28 we provide a checklist and framework for comparing methodologies, and we use the framework to compare methodologies discussed in the book. The framework consists of philosophy, model, techniques and tools, scope, output, practice, and product. Again, we do not suggest an objective comparison, indeed our discussion is sometimes philosophical in nature, our purpose is more to raise issues and engender debate.

27 Issues

27.1 What is a methodology?

In Chapter 2 we provided a working definition of the term 'methodology', and, although this has been adequate for our purposes, we will now look in more depth at the question 'what is a methodology?'. The term is not well defined either in the literature or by practitioners. There is very little agreement as to what it means other than at a very general level. The term is used very loosely but also very extensively.

This loose use of the term does not, of course, mean that there are no definitions, simply that there are no universally agreed definitions. At the general level, a methodology is regarded as a recommended series of steps and procedures to be followed in the course of developing an information system. In a brief *ad hoc* survey, this proves to be about the maximum that people will agree to, and, of course, such a definition raises many more questions than it answers. For example:

- What is the difference between a methodology and a method?
- Does a methodology include a specification of the techniques and tools which are to be used?
- Does a collection of techniques and tools constitute a methodology?
- Should the use of a methodology produce the same results each time?

The questions that arise are fundamental as well as numerous. Unfortunately the problem will not be solved here; the most that can be achieved is that the issues will be aired. The information systems community has regularly debated the meaning of the term methodology in an information systems context, and as yet it has not come up with any universal definition.

One of the most useful definitions for the authors is that provided by the British Computer Society (BCS) Information Systems Analysis and Design Working Group as long ago as 1983. They defined an information systems methodology as:

> a recommended collection of philosophies, phases, procedures, rules, techniques, tools, documentation, management, and training for developers of information systems.
>
> (Maddison, 1983)

Utilizing this definition, we suggest that a methodology has a number of components which specify:

- what tasks are to be carried out at each stage;
- what outputs are to be produced;
- when, and under what circumstances, they are to be carried out;
- what constraints are to be applied;
- which people should be involved;
- how the project should be managed and controlled;
- what support tools may be utilized.

In addition, a methodology should specify the training needs of the users of the methodology. On the other hand, the frameworks discussed in Chapter 26 are far less prescriptive.

Apart from the above, we believe that a methodology should also specifically address the critical issue of 'philosophy'. We mean by this the underlying theories and assumptions that the authors of the methodology believe in and that have shaped the development of the methodology. This identifies those sometimes unwritten aspects and beliefs that make a methodology an effective approach to the development of information systems in the eyes of their authors. We believe that the definition of a methodology should include specific reference to its philosophy as this has a critical bearing on the understanding of a particular methodology. An information systems development methodology is therefore much more than just a series of techniques along with the use of software tools.

Utilizing these ideas and beliefs we extend the BCS definition of a **methodology** as follows:

> A systems development methodology is a recommended means to achieve the development, or part of the development, of information systems based on a set of rationales and an underlying philosophy that supports, justifies and makes coherent such a recommendation for a particular context. The recommended means usually includes the identification of phases, procedures, tasks, rules, techniques, guidelines, documentation and tools. They might also include recommendations concerning the management and organization of the approach and the identification and training of the participants.

Some methodologies are, of course, more comprehensive than others, which is why we include the statement about 'development or part development' and, as mentioned above, the 'rationales and underlying philosophy' are not always made clear but we believe that they always exist and are the key to understanding a particular methodology. For example, they will shed light on the ethical stance that a methodology incorporates. We include 'a particular context' in the definition because it is clear that methodologies are not universal in their applicability, despite some methodology authors seemingly thinking they are. For us there are always limitations to the applicability, and thus we believe that those contexts and limits should be addressed. Together the philosophy and the context help illuminate the assumptions that the authors of the methodology are making, which is why we do not specifically include a reference to 'assumptions' in our definition. In practice, many methodologies, particularly commercial ones, are products that are 'packaged' and might include:

- manuals;
- education and training (including videos);
- consultancy support;
- tools and toolsets;
- *pro forma* documents;
- model-building templates, and so on.

Some have argued (e.g. Flynn, 1992) that the term 'methodology' is not apt in the context of systems development and that the term 'method' is perfectly adequate to cover everything that we mean by a methodology. Indeed Flynn states that 'the term methodology was popular for a time in the 1980s' implying that it is no longer much used. This is contrary to our experience, although it is true that the term 'method' is also used. For us, this might be argued to substitute one ill-defined word for, another, but more importantly we believe that the term 'methodology' has certain characteristics that are not implied by 'method', for example, the inclusion of 'philosophy' which we have already suggested is key. Methodology is thus a wider concept than method.

Checkland (1981) has distinguished between the two terms and says that a methodology:

> . . . is a set of principles of method, which in any particular situation has to be reduced to a method uniquely suited to that particular situation.

In Checkland (1985), he argues that information systems development must be seen as a form of enquiry in the context of the general model of organized enquiry, which consists of three components: an intellectual framework, a methodology, and an application area. This suggests a hierarchy of elements to enquiry.

The first element is the **intellectual framework**, which comprises the ideas that we use to make sense of the world. This is described as the philosophy that guides and constrains the enquiry. It consists of ontological assumptions, that is, beliefs about the fundamental nature of the physical and social world and the way it operates, and epistemological assumptions, that is, the theory of the method or grounds of knowledge. These terms are discussed in Section 28.2. It also consists of ethical values, which should be articulated, that may serve to guide or constrain the enquiry.

The second element is the methodology. This is the operationalization of the intellectual framework of ideas into a set of prescriptions or guidelines for investigation which require, or recognize as valid, particular methods and techniques.

The third element is the application area; that is, some part of the real world that is deemed to be problematical and worthy of investigation.

This is a useful framework for discussions of research and enquiry, but we need to know how well it relates to the world of information systems development methodologies. Working backward, it would seem that the application area is that of information systems development in general. The methodology element is equivalent to the collection of phases, procedures, rules, methods, and techniques that are usually considered to be a methodology in the information systems world. The intellectual framework is the element that is often missing from methodologies, or rather it is not missing, for it exists, but is not explicitly articulated. In our definition, much of this intellectual framework element is included in what we term the underlying philosophy, which we include in the definition of methodology itself.

27.2 Rationale for a methodology

It is important to discover why people adopt a particular methodology. Obviously this varies substantially between organizations and individuals, but we can identify three main categories of rationale: a better end product, a better development process, and a standardized process.

1 A better end product

People may want a methodology to improve the end product of the development process, that is, they want better information systems. This should not be confused with the quality of the development process, addressed below. It is difficult to assess the quality of information systems produced as a result of using a particular methodology. For example, we cannot know that the use of the methodology produced the particular results. The same results might have occurred if the system had been developed using another methodology or without using any methodology at all.

Even if we had some way of comparing the results of using different methodologies, the elements that are perceived to constitute measures of quality differ considerably from person to person, and there is little agreement within the information systems community on this issue. The following represents our attempt to address some of the components of quality of an information system:

- *Acceptability* – whether the people who are using the system find the system satisfactory and whether it fulfils their information needs. This includes business users and managers and their requirements.
- *Availability* – whether it is accessible, when and where required.
- *Cohesiveness* – whether there is interaction between components (subsystems) so that there is overall integration of both information systems and associated manual and business systems.
- *Compatibility* – whether the system fits with other systems and other parts of the organization.
- *Documentation* – whether there is good documentation to help communications between operators, users, developers, and managers.
- *Ease of learning* – whether the learning curve for new users is short and intuitive.
- *Economy* – whether the system is cost-effective and within resources and constraints.
- *Effectiveness* – whether the system performs and operates in the best possible manner to meet its overall business or organizational objectives.

- *Efficiency* – whether the system utilizes resources to their best advantage.
- *Fast development rate* – whether the time needed to develop the project is quick, relative to its size and complexity.
- *Flexibility* – whether the system is easy to modify and whether it is easy to add or delete components.
- *Functionality* – whether the system caters for the requirements.
- *Implementability* – whether the changeover from the old to the new system is feasible, in technical, social, economic, and organizational senses.
- *Low coupling* – whether the interaction between subsystems is such that they can be modified without affecting the rest of the system.
- *Maintainability* – whether it needs a lot of effort to keep the system running satisfactorily and continuing to meet changing requirements over its lifetime.
- *Portability* – whether the information system can run on other equipment or in other sites.
- *Reliability* – whether the error rate is minimized and outputs are consistent and correct.
- *Robustness* – whether the system is fail-safe and fault-tolerant.
- *Security* – whether the information system is robust against misuse.
- *Simplicity* – whether ambiguities and complexities are minimized.
- *Testability* – whether the system can be tested thoroughly to minimize operational failure and user dissatisfaction.
- *Timeliness* – whether the information system operates successfully under normal, peak, and every condition, giving information when required.
- *Visibility* – whether it is possible for users to trace why certain actions occurred.

Maximization of all these criteria is, of course, not attainable; indeed, some actually work against each other. Ideally, the most appropriate information systems methodology can be chosen and then 'tuned' so that emphasis can be given to those criteria which are particularly important in the problem situation. Such tuning, or even any consideration of tuning, is rare in methodologies.

2 A better development process

Under this heading comes the benefits that accrue from tightly controlling the development process and identifying the outputs (or deliverables) at each stage. This results in improved management and project control. It is usually argued that productivity is enhanced with the use of a methodology; that is, we can either build systems faster, given specific resources, or use fewer resources to achieve the same results. It is sometimes also argued that the use of a methodology reduces the level of skills required of the analysts, which improves the development process by reducing its cost.

For some, the problems of developing information systems can be improved by adopting the quality standards that have proved popular in manufacturing and industrial processes. These standards are designed to ensure the quality of processes rather than that of the end product, and this sometimes leads to emphasis on conformity to the standard irrespective of whether this helps the quality of the system.

Another problem argued in Avison et al. (1994) is that the traditional manufacturing process is quite different from the process of developing software products. For example, the product of software is a 'one-off' rather than a mass replication of a design. There are schemes which recognize these difficulties and attempt to interpret the standards in methods applicable to software development. Although such standards have not yet made a large impact, either in terms of coverage or improved quality, they are helping to raise the issues and level of debate concerning quality in the development of information systems.

3 A standardized process

The needs associated with this category relate to the benefits of having a common approach throughout an organization. This means that more integrated systems can result, that staff can easily change from project to project without retraining being necessary, and that a base of common experience and knowledge can be achieved. In short, all the normal benefits of standardization can be achieved, including the specific benefit of easier maintenance of systems.

All the reasons contained in the above categories have been specified, in some form or other, by the authors or vendors of methodologies as being benefits of adopting their particular approach. In contrast purchasers might well be more interested in:

1. Improved systems specifications.
2. Easier maintenance and enhancement.

Methodologies rarely directly address improvement of the maintenance task (but see Section 25.5), although vendors often claim it as an indirect benefit as the information system developed will require less maintenance through use of the methodology. Further, some approaches – particularly those discussed in Chapter 26 – suggest flexibility within a framework, rather than 'a standardized process'.

27.3 Adopting a methodology in practice

An organization thinking of adopting or purchasing a methodology has a number of concerns. These relate to what they get and whether they are guaranteed successful information systems as a result.

What do they get? To address the first question, they get what the vendor or methodology author gives them, and this differs greatly:

- a methodology can range from being a fully fledged product detailing every stage and task to be undertaken to being a vague outline of the basic principles in a short pamphlet;
- a methodology can cover widely differing areas of the development process, from high-level strategic and organizational problem solving to the detail of implementing a small computer system;
- a methodology can cover conceptual issues or physical design procedures or the whole range of intermediate stages;

- a methodology can range from being designed to be applicable to specific types of problem in certain types of environment or industry to an all-encompassing general-purpose methodology;
- a methodology may imply a number of assumptions, for example, that the environment is one where an old system is replaced by a new one rather than development being incremental and integrative (see Section 3.6 and the assumption of 'green field' development);
- a methodology may be potentially usable by anybody or only by highly trained specialists or be designed for users to develop their own applications;
- a methodology may require an army of people to perform all the specified tasks or it may not even have any specified tasks;
- a methodology may or may not include tools and toolsets.

The variations on this theme are numerous. It is clear that methodology adopters should be aware of what their needs are and choose their methodology accordingly. This does not always seem to be the case. One organization that adopted a particular methodology found that they had to write detailed manuals themselves to specify what was required of their development staff. What they purchased was a management overview without any detail.

Some methodologies are purchased as a product, others are available by purchasing a licence, others are obtained through a contract for consultancy work, some come as part of the purchase of a tool, and some by a combination of the above. As can be imagined from this, the cost of methodologies varies considerably. Some are effectively free, and this is the case for Merise. Often the initial purchase of the methodology product is the least of the investment, and it is the training, additional hardware and software, and ongoing consultancy costs that mount up and eventually dwarf the initial purchase price. What is also clear is that the potential organizational costs of adopting a methodology, particularly if it is adopted as a company standard, are enormous. These organizational costs are not the purchase price, but the costs of embedding the methodology in the culture of the organization, the opportunity costs of not doing something else, the costs of training and educating users and managers to participate in the use of the methodology, and so on. There may also be large costs associated with changing methodologies, as once a particular methodology becomes embedded in an organization it is not easy to change the development culture.

These costs are usually seriously underestimated in the evaluation of the cost of a methodology. It also indicates that the IT or systems development department should not necessarily be the ones to make decisions concerning the adoption of a particular methodology. It appears that in the majority of cases the decision to adopt a standard methodology and the choice of the particular methodology is made by the IT/IS department. This is often because this department assumes it as its role and prerogative. Indeed, it has been suggested that this assumption of technical expertise is the structural basis of IS professionals' power in organizations (Markus and Bjørn-Andersen, 1987). It can legitimately be argued that the users, and business managers in general, should be making such decisions, as they are the ones who have to make the actual investment, not just in terms of money, but in time and effort, and ultimately business success, as a result of methodology decisions. However, the assumption that it is an IT decision is frequently accepted and encouraged by the rest of the organization. Information systems development is often seen as a technical issue that the technicians should decide upon. After all that is what they are paid for! We have not taken this view in this book.

However, we know of instances where IT departments have attempted to involve users and business managers in decisions concerning methodologies without success. The users and managers preferred to leave it to the IT department despite the potential significant implications on the degree of user involvement in development that the different methodologies under consideration embodied, and ultimately on the overall future success of the business. Equally, there are some cases where users and managers have not been consulted in the choice of methodology at all, and they have refused to co-operate in developments as a result.

Markus and Bjørn-Andersen (1987) argue that the choice of methodology is in practice made by corporate IS rather than by involving individual analysts who have to work with the methodology following the decision. Therefore, the power structure is argued to be conspiring against the users and the business managers, as discussed above, and against the information systems worker as well.

The answer to the second question posed above, *'are they guaranteed successful information systems as a result of using the methodology?'*, is clearly 'no'. However, this does raise the question of what a methodology is supposed to produce. If two teams of developers are using the same methodology on a similar project, in the same type of environment, can the same results be expected, and if not, why not?

Clearly the developers may interpret the demands of the methodology differently and thus end up with different results. The methodology may give a lot of leeway to the developers in terms of how they perform the specified tasks, and so the results will be different. The developers may have varying skill levels, which will also produce differing results. However, it is sometimes argued that, given these variations, multiple uses of the same methodology in the same circumstances should yield roughly the same results. The tighter, more specific the methodology, the more reproducible the results, particularly where the methodology specifies the exact techniques and tools to be employed under each circumstance.

This is a highly contentious area, but the implication is that if the results are to some degree reproducible, then we must be sure that the methodology specifies 'a best' way of producing information systems. We cannot say *the* best way, because there may be trade-offs between quality, quantity, skill levels, reliability, generality, and so on. But we want a methodology that will produce good results. This implies that a methodology is not just a helpful set of guidelines that enables the developer to organize the development process more effectively and efficiently, but that it embodies a good way of developing systems. If it is reproducible then it leads to the development of particular solutions, and thus if we adopt a methodology, then this methodology must be, to our minds, a good way of doing things. If it is not, it will lead to problems, because the nature of the methodology will exclude other ways of doing things. For example, the adoption of a data flow diagram-based methodology will, if it is rigidly followed, exclude the kind of analysis that the SSM methodology recommends. There will, for example, be no analysis of conflict between various actors in the existing system.

The question of repeatability or reproducibility is obviously not one that can be easily tested in practice. It is impossible to create exactly the same environment in an organization for two sets of developers to develop systems which we can then compare. Checkland (1987) highlights the problem by challenging developers of systems perceived to be successful to prove that another methodology would not have been better. And, of course, they cannot. He asks the developers of systems, where there have been problems, to *prove* it was not their incompetence that led to the problems rather than the methodology. Once again they cannot. It has been suggested that two development teams could work independently on developing a system to the same requirements in the same environment enabling a legitimate comparison of the systems at the end. However, the very fact that there are two sets of developers will undoubtedly influence the results. For example, the fact that one set will have talked to a user first will influence the results when the second set of developers talks to that same user. Even if flawed, the results of such experimentation would be interesting, but it seems the practical problems are insurmountable as the authors have no knowledge of any such attempt, let alone a systematic study.

This lack of repeatability, or rather the ability to demonstrate repeatability, is often used to suggest that information systems is not really an engineering discipline. However, engineering is also a creative profession. Will two engineers design the same bridge or two electronic engineers design the same circuit for a particular function? Probably not, but the degrees of freedom in designing a bridge are greater than that of designing a particular circuit; that is, the two designs of the circuit will probably be more similar than the two designs of the bridge. With information systems, the degrees of freedom are typically even greater than for the bridge. The adoption of an information systems methodology can thus be argued to be an attempt to reduce the degrees of freedom.

If this is the case, then the implication still holds that the methodology should embody a best way of developing systems. This is not always appreciated when methodologies are being selected and may result in the development of inappropriate systems. The question whether the methodology embodies a best way of developing systems is rarely asked. The more usual questions are:

- whether it fits in with the organization's way of working;
- whether it specifies what deliverables (or outputs) are required at the end of each stage;
- whether it enables better control and improved productivity;
- whether it supports a particular set of tools.

A further important consideration when adopting a methodology is that often the practice, behaviour and outcomes are somewhat different from what might be expected or to the theory implicit in the methodology. This is what Fitzgerald et al. (2002) have termed 'methods in action'. They suggest that methods 'are rarely applied in their entirety, nor as originally intended by their creators'. The authors of Multiview and RUP also do not like the term methodology for similar reasons; they believe that the methods as described are quite different from the way that they are used in practice (see Section 22.2). Such differences happen for a variety of reasons, including:

- developers interpret the strictures of the methodology in different ways;
- developers are individuals and act in different ways;
- developers only apply parts of the methodology as they think appropriate or expedient;
- developers may react to the controlling nature of a methodology by attempting to subvert it;
- more experienced developers may be more prepared to be flexible with the methodology, whereas inexperienced developers may follow a methodology more closely;
- developers use the methodology for political reasons, for example, to enhance their power base as 'experts' or gatekeepers of the methodology;
- users may refuse to participate, or obstruct, the use of the methodology;
- organizations (as well as developers) are different in their style and culture resulting in very different approaches to a methodology;
- managers may or may not enforce adherence to the methodology;
- senior managers may have other agendas and force/persuade developers not to adhere to the methodology;
- time and/or budget pressures may lead to the abandonment of a methodology or to the taking of short-cuts;
- different methodologies in the organization may come into conflict and one or other may have to give way or adapt;
- contingency and judgement is applied dynamically as situations or problems develop over time.

As has been observed earlier, there exist a large number of different methodologies. This implies that there are large numbers of 'best' ways of developing information systems. Many of them are probably quite similar when closely examined, but many of them do differ substantially on fundamentals. It is the identification of these fundamentals that are of importance in our opinion, and Section 28.2 will discuss the comparison of methodologies. But, first, we categorize a number of eras of methodology evolution.

27.4 Evolution and development of methodologies

In considering the evolution and development of methodologies we have identified a number of methodology 'eras' as follows.

1 Pre-methodology era

The first era is known as the pre-methodology era and information systems, say until the early 1970s, were developed and implemented without the use of an explicit or formalized development methodology. In these early days, the emphasis of computer applications development was on programming and solving various technical problems. Indeed most of the problems were perceived to be in the technical arena, particularly those resulting from the rather limited hardware of the time. Developers were technically trained but rarely fully understood the business or the organizational context in which the systems were being implemented, and they were typically not very good at communicating with non-technical people. The needs of the users were rarely well established with the consequence that the system designs were often inappropriate to the real requirements of the application and the business. The approach to development was usually individualistic, depending on the experience and skills of the programmer, and based on some simple rules of thumb. This approach often resulted in poor control and management of projects. Emphasis was also placed on maintaining operational systems to get them working properly rather than on developing new systems and responding to the needs of users.

However, despite these problems, the demand for computer-based business systems was steadily increasing and management was demanding more appropriate systems for their expensive outlay. This led to a number of changes. First, there was a growing appreciation that analysis and design required different people with different skills to that of programmers. This led to the systems analyst as the key role in systems development. Second, there was a growing appreciation of the desirability for standards and a more disciplined approach to the development of information systems in organizations. Therefore, the first information systems development methodologies were established.

2 Early-methodology era

The early-methodology era of the 1970s and early 1980s was characterized by an approach to building computer-based applications that focused on the identification of phases and stages that it was thought would help control and enable the better management of systems development and bring a discipline to bear. This approach has come to be known as the Systems Development Life Cycle (SDLC) or the 'waterfall model'. It consisted of a number of stages of development that had to be followed sequentially (it has been discussed extensively in Chapter 3). There were a number of variants of the SDLC but usually the major stages were: feasibility study, systems investigation, analysis, design, development, implementation, and maintenance. One phase had to be completed before the next one began (hence the term waterfall), and each phase had a set of defined outputs or deliverables that had to be produced before the phase could be deemed complete. Therefore, it was also a mechanism for project control. In systems development it also became associated with a set of techniques, such as flowcharting, that were applied to particular phases. There was also the notion of iteration around the phases, as things changed and problems were encountered, although this was often ignored in practice.

The SDLC has a number of features to commend it (Chapter 3). It is tried and tested and, for some, it is well 'proven'. The use of documentation standards helps to ensure that proposals are complete and that they are communicated to both users and developers. There are controls and these, along with the division of the project into phases of manageable tasks with deliverables, help to avoid missed cutover dates and disappointments regarding what is delivered.

Unexpected cost overruns and unrealized benefits are also less likely. The approach also ensures that developers are trained and follow a well-formed, standard process and that users are also trained to enable them to use the systems effectively.

However, as we have seen in Section 3.6 there are a number of serious weaknesses or limitations to the approach, as well as limitations in the way it is used.

3 Methodology era

As a response to one or more of the limitations or criticisms of the SDLC a number of different approaches to systems development began to emerge in the mid to late 1980s, running through till the mid to late 1990s, and what we term 'the methodology era' began. In this era the term methodology was probably used for the first time to describe these different approaches, and methodologies proliferated.

Broadly speaking these methodologies emerged from one of two sources:
- those developed from practice;
- those developed from theory.

The methodologies in the first category have typically evolved from usage in an organization and then been developed into a commercial product. The second category of methodologies have typically been developed in universities or research institutions. These are usually written up in books and journals, although occasionally may have evolved into commercial products.

The commercial methodologies evolving from practice are the most widely known and some claim large numbers of users. They each have a different history, but it was often typically the case that system developers in an organization found that particular techniques they were using, or had helped to develop, were more useful than others, and they then concentrated on improving the use and effectiveness of these techniques. Typically, the people concerned would find that the organization for which they were working was not interested in investing in the development of the technique. Sometimes this resulted in the developers leaving the organization, and either setting up their own company or joining an existing consultancy company, where the opportunity to develop the techniques and methods was greater, using the clients as guinea pigs. At this stage, it was not the methodology itself, but the consultancy work developing information systems that was sold to clients.

Most of the early methodologies relied on one technique, or on a series of closely related techniques, as the foundation stone of the methodology. Commonly, these techniques were either entity modelling or data flow diagramming, but usually not both. It was only later that methodology authors began to include other techniques and to widen the scope and include prescriptions and phases or stages for development. Slowly these informal and somewhat *ad hoc* procedures or 'cookbooks' evolved into the early methodologies. From time to time the development of the methodologies would go through periods of introspection, where various aspects would be added and others dropped, and then the revised methodology would be put to the test, again by usage.

Sometimes consultants using the same methodology in a consultancy company would discover that they were doing things quite differently from their colleagues. They had their own styles and favourite elements, and yet they were supposed to be applying the same methodology. It was at this stage that some of these consultancy companies realized that it was no longer possible to rely on an ill-defined, inadequately researched, often conflicting set of procedures and techniques. Rather than selling consultancy, the realization began to dawn that the methodology itself had to be the product. In one organization, this happened when it was discovered that no one person had responsibility for the methodology and its content. People in the organization could add things to it as they thought fit, and they did. The main reason for this state of affairs was the nature of consultancy business. Consultants were charged out on the basis of the amount of time they spent on a client's project, and there was no mechanism for accounting for time spent developing a methodology. Therefore, nobody was responsible for the methodology because at that stage it was not something that was sold.

Eventually most organizations with a potential methodology product grasped the nettle and invested resources (people and time) into developing the methodology as a product. This involved ensuring that the methodology was:

- written up;
- made consistent;
- made comprehensive;
- made marketable;
- updated as needed;
- maintained;
- researched and developed;
- evolved into training packages;
- provided with supporting software.

The consultancy houses had finally invested in their methodologies. As a result of this investment, a number of things have happened.

Filling the gaps

It was realized that most of the methodologies had some gaps in them or, if not complete gaps, they had areas that were treated much less thoroughly than others. This was usually as a direct result of their background of development that had typically involved a concentration on a single, or small set, of development techniques. These gaps needed to be filled because their clients assumed that the methodologies covered the whole spectrum of activities necessary. It was often quite a surprise for users to find that this was not the case. For example, a methodology based on entity modelling techniques might have been very powerful for data analysis and database design, but not so comprehensive when it came to specifying functions and designing the applications, and might not provide any support for dialogue design. Almost all information systems methodologies went through a process of filling the gaps to make the methodology more comprehensive.

Expanding the scope

Another process was that of expanding the scope of methodologies. This occurred because the methodologies did not address the whole of the life cycle of systems development. Frequently, the implementation phase was not included, some did not include design, others did not address analysis. So, decisions were taken to expand the scope of the methodologies.

One of the most important aspects of this expansion of scope was for methodologies to expand into the areas of strategy and planning. The traditional life cycle of systems development usually started with a stage termed 'project selection' or 'problem identification', which was characterized by the identification of some problem to be solved, some area of the business where computerization was a possible option, or some application that needed building. The development process was viewed as a one-off, *ad hoc* solution to the identified problem, and, while this may have been a reasonable approach in the early days of systems development, it was now no longer valid. Organizations had had many such 'identified problems' solved in this one-off manner and found that, although the individual problem may have been solved to some extent, the existence of a variety of different one-off systems in a business did not lead to harmony nor any general improvement to business processes.

Further, it was realized that these individual problems were not so 'individual', and that almost all areas of the business needed to interact and integrate in some way. In particular, the requirements of tactical and strategic levels of management needed integration across traditional boundaries. A series of systems, developed as individual solutions, at different times, and without reference to each other, is unlikely to be the ideal starting point for such integration. Yet, for many organizations, this is just what existed, and methodologies were forced to address themselves to this situation if they were to provide more than improved implementations of one-off systems.

Therefore, in order to achieve this integration, and because the market was demanding it, some methodologies turned themselves to the topic of information systems strategy. This involved the recognition that:

1. Information systems were becoming a fundamental part of the organization, and that they could contribute significantly to the success, or otherwise, of the enterprise.
2. Information was increasingly being regarded by organizations as an important, and previously neglected, resource, in the same way as the more traditional resources of land, labour, and capital.

It was realized that such a resource was not free, as had been previously assumed, but needed to be controlled, co-ordinated, and planned. Further, the controlling and planning had to take place at the highest level in an organization in order for these resources to contribute effectively to fulfilling the organization's objectives. So, the starting point for effective information system development methodologies was now seen as a strategic plan for the organization, including a specification of the way in which information systems would contribute to the achievement of that plan.

In practice, it was found that, although most organizations, but by no means all, had some kind of strategic or corporate plan, this plan did not usually address the role of the information systems. For this reason, some methodologies incorporated the development of the required strategy plan into their scope. This not only helped to ensure that the information systems met the high-level needs of the organization, but effectively pushed the information system function up the hierarchy of importance in the organization.

Gaining competitive advantage

Another reason for addressing information systems strategy at a high level in an organization was the developing management belief that not only could the information system make the running of the business easier and more efficient, but that information systems and information technology could change the position of the organization in relation to its business competitors. The idea that information systems could enable business change and advantage was prevalent, and this meant that most commercial methodologies introduced strategic and planning phases. This, it was argued, not only ensured that the business and information systems strategy was in alignment, but it also often led to the influencing and determination of business strategy by the identification of new opportunities that information systems could provide.

The process of expanding their scope and 'filling the gaps' has continued for most commercial methodologies. After the introduction of strategic and business planning phases and tasks, the next development was to integrate new and evolving techniques and approaches, such as object-oriented techniques and the introduction of support tools into the methodology package.

As existing methodology vendors sought to fill the gaps and expand the scopes of their products, a number of other organizations entered the marketplace with new methodology products. These were only new in the sense that they sought to blend together what were seen as strong features of a variety of existing methodologies, in particular, the combination of entity modelling techniques with data flow diagramming techniques (see Chapter 21 on Blended Methodologies and SSADM in particular).

There were also other pressures that led to the development of more formalized methodologies, for example, the requirements of certain large organizations or standards bodies. Fitzgerald (1996), for example, suggests that the ISO (International Standards Organization) and the SEI (Software Engineering Institute) were influential in this respect, as was the perceived wisdom in some quarters that 'better methods will solve the problems of IS development'.

Fitzgerald et al. (2002) suggest that there has been a widespread bias in the literature that suggests that practitioners are steadily moving towards more formalized methods but, they argue, this is not borne out by empirical evidence. Yet, despite this, the literature has been highly influential in promoting the move to formalized methods and the methodology era. The literature frequently castigates IS practitioners for being undisciplined and individualistic in their approach to IS development and there is a 'prevalent view' that the many problems of IS development can be successfully addressed by the adoption of a methodology of some kind (Fitzgerald et al., 2002).

This process of expanding the scope and filling the gaps has sometimes resulted in methodologies becoming extremely large and often cumbersome. They seek to be comprehensive and provide all things to all people, but in doing so they have perhaps lost their original specialist focus and sown the seeds for some of a growing discontent with methodologies that we shall examine later. This expansion and filling of gaps also tended to result in some methodologies adopting a process that looks not unlike the conventional SDLC. This is relatively easy to understand as the SDLC, as its name implies, addresses the whole process and even with the advent of new techniques there was still seen to be a need for a life cycle type of process in some cases.

The second basic category of methodology has evolved from an academic or theoretical background. These are generally less well known than the commercial methodologies and may have relatively few users, although their influence is often substantially greater than their user base. Academic methodologies were usually developed by individuals who evolved and popularized the methodologies by means of action research and consultancy.

Typically these methodologies started life as research projects in universities or research institutions. Here the researchers took a more theoretical viewpoint and were less concerned by commercial considerations, although they often wanted access to real situations in order to test and experiment. The income from consultancy was also no deterrent. What intrigued the academic researcher, in particular, was that there did not seem to be any standard techniques or approaches based upon sound theoretical concepts. This was clearly a challenge which was taken up by a number of people from a variety of different background disciplines. It was sometimes felt that methods were already available and successfully being used in other disciplines and that these only needed a small amount of adaptation to be useful in the area of information systems development. Mathematicians, psychologists, linguists, social scientists, engineers, sociologists, and others turned their attention to the challenge. This did not, of course, happen all at the same time but some of the approaches have proved interesting and useful.

In areas other than information systems, the development of techniques and methods by academics has been very influential on practice, particularly in the areas of database design and software engineering. But in the case of information systems development methodologies, the impact was relatively low. Some argued that the research-based methods were not good enough, nor practical enough, or that they were no better than the new methods that the practitioners were already developing themselves. Slowly academics persuaded organizations to try their methods under their guiding hands on a consultancy basis. Others adopted a practice known as action research, which includes the experimentation with and testing of the methodology in a practical situation with the academic playing dual roles of participant and researcher.

As this process evolved, the methods became more practical, and the academic methodologies played an increasingly important role. The adoption of some elements from academic methodologies into commercial methodologies sometimes happened, while others attempted to combine the data and process techniques of entity modelling and data flow diagramming from commercial methodologies with academic methodologies, such as SSM and Multiview.

We characterize the above as the methodology era because of the apparent proliferation of methodologies during this period. This does not mean that every organization was using a methodology, particularly a named methodology purchased from a vendor, and the claims of methodology vendors at the time certainly need to be viewed with some suspicion. Nevertheless, many organizations were using a methodology of some kind, albeit frequently an in-house-developed methodology, which might or might not be based on a named commercial methodology.

A survey conducted in the UK by Fitzgerald et al. (1999) found that 57 per cent of the sample were claiming to be using a methodology for systems development, but, of these, only 11 per cent were using a commercial development methodology unmodified, whereas 30 per cent were using a commercial methodology adapted for in-house use, and 59 per cent a methodology which they claimed to be unique to their organization (i.e. one that was internally developed but not solely based on a commercial methodology). So, the picture seems to emerge that the majority of organizations were using some kind of methodology, but that most of these were developed or adapted to fit the needs of the developers and the organization. Therefore, although large-scale use of commercial methodologies is not the case, we justify the term methodology era on the number of organizations using a methodology of some kind, albeit often home-grown or adapted. Additionally, we argue that the influence of commercial methodologies is considerably larger than their use.

4 Era of methodology reassessment

The most recent era, from the mid to late 1990s onward, is characterized by a serious reappraisal of the concepts and practicalities of the methodologies of the methodology era. As a result some organizations have turned to yet different methodologies and approaches while others have abandoned their use of methodologies completely.

Methodologies were often seen as a panacea for the problems of traditional development approaches, and, as we have also seen, they were often chosen and adopted for the wrong reasons. Some organizations simply wanted a better project control mechanism, others a better way of involving users, still others wanted to inject some rigour or discipline into the process. For many of these organizations, the adoption of a methodology has not always worked or been the total success its advocates expected. Indeed, it was very unlikely that methodologies would ever achieve the more overblown claims made by some vendors and consultants. Some organizations have found their chosen methodology has not been successful or appropriate for them and have adopted a different one. For some this second option has been more useful, but others have found the new one not to be successful either, and some organizations appear to cycle through methodologies on a regular basis. This has led some developers and organizations to reject methodologies in more general terms and attack the concepts upon which they are based. This has been termed a 'backlash against methodologies', which we now discuss. The criticisms of methodologies are deliberately couched in generic terms and are not related to specific methodologies:

- *Productivity*. The first general criticism of methodologies is that they fail to deliver the suggested productivity benefits. It is said that they do not reduce the time taken to develop a project, rather their use increases systems development lead-times when compared with not using a methodology. This is usually because the methodology specifies many more activities and tasks that have to be undertaken. It specifies the construction of many more diagrams and models, and in general the production of considerably more documentation at all stages. Much of this may be felt by users to be unnecessary. As well as being slow, they are resource-intensive. This is true, first, in terms of the number of people required, from both the development and user side, and, second, from the point of view of the costs of adopting the methodology, for example, the purchase costs, training, tools, organizational costs, and so on.

- *Complexity*. Methodologies have been criticized for being over-complex. They are designed to be applied to the largest and most comprehensive development project and therefore specify in great detail every possible task that might conceivably be thought to be relevant, all of which is expected to be undertaken for every development project.

- *'Gilding the lily'*. Methodologies develop any requirements to the ultimate degree, often over and above what is legitimately required. Every requirement is treated as being of equal weight and importance which results in relatively unimportant aspects being developed to the same degree as those that are essential. It is sometimes said that they encourage the creation of 'wish lists' by users.

- *Skills*. Methodologies require significant skills in their use and processes. These skills are often difficult for methodology users and end-users to learn and acquire. It is sometimes also argued that the use of the methodology does not improve system development skills or organizational learning.

- *Tools*. The tools that the methodology advocates are difficult to use and do not generate enough benefits. They increase the focus on the production of documentation rather than leading to better analysis and design.

- *Not contingent*. The methodology is not contingent upon the type of project or its size. Therefore the standard becomes the application of the whole methodology, irrespective of its relevance.

- *One-dimensional approach*. The methodology usually adopts only one approach to the development of projects and while this may be a strength it does not always address the underlying issues or problems. In some cases the approach needs a more political, organizational, or other dimension.

- *Inflexible*. The methodology might be inflexible and does not allow changes to requirements during development. This is problematic as requirements, particularly business requirements, frequently change during the long development process.

- *Invalid or impractical assumptions*. Most methodologies make a number of simplifying yet invalid assumptions, such as a stable external and competitive environment. Many methodologies that address the alignment of business and information systems strategy assume the existence of a coherent and well-documented business strategy as a starting point for the methodology. This may not exist in practice.

- *Goal displacement*. It has frequently been found that the existence of a methodology standard in an organization leads to its unthinking implementation and to a focus on following the procedures of the methodology to the exclusion of the real needs of the project being developed. In other words, the methodology obscures the important issues. De Grace and Stahl (1993) have termed this 'goal displacement' and talk about the severe problem of 'slavish adherence to the methodology'. Wastell (1996) talks about the 'fetish of technique' which inhibits creative thinking. He takes this further and suggests that the application of a methodology in this way is the functioning of methodology as a social defence which he describes 'as a highly sophisticated social device for containing the acute and potentially overwhelming pressures of systems development'. He is suggesting that systems development is such a difficult and stressful process that developers often take refuge in the intense application of the methodology in all its detail as a way of dealing with these difficulties. Developers can be seen to be working hard and diligently, but this is in reality goal displacement activity because they are avoiding the real problems of effectively developing the required system.

- *Problems of building understanding into methods.* Introna and Whitley (1997) argue that some methodologies assume that understanding can be built into the method process. They call this 'method-ism' and believe it is misplaced. Method-ism assumes that the developers need to understand little or nothing about the problem situation, and that the method will somehow 'bring to light' all the characteristics that need to be discovered. Therefore, all that needs to be understood is the method itself. This, it is argued, is far too constraining and prevents real understanding of the problem situation emerging and being acted upon. It also inhibits the contingent use of methodologies. Introna and Whitley are not against methods as such, just this underlying assumption and its implications.

- *Insufficient focus on social and contextual issues.* The growth of scientifically based, highly functional methodologies has led some commentators to suggest that we are now suffering from an overemphasis on the narrow, technical development issues, and that not enough emphasis is given to the social and organizational aspects of systems development. Hirschheim et al. (1996), for example, argue that changes associated with systems development are emergent, historically contingent, socially situated, and politically loaded and that as a result sophisticated social theories are required to understand and make sense of IS development. They observe that these are sadly lacking in most methodologies.

- *Difficulties in adopting a methodology.* Some organizations have found it hard to adopt methodologies in practice. They have found resistance from developers who are experienced and familiar with more informal approaches to systems development and see the introduction of a methodology as restricting their freedom and a slight on their skills. One organization experienced these problems to the extent that they had to introduce the methodology by setting up a new development team from scratch, recruited from new graduates not tainted with the old ways of doing things. In other cases it has been the users who have objected to a methodology, because it did not embody the way they wished to work and included techniques for specifying requirements that they were not familiar with and did not see a good reason to adopt.

- *No improvements.* Finally in this list, and perhaps the acid test, is the conclusion of some that the methodology has not resulted in better systems, for whatever reasons. This is obviously difficult to prove, but nevertheless the perception of some is that 'we have tried it and it didn't help and it may have actively hindered'.

We thus find that for some the great hopes in the 1980s that methodologies would solve most of the problems of information systems development have not come to pass.

Strictly speaking, a distinction should be made in the above criticisms of methodologies between an inadequate methodology itself and the poor application and use of a methodology. Sometimes a methodology vendor will argue that the methodology is not being correctly or sympathetically implemented by an organization. While this may be true to some extent, it is not an argument that seems to hold much sway with methodology users. They argue that the important point is that they have experienced disappointments in their use of methodologies.

So, for the reasons above, some organizations have rejected the use of a methodology altogether and are returning to a less formal, more off the cuff, perhaps more flexible approach. While for others, it is not the concept of a methodology that is the problem, it is simply the inadequacy of the current methodologies, and they continue to seek a different and better methodology. Still others seek, not a better methodology, but an alternative to traditional in-house systems development. The directions we see these organizations moving in this era of methodology reappraisal are as follows:

1. *Ad hoc development.* This might be described as a return to the approach of the pre-methodology days in which no formalized methodology is followed. The approach that is adopted is whatever the developers understand and feel will work. It is driven by, and relies heavily on, the skills and experiences of the developers. This is perhaps the most extreme reaction to the backlash against methodologies and in general terms it runs the risk of repeating the problems encountered prior to the advent of methodologies. One area where, in the authors' experience, methodologies are not being used is in the development of web-based applications. Many of these applications are currently being developed in a very ad-hoc, trial-and-error manner relying on the skills and experience of a few key personnel in organizations. Not unlike the development approach of the pre-methodological era! We have chosen to illustrate WISDM (Section 23.4) but it is true that no methodology has become a standard for web development. Another group of organizations are pinning their faith on the evolution of toolsets to increasingly guide and automate the development process. Truex et al. (2000) have formalized ad-hoc development into what they refer to as **amethodical systems development**, an opposing viewpoint to methodical systems development. Amethodical systems development rejects structure; it occurs without a predefined sequence, control, rationality, or claims to universality. They argue that conventional descriptions of methodologies do not represent what happens in practice. In other words, it is an espoused theory, but does not reflect theories in use. Although we agree with the authors on this point, and some flexibility and interpretation of a methodology according to the problem situation at hand is

inevitable, there does need to be some support for developers for the purposes of training, control, maintenance and the development process itself.

2. *Further developments in the methodology arena.* For some there is the continuing search for the methodology holy grail. Methodologies will continue to be developed and existing ones evolve. For example, object-oriented techniques and methodologies have been gaining ground over process and entity modelling approaches for some time, although whether this is a fundamental advance is debatable. It may be that component-based development, which envisages development from the combination and recombination of existing components, will make a long-term impact. But this may simply be the current fashion to be overtaken by the next panacea at some point in the future. It may be that the RAD approaches will prevail, or perhaps the need for flexibility will favour prototyping approaches. The current emphasis on knowledge, rather than information, may make approaches like CommonKADS popular. With the importance of web applications, focusing on customers as stakeholders might make Customer Relationship Management (CRM) even more important. But these are conjectures made at the time of writing and it is difficult to predict the future. What we do know, based on past experience, is that proposed new solutions will come and go, some will be easily forgotten while others will probably stand the test of time and make genuine contributions. However, we believe it unlikely that any single approach will provide the solution to all the problems of information systems development.

3. *Contingency.* Most methodologies are designed for situations which follow a stated or unstated 'ideal type'. The methodology provides a step-by-step prescription for addressing this ideal type. However, situations are all different and there is no such thing as an 'ideal type' in reality. We therefore see a contingency approach to information systems development, where a structure is presented but tools and techniques are expected to be used or not (or used and adapted), depending on the situation, as being a third movement of this present era. Situations might differ depending on, for example, the type of project and its objectives, the organization and its environment, the users and developers, and their respective skills. The type of project might also differ in its purpose, complexity, structuredness, and degree of importance, the projected life of the project, or its potential impact. The organization might be large or small, mature or immature in its use of IT. Different environments might exhibit different rates of change: the number of users affected by the system, their skills, and those of the analysts. All these characteristics could affect the choice of development approach that is required. A contingent methodology allows for different approaches depending on situations. This is a reaction to the 'one methodology for all developments' approach that some companies adopted, and is a recognition that different characteristics require different approaches. There are, however, potential problems of the contingent approach as well. First, some of the benefits of standardization might be lost. Second, there is a wide range of different skills that are required to handle many approaches. Third, the selection of approach requires experience and skills to make the best judgements. Finally, it has been

suggested that certain combinations of approaches are untenable because each has different philosophies that are contradictory. Multiview aims to provide a framework which helps people make such contingent decisions (Section 26.1).

4. *External development*. We also see a movement toward external development in a variety of ways (Chapter 9). In particular we discuss the use of packages and outsourcing. Some organizations are attempting to satisfy their systems needs by buying packages from the marketplace. Clearly the purchasing of packages has been commonplace for some time, but the present era is characterized by some organizations deciding not to embark on any more in-house system development activities but to buy in all their requirements in the form of package systems. This is regarded by many as a quicker and cost-effective way of implementing systems for organizations that have fairly standard requirements. Only systems that are strategic or for which a suitable package is not available would be considered for development in-house. The package market is becoming increasingly sophisticated, and more and more highly tailorable packages are becoming available. Integrated packages which address a wide range of standard business functions, purchasable in modular form, known as Enterprise Resource Packages (ERPs) have emerged in the last few years and have become particularly popular with large corporations. The key for these organizations is ensuring that the correct trade-off is made between a standard package, which might mean changing some elements of the way the business currently operates, and a package that can be modified to reflect the way they wish to operate. There are dangers of becoming locked in to a particular supplier and of not being in control of the features that are incorporated in the package, but many companies have taken this risk. For others, the continuing problems of systems development, and the perceived failure of methodologies to deliver, has resulted in them outsourcing systems development to a third party. The client organization is no longer so concerned with how a system is developed and what development approach or methodology is used, but with the end results and the effectiveness of the system that is delivered. This is different to buying in packages or solutions, because normally the management and responsibility for the provision and development of appropriate systems is given to a vendor. The client company has to develop skills in selecting the correct vendor, specifying requirements in detail, and writing and negotiating contracts rather than thinking about system development methodologies.

The above features of the present era of methodology reappraisal as we see it are not mutually exclusive, and some organizations are moving to a variety of these approaches. Some aspects are being absorbed or incorporated into some existing methodologies (i.e. the 'filling the gaps' and 'blending' process is still continuing). This present era is not one where all methodologies have been abandoned. It is an era where there is diversity and perhaps a more realistic view of the limitations of methodologies. For some, however, it is about the abandonment of methodologies altogether. For others, it is seeking improved methodologies, but moving away from the highly bureaucratic types of the methodology era. For still others it is about moving out of in-house systems development altogether. But it should also not be forgotten that even in the post-methodology era some are still using methodology-era methodologies effectively and successfully.

Our identification and characterization of these methodology eras has been done to provide a more categorized view of the history and evolution of methodologies and to make such a history more understandable. However, it has been criticized by some because they do not recognize the concept of the methodology era itself. They argue there was never a period when methodologies proliferated, particularly in terms of their use; we disagree but as with any historical categorization it is open to debate and interpretation. Our hope is that we have engendered, and contributed to, such a debate.

Summary

- In this chapter we have further refined the definition of the term 'methodology' to include the critical issue of 'philosophy', or underlying theory behind a methodology. The philosophy is key to understanding the methodology.

- We also looked at the rationale for adopting a methodology under the headings: a better end product, a better development process, and a standardized process.

- The issue of what you get when you buy or adopt a methodology in practice was addressed. This varies considerably from a fully comprehensive product, including manuals and software, covering the range of the life cycle, to a short paper or a book.

- Next an historical examination of the evolution of methodologies was undertaken which identified four methodology eras: a pre-methodology era, an early-methodology era, a methodology era, and the current era which is described as an era of methodology reassessment. This means that we are now in an era which is perhaps more critical and demanding of methodologies and that does not see them as a panacea for the problems of systems development. In this era a range of approaches exist, including the use of a methodology through to *ad hoc* development without a methodology or outsourcing of development.

Questions

1. How would you define a methodology and why?
2. In the context of a company with which you are familiar, choose a methodology, and identify a convincing rationale for adopting it that might persuade the Chief Executive.
3. What would you expect to find in the box if you purchased a methodology from a well-known source?
4. Why is the current methodology era described as one of reassessment?

Further reading

Fitzgerald, B., Russo, N.L., and Stolterman, E. (2002) *Information Systems Development: Methods in Action*, McGraw-Hill, Maidenhead, UK.

Introna, L.D. and Whitley, E.A. (1997) Against method-ism: Exploring the limits of method, *Information Technology and People*, Vol. 10, No. 1, 31–45 (MCB University Press).

28 Methodology comparisons

28.1 Comparison issues

There are two main reasons for comparing methodologies:

- *an academic reason* – to better understand the nature of methodologies (their features, objectives, philosophies, and so on) in order to perform classifications and to improve future information systems development;
- *a practical reason* – to choose a methodology, part of one, or a number of methodologies for a particular application, a group of applications, or for an organization as a whole.

The two reasons are not totally separate, and it is hoped that the academic studies will help in the practical choices and that the practical reasons will influence the criteria applied in the academic studies. In this section we look at a number of different approaches to the comparison of methodologies which have been attempted, and discuss some of the issues that arise from such comparisons.

We provide the following list of 'ideal-type' criteria that might be considered in assessing methodologies:

- *Rules.* Methodologies should provide formal guidelines to cover phases, tasks, and deliverables, and their ordering, techniques and tools, documentation and development aids, and guidelines for estimating time and resource requirements.
- *Total coverage.* A methodology should cover the entire systems development process from strategy to cutover and maintenance.
- *Understanding the information resource.* A methodology should ensure effective utilization of the corporate information resource, in terms of the data available and the processes which need to make use of the data.
- *Documentation standards.* There should be agreed standards, and all output, using the methodology, should be easily understandable by both users and analysts.
- *Separation of logical and physical designs.* There should be a separation of logical descriptions and requirements (e.g. what an application does, what the interactions are, and what data are involved).
- *Validity of design.* There should be a means of checking for inconsistencies, inaccuracies, and incompleteness.

- *Early change.* Any changes to a system design should be identifiable as early as possible in the development process.
- *Inter-stage communication.* The full extent of work carried out must be communicable to other stages, with each stage being consistent, complete, and usable.
- *Effective problem analysis.* The methodology should provide a suitable means for expressing and documenting the problems and objectives of an organization.
- *Planning and control.* Careful monitoring of an information system is required, and a methodology must support development in a planned and controlled manner to contain costs and timescales.
- *Performance evaluation.* The methodology should support a means of evaluating the performance of operational applications developed using it.
- *Increased productivity.* The use of a methodology should lead to increases in productivity.
- *Improved quality.* A methodology should improve the quality of analysis, design, and programming products and hence the overall quality of the information system.
- *Visibility of the product.* A methodology should maintain the visibility of the emerging and evolving information system as it develops.
- *Teachable.* Users as well as technologists should understand the various techniques in a methodology in order that they can verify analysis and design work and train others to use them.
- *Information systems boundary.* A methodology should allow definition of the areas of the organization to be covered by the information system. These may not all be areas for computerization.
- *Designing for change.* The logical and physical designs should be easily modified.
- *Effective communication.* The methodology should provide an effective communication medium between analysts and users.
- *Simplicity.* The methodology should be simple to use.
- *Ongoing relevance.* A methodology should be capable of being extended so that new techniques and tools can be incorporated as they are developed, but still maintain overall consistency and framework.
- *Automated development aids.* Where possible software tools should be used since they can enhance productivity.
- *Consideration of user goals and objectives.* The goals and objectives of potential users of a system should be noted, so that when an information system is designed it can be made to satisfy these users and assist them in meeting goals and objectives.
- *Participation.* The methodology should encourage participation through simplicity and the ability to facilitate good communications.
- *Relevance to practitioner.* The methodology has to be appropriate to the practitioner using it, in terms of level of technical knowledge, experience with computers, and social and communications skills.
- *Relevance to application.* The methodology must be appropriate to the type of system being developed, which might be real-time, distributed, web-based, etc.

- *The integration of the technical and the non-technical systems.* The methodology should not only address the technical and non-technical aspects of a system but should make provision for their integration.
- *Scan for opportunity.* The methodology should enable the system to be thought about in new ways. Rather than being viewed as simply a solution to existing problems it should be seen as an opportunity to address new areas and challenges.
- *Separation of analysis and design.* This separation ensures that the analysis of the existing system and the user requirements are not influenced by design considerations.

Of course, the above 'ideal-type' requirements will not be found in any one methodology. However, the above criteria form a useful checklist but clearly need to be tailored for the particular purpose. Other commentators have taken these debates further and suggested a much broader range of issues that they feel are relevant in the comparison of methodologies. Bjørn-Andersen (1984), for example, identifies a checklist that includes criteria relating to values and society:

1. What research paradigms/perspective form the foundation for the methodology?
2. What are the underlying value systems?
3. What is the context where a methodology is useful?
4. To what extent is modification enhanced or even possible?
5. Does communication and documentation operate in the users' dialect, either expert or not?
6. Does transferability exist?
7. Is the societal environment dealt with, including the possible conflicts?
8. Is user participation 'really' encouraged or supported?

This checklist is useful as it focuses attention on some wider issues that are often ignored.

It is, of course, a subjective list and makes a number of assumptions, for example, that user participation is a desirable feature.

Another comparison framework is NIMSAD (Normative Information Model-based Systems Analysis and Design) (Jayaratna, 1994). This is based on the models and epistemology of systems thinking and to a large degree evaluates and measures a methodology against these criteria. The evaluation has three elements:

- the 'problem situation' (the methodology context);
- the intended problem solver (the methodology user);
- the problem-solving process (the methodology).

The evaluation of the elements is wide-ranging and expressed in terms of the kinds of question that require answers. The questions concerning the first element (the problem situation) deal with the way in which the methodology helps the understanding and identification of the following:

- the clients and their understandings, experiences, and commitments;
- the problem owners, their concerns, and problems;

- the situation that the methodology users are facing, its diagnosis as structured or unstructured;
- the ways in which the methodology might help the situation;
- the culture and politics of the situation, including the risks associated with using the methodology in various circumstances;
- the views of the stakeholders concerning 'reality', for example, is there an objective real-world problem out there, and what is the relationship of this to the methodology's philosophical assumptions about reality?;
- the dominant perceptions in the problem situation; for example, are they technical, political, social, and so on?

The questions concerning the second element (the intended problem solver) ask about:

- the methodology users' beliefs, values, and ethical positions;
- the relationship of the above to that assumed or demanded by the methodology;
- the way in which mismatches in the above two may be handled or reconciled; for example, can the methodology processes be changed, and does the methodology help to achieve this?;
- the methodology users' philosophical views, for example, science or systems-based;
- the methodology users' experience, skills, and motives, in relation to those required by the methodology.

The questions concerning the third element (the methodology itself) ask about the way in which the methodology provides specific assistance for:

- understanding the situation of concern and the setting of boundaries;
- performing the diagnosis; for example, the models, tools, techniques, and the levels at which they operate, how they interact, what information they capture, what is not captured, what happens when people disagree, and so on;
- defining the prognosis outline, for example, the desired states, what constitutes legitimate states, and the handling of conflict;
- defining problems;
- deriving notional systems; for example, are they derived at all, and if so how, and in what ways are they recorded?;
- performing conceptual/logical and physical design, including who is involved and what are the implications; for example, does it lead to systems improvements or systems innovation?;
- implementing the design; for example, how does it handle alternatives and how does it ensure success?

One feature of this framework is that it recommends that the evaluation be conducted at three stages. First, before intervention (i.e. before a methodology is adopted), second, during intervention (i.e. during its use), and finally, after intervention (i.e. after an assessment of the success or otherwise of the methodology). These stages are an important feature of the framework and introduce the important element of organizational learning.

A number of other commentators have suggested alternative approaches to an overall feature analysis when selecting methodologies. They adopt a more pragmatic approach and suggest that there is no benefit to be gained from attempting to find, in isolation, a 'best' methodology, because the approaches are not necessarily mutually exclusive. One or more approaches may be suitable, depending on the circumstances. Therefore, there should be a search for an appropriate methodology in the context of the problems being addressed, the applications, and the organization and its culture.

Davis (1982) advocates the contingency approach, that is, the selection of an appropriate approach as part of the framework or methodology itself. Davis offers guidelines for the selection of an appropriate approach to the determination of requirements, rather than to the selection of a methodology itself (although the same principles may well apply). Davis suggests measuring the level of uncertainty in a system. This will help ascertain the appropriate methodology. There are four measures:

1. System complexity or ill structuredness.
2. The state of flux of the system.
3. The user component of the system, for example, the number of people affected and their skill level.
4. The level of skill and experience of the analysts.

Once the level of uncertainty has been ascertained, the appropriate approach to determining the requirements can be made. For low uncertainty, the traditional method of interviewing users would be appropriate. For high levels of uncertainty, a prototype or an evolutionary approach would be better. For intermediate levels of uncertainty, a process of synthesizing from the characteristics of the existing system might be appropriate.

Avison and Taylor (1996) identify five different classes of situation and appropriate approaches as follows:

1. Well-structured problem situations with a well-defined problem and clear requirements. A traditional SDLC approach might be regarded as appropriate in this class of situation.
2. As above but with unclear requirements. A data, process modelling, or a prototyping approach is suggested as appropriate here.
3. Unstructured problem situation with unclear objectives. A soft systems approach would be appropriate in this situation.
4. High user-interaction systems. A people-focused approach, for example, ETHICS, would be appropriate here.
5. Very unclear situations, where a contingency approach, such as Multiview, is suggested.

In addition to some of the conceptual problems of comparing methodologies discussed above, we also discovered a number of practical problems in attempting to compare methodologies themselves. First, methodologies are not stable, but are in fact moving targets that are continuing to develop and evolve. Therefore, there exists a version problem, and it is often difficult to know which version of a methodology is being applied in a particular situation or which is the latest version. Second, for commercial reasons, the documentation is not always published or readily available to people or organizations not purchasing the methodology. Third, the practice of a methodology is sometimes significantly different from that prescribed by the documentation or the authors of the methodology. This is sometimes talked about in terms of the espoused version of the methodology and the way that it is actually used in practice. Fourth, consultants or developers using the methodology often interpret aspects of the methodology in quite different ways.

A further problem in undertaking comparisons concerns terminology, in particular, the use of different terms for the same phenomena and similar terms for different phenomena. Information systems continues to exhibit rather more than its fair share of these. It is unhealthy, as it not only leads to confusion and poor communication, but to a restriction of development, due to the inability to enhance and expand upon earlier research work. Progress in most successful disciplines is usually a process of evolutionary development because 'out of the blue' quantum leaps are rare. Any restriction in evolution is therefore very serious. One problem is that sometimes 'new' approaches only express a new terminology and make no substantial contribution to the state of the art.

A common approach to methodology comparison attempts to identify a set of idealized 'features', followed by a check to see whether different methodologies possess these features or not, the implication being that those that do possess them or at least score highly on a features rating are 'good', and that those that do not are 'less good'. The set of features must be chosen by somebody and are thus subjective, although often purported to be objective. The most obvious indication of this is the kind of comparison conducted by particular methodology vendors in which their methodology scores highly and the competition poorly. The comparison has been designed to give this result. We are more familiar with this kind of comparison in relation to cars or soap powders.

We identify our own set of comparison criteria below, in the full knowledge that they are subjective and can be criticized in exactly the same way as all the other attempts discussed above:

1. What aspects of the development process does the methodology cover?
2. What overall framework or model does it utilize? For example, is it systems development life cycle based, linear, or spiral?
3. What representation, abstractions, and models are employed?
4. What tools and techniques are used?
5. Is the content of the methodology well defined and described, such that a developer can understand and follow it? This applies not only to the stages and tasks but also to the philosophy and objectives of the methodology.
6. What is the focus of the methodology? Is it, for example, people-, data-, process-, and/or problem-oriented? Does it address organizational and strategic issues?

7. How are the results at each stage expressed?

8. What situations and types of application is it suited to?

9. Does it aim to be scientific, behavioural, systemic, or whatever?

10. Is a computer solution assumed? What other assumptions are made?

11. Who plays what roles? Does it assume professional developers, require a methodology facilitator, involve users and managers, and, if so, how and to what degree?

12. What particular skills are required of the participants?

13. How are conflicting views and findings handled?

14. What control features does it provide and how is success evaluated?

15. What claims does it make as to benefits? How are these claims substantiated?

16. What are the underlying philosophical assumptions of the methodology? What makes it a legitimate approach?

Perhaps the most important aspects of this list in terms of comparing methodologies has been found to be the final one. The features of a methodology are highly dependent upon the philosophy, and without this understanding the methodology is difficult to explain. Some methodologies, especially the more commercial ones, do not always explicitly state their underlying philosophy: it has to be searched for and interpreted, and this makes analysis difficult. Others are more explicit; some of the object-oriented methodologies make great play of explaining the reasoning behind the concept of objects. However, relatively few explain their philosophy or in Checkland's terms the 'intellectual framework of enquiry'.

Despite all the difficulties identified, which may imply failure from the outset, comparisons continue to be made, because it is becoming increasingly important in a world where large numbers of widely differing methodologies claim the same promises of universal applicability and overall usefulness.

28.2 Framework for comparing methodologies

The reader should now be aware that comparing methodologies is a very difficult task, and the results of any such work are likely to be criticized on many counts. There are as many views as there are writers on methodologies. The views of analysts do not necessarily coincide with users, and these views are often at variance with those of the methodology authors. Therefore, the following is simply another set of views and is unlikely to satisfy all the players in the methodologies' game.

There are seven basic elements to our framework as follows; some elements are broken down into a number of subelements (see Figure 28.1).

The framework is not supposed to be fully comprehensive, and one could envisage a number of additional features that might usefully be compared for particular purposes, for example:

- the speed at which systems can be developed;
- the quantity of the specifications and documentation produced;
- the potential for modification by users to suit their own environment.

However, we believe that this gives a set of features that proves to be a reasonable guide when examining an individual methodology and when used as a basis for comparing methodologies. The headings are not mutually exclusive and there are obviously interrelationships between them. For example, aspects of philosophy are reflected in some senses in all the other elements.

Each of the above-listed elements in the framework will be described, and then a sample application of the framework is made to some of the methodologies of Part VI. The reader is also invited to apply the framework to methodologies of his or her own choice.

1 Philosophy

The question of philosophy is an important aspect of a methodology because it underscores all other aspects. As discussed in Section 27.1, it distinguishes, more than any other criterion, a 'methodology' from a 'method'. The choice of the areas covered by the methodology, the systems, data or people orientation, the bias or otherwise toward a purely IT solution, and other aspects are made on the basis of the philosophy of the methodology. This philosophy may be explicit but in most methodologies the philosophy is implicit. Indeed, many feature analyses have neglected this aspect completely, partly because methodology authors seldom stress their philosophy.

In this context we regard 'philosophy' as a principle, or set of principles, that underlie the methodology. It is sometimes argued that all methodologies are based on a common philosophy to improve the world of information systems development. While this is true to some extent, this is not what is meant here by philosophy.

As a guide to philosophy the four factors of paradigm, objectives, domains, and applications are highlighted.

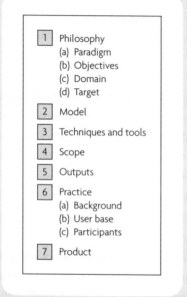

1 Philosophy
   (a) Paradigm
   (b) Objectives
   (c) Domain
   (d) Target

2 Model

3 Techniques and tools

4 Scope

5 Outputs

6 Practice
   (a) Background
   (b) User base
   (c) Participants

7 Product

Figure 28.1: The seven elements of the framework for comparing methodology

(a) Paradigm

We identify two paradigms of relevance. The first is the science paradigm, which has characterized most of the hard scientific developments of recent times, and the second is the systems paradigm, which is characterized by a holistic approach.

The term paradigm is defined here, in the sense that Kuhn (1962) uses it, as a specific way of thinking about problems, encompassing a set of achievements which are acknowledged as the foundation of further practice. A paradigm is usually regarded as subject-free, in that it may apply to a number of problems regardless of their specific content.

As Checkland (1981) summarizes it, the science paradigm consists of reductionism, repeatability, and refutation:

> We may reduce the complexity of the variety of the real world in experiments whose results are validated by their repeatability, and we may build knowledge by the refutation of hypotheses.

The science paradigm has a long and successful history and is responsible for much of our current world. The systems paradigm has a much shorter and less successful history, but was evolved as a reaction to the reductionism of science and its perceived inability to cope with living systems and those categorized as human activity systems.

Science copes with complexity through reductionism, breaking things down into smaller and smaller parts for examination and explanation. This implies that the breakdown does not disrupt the system of which it is a part. Checkland argues that human activity systems are systems which do not display such characteristics; they have **emergent properties** (i.e. the whole is greater than the sum of the parts) and perform differently as a whole or as part of a system than when broken down to their individual components. This led directly to the development of the systems paradigm which is characterized by its concern for the whole picture, the emergent properties, and the interrelationships between parts of the whole. The science and systems paradigms are closely related to concepts of hard and soft thinking discussed in Section 4.1.

In a series of papers, Iivari et al. (2000/2001) extend the debate, distinguishing between ontology and epistemology. Ontology is concerned with the essence of things and the nature of the world, and two extreme positions of realism and nominalism are identified. Realism, according to Hirschheim (1985), 'postulates that the universe comprises objectively given, immutable objects and structures. These exist as empirical entities, on their own, independent of the observer's appreciation of them.' On the other hand, according to Hirschheim and Klein (1989), nominalism is where:

> reality is not a given immutable 'out there', but is socially constructed. It is the product of the human mind. Social relativism is the paradigm adopted for understanding social phenomena and is primarily involved in explaining the social world from the viewpoint of the organizational agents who directly take part in the social process of reality construction.

Epistemology is the grounds of knowledge. The term relates to the way in which the world may be legitimately investigated and what may be considered as knowledge and progress. It includes elements concerned with sources of knowledge, structure of knowledge, and the limits of what can be known. Again, two extreme positions are frequently identified: positivism and interpretivism. Positivism implies the existence of causal relationships which can be investigated using scientific methods whereas interpretivism implies that there is no single truth that can be 'proven' by such investigation. Different views and interpretations are potentially legitimate and the way to progress is not to try to discover the one 'correct' view but to accept the differences and seek to gain insight by a deep understanding of such complexity.

Lewis (1994) brings these elements together in a framework, shown in Figure 28.2, that identifies objectivist and subjectivist approaches as positions between the ontology of realism and nominalism and the epistemology of positivism and interpretivism. We find this a helpful framework for thinking about and identifying the underlying philosophies of methodologies. For example, if we believe in an ontology of nominalism we should not adopt a methodology

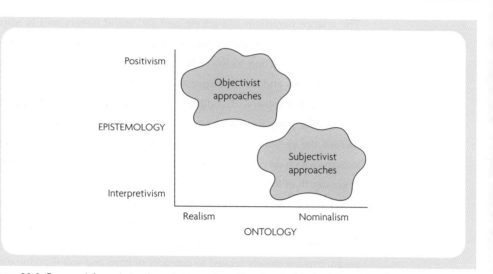

Figure 28.2: Framework for analysing the underlying philosophies of methodologies (adapted from Lewis, 1994)

based on an assumption of realism. A subjectivist approach might help us think about data collected in the analysis of a system not as 'facts' but more as perceptions from a particular point of view. Sales targets are not necessarily a set of facts, but perhaps part of a political process which is negotiated between sales personnel, management, and directors. This negotiation may have far-reaching implications, relating to people's lives, remuneration, job satisfaction, self-esteem, and so on. Furthermore, we need ways of handling these different perceptions. This may mean, for example, that we need a methodology that focuses on the highlighting of different opinions and interpretations or one that allows a series of different designs in order to accommodate different perceptions.

It is sometimes argued that not only do methodologies adopt particular philosophical assumptions that we should be aware of but that individual techniques and tools also embody and reflect these concepts. Entity, process, and object modelling are usually all seen as objectivist techniques, but Lewis (1994) argues that 'this does not rule out the possibility that they could be employed as part of an alternative (subjectivist) paradigm'. He suggests that some analysts use the technique in this way in practice. They do not think of particular entities and relationships as representations of reality, but as representations of interpretations of reality. They use it as a way of negotiating a shared understanding or interpretation of the problem domain rather than as a way of correcting the user's ambiguities and misunderstandings.

This raises some interesting issues, for if a generally perceived objectivist technique can be used in a subjectivist manner, then this implies that the techniques themselves are not such an important feature in determining the characteristics of a methodology as usually thought. However, we could argue that entity models are not generally used in a subjectivist way but, even if they can be, the technique is still objectivist because it pushes the user of the technique in that general direction. It is a sophisticated developer that does not seek to resolve differing views to one version of reality. It can be argued, however, that it is not so much whether a technique assumes an independent existence of reality or a consensus interpretation of reality but whether it is able to handle differing perceptions of reality that cannot be resolved by consensus.

Another issue is that the content and expression of a technique can 'frame' and limit the user's perception of the problem situation and potential solution to that allowed by the technique. This framing effect can also apply to tools and even methodologies themselves.

This is a complex area, but the purpose of the discussion has been to indicate the importance of a methodology's underlying philosophies and assumptions and that developers need to be aware that they should match their beliefs to that of the authors of the methodology. As may be appreciated, this discussion of ontology and epistemology is an oversimplification of a long history of philosophical debate, and the reader is referred to Burrell and Morgan (1979), Searle (1995), Walsham (1993), and Iivari et al. (2000) for more detailed treatments.

(b) Objectives

One fairly obvious clue to the methodology philosophy is the stated objective or objectives. Some methodologies state that the objective is to develop a computerized information system. Others, for example, ISAC, have as an objective to discover if there really is a need for an information system. So, there exists a difference in that some methodologies are interested only in aspects that are 'computerizable' while others take a wider view and direct their attention to achieving solutions or improvements no matter what this implies. Such improvements might involve manual, procedural, managerial, organizational, educational, or political change. This is an important difference, because it determines the boundaries of the area of concern. The problem with concentrating only on aspects to be computerized is that this is an artificial boundary in terms of the logic of the business. There is no reason why the solution to a particular problem should reside only in the area that can be automated. More likely the problem needs to be examined in a wider context.

It has often been found that 'computerization' *per se* is not the answer. What is required is a thorough analysis and redesign of the whole problem area. It may be viewed as 'putting the cart before the horse' to decide that an IT system is the solution to a particular problem. Clearly a methodology that concerns itself solely with providing an IT solution is quite a different methodology from one that does not. In choosing and understanding a methodology, it is important to ask the question: 'could the use of this methodology lead to the implementation of a purely organizational or non-IT solution?' If the answer is 'no' then the methodology is not addressing the same problems as one to which the answer is 'yes'. It is interesting to note that many of the most widely used commercial methodologies would seem to answer 'no' to this question, whereas most of the academic methodologies would probably answer 'yes'. An exception would be the 'new' BPR methodologies, whose focus is not primarily on IT but on generalized improvements to business processes.

(c) Domain

The third factor relating to philosophy is the domain of situations that methodologies address. Early methodologies, such as the conventional life cycle approach discussed in Chapter 3, saw their task as overcoming a particular and sometimes narrow problem. Obviously the solving of the problem, often through the introduction of a computer system of some kind, might be beneficial to the organization. However, the solution of a number of these kinds of problem on an *ad hoc* basis at a variety of different points in time can lead to a mishmash of different physical systems being in operation at the same time.

Even if the developments of solutions to the different problems have been well co-ordinated, and the later systems have been designed with the earlier systems in mind, it is often found that the systems and problems interrelate and the solution to a number of interrelated problems is different to the sum of the solutions to the individual problems viewed in isolation. This has led to a number of methodologies adopting a different development philosophy. They take a much wider view of their starting point and are not looking to solve, at least in the first instance, particular problems. The argument is a basis of a number of approaches, for example, the systems and strategic approaches described in Sections 4.1 and 4.2. In other words, it is argued that in order to solve individual problems, it is necessary to analyse the organization as a whole, devise an overall information systems strategy, sort out the data and resources of the organization, and identify the overlapping areas and the areas that need to be integrated. In essence, it is necessary to perform a top-down analysis of the organization, sort out the strategic requirements of the business, and in this way ensure that the information systems are designed to support these fundamental requirements.

(d) Target

The last aspect of philosophy is the applicability of the methodology. Some methodologies are specifically targeted at particular types of problem, environment, or type or size of organization, while others are said to be general purpose.

2 Model

The second element of the framework concerns an analysis of the model that the methodology adheres to. The model is the basis of the methodology's view of the world, it is an abstraction and a representation of the important factors of the information system or the organization.

The model works at a number of levels: first, it is a means of communication; second, it is a way of capturing the essence of a problem or a design, such that it can be translated or mapped to another form (e.g. implementation) without loss of detail; and, third, the model is a representation which provides insight into the problem or area of concern.

Models have been categorized into four distinct types:

1. Verbal.
2. Analytic or mathematical.
3. Iconic, pictorial, or schematic.
4. Simulation.

The models of concern in the field of information systems methodologies are almost exclusively of the third type. The reason for the current dominance of iconic, pictorial, or schematic models is the perceived importance of using the models as a means of communication, particularly between users and analysts. Further important aspects of the models in information systems development are to ensure that the information necessary is captured, at the appropriate stage, and that this information is that required to be able to develop a working system.

3 Techniques and tools

A key element of the framework is the identification of the techniques and tools used in a methodology, but as these have been extensively discussed in Parts IV and V they will not be further elaborated here.

4 Scope

The scope of a methodology is the next element in the framework. Scope is an indication of the stages of the life cycle of systems development which the methodology covers (Chapter 3). Further, an analysis of the level of detail at which each stage is addressed is useful. The problem with using the stages of the life cycle as a basis for the examination of scope is that there are methodologies that do not follow a life cycle; they may adopt a more iterative, evolutionary, or spiral model (Section 7.1). Nevertheless we think an examination of the scope of a methodology in relation to the life cycle is still useful.

5 Outputs

The next element in the framework concerns the outputs from the methodology. It is important to know what the methodology is producing in terms of deliverables at each stage and, in particular, the nature of the final deliverable. This can vary from being an analysis specification to a working implementation of a system and all its related procedures.

6 Practice

The next element of the framework is termed the practice and is measured according to: (a) the methodology background (discussed in Section 27.4 in terms of commercial or academic); (b) the user base (including numbers and types of user); (c) the participants in the methodology (e.g. can it be undertaken by users themselves or must professional analysts be involved); and what skill levels are required. The practice should also include an assessment of difficulties and problems encountered, and perceptions of success and failure. This should be undertaken by investigating the experiences of prior users of the methodology. This will inevitably be subjective, depending on who is consulted, but it can be a revealing exercise. In examining the practice, the degree to which the methodology can be, and is, altered or interpreted by the users according to the requirements of the particular situation should be assessed. It is also important to assess any differences there appear to be between the practice and the theory of the methodology.

7 Product

The last element of the framework is the product of the methodology, that is, what purchasers actually get for their money. This may consist of software, written documentation, an agreed number of hours' training, a telephone help service, consultancy, and so on.

28.3 Comparison

In this section we offer a discussion of some of the methodologies we have described in this text in the context of the comparison framework identified in Section 28.2 above. This discussion is illustrative, rather than comprehensive, so we do not include all methodologies from Part VI. For example, we exclude a number of methodologies because of their restricted application domain, such as KADS (expert systems development), Welti ERP, and Renaissance (legacy systems). We also exclude those that are frameworks, such as Euromethod, SODA, Multiview, and CMM. We hope readers will regard our comparison as a basis for stimulating debate rather than a statement of 'facts' about methodologies. It represents only our subjective view.

1 Philosophy

Element 1 of the framework concerns the identification of the philosophy of the methodology. There are a number of subelements to philosophy which we will examine in turn.

The first subelement is paradigm, and it provides ample illustration of the difficulties in attempting to compare methodologies. In the discussion of the framework, we identified the science paradigm and the systems paradigm to be of critical importance. We suggest that SSM and ETHICS belong to the systems paradigm and that STRADIS, YSM, IE, SSADM, Merise, JSD, OOA, RUP, DSDM, and ISAC belong to the science paradigm.

(a) Paradigm

It is clear that SSM adopts the systems paradigm, indeed it is explicitly stated to do so by the methodology author. This is one of the few cases where the issue is addressed as part of the methodology itself. But even if we were not so told, it is clear that the methodology uses many systems concepts and avoids a reductionist approach.

In ETHICS it is the belief in the interaction of the social and the technical subsystems (the socio-technical approach) that leads to an advocacy of the participative design philosophy. The work system is analysed for variances or weaknesses which prevent the systems objectives being realized. These variances are often discovered at subsystems boundaries, particularly where the social and technical subsystems meet. The ideas of job enrichment and participative design are particular solutions to the more common variances which are encountered. In addition, ETHICS makes no attempt to break down the system into its constituent parts for the purpose of understanding the problems. Therefore, the underlying paradigm for ETHICS is, we believe, the systems paradigm.

In the analysis of paradigm, ISAC generates the most discussion. ISAC is probably one of the less well-known and used methodologies we describe, but it is important in relation to its philosophical contribution. While it is clear that ISAC is firmly in the participative tradition, we believe that this does not mean it automatically incorporates the systems paradigm. The ETHICS methodology is also highly participative, but it is more the socio-technical aspects which lead us to classify that as being of the systems paradigm. ISAC adopts a more reductionist approach to the understanding of systems. Its authors state (Lundeberg et al., 1982) that:

The only way to solve complex problems is to divide them into sub-problems until they become manageable. A requirement for this to work is that the solution to the sub-problems gives the solution for the problem as a whole, that is, that the division in sub-problems is coherent.

This would appear to be a categorical endorsement of the science paradigm, and despite its socio-technical spirit we classify ISAC to be in the science paradigm, although if we accept the notion of a continuum it would not be at the extreme end. We also, on the basis of their clear reductionist approaches and their acceptance of the ontological position of realism, identify STRADIS, YSM, IE, SSADM, Merise, JSD, RUP, DSDM, and OOA to belong to the science paradigm.

IE, for example, adopts what is termed the 'divide and conquer' approach which is clearly reductionist, and JSD describes its approach to modelling as attempting to reflect the real-world situation; for example, in the entity action step, 'real-world entities are defined' without any discussions of the real world being socially constructed or any problems that might be encountered concerning differing perceptions of the real world.

One of the interesting aspects of applying the comparison framework is not so much whether the classification is right, but the discussion and debate that it generates. The debate proves insightful and causes many significant questions to be asked of the methodologies, some of which prove difficult to answer.

(b) Objectives

The second subelement of philosophy concerns the objectives of the methodology. There are many objectives that could be discussed, but for the purposes of this framework it is of prime importance whether the objective is to build a computerized information system or whether wider improvements and more general problem solving are involved. Some of our methodologies indicate their position more clearly than others. ISAC, for example, decides on information systems development as the suitable development measure only if the change analysis indicates that there are problems and needs in the information systems area. In other situations, other development measures are chosen, for example: (i) development of the direct business activities, such as production development, product development, or a development of distribution systems; (ii) organizational development; or (iii) development of personal relations (communication training). We therefore see that ISAC is very much more than the development of IT systems, as are SSM and ETHICS.

PI also has objectives that are much wider than the development of computerized systems. Its objectives focus on improving and redesigning business processes for an organization as a whole, and, although IT is usually regarded as an important enabler of process innovation, many of the specified improvements and redesigned processes are achieved without the construction of computer systems. DSDM, although often resulting in the design of computer systems, is sometimes used to address organizational or general problem-solving issues.

On the other hand we classify STRADIS, YSM, IE, SSADM, Merise, JSD, RUP, and OOA not as general problem-solving methodologies, but as having clear objectives to develop computerized information systems. Some methodologies claim that they are applicable whether the system is going to be automated or not, for example, STRADIS, but we can find no evidence that this is ever put into practice, and an examination of the activities of these methodologies illustrates that their main focus clearly embodies an assumption that a computerized system is to be constructed.

Apart from objectives concerning whether a computerized system is the goal or not, there are other objectives of importance in comparisons. For example, in ETHICS there are objectives relating to improving the quality of working life and enhancing the job satisfaction of users, and a clear ethical position implying that it should not be used in circumstances where an IT system might dehumanize work.

(c) Domain

The next subelement for analysis is that of domain. This is related to the above subelement of objective but focuses on what aspects or domain the methodology seeks to address. Of particular interest is the distinction between those methodologies that seek to identify business or organizational need for an information system, that is, those which address the general planning, organization, and strategy of information and systems in the organization, and those concerned with the solving of a specific, pre-identified problem, for example, the need to provide a wider range of marketing information to the salesforce.

Often the key to this distinction is the starting point of the methodology. IE, PI, and SSM are identified as being of the planning, organization, and strategy type. In PI the development of information systems is clearly driven by the identified improvements to the processes required for the benefit of the business and organization. IE has as its first stage information strategy planning. Here an overview is taken of the organization in terms of its business objectives and related information needs and an overall information systems plan is devised for the organization. This clearly implies that it is an approach adopting the philosophy that an organization needs such a plan in order to function effectively, and that success is related to the identification of information systems that will benefit the organization and help achieve its strategic objectives.

SSM is quite different to IE, and yet we also classify it as a methodology of the planning, organization, and strategy type. Such terminology might not easily be associated with SSM, but it is clearly not a specific problem-solving methodology in the sense that it does not assume that a well-defined, structured problem already exists. Much of its focus is on the understanding of these wider issues and the context in which the problem situation exists. The term 'problem situation' in SSM is not meant to imply the existence of a well-defined problem, quite the reverse. However, SSM is not usually thought of as a methodology that addresses planning, organization, and strategy, but if we remove the managerial implications from these terms, this is fundamentally what it is about. It is attempting to identify the underlying issues that help in the understanding of the problem situation, including the purpose of the organization. Later stages of the methodology assess feasible and desirable change and recommend action to improve the situation, the results of which can be the development of information systems.

STRADIS, YSM, SSADM, Merise, JSD, OOA, RUP, DSDM, ISAC, and ETHICS are classified as specific problem-solving methodologies; that is, they do not focus on identifying the systems required by the organization but begin by assuming that a specific problem is to be addressed.

(d) Target

The final subelement of philosophy in the framework is concerned with the target system to be developed, that is, whether the methodology is aimed at particular types of application, types of problem, size of system, environment, and so on. This is also a difficult area, because most methodologies appear to claim to be general purpose. Such a claim is clearly made within certain assumptions. In the majority of cases, a large organization with an in-house data processing department is assumed. Further, it is often assumed that bespoke (tailor-made) systems are going to be developed, rather than, for example, the use of application packages. An exception is IE, where alternative approaches are envisaged. OOA and RUP are considered to be general purpose, although it is suggested that they are not very helpful for simple, limited systems or systems with only a few class-&-objects. STRADIS is also stated to be general purpose and applicable to any size of system, yet the main technique is data flow diagramming, which is not particularly suitable for all types of application, for example, the development of management information systems or web-based systems. Therefore, the claims of the methodology authors have to be tempered by an examination of the methodology itself. JSD, for example, has been described as most suitable for real-time processing applications. SSM has been developed to be applicable in human activity situations where very complex (wicked) problem situations exist.

The size of organizations that the methodologies address is also an important aspect of target. Whereas STRADIS, YSM, IE, SSADM, Merise, OOA, and RUP have all been designed primarily for use in large organizations, Multiview has been designed to be applicable in relatively small organizations. There is, however, a version of SSADM, called MicroSSADM, which is intended to help develop information systems in smaller environments or where the target system is PC-based.

2 Model

The second element of the framework deals with the model or models that the methodology uses. This can be investigated in terms of the type of model, the levels of abstraction of the model, and the orientation or focus of the model.

In this subsection, we will concentrate on examining the various models of process that methodologies use. The primary process model used is the data flow diagram. In STRADIS it is the primary model of the methodology. It also appears in YSM, SSADM and ISAC as an important model, although not the only one. It is also used in Wilson's (1990) description of SSM (but not in Checkland's version), referred to as a "Gane and Sarson" type diagram'. It also features in IE, though in a slightly different form, termed a 'process dependency diagram', but it plays a less significant role than, for example, in STRADIS. The data flow diagram is predominantly a process model, and data are only modelled as a by-product of the processes.

The models in JSD, ETHICS, SSM (Checkland), and PI are also, in their various ways, process-oriented, but they do not use data flow diagrams. The structure diagram is used in JSD to model aspects of process, and we see that this depicts the structure of processes rather than the flow of data between processes, which dramatically changes the focus of what is important in JSD and helps to explain why it is regarded as more suitable for real-time process applications than data flow diagram-based methodologies. In SSM, the rich picture, which is a model of the problem situation, is also, in part, a model of processes, structures, and their relationships. ETHICS uses a socio-technical model which involves the interaction of technology and people and the processes performed. It is interesting to note that of the identified model types, this is not a pictorial model-type but a 'verbal' or narrative model. In OOA and RUP the basic models are an integration of both process and data orientation, often in the same diagram, which as we have seen is a key element of the object-oriented approach.

3 Techniques and tools

The third element of the framework is that of the techniques and tools that a methodology employs. These have been examined in Parts 4 and 5, respectively. The techniques to be used in a methodology are usually made explicit by the authors which makes it relatively easy to compare and contrast. Many of the models discussed above are closely reflected in the techniques used but there are sometimes differences in the way the models are used and their importance in the methodology. Many methodologies appear to include the techniques as part of the methodology. STRADIS is an example of a methodology which is largely described in terms of its techniques, but others, such as YSM, SSADM, JSD, OOA, and RUP, also have specific techniques which are regarded as fundamental to the methodology. Other methodologies, for example, ISAC, do not rely on particular techniques quite so much, and it is relatively easy to envisage similar but alternative techniques being used without affecting the essence of the methodology.

Some methodologies, for example, IE, explicitly suggest that the techniques are not a fundamental part of the methodology and that the current recommended techniques can be replaced, or substituted, as better techniques become available, providing, of course, they address the same fundamentals. This is potentially an important conceptual difference between methodologies, and it is a useful exercise to strip a methodology of its techniques and see what, if anything, is left. For example, are the phases and tasks of the methodology described in terms of when and how to use the techniques? Obviously those methodologies that allow new techniques to be incorporated are somewhat more flexible, but achieving this in practice is not so easy. For example, in a methodology which advocates the clear separation of the modelling of data and processes, such as SSADM, Merise, or IE, it would be quite difficult to accommodate an object-oriented modelling technique, which integrates the two without amending the fundamentals of the methodology. Similarly it would be difficult to imagine OOA or RUP accommodating anything other than object-oriented techniques and models, particularly ones which did not incorporate the essential combination of data and process. Therefore, the identification of the fundamentals is an important part of any comparison. This also raises interesting issues about how methodologies can legitimately develop and evolve over time without losing the essence of the methodology.

The comparison of the tools of methodologies begins with whether any are specifically advocated. SSM, for example, does not advocate, or even mention, any tools, but most methodologies, such as YSM, IE, SSADM, Merise, JSD, OOA, and PI, recommend the use of tools to some degree. These range from simple drawing tools through to tools supporting the whole development process, including prototyping, project management, code generation, simulation, and so on. The degree of recommendation varies considerably. Some, such as IE and RUP, suggest that the process should not be contemplated without the use of tools, the process being too complicated and time consuming. Others, for example, SSADM and OOA, argue that they might be helpful but are not necessarily essential. An important element of comparison is whether the methodology, like IE, implies the use of a specific brand-name tool, or whether appropriate tools from any vendor can be used. In practice, there is often a trade-off to be made between the degree that the tool is specific to the methodology and the degree of lock-in to a particular vendor.

4 Scope

The fourth element of the framework is scope. For the purposes of this text we examine methodologies in terms of the stages of the life cycle they address. We identify nine stages: strategy (planning), feasibility, analysis, logical design, physical design, programming, testing, implementation, and maintenance. This is not the only possible approach to the analysis of scope and, depending on the methodologies being compared and the purpose of the comparison, other dimensions may be more appropriate. Any analysis of the scope of a methodology is difficult and subjective. Using the stages of the life cycle may misrepresent those methodologies that are not designed to follow such a structure, for example, prototyping methodologies, in which case a definition of scope based on the spiral model might be more appropriate. It is therefore important that scope is not viewed in isolation from the rest of the framework and that the earlier warnings concerning comparisons are heeded.

Figure 28.3 summarizes the results of the analysis of scope. Shaded areas indicate that the methodology covers the stage in some detail, which may include the provision of specific techniques, and tools of support. An unshaded area means that the methodology addresses the area, but in less detail and depth. In this case there is less guidance from the methodology and more is left for the developers to interpret and perform for themselves. The broken lines indicate areas that are only briefly mentioned in the methodology, but no procedures, techniques, or rules are provided although the methodology recognizes that the area should be addressed. The identification that a methodology covers a particular stage does not necessarily mean that there exists a defined stage of that name but that within the methodology there are elements that can be construed to be equivalent to that stage.

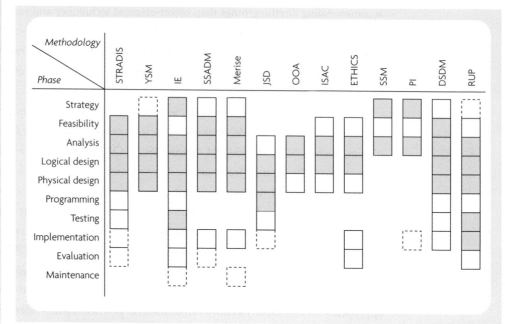

Figure 28.3: Scope of methodologies

In our analysis of scope, *strategy* is used to indicate whether the methodology addresses any aspects which relate to an organization-wide context, and that deals with overall information systems strategy, purpose, and planning, rather than just that of the particular system or area of concern. IE and PI are identified as addressing this in some detail, as is SSM which also deals with the wider context, although in a significantly different way to the others. SSADM, Merise, DSDM, and to a lesser extent YSM and RUP, also include aspects of a strategic nature.

The next stage is *feasibility* which is defined as the economic, social, and technical evaluation of the system under consideration. STRADIS, YSM, SSADM, DSDM, and Merise include detailed aspects for the evaluation of feasibility in their methodologies. IE, ISAC, PI, and RUP also address feasibility, but less comprehensively. However, it should be noted that the way in which they deal with feasibility differs considerably. For example, ISAC does not identify a specific feasibility phase (it is contained in some of the analysis steps) and STRADIS checks and rechecks feasibility at many stages throughout the methodology. ETHICS and SSM are identified as dealing in detail with feasibility, although in quite a different way from the others as it is not financial but social feasibility of change which is addressed. ETHICS focuses on social and technical 'fit' and SSM on feasible and desirable change. In Figure 28.3, these methodologies are depicted as open boxes rather than shaded boxes because, although their social focus is comprehensive, the other aspects of feasibility are not.

The *analysis* stage, which includes user requirements analysis, is covered in detail, although in a wide variety of different ways by all the methodologies except JSD. JSD does not specifically address user requirements analysis but begins with an assumption that they are given. The *logical design* stages are covered in detail by all the methodologies except SSM and PI which do not cover the stage or indeed most subsequent stages. *Physical design* is covered in detail by all except OOA, ISAC, and ETHICS, which are less explicit. ISAC suggests the use of JSP in the data systems design phase.

The next stage we identify is that of the physical development of the system, characterized as *programming*. Only STRADIS, IE, JSD, DSDM, and RUP cover programming, with JSD judged to address it in most detail. Some methodologies suggest or assume that other approaches are used for the physical development of the system. ISAC, for example, recommends the use of JSP, and OOA recommends an object-oriented language. In such cases we do not include this as being covered in the scope of the methodology. *Testing,* which includes the planning as well as the testing of systems, programs, and procedures, is addressed by STRADIS, IE, JSD, and RUP, but only in detail by IE and RUP.

Implementation, in which we include planning and implementation of technical, social, and organizational aspects, is covered by IE, SSADM, Merise, DSDM, and ETHICS, although not in as much detail as earlier stages, with STRADIS and PI addressing it, but in less detail. JSD is a special case, because its interpretation of implementation is simply technical, covering the sharing of processors among processes. RUP deals in detail with the technical issues of implementation but does not address the social and organizational aspects to the same extent. Nevertheless, we classify this as detailed coverage. Post-implementation evaluation and review concerns the measurement and evaluation of the implemented system and a comparison with the original objectives. This is addressed by IE, DSDM, RUP, and ETHICS, with ETHICS particularly focusing on the *evaluation* of 'fit'; STRADIS and SSADM cover evaluation in less detail.

The final stage in the analysis of scope is *maintenance*, which we regard as being covered by a methodology only if it is specifically addressed in terms of tasks. The fact that maintenance may in general be improved by the use of the methodology in earlier stages is not regarded as coverage of maintenance for the purposes of this analysis. With this definition we find that only IE and Merise mention maintenance in any detail.

A glance at Figure 28.3 illustrates that the main focuses of most of these methodologies are at the analysis and design stages, but that there are wide scope variations, and that to assume all methodologies are the same in this respect, as people often do, is obviously incorrect.

5 Outputs

The fifth element of the framework is concerned with outputs. This is an investigation of what is actually produced in terms of deliverables at the end of each stage of the methodology. The outputs of methodologies differ quite substantially, not only in terms of what should be produced but also in the level of detail that the methodology specifies. This is closely related to the degree that the methodology is seen as a blueprint for action; that is, how detailed are the rules about how to proceed and how much is left to the discretion of the analyst.

A related issue concerns how, and to what extent, the analysts know that they are proceeding correctly. As an example, ISAC specifies in some detail the outputs of the change analysis stage (A-graphs), but the process of generating change alternatives is not described in any detail. This is regarded as the creative part of the methodology and not amenable to specification. The identification of such areas in a methodology is regarded as an important contribution to an understanding of that methodology.

6 Practice

The penultimate element of the framework is the practice or use of the methodology. It contains the subelements of background, user base, applications, and players. The background of the methodology broadly identifies its origins in terms of academic or commercial. STRADIS, YSM, IE, SSADM, DSDM, and OOA lie in the commercial sphere, whereas Merise, JSD, ISAC, ETHICS, SSM, RUP, and PI have academic backgrounds, at least in part, though this does not mean that they are not now commercial methodologies.

The user base, perhaps surprisingly, is often difficult to discover and is possibly less widespread than vendors would have us believe. Vendors have a habit of suggesting that any company that has expressed interest in their methodology, or has requested an evaluation version, is a user. There is a view that commercial methodologies are not in as widespread use as is claimed but finding evidence is difficult. Fitzgerald et al. (1999) found that 57 per cent of organizations were using a systems development methodology, but, of these, only 11 per cent used a commercial development methodology, 30 per cent used a commercial methodology adapted for in-house use, and 59 per cent a methodology which was claimed to be unique to the organization (i.e. internally developed and not based on a commercial methodology). A similar finding concerning the adaptation of methodologies was found in a study conducted in the Republic of Ireland, although in this case the proportion using methodologies was only 40 per cent (Fitzgerald, 1996). In the USA Russo and Wynekoop (1995) reported on a survey that identified generic types of methodology and found that, of 132 organizations using a methodology, 77 per cent were using a structured approach, with 8 per cent using a prototyping/iterative approach, and 5 per cent using a RAD approach. Therefore, it would appear that commercial methodologies are not as well used as some might believe but nevertheless they are clearly influential on the practice of systems development.

The last subelement of practice requires an analysis of the participants involved; in some contexts these are referred to as 'actors' or 'stakeholders'. This requires answers to the questions 'who is supposed to use the methodology?' and 'what roles do they perform?', 'who are the other stakeholders', etc. It also attempts to identify the skill levels required. The traditional view of information systems development is that a specialist team of professional systems analysts and designers perform the analysis and design aspects and professional programmers design the programs and write the code. The system is then implemented by the analysts. Although the exact roles have different names, in general systems development is undertaken by professional technical developers. This view is taken by STRADIS, YSM, IE, SSADM, Merise, JSD, OOA, and RUP. ETHICS takes a different view, and users have a much more proactive role. In ETHICS, the users perform the analysis and design themselves and the data processing professionals are used as consultants as and when required. In addition, ETHICS incorporates a facilitator role, whose task is to guide the users in the use of the methodology. Facilitators do not actually perform the tasks themselves, but smooth the path and ensure that the methodology is followed correctly. The facilitator should be expert and experienced in the use of the methodology. Introna and Whitley (1997) advocate what they term 'a formal apprenticeship' in the use and applicability of a range of methodologies and the context in which they might be used to overcome the problems of method-ism discussed earlier.

The levels of skill required by the players varies considerably. In almost all methodologies considerable training and experience is necessary for at least some of the players. Further, many make significant demands on the users, in which case the methodology would be expected to include training aspects. This may significantly increase the time and costs required to develop a project. ETHICS, which makes heavy demands on the users, recognizes and addresses this problem. Even where methodologies adopt the more traditional roles, professional analysts and designers can find them quite difficult to learn at first (and some complain laborious as well), with new vocabularies, techniques, and tools to work with. Other methodologies, while not going as far as ETHICS, have a much wider range of stakeholders involved, particularly from the business side as well as the professional systems developers, for example, SSM, ISAC, PI, and DSDM.

7 Product

The final element of the framework is that of product. This describes what is supplied when purchasing a methodology and at what cost. Most methodologies have a range of products and services available, which can be taken or not, although there is likely to be a minimum set. The product is also likely to be changing rapidly, mainly due to the increasing numbers of support tools available. The product can range from large and copious sets of manuals, for example, SSADM, to a set of academic papers and a book, as in the case of SSM. RUP has a range of documents, books, and specifications but also has a multimedia website, and indeed claims that the methodology product is actually delivered using this web technology, 'so it's not in books or binders but literally at the developers' fingertips'. Some methodologies require consultants, facilitators, and/or training courses to be used as part of the product. Some methodologies offer certification of competency for developers. There is, for example, a certificate of proficiency scheme for SSADM.

This discussion of methodologies using the framework is by no means comprehensive, but is intended to be illustrative of the issues that the framework raises. It is likely to be contentious. It is meant to stimulate debate, to open discussions rather than be taken as a statement of facts.

Summary

- In this chapter we examine the comparison of methodologies. We identify two main reasons for wanting to compare methodologies. First, an academic reason, to understand the nature of methodologies better, and, second, a practical reason, to choose a methodology for use in an organization.

- A general framework for comparing methodologies is discussed (NIMSAD). This is a broader framework than most and has three main elements as follows: the problem situation (context); the problem solver (methodology user); and the problem-solving process (the methodology itself).

- A more specific framework for comparing methodologies is proposed. This framework has seven elements. The first element relates to the identification of the underlying philosophy behind the methodology (i.e. what gives it its particular flavour and why the authors think it is an appropriate way to develop information systems). The second element relates to the models that the methodology uses (i.e. the way that the 'real world' is modelled and abstracted). The third element is the specific techniques and tools that are recommended. The fourth element suggests comparing the scope of the methodologies, the fifth the outputs of the methodology, the sixth the practice (e.g. number and experience of existing users), and finally the seventh element is the nature of the product itself (i.e. what you actually get when you procure a methodology).

- The final section of the book takes many of the methodologies described in Part 6 and uses the framework, described above, to compare them. The value of such a comparison is not necessarily the outcome but the enhanced understanding that comes from asking questions of methodologies and discussing the issues that arise. For example, the comparison of scope reveals that quite different aspects are addressed in very different ways by different methodologies.

Questions

1. Why do you think we attach great importance to the 'philosophy' of a methodology?

2. What elements do you regard as most important in the comparison of methodologies?

3. What would be your criteria for comparison if you were making a choice of a methodology for potential adoption in a public service organization, such as a hospital?

4. Undertake a comparison of three methodologies with which you are familiar using just three criteria. What did you learn from the comparison?

5. Why are some methodologies essentially a set of recommended techniques while others say that in a particular stage or phase any 'appropriate' techniques may be used?

Further reading

Checkland, P. (1981) *Systems Thinking, Systems Practice,* John Wiley & Sons, Chichester, UK.

Hirschheim, R., Klein, H.K., and Lyytinen, K. (1996) Exploring the intellectual structures of information systems development: A social action theoretical analysis, *Accounting Management & Information Technology,* Vol. 6, Nos 1/2, 1–64.

Jayaratna, N. (1994) *Understanding and Evaluating Methodologies, NIMSAD: A Systemic Framework,* McGraw-Hill, Maidenhead, UK.

Olle, T.W., Sol, H.G., and Verrijn-Stuart, A.A. (1982) *Information Systems Design Methodologies: A Comparative Review,* North Holland, Amsterdam.

Russo, N.L. and Stolterman, E. (2000) Exploring the assumptions underlying information systems methodologies: Their impact on past, present and future ISM research, *Information Technology & People,* Vol. 13, No. 4.

Bibliography

Abrahamsson, P., Outi, S., Ronkainen, J. and Warsta, J. (2002) *Agile Software Development Methods: Review and Analysis*, VTT Publications 478, www.inf.dtt.fi/pdf/.

Ackerman, F. and Eden, C. (2001) SODA – journey making and mapping in practice, in J. Rosenhead and J. Mingers (eds) *Rational Analysis for a Problematic World Revisited – Problem Structuring Methods for Complexity, Uncertainty and Conflict*, 2nd edn, John Wiley & Sons, Chichester, UK.

Ackoff, R. L. (1971) Towards a system of system concepts, *Management Science*, Vol. 17.

ACM (2005) www.acm.org

ACS (2005) www.acs.org.au

Adams, C. and Avison, D. (2003) Dangers Inherent in the Use of Techniques: Identifying Framing Influences, *Information Technology & People*, Vol. 16, No. 2, 203–34.

Adams, J. (1987) *Conceptual Blockbusting, a Guide to Better Ideas*. Penguin, Harmondsworth, Middlesex.

Ambler, S. (2002) *Agile Modelling: Effective Practices for eXtreme Programming and the Unified Process*, John Wiley, Chichester.

Amoroso, D. L. and Cheney, P. H. (1991) Testing a causal model of end-user application effectiveness, *Journal of Management Information Systems*, Vol. 8, No. 1, 63–89.

Andrews, D. C. (1991) JAD: A crucial dimension for rapid application development, *Journal of Systems Management*, March, 23–31.

Angell, I. O. and Smithson, S. (1991) *Information Systems Management: Opportunities and Risks*, Macmillan, Basingstoke, UK. This text emphasizes the management of information systems within business organizations, suggesting that the real problems associated with information systems are human, managerial, social, and organizational, rather than technological.

Ansoff, H. I. (1965) *Corporate Strategy*, McGraw-Hill, New York.

Ante, S. E. (2004) Shifting Work Offshore? Outsourcer beware; quality and security woes can eat expected savings, *Business Week*, New York, 12 January.

Aonix (2005) http://www.pugh.co.uk/Products/aonix_europe/select-enterprise.htm

Argyris, C. (1993) *Knowledge for Action: A Guide to Overcoming Barriers to Organization Change*, Jossey-Bass, San Francisco.

Argyris, C. and Schön, D. (1978) *Organizational Learning: A Theory of Action Perspective*, Addison-Wesley, Reading, Massachusetts.

Arnett, K. P. and Litecky, C. R. (1992) Retooling systems analyst skills for small hospitals, *Journal of Systems Management*, June.

Augustine, S. and Woodcock, S. (2003) Agile Project Management, http://www.ccpace.com/Resources/documents/AgileProjectManagement.pdf

Avison, D. E. and Catchpole, C. P. (1987) Information systems for the community health services. *Medical Informatics*, Vol. 13, 2.

Avison, D. E. and Myers, M. D. (1995) Information systems and anthropology: An anthropological perspective on IT and organizational culture, *Information Technology & People*, Vol. 8, No. 3, 43–56.

Avison, D. E. and Taylor, A. V. (1996) Information systems development methodologies: A classification according to problem situations, *Journal of Information Technology*, Vol. 11.

Avison, D. E. and Truex, D. (2003) *Method Engineering: Reflections on the past and ways forward*, Americas Conference in Information Systems (AMCIS), Tampa, Florida (August).

Avison, D. and Wilson, D. N. (2002) IT Failure and the Collapse of One.Tel. in Traunmuller, R. (ed.) *Information Systems: The e-Business Challenge*, Kluwer, Boston, 31–46.

Avison, D. E. and Wood-Harper, A. T. (1986) Multiview – an exploration in information systems development, *Australian Computer Journal*, Vol. 18, 4.

Avison, D. E. and Wood-Harper, A. T. (1990) *Multiview: An Exploration in Information Systems Development*, McGraw-Hill, Maidenhead, UK.

Avison, D. E., Powell, P., Keen, P., Klein, J. H., and Ward, S. (1995) Addressing the need for flexibility in information systems, *Journal of Management Systems*, Vol. 7, No. 2, 43–60.

Avison, D. E., Powell, P. L., and Adams, C. (1994) Identifying and incorporating change in information systems, *Systems Practice*, Vol. 7, No. 2, 143–159.

Avison, D. E., Wood-Harper, A. T., Vidgen, R. T., and Wood, J. R. G. (1998) A further exploration into information systems development: The evolution of Multiview2, *Information Technology & People*, Vol. 11, No. 2, 124–139.

Bahrami, A. (1999) *Object-oriented Systems Development*, McGraw-Hill, Boston.

Bardoloi, S. (2002) *Quality: A Health Capsule to Retain Growth*, The Project Perfect White Paper Collection, available at http://www.projectperfect.com.au/downloads/info_cmm.pdf

Baskerville, R., and Pries-Heje, J. (2001) Racing the e-bomb: How the Internet is redefining information system development methodology, in L. Russo, B. Fitzgerald, and J. DeGross (eds) *Realigning Research and Practice in Information System Development*, Proceedings of the IFIP TC8/WG8.2 Working Conference, 27–29 July, Boise, Idaho.

BCS (2005a) http://www.bcs.org/bcs

BCS (2005b) http://www.bcs.org/BCS/AboutBCS/codes/conduct/

BCS (2005c) http://www.bcs.org/BCS/AboutBCS/codes/cop/

BCS (2005d) http://www.bcs.org/BCS/AboutBCS/codes/cop/

Beck, K. (2000) *Extreme Programming Explained: Embrace Change*, Addison-Wesley Longman, Boston.

Beck, K. et al. (2001) *Agile Manifesto*, available at http://agilemanifesto.org/

Benbasat, I., Dexter, A. S., Drury, D. H., and Goldstein, R. C. (1984) A critique of the stage hypothesis: Theory and empirical evidence, *Communications of the ACM*, Vol. 27, No. 5, 476–485.

Bennett, S., McRobb, S., and Farmer, R. (1999) *Object-oriented Systems Analysis and Design Using UML*, McGraw-Hill, London.

Bentley, C. (2002) *PRINCE2: A Practical Handbook*, Butterworth-Heinemann, Oxford.

Benyon, D. (1990) *Information and Data Modelling*, McGraw-Hill, Maidenhead, UK.

Bergeron, F. and Berube, C. (1988) The management of the end-user environment: An empirical investigation, *Information and Management*, Vol. 14, No. 2, 107–113.

Beynon-Davies, P. (1995) Information systems 'failure': The case of the London Ambulance Service's Computer Aided Despatch project, *European Journal of Information Systems*, No. 4, 171–184.

Beynon-Davies, P., Mackay, H. and Tudhope, D. (2000) 'It's lots of bits of paper and ticks and post-it notes and things . . .' a case study of a rapid application development project, *Information Systems Journal*, Vol. 10, No. 3, 195–216.

Bikson, T. K. (1996) Groupware at the World Bank, in C.U. Ciborra (ed.) *Groupware and Teamwork: Invisible Aid or Technical Hindrance?*, Wiley, Chichester, UK, pp. 145–183.

Bisp, S., Sorensen, E., and Grunert, K. G. (1998) Using the key success factor concept in competitor intelligence and benchmarking, *Competitive Intelligence Review*, Vol. 9, No. 3, 55–67.

Bjørn-Andersen, N. (1984) Challenge to certainty, in T.M.A. Bemelmans (ed.) *Beyond Productivity: Information Systems Development for Organizational Effectiveness*, North Holland, Amsterdam.

Boehm, B. (1999) Making RAD work for your project, *Computer*, Vol. 32, No. 3, 113–114.

Boehm, B. (2002) Get ready for agile methods with care, *Computer*, January, 64–69.

Boehm, B. W. (1988) A spiral model for software development and enhancement, *IEEE Computer*, Vol. 21, No. 5.

Booch, G. (1991) *Object-oriented Design with Applications*, Benjamin/Cummings, Redwood City, California.

Booch, G. (1994) *Object-oriented Analysis and Design with Applications*, 2nd edn, Benjamin/Cummings, California.

Booch, G. (1995) *Object Solutions: Managing the Object-oriented Project*, Addison-Wesley, Boston.

Booch, G., Jacobson, I., and Rumbaugh, J. (1997) The Unified Modeling Language, Notation Guide Version 1.1, available at http://www.rational.com/uml/index.jsp

Booch, G., Jacobson, I., and Rumbaugh, J. (1998) *Unified Modeling Language – User's Guide*, Addison-Wesley, Boston.

Booch, G., Jacobson, I., and Rumbaugh, J. (1999) *Unified Modeling Language 1.3*, White paper, Rational Software Corp., available at http://www.rational.com/uml/index.jsp

Boynton, A. C. (1984) An assessment of critical success factors, *Sloan Management Review*, Summer.

Boynton, A. C. and Zmud, R. W. (1984) An assessment of critical success factors, *Sloan Management Review*, 25 (Summer), 17–27.

Braa, K., Sorensen, C., and Dahlbom, B. (eds) (2000) *Planet Internet*, Studentliteratur, Lund.

Brancheau, J. C. and Wetherbe, J. C. (1990) The adoption of spreadsheet software: Testing innovation diffusion theory in the context of end-user computing, *Information Systems Research*, Vol. 1, No. 2.

Braue, D. (2004) Certification: What's in a name?, *Technology and Business Magazine*, 20 October, available at http://www.builderau.com.au/

Brinkkemper, S., Lyytinen, K., and Welke, R. J. (eds) (1996) *Method Engineering: Principles of Method Construction and Tool Support*, Chapman & Hall, London.

Broadman, J. G. and Johnson, D. I. (1996) Realities and rewards of software process improvement, *IEEE Software*, Vol. 29, No. 11, 99–101.

Brooks, F. P. (1995) *The Mythical Man Month and Other Essays on Software Engineering* (anniversary edition with four new chapters), Addison-Wesley, Reading, Mass.

Brown, A. W. (ed.) (1996) *Component-based Software Engineering*, IEEE Computer Society, Los Alamitos, California, p. 140.

Brunelli, M. (2005) Sleepycat CEO talks Microsoft open source tactics, SearchEnterpriseLinux.com, item dated 28 March 2005.

Bubenko, J. A. Jr (1986) Information system methodologies – a research view, in T. W. Olle, H. G. Sol, and A. A. Verrijn-Stuart (eds) *Information Systems Design Methodologies: A Comparative Review*, North Holland, Amsterdam.

Bucki, L.A. (2000) *Managing with Microsoft Project 2000*, Prima, Rocklin, California.

Buckingham, R. A., Hirschheim, R. A., Land, F. F., and Tully, C. J. (eds) (1987) Information systems curriculum: A basis for course design, *Information Systems Education: Recommendations and Implementation*, Cambridge University Press, Cambridge.

Budde, R., Kautz, K., Kuhlenkamp, K., and Züllighoven, H. (1992) *Prototyping – An Approach to Evolutionary Systems Development*, Springer-Verlag, Berlin.

Burke, G. and Peppard, J. (1995) *Examining Business Process Re-engineering: Current Perspectives and Research Directions*, Kogan Page, London

Burrell, G. and Morgan, G. (1979) *Sociological Paradigms and Organisational Analysis*, Heinemann, London.

Buttle, F. (1992) The marketing strategy worksheet: A practical tool, *Cornell Hotel Restaurant Administration Quarterly*, Vol. 33, No. 3.

CA (2005) *www3.ca.com*

Cadle, J. and Yeates, D. (2001) *Project Management for Information Systems*, Prentice Hall, Harlow, UK.

Cameron, J. R. (1989) *JSP & JSD: The Jackson Approach to Software Development*, IEEE Computer Society Press, Los Angeles.

Card, D. (1995) Is timing really everything? *IEEE Software*, September.

Cardenas, A. F. (1985) *Database Management Systems*, 2nd edn, Allyn and Bason, Boston.

Carmel, E. (1995) Does RAD live up to the hype? *IEEE Software*, September, 25–26.

Carmel, E. and Nunamaker, J. (1992) Supporting joint application development (JAD) with electronic meeting systems: A field study, in J. I. DeGross, J. D. Becker, and J. J. Elam (eds) *Proceedings of the Thirteenth International Conference on Information Systems (ICIS)*, ACM, Baltimore.

Carr, N. G. (2003) IT doesn't matter, *Havard Business Review*, Vol. 81, No. 5, 41–49.

CCTA (1990) *Euromethod Project, Phase 2, Deliverable 1, State of the Art Report*, Eurogroup, CCTA, Norwich, UK.

CCTA (1994a) *Euromethod Overview*, CCTA, Norwich, UK.

CCTA (1994b) *Using Euromethod in Practice*, CCTA, Norwich, UK.

Checkland, P. (1981) *Systems Thinking, Systems Practice*, John Wiley & Sons, Chichester, UK.

Checkland, P. (1985) From optimising to learning: A development of systems thinking for the 1990s, *Journal of the Operational Research Society*, Vol. 36, 9.

Checkland, P. (1987) *Systems Thinking and Computer Systems Analysis: Time to Unite*, DEC Seminar Series in Information Systems, presentation given at the London School of Economics, 5 November 1987.

Checkland, P. (1999) Soft systems methodology: A 30-year retrospective, in *Systems Thinking, Systems Practice*, John Wiley & Sons, Chichester, UK.

Checkland, P. and Scholes. J. (1990) *Soft Systems Methodology in Action*, John Wiley & Sons, Chichester, UK.

Checkland, P. and Scholes, J. (1999) *Soft Systems Methodology in Action* (edition including a 30-year perspective), Wiley, Chichester.

Chen, J. (2001) Building web applications, *Information Systems Management*, Winter, Vol. 18, No. 1.

Churchman, C. W. (1971) *The Design of Inquiring Systems, Basic Concepts of Systems and Organization*, Basic Books, New York.

Ciborra, C. U. (1996) The platform organization: Recombining strategies, structures, and surprises, *Organization Science*, Vol. 7, No. 2, 103–118.

CIO (2005) www.cio.com

Coad, P. and Argila, C. (1996) *Case Studies in Object Oriented Analysis and Design*, Yourdon Press, New Jersey.

Coad, P. and Yourdon, E. (1991) *Object Oriented Analysis*, 2nd edn, Prentice Hall, Englewood Cliffs, New Jersey.

Cockburn, A. (2002) *Agile Software Development*, Pearson-Longman, Harlow, UK.

Cockburn, A. (2004) *Crystal Clear: A Human-Powered Methodology for Small Teams*, Addison-Wesley, Reading, Mass.

Computerworld (2004), *Computerworld*, 17 May.

Conn, S. S. (2003) A New Teaching Paradigm in Information Systems Education: An Investigation and Report on the Origins, Significance, and Efficacy of the Agile Development Movement, Proceedings of ISECON, San Diego. Available at http://isedj.org/isecon/2003/3212/ISECON.2003.Conn.pdf

Cooper, D. F. and Chapman, C. B. (1987) *Risk Analysis for Large Projects – Models, Methods and Cases*, John Wiley & Sons, Chichester, UK.

Cooper, R. B. and Zmud, R. W. (1990) Information technology implementation research: A technology diffusion approach, *Management Science*, Vol. 36, No. 2, 123–139.

Couger, D. (1995) *Creative Problem Solving and Opportunity*, Boyd and Fraser, Massachusetts.

Coughlan, J. and Macredie, R. D. (2002) Effective communication in requirements elicitation: a comparison of methodologies, *Requirements Engineering*, Vol. 7, No. 2, 47–60.

Currie, W. L. and Galliers, R. (1999) *Rethinking Management Information Systems*, Oxford University Press, Oxford.

Currie, W. L. (2001) Application outsourcing: A new business model for enabling competitive electronic commerce, *International Journal of Technology Management*, Special Issue on Enabling Organizational Competitiveness Using Electronic Commerce (forthcoming).

Cysneiros, L. M. and Leite, J. C. S. P. (2001) Using UML to Reflect Non-Functional Requirements, Proceedings of the 11th CASCON (IBM Center for Advanced Studies CONference), IBM Canada, Toronto, November, 202–216.

Daniel, R. H. (1961) Management data crisis, *Harvard Business Review*, September–October, 111–112.

Daniels, J. and Cook, S. (1992) Making objects stick, in R. Tagg and J. Mabon (eds) *Object Management*, Aldgate Publishing, Aldershot, UK.

Daniels, J. (1996) *Why RAD Is Bad*, Presentation to the BCS Specialist Group, Imperial College, 10 July. A version of this talk is available from the Requirements Engineering Specialist Group of the British Computer Society (BCS) at http://www.resg.org.uk/newsletter.html, Requirements Quarterly Volume 8.

Davenport, T. H. (1993) *Process Innovation*, Harvard Business School Press, Cambridge, Massachusetts.

Davenport, T. H. and Prusak, L. (2000) *Working Knowledge: How Organizations Manage What They Know*, Harvard Business School Press, Boston.

Davenport, T. H. and Short, J. (1990) The new industrial engineering: Information technology and business process redesign, *Sloan Management Review*, Vol. 31, No. 4, 11–27.

Davenport, T. H. and Stoddard, D. (1994) Re-engineering: Business change of mythic proportions, *MIS Quarterly*, Vol. 18, 121–127.

Davidson, E. J. (1999) Joint Application Design (JAD) in practice, *Journal of Systems and Software*, Vol. 45, No. 3, 215–223.

Davis, G. B. (1982) Strategies for information requirements determination, *IBM Systems Journal*, Vol. 21, 2.

Davis, L. E. (1972) *The Design of Jobs*, Penguin, Harmondsworth, UK.

De Bono E. (1977) *Lateral Thinking: A Textbook of Creativity*, Penguin, Harmondsworth.

De Bono, E. (1990) *Lateral Thinking for Management*, Penguin, Harmondsworth, UK.

De Grace, P. and Stahl, L. (1993) *The Olduvai Imperative: CASE and the State of Software Engineering Practice*, Prentice Hall, Englewood Cliffs, New Jersey.

De Greef, P. and Breuker, J. A. (1992) Analysing system-user cooperation, *Knowledge Acquisition*, Vol. 4.

DeMarco, T. (1979) *Structured Analysis and System Specification*, Prentice Hall, Englewood Cliffs, New Jersey.

Diaz, M. and Sligo, J. (1997) How software process improvement helped Motorola, *IEEE Software*, Vol. 30, No. 5, 75–81.

Dibbern, J., Goles, T., Hirschheim, R. and Jayatilaka, B. (2004) Information Systems Outsourcing: a survey and analysis of the literature, *ACM SIGMIS Database*, Vol. 35, No. 4, 6–102.

Donaldson, T. and Preston, L. E. (1995) The stakeholder theory of the corporation: Concepts, evidence, and implications, *Academy of Management Review*, Vol. 20, No. 1, 65–91.

Dreamweaver (2005) www.Macromedia.com

DSDM (2005) http://www.dsdm.org/

DSDM Manual Version 3 (1998) DSDM Consortium, Tesseract Publishing, Surrey, UK.

Dyché, J. (2001) *The CRM Handbook: A Business Guide to Customer Relationship Management*, Addison-Wesley, Boston.

Eardley, A., Avison, D. E. and Powell, P. (1996) How strategic are strategic information systems?, *Australian Journal of Information Systems*, Vol. 4, No. 1, 11–20.

Earl, M. and Khan, B. (1994) How new is business process re-design?, *European Management Journal*, Vol. 12, No. 2, 20–30.

Earl, M. J. (1989) *Management Strategies for Information Technology*, Prentice Hall, Hemel Hempstead, UK.

Eden, C. and Ackerman, F. (2001) SODA – the principles, in J. Rosenhead and J. Mingers (eds) *Rational Analysis for a Problematic World Revisited – Problem Structuring Methods for Complexity, Uncertainty and Conflict*, 2nd edn, John Wiley & Sons, Chichester, UK.

Espejo, R. and Harnden, R. (eds) (1989) *The Viable System Model: Interpretations and Applications of Stafford Beer's VSM*, John Wiley & Sons, Chichester, UK.

Essex, P. and Magal, S. R. (1998) Determinants of information center success, *Journal of Management Information Systems*, Fall, Vol. 15, No. 2.

Eva, M. (1994) *SSADM Version 4: A User's Guide*, 2nd edn, McGraw-Hill, Maidenhead, UK.

Eva, M. and Guilford, S. (1996) Committed to a RADical solution, paper given at *4th BCS ISM Conference*, edited by Fitzgerald and Jayaratne, N. (eds) *Proceedings of the Fourth BCS Conference on Information Systems Methodologies*, University College, Cork.

Evan, W. and Freeman, R. E. (1988) A stakeholder theory of the modern corporation: Kantian capitalism, in T. Beauchamp and N. Bowie (eds) *Ethical Theory and Business*, 3rd edn, Prentice Hall, Englewood Cliffs, New Jersey.

Evans, J. S. (1991) Strategic flexibility for high technology manoeuvres: A conceptual framework, *Journal of Management Studies*, Vol. 28, No. 1, 69–89.

Fagin, R. (1977) Multivaried dependencies and a new normal form for relational databases, *ACM Transactions on Database Systems*, Vol. 2, 3.

Failla, A. (1996) Technologies for co-ordination in a software factory, in C.U. Ciborra (ed.) *Groupware and Teamwork: Invisible Aid or Technical Hindrance?*, John Wiley & Sons, Chichester, UK, pp. 61–88.

Feeny, D., Willcocks, L., Rands, T., and Fitzgerald, G. (1993). Strategies for IT management: When outsourcing equals rightsourcing, in S. Rock (ed.) *Directors Guide to Outsourcing*, Institute of Directors/IBM, London.

Feeny, D. F., Earl, M. J., and Edwards, B. (1996) Organisational arrangements for IS: Roles of users and specialists, in M.J. Earl (ed.) *Information Management: The Organisational Dimension*, Oxford University Press, Oxford.

Feigenbaum, E. A. (1982) *Knowledge Engineering in the 1980's*, Department of Computer Science, Stanford University, Stanford, California. [Quoted in Giarratano, J. and Riley, G. (1994) *Expert Systems, Principles and Programming*, PWS Publishing, Boston.]

Feller, J. and Fitzgerald, B. (2000) A framework analysis of the open source software development paradigm, Proceedings of ICIS 2000, 14–16 December.

Feller, J. and Fitzgerald, B. (2002) *Understanding Open Source Software Development*, Addison-Wesley, Harlow, UK.

Fine, C. H. (1998) *Clockspeed: Winning Industry Control in the Age of Temporary Advantage*, Perseus Books, Reading, MA.

Fine, E. (2001) Where heros go, *IIE Solutions*, Vol. 33, No. 5, 26–31.

Fitzgerald, B. (1994) The systems development dilemma: whether to adopt formalised systems development methodologies or not?, in W.R.J. Baets (ed.) *Proceedings of the Second European Conference on Information Systems*, Nijenrode University Press, Breukelen, The Netherlands.

Fitzgerald, B. (1996) Formalized systems development methodologies: A critical perspective, *Information Systems Journal*, Vol. 6, No. 1, 3–23.

Fitzgerald, B. (1999) An empirical investigation of RAD in practice, in T. Wood-Harper, N. Jayaratne, and B. Woods (eds) *Proceedings of the BCS Information Systems Methodologies Conference*, Springer-Verlag, London, pp. 77–87.

Fitzgerald, G. (1990) Achieving flexible information systems: The case for improved analysis, *Journal of Information Technology*, Vol. 5, No. 1, 5–11.

Fitzgerald, G. (1994) Strategic outsourcing of IT in the UK, Keynote address given at *Proceedings of the 5th Australasian Conference on Information Systems*, edited by G. Shanks and D. Arnott, published by Monash University, Melbourne, Australia, pp. 27–40.

Fitzgerald, G. (1998) Evaluating information systems projects: A multidimensional approach, *Journal of Information Technology*, Vol. 13, No. 1, 15–28.

Fitzgerald, G. (2000) *IT at the Heart of Business*, British Computer Society, Swindon, UK.

Fitzgerald, G. and Fitzgerald, B. (1999) Categories and contexts of ISD: Making sense of the mess, in J. Pries-Heje et al. (eds) *European Conference on Information Systems (ECIS)*, Copenhagen Business School, Copenhagen.

Fitzgerald, G. and Russo, N. L. (2005) The turnaround of the London Ambulance Service Computer-Aided Despatch system, *European Journal of Information Systems*, Vol. 14, 244–257.

Fitzgerald, G. and Willcocks, L. (1994) *A Business Guide to Outsourcing Information Technology, A Study of European Best Practice in the Selection, Management and Use of External IT Services*, Business Intelligence, Wimbledon, UK, p. 372.

Fitzgerald, G., Philippides, A., and Probert, P. (1999) Information systems development, maintenance and enhancement: Findings from a UK study, *International Journal of Information Management*, Vol. 40, No. 2, 319–329.

Fitzgerald, B., Russo, N. L., and Stolterman, E. (2002) *Information Systems Development: Methods in Action*, McGraw-Hill, Maidenhead.

Flynn, D. J. (1992) *Information Systems Requirements – Determination and Analysis*, McGraw-Hill, Maidenhead, UK.

Forrester Research (2005) see http://www.forrester.com

Fowler, K. M. (1997) SEI CMM Level 5: A practitioner's perspective, *Crosstalk*, Vol. 10, No. 9.

Fowler, M. (2002) Agile Development: What, Who, How, and Whether, http://www.fawcette.com/resources/managingdev/interviews/fowler/

Fowler, M. (with Scott, K.) (2000) *UML Distilled: A Brief Guide to the Standard Object Modelling Language*, 2nd edn, Addison-Wesley, Reading, Massachusetts.

Freeman, R. E. (1984) *Strategic Management: A Stakeholder Approach*, Harper Collins, Boston.

Gacek, C., Lawrie, T. and Arief, B., (2004) The many meanings of open source, *IEEE Software*, Vol. 21, No. 1, 34–40.

Galliers, R. D. and Sutherland, A. R. (1991) Information systems management and strategy formulation: the 'stages of growth model' revisited, *Journal of Information Systems*, Vol. 1, No. 2.

Gane, C. (1989) *Rapid Systems Development*, Prentice Hall, Englewood Cliffs, New Jersey.

Gane, C. and Sarson, T. (1979) *Structured Systems Analysis: Tools and Techniques*, Prentice Hall, Englewood Cliffs, New Jersey.

Gibson, C. and Nolan, R. (1974) Managing the four stages of EDP growth, *Harvard Business Review*, Vol. 52, No. 1, 76–78.

Gilb, T. (1988) *Principles of Software Engineering Management*, Addison-Wesley Longman, Harlow, UK.

Goldsmith, R. F. (1994) Confidently outsourcing software development, *Journal of Systems Management*, Vol. 45, No. 4.

Greiner, L. (2000) Tapping into the best tools, *Computer Dealer News*, Vol. 16, Issue 12, 25–27.

Greiner, M. E. and Goodhue, D. L. (2005) Make or buy in the age of open source: A transaction cost analysis, Proceedings of ICIS, Las Vegas, 11–14 December.

Groupsystems (2005) www.Groupsystems.com

Grover, V. (2001) Power of Modern Information Technology is Impetus for Business Process Change, *Information Management*, Vol. 14, No. 1/2.

Grover, V. and Malhotra, M. (1997) Business Process Re-engineering: A tutorial on the concept, evolution, method, technology and application, *Journal of Operations Management*, Vol. 15, 193–213.

Grundén, K. (1986) Some critical observations on the traditional design of administrative information systems and some proposed guidelines for human-oriented system evolution, in H-E. Nissen and G. Sandsröm (eds) *Quality of Work versus Quality of Information Systems*, Lund University, Sweden.

Guimaraes, T. (1996) Assessing the impact of information centers on end-user computing and company performance, *Information Resources Management Journal*, Vol. 9, No. 1, Winter, 6–15.

Haag, M. (2004) *Cairo Luxor Aswan*, CadoganGuides, London.

Hammer, M. (1990) Re-engineering work: don't automate, obliterate, *Harvard Business Review*, Vol. 90, No. 4, 104–112.

Hammer, M. and Champy, J. (1990) *Re-engineering the Corporation: A Manifesto for Business Revolution*, Harper Business, New York.

Hammer, M. and Champy, J. (1993) *Re-engineering the Corporation: A Manifesto for Business Revolution*, 2nd edn, Harper Business, New York.

Hardgrave, B. C. (1995) When to prototype: Decision variables used in industry, *Information and Software Technology*, Vol. 37, No. 2, 113–118.

Henderson, J. C. and Treaty, M. E. (1986) Managing end-user computing for competitive advantage, *Sloan Management Review*, Vol. 27, No. 2, Winter, 3–14.

Herbsleb, J. et al. (1994) Software process improvement: State of the payoff, *American Programmer*, September.

Herzberg, F. (1966) *Work and the Nature of Man*, Staple Press, New Hope, Minnesota.

Highsmith, J. (2002) *Agile Software Development Ecosystems*, Addison-Wesley, Harlow, UK.

Highsmith, J. and Cockburn, A. (2001) Agile software development: the business of innovation, *IEEE Computer*, Vol. 34, No. 9, 120–122.

Highsmith, J. A., III (2000) *Adaptive Software Development: A Collaborative Approach to Managing Complex Systems*, Dorset House, New York.

Hill, T. and Westbrook, R. (1997) SWOT analysis: It's time for a product recall, *Long Range Planning*, Vol. 30, No. 1, 46–52.

Hirschheim, R. (1985) Information systems epistemology: An historical perspective, in E. Mumford, R.A. Hirschheim, G. Fitzgerald, and A.T. Wood-Harper (eds) *Research Methods in Information Systems*, North-Holland, Amsterdam.

Hirschheim, R. and Klein, H. K. (1989) Four paradigms of information systems development, *Communications of the ACM*, Vol. 32, No. 10.

Hirschheim, R., Earl, M., Feeny, D., and Lockett, M. (1988) An exploration into the management of the information systems function: Key issues and an evolutionary model, paper given at *Information Technology Management for Productivity and Strategic Advantage, IFIP TC-8 Open Conference, Singapore*, March.

Hirschheim, R., Klein, H. K., and Lyytinen, K. (1996) Exploring the intellectual structures of information systems development: A social action theoretical analysis, *Accounting Management & Information Technology*, Vol. 6, Nos 1/2, 1–64.

Holck, J. (2003) 4 Perspectives on web information systems, *HICSS*, Vol. 8, No. 8, 265–272.

Holland, C. P. and Light, B. (2001) A stage maturity model for enterprise resource planning systems use, *Database for Advances in Information Systems*, Vol. 32, No. 2.

Hopkins, J. (2000) A component primer, *Communications of the ACM*, 43 (10) 27–30.

Hough, D. (1993) Rapid delivery: An evolutionary approach for application development, *IBM Systems Journal*, Vol. 32, No. 3.

Howard, A. (1997) A new RAD-based approach to commercial information systems development: The dynamic systems development method, *Industrial Management & Data Systems*, Vol. 5, 175–177.

Huff, S. L., Munro, M. C., and Martin, B. H. (1988). Growth stages of end-user computing, *Communications of the ACM*, Vol. 31, No. 5, 542–550.

Hume, S., DeVane, T., and Slater, J. S. (1999) Transforming an organization through prototyping: A case study, *Information Systems Management*, Fall, Vol. 16 No. 4; *IBM Systems Journal*, Vol. 32, No. 3.

Humphrey, W. S. (1989) *Managing the Software Process*, Addison-Wesley, Reading, Massachusetts.

Iansiti, M. and MacCormack, A. (1997) Developing production on internet time, *Harvard Business Review*, September–October.

IBM (2005) *www14.software.ibm.com*

Iivari, J., Hirschheim, R., and Klein, H. K. (2000) A dynamic framework for classifying information systems development methodologies and approaches, *Journal of Management Information Systems*, Vol. 17, No. 3, 179–218.

Introna, L. D. and Pouloudi, A. (1999) Privacy in the information age: Stakeholders, interests and values, *Journal of Business Ethics*, Vol. 22, No. 1, 27–38.

Introna, L. D. and Whitley, E. A. (1997) Against method-ism: Exploring the limits of method, *Information Technology and People*, Vol. 10, No. 1, 31–45.

Jackson, M. (1983) *Systems Development*, Prentice Hall, Hemel Hempstead, UK.

Jackson, M. A. (1975) *Principles of Program Design*, Academic Press, New York.

Jacobsen, J. J. (2000) Case study: Selling the CMM, *Software Testing and Quality Engineering*, Vol. 2, No. 2, 32–39.

Jacobson, I. (updated by Bylund, S.) (2000) *The Unified Software Development Process*, Cambridge University Press, Cambridge.

Jacobson, I., Booch, G., and Rumbaugh, J. (1999) *The Unified Software Development Process*, Addison-Wesley, Boston.

Jacobson, I., Christerson, M., Jonsson, P. and Øvergaard, G. (1992) *Object-oriented Software Engineering: A Use Case Driven Approach*, Addison-Wesley, Wokingham, UK, 582 pp.

Jacobson, I., Griss, M., and Jonsson, P. (1997) *Software Reuse Architecture, Process and Organization for Business Success*, AWL, Harlow, UK.

Jalote, P. (2000) *CMM in Practice: Processes for Executing Software Projects in Infosys*, Addison-Wesley, Reading, Massachusetts.

Janson, M. A. and Smith, L. D. (1985) Prototyping for systems development: A critical appraisal, *MISQ*, December.

Jayaratna, N. (1994) *Understanding and Evaluating Methodologies, NIMSAD: A Systemic Framework*, McGraw-Hill, Maidenhead, UK.

Jeffries, R. (2001) *Extreme Programming Installed*, Pearson Education, Harlow, UK.

Jenkins, T. (1994) Report back on the DMSG sponsored UK Euromethod Forum '94, *Data Management Bulletin*, Summer Issue, Vol. 11, 3.

Johnson, D .I. and Broadman, J. G. (1996) Realities and rewards of software process improvement, *IEEE Software*, Vol. 29, No. 11, 99–101.

Jones, C. (1986) *Programming Productivity*, McGraw-Hill, New York.

Jones, C. (1993) *Assessment and Control of Software Risks*, Prentice Hall, Englewood Cliffs, New Jersey.

Kambil, A. and van Heck, E. (1998) Re-engineering the Dutch flower auctions: A framework for analyzing exchange organizations automation, *Information Systems Research*, Vol. 9, No. 1, 1–19.

Kattanjian, S. (2005) Private Discussions with Samuel Kattanjian, PreSales Engineer, SPSS, Paris.

Kettelhut, M. C. (1992) Supporting end-user database development, *Data Resource Management*, Vol. 3, No. 3, 29–39.

King, J. L. and Kraemer, K. L. (1984). Evolution and organizational information systems: An assessment of Nolan's stage model, *Communications of the ACM*, Vol. 27, No. 5, 466–475.

King, W. R. and Thompson, S. H. T. (1997) Integration between business planning and information systems planning: Validating a stage hypothesis, *Decision Sciences*, Vol. 28, No. 2.

Klein, H., Hirschheim, R. A. and Lyytinen, K. (1995) *Information Systems Development and Data Modeling: Conceptual and Philosophical Foundations*, Cambridge Tracts in Theoretical Computer Science, Cambridge University Press, Cambridge.

Knights, D. and Murray, F. (1994) *Managers Divided: Organization Politics and Information Technology Management*, John Wiley & Sons, Chichester, UK.

Kraushaar, J. M. and Shirland, L. E. (1985) A prototyping method for applications development by end-users and information systems specialists, *MISQ*, September.

Kripalani, M., Hamm, S., Ante, S. E. and Reinhardt, A. (2004) *Business Week*, New York. 26 January.

Kruchten, P. (1995) The 4+1 view model of architecture, *IEEE Software*, Vol. 12, No. 6, November, 42–50.

Kruchten, P. (2000) *The Rational Unified Process An Introduction*, 2nd edn, Addison-Wesley, Reading, Massachusetts.

Kuhn, T. S. (1962) *The Structure of Scientific Revolutions*, 2nd edn, University of Chicago Press, Chicago.

Kumra, G. and Sinha, J. (2003) The next hurdle for Indian IT, *McKinsey Quarterly*, Special Edition: Global directions, 43–53.

Lacity, M. and Hirschheim, R. (1993) *Information Systems Outsourcing: Myths, Metaphors and Realities*, John Wiley & Sons, Chichester, UK.

Land, F. (1982) Adapting to changing user requirements, *Information and Management*, Vol. 5.

Lawlis, P. K., Flowe, R. M., and Thordahl, J. B. (1995) A correlation study of the CMM and software development performance, *Crosstalk*, Vol. 8, No. 9, 21–25, September.

Lederer, A. L. and Mendelow, A. L. (1989) Information systems planning: Incentives for effective action, *Data Base*, Fall.

Leffingwell, D. (1997) Calculating the return on investment from more effective requirements management, *American Programmer*, Vol. 10, No. 4, 13–16 (now called *Cutter IT Journal*).

Lewis, P. J. (1994) *Information Systems Development: Systems Thinking in the Field of IS*, Pitman, Lonoon.

Liebenau, J. and Backhouse, J. (1990) *Understanding Information: An Introduction*, Macmillan, Basingstoke, UK.

Lincoln, T. J. and Shorrock, D. (1992) Cost justifying current use of information technology, in T. J. Lincoln (ed.) *Managing Information Systems for Profit*, John Wiley & Sons, Chichester, UK.

Ljubic, T. and Stefancic, S. (1994) Problems in the introduction of RAD principles put to praxis, *Proceedings of the Fourth International Conference on Information Systems Development (ISD94), Methods & Tools, Theory & Practice*, Moderna Organizacija, Bled, Slovenia.

Lundeberg, M., Goldkuhl, G., and Nilsson, A. (1982) *Information Systems Development − A Systematic Approach*, Prentice Hall, Englewood Cliffs, New Jersey.

Lycett, M. (2001) Understanding 'variation' in component-based development: case findings from practice, *Information and Software Technology*, Vol. 43, No. 3, 203–213.

MacCormack, A., Verganti, R. and Iansiti, M. (2001) Developing products on 'internet time': the anatomy of a flexible development process, *Management Science*, Vol. 47, No. 1, 133–150.

Maddison, R. N. (ed.) (1983) *Information System Methodologies*, Wiley Heyden, Chichester, UK.

Mansuy, J. E. (1989) Evolutionary development srategy for MIS, *Journal of Systems Management*, July.

Marchand, D. A., Davenport, T. H., and Dickson, T. (2000) *Mastering Information Management*, Prentice Hall, Harlow, UK.

Markus, M. L. and Bjørn-Andersen, N. (1987) Power over users: Its exercise by systems professionals, *Communications of the ACM*, Vol. 30, 6.

Martin, J. (1989) *Information Engineering*, Prentice Hall, Englewood Cliffs, New Jersey.

Martin, J. (1991) *Rapid Application Development*, Prentice Hall, Englewood Cliffs, New Jersey.

Martin, J. and Finkelstein, C. (1981) *Information Engineering*, Vols 1 and 2, Prentice Hall, Englewood Cliffs, New Jersey.

Martin, J. and Odell, J. J. (1992) *Object Oriented Information Engineering (OOIE) − Object Oriented Methods A Foundation*, Prentice Hall, Englewood Cliffs, New Jersey.

Maskell, B. H. (1996) Agile Manufacturing, article available at http://www.maskell.com/AgileArticle.htm

Mason, D. and Willcocks, L. (1994) *Systems Analysis, Systems Design*, McGraw-Hill, Maidenhead, UK.

Mason, R. and Mitroff, I. (1981) *Challenging Strategic Planning Assumptions*, John Wiley & Sons, New York.

Mathiassen, L., Munk-Madsen, A., Nielsen, P. A., and Sage, J. (2000) *Object Oriented Analysis and Design*, Marko Publishing, Aalborg, Denmark.

Mayer, R. E. (1996) *Thinking, problem solving, cognition*, 2nd edn. Freeman, New York.

McConnell, S. (1996) Software quality at top speed, *Software Development*, August, 38–44, available at http://www.stevemcconnell.com/articles/art04.htm.

McCracken, D. D. and Jackson, M. A. (1982) Lifecycle concept considered harmful, ACM SIGSOFT, *Software Engineering Notes*, Vol. 17, 2.

McFarlan, F. W. (1984), Information technology changes the way you compete, *Harvard Business Review*, May–June, 98–103.

McGrew, J. F., Bilotta, J. G., and Deeney, J. M. (1999) Software team formation and decay, *Small Group Research*, April, Vol. 30, No. 2, 209–234.

McGuire, E. G. (2000) The Culture Side of CMM, Research Briefs, Cutter Consortium, www.cutter.com/consortium/research/2000/crb000613.html

McKeen, J. D. and Smith, H. A. (1996) *Management Challenges in IS*, John Wiley & Sons, Chichester, UK.

McLeod, G. and Smith, D. (1996) *Managing Information Technology Projects*, Boyd & Fraser, Danvers, Massachusetts.

McManus, J. (2001) Risk in software projects, *Management Services*, Vol. 45, No. 10.

Melao, N. and Pidd, M. (2000) A conceptual framework for understanding business processes and business process modelling, *Information Systems Journal*, Vol. 10, No. 2, 105–129.

Michell, V. A. (1994) *IT Outsourcing: The Changing Outlook*, International Data Corporation, London, pp. 13–21.

Microsoft (2005) www.Microsoft.com

Milburn, M. (1978) Sources of bias in the prediction of future events, *Organisational Behaviour and Human Performance*, Vol. 21, 17–26.

Miller, G. G. (2001) The characteristics of agile software processes, 39th International Conference of Object-Oriented Languages and Systems (TOOLS 39), Santa Barbara, CA.

Mills, S. (2002) http://www.ibm.com/news/us/2002/12/061.html

Mirani, R. and King, W. R. (1994) Impacts of end-user and information center characteristics on end-user computing support, *Journal of Management Information Systems*, Summer, Vol. 11, No. 1.

Mitroff, I. and Linstone, H. (1993) *The Unbounded Mind, Breaking the Chains of Traditional Business Thinking*, Oxford University Press, Oxford.

Moad, J. (1993) Does reengineering really work?, *Datamation*, Vol. 39, 15.

Moran, N. (2003) Looking for saving on distant horizons, *Financial Times*, 2 July.

Moynihan, E. (1993) *Business Management and Systems Analysis*, McGraw-Hill, Maidenhead, UK. This text provides a hybrid approach to the subject, treating the business and computer sides with equal importance and emphasizing the necessary integration of the two in the information systems development process.

Mumford, E. (1983a) *Designing Human Systems*, Manchester Business School, Manchester.

Mumford, E. (1983b) *Designing Participatively*, Manchester Business School, Manchester.

Mumford, E. (1986) *Using Computers for Business*, Manchester Business School, Manchester.

Mumford, E. (1989) *XSEL's Progress*, John Wiley & Sons, Chichester, UK.

Mumford, E. (1995) *Effective Requirements Analysis and Systems Design: The ETHICS Method*, Macmillan, Basingstoke, UK.

Mumford, E. and Weir, M. (1979) *Computer Systems in Work Design – The ETHICS Method*, Associated Business Press, London.

Myers, G. (1975) *Reliable Software Through Composite Design*, Petrocelli/Charter, New York.

Myers, G. (1978) *Composite/Structured Design*, Van Nostrand Reinhold, New York.

National Audit Office (1999) *Government on the Web*, A Report by the Comptroller and the Auditor General, HC 87, 13 December.

National Audit Office (2002) *Government on the Web II*, A Report by the Comptroller and the Auditor General, 25 April.

Nauman, J. D. and Jenkins, A. M. (1982) The new paradigm for systems development, *MISQ*, September.

Nolan, R. L. (1973) Managing the computing resource: A stage hypothesis, *Communications of the ACM*, Vol. 16, No. 7, 399–405.

Nolan, R. L. (1979) Managing the crises in data processing, *Harvard Business Review*, Vol. 57, 2.

Nonaka, I. and Takeuchi, H. (1986) The new product development game, *Harvard Business Review*, January, 137–146.

Norris, G., Wright, I., Hurley, J. R., Dunleavy, J., and Gibson, A. (1998) *SAP: An Executive's Comprehensive Guide*, John Wiley & Sons, New York.

NTA (2003) http://www.nta-monitor.com/fact-sheets/pwd-main.htm, accessed July 2005.

Nunamaker, J., Dennis, A., Valacich, J., Vogel, D., and George, J. (1991) Electronic meeting systems to support group work, *Communications of the ACM*, July, 41–61.

Oliga, J. (1988) Methodological foundations of systems methodologies, in R. L. Flood and M. C. Jackson (eds) *Critical Systems Thinking: Directed Readings*, John Wiley & Sons, Chichester, UK.

Olle, T. W., Sol, H. G., and Verrijn-Stuart, A. A. (1986) *Information Systems Design Methodologies: A Comparative Review*, North Holland, Amsterdam.

OMG (2002) Introduction to OMG's Unified Modeling Language, http://www.omg.org/gettingstarted/what_is_uml.htm

Open Source Initiative (2005) see http://www.opensource.org/

Orman, L. (1998/9) Evolutionary development of information systems, *Journal of Management Information Systems*, Vol. 5, No. 3.

Page-Jones, M. (2000) *Fundamentals of Object-oriented Design in UML*, Addison-Wesley, Reading, Massachusetts.

Palmer, S. R. and Felsing, J. M. (2002) *A Practical Guide to Feature-Driven Development*, Prentice-Hall, New Jersey.

Pankaj, J. (2000) *CMM in Practice: Processes for Executing Software Projects at Infosys*, Addison-Wesley.

Parsons, T. and Shils, E. (1951) *Towards a General Theory of Action*, Harvard University Press, Massachusetts.

Patane, J. R. and Jurison, J. (1994) Is global outsourcing diminishing the prospects for American programmers? *Journal of Systems Management*, Vol. 45, No. 6.

Paulk, M. C., Curtis, B., Chrissis, M. B., and Weber, C. V. (1993a) Capability Maturity Model, Version 1.1, *IEEE Software*, Vol. 10, No. 4, July, 18–27.

Paulk, M. C., Weber, C. V., Curtis, B., and Chrissis, M. B. (1995) *The Capability Maturity Model, Guidelines for Improving the Software Process*, Addison-Wesley, Reading, Massachusetts, 441 pp.

Paulk, M. C., Weber, C. V., Garcia, S. M., Chrissis, M. B., and Bush, M. (1993b) *Key Practices of the Capability Maturity Model, Version 1.1*, Software Engineering Institute, CMU/SEI-93-TR-25.

Peppard, J. and Preece, I. (1995) The content, context and process of business process re-engineering, in G. Burke and J. Peppard (eds) *Examining Business Process Re-engineering: Current Perspectives and Research Directions*, Kogan Page, London, pp. 157–185.

Pfahl, D., Klemm, M. and Ruhe, G. (2000) Using System Dynamics Simulation Models for Software Project Management Education and Training, Extended Abstract, the Software Process Simulation Modeling Workshop (ProSim2000), London, 10–12 July.

Phan, D. D. (2001) Software quality and management, *Information Systems Management*, Winter, Vol. 18, No. 1, 56–68.

Plato, J. J. (1995) Prototyping: Proceed with caution, *Information Systems Management*, Fall, Vol. 12, No. 4.

Ploskina, B. (1999) Rational adds quality to rapid application development, *ENT*, Vol. 4, Issue 21, August, 25–27

Porter, M. E. (1980) *Competitive Strategy*, Free Press, New York.

Porter, M. E. (1985) *Competitive Advantage*, Collier-Macmillan, New York.

Porter, M. E. (1991) Toward a dynamic theory of strategy, *Strategic Management Journal*, Vol. 12, 95–117.

Prahalad, C. K. and Hamel, G. (1994) Strategy as a field of study: Why search for a new paradigm? *Strategic Management Journal*, Vol. 15, 5–16.

Pralahad, C. K. and Hamel, G. (1990) The core competence of the corporation, *Harvard Business Review*, May/June, 79–91.

Pressman, R. S. (2004) *Software Engineering: A Practitioner's Approach*, 6th edn, McGraw-Hill, London.

Prince2 (2005) www.prince2.com

Probasco, L. (2000) *The Ten Essentials of RUP: The Essence of an Effective Development Process*, Rational White Paper.

Pyron, T. (2000) *Using Project 2000*, Que Macmillan, Indianapolis, Indiana.

Quang, P .T. and Chartier-Kastler, C. (1991) *Merise in Practice. Macmillan*, Basingstoke (translated by D. E. and M. A. Avison from the French: *Merise Appliquée*, Eyrolles, Paris, 1989).

Quinn, J. (1992) *The Intelligent Enterprise*, Free Press, New York, 1992.

Radding, A. (1999) Enterprise RAD tools: Can they do the job? *InformationWeek*, Issue 716, November, 1–5.

Rakitin, S. (2001) Manifesto elicits cynicism, *Computer*, December, p.4 (quoted by Boehm, 2002).

Randall, D., Hughes, J., and Shapiro, D. (1994) Steps towards a partnership: Ethnography and system design, in M. Jirotka and J. Goguen (eds) *Requirements Engineering: Social and Technical Issues*, Academic Press, London, pp. 241–258.

Ravichandran, T. and Rothenberger, M. (2003) Software reuse strategies and component markets, *Communications of the ACM*, Vol. 46, No. 8, 109–114.

Raymond, E. S. (2001) *The Cathedral and the Bazaar: Musings on Linux and Open Source by an Accidental Revolutionary*, O'Reilly, Sebastopol, California.

Ringland, G. (1998) *Scenario Planning: Managing for the Future*, John Wiley & Sons, Chichester, UK.

Robertson, S. and Robertson, J. (1999) *Mastering the Requirements Process*, Addison-Wesley, Boston, MA.

Rockart, J. F. (1979) Chief executives define their own data needs, *Harvard Business Review*, Vol. 57, No. 2, 238–241.

Rockart, J. F. (1982) The changing role of the information systems executive: A critical success factors perspective, *Sloan Management Review*, Fall.

Rockart, J. F. and Flannery, L.S. (1983) The management of end-user computing, *Communications of the ACM*, Vol. 26, 10.

Rogers, E. M. (1995) *Diffusion of Innovations*, 4th edn, The Free Press, New York.

Rosenhead, J. (1989) *Rational Analysis for a Problematic World: Problem Structuring Methods for Complexity, Uncertainty and Conflict*, edited by J. Rosenhead, John Wiley & Sons, Chichester, UK.

Rosenhead, J. and Mingers, J. (2001) *Rational Analysis for a Problematic World Revisited: Problem Structuring Methods for Complexity, Uncertainty and Conflict*, 2nd edn, Wiley, Chichester.

Ross, J. W. and Feeny, D. F. (2000) The evolving role of the CIO, in R. Zmud (ed.) *Framing the Domain of IT Management: Glimpsing the Future through the Past*, Pinnaflex, Cincinnati, Ohio.

Rumbaugh, J., Blaha, M., Premerlani, W., Eddy, F., and Lorensen, W. (1991) *Object Oriented Modelling and Design*, Prentice Hall, Englewood Cliffs, New Jersey.

Russell, M. (2004) The importance of non-functional and operational requirements, IBM developerWorks, at http://www-106.ibm.com/developerworks/web/library/wa-qualbust1/

Russo, N. L. and Wynekoop, J. L. (1995) The use and adaptation of system development methodologies, *Proceedings of the 6th Information Resources Management Association International Conference*, Idea Group Publishing, Hershey, Pennsylvania.

Sachs, P. (1995) Transforming work: Collaboration, learning, and design, *Communications of the ACM*, Vol. 38, No. 9, 36–44.

Saiedian, H. and Dale, R. (2000) Requirements engineering: making the connection between the software developer and the customer, *Information and Software Technology* Vol. 42, No. 6, 419–428.

Scacchi, W. (2004) Socio-technical interaction networks in free/open source software development processes, in *Peopleware and the Software Process*, Silvia Teresita Acuña and Natalia Juristo (eds.), World Scientific Press.

Schreiber, G., Akkermans, H., Anjewierden, A., de Hoog, R., Shadbolt, N., Van de Velde, W., and Wielinga, B. J. (2000) *Knowledge Engineering and Management: The Common KADS Methodology*, MIT Press, Cambridge, Massachusetts.

Schwaber, K. and Beedle (2002) *Agile Software Development with SCRUM*, Prentice Hall.

Schware, R. (1987) Software industry in the Third World: policy guide lines, institutional options and constraints, *World Development*, Vol. 15, No. 10/11, 1249–1267.

Searle, J. R. (1995) *The Construction of Social Reality*, Allen Lane, Penguin Press, Harmondsworth, UK.

SEI (2005) Software Engineering Institute website, http://www.sei.cmu.edu/sei-home.html

SEIa (2001) SEI Mission, http://www.sei.cmu.edu/about/overview/sei/mission.html

Silk, D. (1990) Managing information systems benefits for the 1990s, *Journal of Information Technology*, Vol. 5, No. 4.

Smith, P. G. and Pichler, R. (2005) Agile risks/agile rewards, *Software Development*, Vol. 13, No. 4, 50–53.

Smith, A. (1997) *Human Computer Factors: A Study of Users and Information Systems*, McGraw-Hill, London.

Smith, H. J. and Hasnas, J. (1999) Ethics and information systems: The corporate domain, *MIS Quarterly*, Vol. 23, No. 1.

Smith, R. (2000) Defining the UML kernal, *Software Development*, October, Rational Software Corp., available at http://www.sdmagazine.com/articles/2000/0010/

Sparling, M. (2000) Lessons learnt through six years of component development, *Communications of the ACM*, Vol. 43, No. 10, 47–53.

Stapleton, J. (1997) *DSDM Dynamic Systems Development Method, The Method in Practice*, Addison-Wesley, Harlow, UK. This source also contains a number of case studies of DSDM development projects.

Stapleton, J. (1998) Giving RAD a good name, *The Computer Bulletin*, November.

Stapleton, J. (2002) *DSDM: A Framework for Business Centred Development*, Addison-Wesley, Harlow, UK.

Stevens, W. P., Myers, G. J., and Constantine, L. L. (1974) Structured design, *IBM System Journal*, Volume 13, 2.

Stewart, J. (1994) *IS Notice: Euromethod*, CCTA, Norwich, UK, p. 61.

Stewart, K. and Gossain, S. (2001) An exploratory study of ideology and trust in open source development groups, *Proceedings of ICIS 2001*, 507–512.

Stewart, T. A. (2003) Does IT matter? An HBR debate, *Harvard Business Review*, June 2003 (http://harvardbusinessonline.hbsp.harvard.edu/b01/en/files/misc/Web_Letters.pdf)

Strassmann, P. (1990) *The Business Value of Computers*, The Information Economics Press, New Canaan, Connecticut.

Subramanian, G. H. and Zarnich, G. E. (1996) An examination of some software development effort and productivity determinants in ICASE tool projects, *Journal of Management Information Systems*, Vol. 12, No. 4.

Sugumaran, V. and Storey, V. C. (2003) A semantic-based approach to component retrieval, *DATA BASE for Advances in Information Systems*, Vol. 34, No. 3, 8–24.

Sumner, M. and Klepper, R. (1986) The impact of end user computing on information systems development, *Computer Personnel*, Vol. 10, No. 4.

Sutcliffe, A. (1988) *Jackson Systems Development*, Prentice Hall, Hemel Hempstead, UK.

Szalvay, V. (2004) An introduction to agile software development, White Paper, Danube Technologies, available at http://www.danube.com/docs/Intro_to_Agile.pdf.

Takeishi, A. (2002) Knowledge partitioning in the interfirm division of labor: the case of automotive product development, *Organization Science*, Vol. 13, No. 3, 321–338.

Taylor, F. W. (1947) *Scientific Management*, Harper & Row, New York.

Taylor-Cummings, A. (1998) Bridging the user-IS gap: a study of major systems projects, *Journal of Information Technology*, Vol. 13, No. 1, 29–54.

Tayntor, C. B. (1994) New challenges or the end of EUC? *Information Systems Management*, Summer, Vol. 11, No. 3. The DSDM website: http://www.dsdm.org/

Tiwana, A. (2000) *The Knowledge Management Toolkit*, Prentice Hall, Upper Saddle River, New Jersey.

Truex, T., Baskerville, R., and Travis, J. (2000) Amethodical systems development: the deferred meaning of systems development methods, *Accounting Management and Information Technologies*, Vol. 10, 53–79.

Tudhope, D. (2000) Prototyping praxis: Constructing computer systems and building belief, *Human–Computer Interaction*, Vol. 15 No. 4.

Turk, D., France, R. and Rumpe, B. (2002) Limitations of agile software processes, Proceedings of Third International Conference on Extreme Programming and Flexible Processes in Software Engineering, XP2002, Alghero, Italy, 43–46.

Turner, P. and Jenkins, T. (1996) *Euromethod and Beyond*, Thomson, London.

Tversky A. and Kahneman D. (1992) Advances in prospect theory: Cumulative representation of uncertainty, *Journal of Risk and Uncertainty*, Vol. 5, 297–323.

Tversky A. and Koehler D. (1994) Support theory: A nonextensional representation of subjective probability, *Psychological Review*, Vol. 101, 547–567.

Valentin, E. K. (2001) SWOT analysis from a resource-based view, *Journal of Marketing Theory and Practice*, Vol. 9, No. 2, Spring.

Van Slooten, K. (1996) Situated method engineering, *Information Resources Management Journal*, Vol. 9, No. 3.

Vidgen, R. (1994) Research in progress: Using stakeholder analysis to test primary task conceptual models in information systems development, in *Proceedings of the Second Annual Conference on Information System Methodologies*, BCS IS Methodologies Specialist Group, 31 August–2 September, Edinburgh.

Vidgen, R. (1996) A multiple perspectives approach to information system quality. Unpublished PhD thesis, June, University of Salford.

Vidgen, R. and McMaster, T. (1996) Black boxes, non-human stakeholders and the translation of IT through mediation, in W. J. Orlikowski, G. Walsham, M. Jones, and J. I. DeGross (eds) *Information Technology and Changes in Organizational Work*, Chapman & Hall, London, pp. 250–271.

Vidgen, R., Avison, D. E., Wood, R., and Wood-Harper, A. T. (2002) *Developing Web Information Systems*, Butterworth-Heinemann, London.

Vitharana, P. and Jain, H. (2000) Research issues in testing business components, *Information & Management*, Vol. 37, No. 6, 297–309.

Walker, R. (1998) *Software Project Management, A Unified Framework*, Addison-Wesley, Boston.

Walsham, G. (1993) *Interpreting Information Systems in Organisations*, John Wiley & Sons, Chichester, UK.

Ward, J. and Griffiths, P. (1996) *Strategic Planning for Information Systems*, 2nd edn, John Wiley & Sons, Chichester, UK.

Ward, J. M. (1985) Evaluating IS projects and charges for services, *Management Accounting*, Vol. 63, No. 1.

Warren, I. (1999) *The Renaissance of Legacy Systems*, Springer-Verlag, London.

Wastell, D. (1996) The fetish of technique: Methodology as a social defence, *Information Systems Journal*, Vol. 6, 1.

Watson, I. (1997) *Applying Case Based Reasoning*, Morgan Kauffman, San Francisco, California.

Weaver, P. L. (1993) *Practical SSADM Version 4: A Complete Tutorial Guide*, Pitman, London.

Weaver, P. L., Lambrou, N., and Walkley, N. (1998) *Practical SSADM+*, Pitman, London.

Weber, C. V., Paulk, M. C., Wise, C. J., and Withey, J. V. (1991) *Key Practices of the Capability Maturity Model*, Software Engineering Institute, CMU/SEI-91-TR-25.

Weinberg, V. (1978) *Structured Analysis*, Prentice Hall, Englewood Cliffs, New Jersey.

Welti, N. (1999) *Successful SAP R/3 Implementation*, Addison-Wesley, Harlow, UK.

Wiegers, K. E. (2001) License to Hack, StickyMinds, Weekly Column, 24 September, available at http://www.stickyminds.com

Wielemaker, M. W., Elfring, T., and Volberda, H. W. (2000) Strategic renewal in large European firms: Investigating viable trajectories of change, *Organization Development Journal*, Vol. 18, No. 4, 49–68.

Wielinga, B. J., Sterner, Th.A., and Breuker, J. A. (1993) KADS: A modelling approach to knowledge engineering, in B. G. Buchanan and D. C. Wilkins (eds) *Readings in Knowledge Acquisition and Learning, Automating the Construction and Improvement of Expert Systems*, Morgan Kaufmann, San Mateo, California.

Willcocks, L. and Fitzgerald, G. (1993) Market as opportunity? Case studies in outsourcing information technology and services, *Journal of Strategic Information Systems*, Vol. 2, No. 3, 223–242.

Willcocks, L., Fitzgerald, G., and Lacity, M. (1996) To Outsource IT or not?: Recent research on economics and evaluation practice, *European Journal of Information Systems*, Vol. 5, 143–160.

Willis, R. R. (and eight others) (1998) *Hughes Aircraft's Widespread Deployment of a Continuously Improving Software Process*, Software Engineering Institute, CMU/SEI-98-TR-006.

Wilson, B. (1990) *Systems: Concepts, Methodologies and Applications*, 2nd edn, John Wiley & Sons, Chichester, UK.

Wilson, B. (2001) *Soft System Methodology*, John Wiley & Sons, Chichester, UK.

Wirth, N. (1971) Program development by stepwise refinement, *Communications of the ACM*, Vol. 14, 4.

Wood, J. and Silver, D. (1989) *Joint Application Design: How to Design Quality Systems in 40% Less Time*, John Wiley & Sons, New York.

Wood, J. R. G., Vidgen, R. T., Wood-Harper, A. T., and Rose, J. (1995) Business process redesign: Radical change or reactionary tinkering?, in G. Burke and J. Peppard (eds) *Examining Business Process Reengineering: Current Perspectives and Research Directions*, Kogan Page, London.

Wood-Harper, A. T., Corder, S., Wood, J., and Watson, H. (1996) How we profess: The ethical systems analyst, *Communications of the ACM*, Vol. 39, No. 3, 69–77.

Wysocki, R. K. and DeMichiell, R. L. (1997) *Managing Information Across the Enterprise*, John Wiley & Sons, New York.

Ximbiot (2005) see http://ximbiot.com/

XP (2005) www.extremeprogramming.org

Yalaho, A., Nahar, N., Käkölä, T., and Wu, C. (2005) A conceptual process framework for IT-supported international outsourcing of software production, Proceedings of the IEEE Eee05 international Workshop on Business Services Networks (Hong Kong, 29 March 2005), *ACM International Conference Proceeding Series*, 87, IEEE Press, Piscataway, NJ, 3–13.

Yoon, Y., Guimaraes, T., and O'Neal, Q. (1995) Exploring the factors associated with expert systems success, *MIS Quarterly*, Vol. 19, 1.

Yourdon, E. and Argila, C. (1996) *Case Studies in Object-oriented Analysis and Design*, Yourdon Press, Prentice Hall, New Jersey.

Yourdon Inc. (1993) *Yourdon Systems Method: Model-driven Systems Development*, Yourdon Press, Englewood Cliffs, New Jersey.

Yourdon, E. (1989) *Modern Structured Analysis*, Prentice Hall, Englewood Cliffs, New Jersey.

Yourdon, E. (2000) *Computerworld*, 21 August, Vol. 34, Issue 34, 36.

Yourdon, E. and Constantine, L. L. (1978) *Structured Design*, 2nd edn, Yourdon Press, New York.

Zach, M. H. (1999) Developing a knowledge strategy, *California Management Review*, Vol. 41, No. 3, 125–145.

Zuboff, S. (1988) *In the Age of the Smart Machine*, Basic Books, New York.

Index

Page numbers for figures have suffix **f**, those for tables have suffix **t**, and those for major topics are in **bold**

A

Abrahamsson, P., 134, 139
Ackerman, F., 213, 214, 215, 216, 549, 551, 564
Ackoff, R.L., 51
ACM *see* Association of Computing Machinery
ACS *see* Australian Computer Society
Adams, C., 315, 320, 324, 326, 386
Adams, James, 317
adaptive software development (ADS), 144
Advantage Gen, 364, 368
Agile Alliance, 135
agile development, **134–145** *see also* rapid and
    evolutionary development theme
Agile Manifesto, 134–136
Agile Modelling (AM), 144
Akkermans, H., 505
Allen, P., 387
Ambler, S., 144
American Express, 90
American Hospital Supply Corporation, 57
Andrews, D.C., 310
Anjewierden, A., 505
Ante, S.E., 191
Aonix, 372
Apache (computer software), 179
application service provision (ASP), 190
application system asset management (ASAM),
    152
Argila, C., 273, 274, 451, 452
Argyris, C., 501, 547
Arnett, K.P., 298
ASAM *see* application system asset management
ASP *see* application service provision
Association of Computing Machinery (ACM), 15
Augustine, S., 141
Australian Computer Society (ACS), 15
Avison, M.A., 450
Avison, Professor David E., xiv, xxi, 14, 44, 117, 149,
    158, 160, 174, 199, 241, 272, 297, 315, 319, 324,
    326, 386, 450, 485, 537, 545, 546, 547, 549,
    564, 572, 595

B

Bahrami, A., 113
Bardoloi, S., 557
Baskerville, R., 137, 146
BCNF *see* Boyce-Codd normal form

BCS *see* British Computer Society
Beck, Kent, 135, 479, 485
Beer, Stafford, 512
Beer's Viable Systems Model, 53
Begg, C., 241
Bennett, S., 283
Bentley, C., 530, 531, 533, 536
Beynon-Davies, P., 144, 189
Bikson, T.K., 334
Bisp, S., 296
Bjørn-Andersen, N., 573, 574, 593
blended methodologies, **419–449**
    Information Engineering (IE), **434–446**
        business area analysis, 439–443
            confirmation, 442
            current systems analysis, 440–443
                bubble charts, 441**f**, 442**f**
                canonical synthesis, 441
            entity and function analysis, 439
            interaction analysis, 439, 440**f**
            planning for design, 442–443
        construction and cutover, 445–446
            cutover tasks, 445–446
            evaluation and maintenance following
                cutover, 446
            system generation and verification, 445
        data, activity and interaction, 436**f**
        divide and conquer, 437
        four levels, 437, 438**f**
        information strategy planning (ISP), 437–439
            architecture definition, 439
            current situation analysis, 438
            executive requirements analysis, 439
            information architectures, 438
            information strategy plan, 439
        philosophy, 435
        system planning and design, 443–445
            confirmation, 444
            dialogue flow, 444**f**
            planning for technical design, 444
            preliminary data structure design, 443
            procedure design, 443–444
            system structure design, 443
            technical design phase tasks, 445
        top-down approach, 436
    Merise, **426–434**

abstraction cycle, 429–434
    Boyce-Codd normal form (BCNF), 433
    conceptual data model, 430, 431**f**
    conceptual processing model with
        synchronization, 430, 432**f**
    flow diagram, 430**f**
    modelling logic, 429**f**
    organizational rules, 431
    Petri nets, 431
    rules for mapping conceptual data, 433**f**
decision cycle, 427
    decisions involved, 427
    schema, 428**f**
documentation and security, 434
life cycle, 427–429
    main phases, 427–429
use of tools, 434
Structured Systems Analysis and Design
    Method (SSADM), **419–426**
definition of requirements, 422–424
    entity life histories, 422, 423**f**, 424**f**
    Jackson structure diagrams, 422
feasibility, 419–420
    data flow diagrams, 421**f**
logical design, 425
physical design, 425–426
    function component implementation
        map, 425
    quality assurance reviews, 426
    rapid application development (RAD),
        426
requirements analysis, 420–422
    business system options, 421–422
        cost/benefit analysis, 422
        investigation of current environment,
            420–421
            customization factors, 421
            logical data flow model, 421
    seven stages, 419, 420**f**
    technical system options, 424–425
Welti ERP development, **447–449**
    enterprise resource planning (ERP)
        system, 447
    project life cycle, 447**f**
        planning, 447
        preparation, 448

productive, 449
realization, 448
Stanford advanced project (SAP)
management, 449
methodology, 449
Boafu, K., 19
Boehm, B., 121, 132, 145
Boeing, 557
Booch, G., 113, 273, 276, 277, 279, 280, 451, 461, 468
Boyce-Codd normal form (BCNF), 238**f**, 239**f**, 433
BPR *see* business process re-engineering
Braa, K., 19
Brancheau, J.C., 70
Braue, D., 557
Breuker, J.A., 496, 505
Brinkkemper, S., 149, 158
British Airways, 473
British Computer Society (BCS), 14, 473
definition of methodology, 567–568
eight-point Code, 16**f, 17f**
adherence to regulations, 17
use of appropriate methods and tools, 18
expert systems specialist group, 87
Information Systems Analysis and Design
Working Group, 567
Professional Advice Register, 15, 17
Professional Examination, 18
British Telecom, 473
Broadman, J.G., 557
Brooks, F.P., 474, 481
Brown, A., 174
Buchanan, B.G., 505
Bucki, L.A., 359
Buckingham, R.A., 21, 22
Budde, R., 547
Burke, G., 64
Burrell, G., 601
business process re-engineering (BPR), **62–65,** 601
see also organizational themes
Buttle, F., 302

C

Cadle, J., 78, 294
Capability Maturity Model (CMM), 71, **551–558**
see also framework methodologies
Cardenas, A.F., 232
Carmel, E., 310
Carr, N.G., 59, 60
CAS *see* complex adaptive systems
CATWOE mnemonic for criteria, **204–205,** 510, 541
Central Computer and Telecommunications
Agency (CCTA), 419, 530

Champy, J., 62, 63, 64, 78
Chapman, C.B., 304
Chartier-Kastler, C., 450
Checkland, P., 52, 53, 78, 115, 213, 216, 319, 320, 392,
507, 508, 510, 512, 514, 515, 516, 536, 537, 569,
575, 598, 599, 607, 608, 614
Checkland's Soft Systems Methodology, 199 *see*
also organizational-oriented
methodologies
Chen, J., 123, 128, 149, 340
Chief Information Officer (CIO), 11
Chief Knowledge Officer (CKO), 95
Chrissis, M.B., 564
Churchman, C.W., 512, 548
Ciborra, C.U., 10, 359
CIO *see* Chief Information Officer
CKO *see* Chief Knowledge Officer
CMM *see* Capability Maturity Model
Coad, P., 113, 115, 123, 274, 341, 392, 451, 452, 454,
455, 456, 457, 458, 459, 460, 468
Cobol computer program, 83
Cockburn, A., 136, 144, 146
CoCoMo technique, 74, 75, 289
CommonKADS methodology, **501–504,** 587 *see*
also people-oriented methodologies
complex adaptive systems (CAS), 140
computer aided software engineering (CASE) *see*
toolsets
Computer Associates, 364, 368
Computerworld (journal), 191
Conceptual Blockbusting (Adams), 317
concurrent versions system (CVS), 182
Conn, S.S., 135
Connolly, T., 241
Constantine, L.L., 395
context of information systems, **3–18**
change, **10–11**
dynamic effects, 10
legacy systems, 10
digital economy, 8–9
call centres, 8
digital divide, 8
effects of technology change, 8
Prudential and Egg, 8
electronic commerce, **9**
considered to be a revolution, 9
non-commercial impacts, 9–10
e-government initiatives, 10
effect on public sector and local
government, 9–10
National Audit Office, 9

open government, 9
reasons for effects on business, 8–9
environment, 7
globalization, 7–8
Dyson, 7
effect on markets, 7–8
human dimension of information systems,
11–14
airline reservation system, 12
development process with steering
committee, 13
external users are also stakeholders, 12–13
information users, 13
shareholders, 13
society, 13
trusted external users, 13
stakeholders, **11**
business analysts, 11
Chief Information Officer, 11
programmers, 11
project managers, 11
senior IT management, 11
system analysts, 11
training, 11
users, **12**
business users, 12
casual users, 12
end users, 12
external users, 12
managers, 12
occasional users, 12
regular users, 12
information systems, 3–7
computer-based, 3
context, 7
defined, 3–4
environment, 7
examples, 4–7
e-Bay electronic auction house, 5, 6**f**, 7
payroll system, 4–5
formalized, 3
goals, 3
type, 4
organization, 14
information systems strategy group, **14**
project management and control, 13–14
project team leader, 13–14
steering committee, **14**
systems development team, **14**
professionalism and information systems,
14–19

British Computer Society Code, 14–15, 16**f**, 17**f**
  failure of One.Tel in Australia, 14–15
    availability of professional codes of conduct and good practice, 15
    Capability Maturity Model (CMM), 14
    flaws identified in the billing system, 15
    prioritization, 15
conventional systems analysis *see* information systems development life cycle
Cook, S., 277
Cooper, D.F., 304
Cooper, R.B., 70
Coopers and Lybrand, 90
Couger, D., 320
Coughlan, J., 101
critical path analysis, 74, 290
critical success factor (CSF), 60, 67, 296–299 *see also* organizational techniques
CRUD matrix (create, read, update or delete), 264, 365
Crystal methodology, 144
CSF *see* critical success factor
Currie, W.L., 78, 190
Curtis, B., 564
customer relationship management (CRM), 96, 587
Cuthbertson, C., 174, 241
Cysneiros, L.M., 106, 107
D
Dahlborn, B., 19
Dale, R., 101
Daniels, J., 277
data and function networking (DAFNE), 559
data flow diagram (DFD), **243**, 244–251, 397 *see also* process techniques
data techniques, **217–240**
  entity modelling, **217–229**
    attribute, defined, **220**
    cardinality of a relationship, **221, 222f**
    deciding level of detail, 219
    defined, 217
    documentation, 227**f**
    entity model for academic department, 217, 218**f**
    entity-relationship model for hospital, 218**f**, 219
    example of wholesaler, 228**f**, 229**f**
    involuted relationship, 223, 224**f**
    mandatory and optional relationships, 223**f**
    operations on attributes, **224, 225f**

relationship
  defined, **220**
  fixed, 222
  many-to-many, 221, 222**f**, 223**f**
  one-to-many, 221
  one-to-one, 221
  optional, mandatory or exclusive, 223
stages of entity analysis, 225–226
synchronization of an operation, 224, 225**f**
ternary relationship, 223, 224**f**
normalization, **229–240**
  Boyce-Codd normal form (BCNF), 238**f**, 239**f**
  COURSE-DETAIL relationship, 232, 234
  COURSE-MODULE relationship, 234–235
  defined, 231
  ELECTION-RESULT relationship, 229, 230**f**
  entities expressed as relations, 230**f**
  first normal form, 232, 233**f**, 234
  fourth normal form, 238–239, 240**f**
  functional dependency, defined, 232
  key attributes, 230–231
    primary, candidate and foreign keys, 231
  reasons for normalizing, 237
  SALES-ORDER relationship, 229**f**
  second normal form, 232, 234**f**, 235**f**
  STATUS relationship, 237
  and STRADIS, 231
  terminology, 229
  third normal form (TNF), 232, 235, 236**f**
  transitive dependency, 236
Date, C.J., 241
Davenport, Tom H., 62, 63, 64, 93, 108, 392, 526, 528, 529, 536
Davidson, E.J., 312
Davis, G.B., 595
Davis, L.E., 487
De Bono, E., 295, 306, 322
De Grace, P., 585
De Greef, P., 496
de Hoog, R., 505
Deeks, D., 241, 272
DeMarco, T., 110, 138
DeMarco, T, 395
DeMichiell, R.L., 64
development of information systems, 21–26
  information systems development
    methodology, **24–26**
    defined, 24
    objectives, 25
    philosophy, 24

techniques and tools, 24–25
    use of methodologies, 25–26
  key concepts, 21–23
    data, information, and knowledge, defined, **21–22**
    information systems development, defined, **23**
    systems, **22**
      boundary, 22
      subsystems, **22**
    need for methodology, 23–24
      changes with computer development, 24
      use of programmers, 23
DFD *see* data flow diagram
Diaz, M., 557
Dibbern, J., 192
Digital Equipment Corporation (DEC), 90, 495
Donaldson, T., 308
Dreamweaver software, **334–335,** 481, 484 *see also* tools
Dunleavy, J., 193
Dyché, J., 108
Dynamic Systems Development Method (DSDM) *see* rapid development methodologies
Dyson plc, 7
E
e-Bay electronic auctions, 5, 6**f,** 7
Earl, M., 57, 58, 60, 61, 62, 64, 78
Eastman Kodak, 187
The Economist (periodical), 10
Eden, C., 213, 214, 215, 216, 549, 551, 564
effective technical and human implementation of computer-based systems (ETHICS) methodology, 36, 82, 392, **487–496,** 604 *see also* people-oriented methodologies
electronic commerce and information systems, **9**
electronic data interchange (EDI), 54
end-user computing (EUC), 84–87
engineering themes, **151–173**
  automated tools, **156–158**
    documentation support tools, 157
    software tools facilities, 157–158
  component development, **161–163**
    basic concept, 161
    benefits, 162
    component based software development (CBSD), 162
    defined, 161
    reasons for lack of applications, 162–163

data warehouse, **170–173**
business intelligence, 171
data mining, **171**
SPSS data mining tool, 172
examples of use, 173
stages in a data mining project, 172–173
database management, **165–170**
access by programmers, 168
database architecture, 166**f**
external, internal and conceptual views, 166–167
database management system (DBMS), 166, 170
use as a tool to develop applications, 166, 170
improvement of standards, 169–170
database administrator (DBA), **169**
independence of data, 165–170
integrity of data, 168
provide management view, 169
reduce program maintenance, 169
shareability of data, 167–168
speed of implementation, 168
legacy systems, **151–152**
application of enterprise resource planning system, 152
problems with environmental and technology changes, 151–152
reverse engineering and forward engineering, **152**
method engineering (ME), **158–161**
defined, 158
types of method engineering, 158, 159**f**, 160
security issues, **163–165**
documentation, 165
examples of protection, 164
objectives, 163–164
use of passwords, 164–165
software engineering, **152–156**
control, 156
definition of requirements, 155
functional decomposition, **153, 154f**
program design, 154**f**, 155
enterprise application integration (EAI), 148
enterprise resource planning system (ERP), 152, 175–178, 183–186
Ericsson (Sweden), 461
Ernst & Young, 95, 96
Espejo, R., 53
Essex, P., 86

ETHICS *see* effective technical and human implementation of computer-based systems (ETHICS) methodology
EUC *see* end-user computing
Euromethod methodology, **558–563** *see also* framework methodologies
European Commission (EC), 558
European Union ESPRIT project, 496, 533
European Union (EU), 558
Eva, M., 132, 450
Evans, J.S., 72
Everard, J., 19
evolutionary development, **119–123** *see also* rapid and evolutionary development theme
expert system, **87–92** *see also* people themes
external development theme, **175–192**
application packages, **175–178**
evaluation, 176, 177**f**
intangible costs and benefits, 178
question the functional requirements, resources, people, and documentation, 176
requirements shortfall, 177–178
enterprise resource planning (ERP), 175–178, **183–186**
advantages, 184
control of organizational activities, 185
functions included, 183
potential for further improvements, 185–186
scope of human resource management function, 183–184
suppliers of systems, 184
open source software (OSS), **178–182**
Apache and Linux, 179
concurrent versions System (CVS), 182
criticisms, 182
open source initiative (OSI), 179
roles and activities of participants, 181**f**
technical applications, 180
outsourcing, **186–192**
application service provision (ASP), 190
defined, 186
development of large vendors, 188
disadvantages, 189–190
information systems, 188
and information technology, 187
insourcing, 187
offshoring, **190–192**
cost savings for developed countries, 191
dependencies for capacity and/or for knowledge, 191–192

effect on India, 191
reasons for growth, 190
reasons for outsourcing, 187–188
web technology, 190
Extreme Programming Installed (Jeffries), 479
Extreme Programming (XP), xiv, 135, 144, **479–481** *see also* rapid development methodologies

F

Fagin, R., 239
Failla, A., 332, 359
feature driven development, 144
Feeny, D.F., 11, 187
Feigenbaum, E.A., 87
Feller, J., 178, 193
Felsing, J.M., 144
Fine, C.H., 192
Fine, E., 554
Finkelstein, Clive, 434
Fitzgerald, B., 179, 180, 182, 186, 187, 193, 558, 590, 612
Fitzgerald, Professor Guy, xiv, xxi, 11, 53, 106, 115, 127, 178, 301, 306, 558, 575, 581, 583, 612
Flannery, L.S., 84
flexibility in organizations, **71–74** *see also* organizational themes
Flynn, D.J., 569
Forrester Research, 191
Fowler, K.M., 557
Fowler, M., 143, 288
framework methodologies, **537–563**
Capability Maturity Model (CMM), **551–558**
defined level, 554
early development, 552
five maturity levels, 553**f**
initial level, 553–554
integration model, 552
managed level, 554
optimizing level, 555
key process areas and key practices, 555, 556**f**
structure, 555**f**
repeatable level, 554
use in industry, 557–558
Euromethod, **558–563**
applies to adaptations of information systems, 559, 560**f**
defining deliverables, 560**f**
deliverable model, 561–562
properties expressed as a set of views, 562
established by the European Commission, 558

objectives, 559

planning process, 560, 561**f**

profile for target domain deliverables, 563

strategy model, 562

tasks to prepare delivery plan, 562

transaction model, 561

Multiview 1, **537–545**

analysis of human activity, 541–542

conceptual model, 542

rich picture of problem situation, 541

using CATWOE checklist, 541

Weltanschauung as basis for system
requirements, 541

analysis of information, 542

development of functional and entity
models, 542

design of technical aspects, 545

major outputs of the methodology,
545

requirements for technical
specification, 545, 546**f**

framework of five stages, 537, 538**f,**
540**f**

human-computer interface, 544

outputs, 538, 539**f**

questions to be answered, 538–539

socio-technical analysis and design, 542,
543**f**, 544

Multiview 2, **545–549**

changes from Multiview 1, 546, 547**f**

constructing information systems
development methodology, 549**f**

framework, 546, 547**f**

interaction of situation, analysis nd
methodology, 546**f**

O perspective, 548

T perspective, 548**f**

TOP multiple perspective, 547–548

Strategic Options Development and
Analysis (SODA), **549–551**

cognitive mapping, 550

framework, 550, 551**f**

journey making, 550

framing influences, 315, 385–386

free/open source software (F/OSS) *see* open
source software (OSS)

French Ministry of Industry, 426

Frost, S., 387

G

Gacek, C., 181

Galliers, R.D., 70, 78

Gane, C., 110, 231, 244, 339, 342, 390, 396, 402, 418,
607

Gantt charts, 74, **292–293**

General Motors, 557

Gestalt psychology, 318–319

Gibson, A., 193

Gibson, C., 68, 69

Gilb, Tom, 134

Goldkuhl, G., 536

Goldsmith, R.F., 108, 189

Goodhue, D.L., 179

Gores Technology Group, 372

Gosain, S., 180

Greiner, M.E., 179

Griffiths, P., 296

Gross Domestic Product (GDP), 191

GroupSystems software, 331–332, 386

Grover, V., 64, 65

Grundén, K., 83

Guardian (newspaper), 213

Guilford, S., 132

Guimaraes, T., 86

H

Haag, M., 98

Hamel, G., 10

Hammer, M., 62, 63, 64, 78

Hardgrave, B.C., 123

Harnden, R., 53

Hasnas, J., 308, 309

Herbsleb, J., 557

Herzberg, F., 488

Highsmith, J., 134, 135, 136, 140, 141, 142, 144

Hill, T., 303

Hirschheim, R., 69, 187, 193, 488, 585, 599, 614

Holck, J., 145

holistic techniques, **199–215**

cognitive mapping, **213–215**

defined, 213

development of Labour Party map, 214, 215**f**

example of Labour Party, 213, 214**f**

conceptual models, **208–213**

compared with reality, 212

level 1 model for professional association
examination system, 210**f**

level 2 model for registering students, 211**f**

purpose of the model, 209

as a technique for stimulating thought, 212

rich picture diagrams, **199–204**

assumptions involved, 203–204

Checkland's Soft Systems Methodology
(SSM), 199

facts to be included, 199–200

identifying tasks and issues, 202

Multiview, 199

not used generally, 204

picture for professional association
examination system, 200, 201**f**

analysis based on the picture, 200, 201**f**

root definitions, **204–208**

applied to prison system, 208

defined, 204

example of hospital system, 206, 207**f**, 208

mnemonic CATWOE, **204–205**

root definition for professional association
examination system, 205

checked by CATWOE criteria, 206

stages for creating a root definition, 205

examples, 207**f**

Holland, C.P., 70

Hopkins, J., 161

Hough, D., 129, 133

Howard, A., 476

Hughes Aircraft, 557

Hume, S., 125

Humphrey, W.S., 552

Hurley, J.R., 193

hypertext mark-up language (HTML), 334–335,
337**f**

I

Iansiti, M., 137, 138

IBM, 461, 473, 557

ICL, 473

Iivari, J., 599

Imperial Chemical Industries (ICI), 495

Information Engineering Facility (IEF), **363–368**

Information Engineering (IE), 62, 67–68, 217, 363,
368, 386, **434–446**, 559 *see also* blended
methodologies

Information Engineering (Martin and Finkelstein),
435

information systems context *see* context of
information systems

information systems development *see*
development of information systems

Information Systems Development (Avison and
Fitzgerald), xiv

information systems development life cycle
(SDLC), **31–43**, 121

basic structure, 31

and evolutionary development, 121

feasibility study, **31–32**

implementation, **34**

documentation, **34**

education and training, 34

quality control, **34**

security, 34

methodology, 35–36

attributes, 35

phases, 35

philosophy, 35

Merise, 36

National Computing Centre, 36

SSADM, 36

Yourdon Systems, 36

potential weaknesses of SDLC, 38–43

application backlog, 41

assumption of green-field development, 42–43

Renaissance, evolutionary development and legacy systems, 42

computing, 38–43

deadlines, maintenance, staff and user problems, 43

design is output driven, 39, 40**f**

documentation, 40

emphasis on hard thinking, 42

failure to meet management needs, 38, 39**f**

incomplete systems, 41

lack of control, 41

maintenance workload, 41

models unstable, 39

need for iterations, 41–42

top-down process, 41

user dissatisfaction, 39–40

user 'sign-off', 40

review and maintenance, **34–35**

organizational learning, 35

strengths of SDLC, 38, 43

systems

analysis, **33**

design, **33**

investigation, **32–33**

observation, interviewing, questionnaires, sampling, 32–33

techniques, **36–37**

data flow diagramming, 37

flow charts and organization charts, 36

grid charts and records, 37

Soft Systems Methodology, 37

tools, **37–38**

Oracle and Visio, **37–38**

report generating package (RPGII), 37

waterfall model, 31, 577

information systems (IS), 10

Information Systems Journal (ISJ), xxi

information systems planning, **65–68** *see also* organizational themes

information systems strategy, 14, **53–62** *see also* organizational themes

information systems work and analysis of change (ISAC) methodology, 392, 395, **516–526,** 601, 604 *see also* organizational-oriented methodologies

information technology (IT), 10, 54, 73

Infosys, 557

Internal rate of return (IRR), 54

International Standards Organization (ISO), 76, 581

Internet, 8, 91, 145–148

internet addresses:

McGraw-Hill textbooks, xviii

Office of Government Commerce, 530

Online Learning Centre, xv

Oracle, 369

Primis Content Centre, xviii

PRINCE2, 533

Renaissance, 533

Introna, L.D., 309, 585, 590, 613

IS *see* information systems

ISAC *see* information systems work and analysis of change (ISAC) methodology

ISJ *see Information Systems Journal*

ISO *see* International Standards Organization

IT *see* information technology

J

Jackson, M., 121, 390, 395, 407, 410, 413, 416, 417, 418

Jackson, P., 108

Jackson Structured Programming, 407, 422

Jackson Systems Development (JSD), 259, 270–271, 341, 407–417, **407–417** *see also* process-oriented methodologies

Jacobson, I., 279, 451, 460, 461, 462, 468

JAD *see* joint application development

Jain, H., 162

Jalote, P., 557

James Martin Associates, 441

James Martin's rapid application development methodology, 36, 128–134, 392, **469–472** *see also* rapid development methodologies

Janson, M.A., 126, 193

Jayaratna, N., 566, 593, 614

JD Edwards, 369

Jeffries, R., 140, 479, 485

Jenkins, T., 559, 564

JIT *see* Just In Time

JMRAD *see* James Martin's rapid application development methodology

Johnson, D.I., 557

joint application development (JAD), 83, 131–132 *see also* people techniques

joint requirements planning (JRP), 83

Jones, C., 312

JSD *see* Jackson Systems Development

JSP *see* Jackson Structured Programming

Jurison, J., 190

just In time (JIT), 54

K

KADS methodology, **496–501** *see also* people-oriented methodologies

Kambil, A., 308

Kantaris, N., 359

Kendall, J., 27

Kendall, K., 27

Khan, B., 64

King, W.R., 84

Klein, H.K., 599, 601, 614

Knights, D., 548

knowledge-based expert system *see* expert system

knowledge management, **93–96** *see also* people themes

Koehler, D., 319

Kogut, B., 19

Kripalani, M., 191

Kruchten, P., 451, 461, 462, 463, 464

Kuhn, T.S., 598

Kumra, G., 191

L

Lacity, M., 187

Lambrou, N., 450

Lancaster University, 508

Lawlis, P.K., 557

Learmouth and Burchett Management Systems (LBMS), 419

Lederer, A.L., 67

Leffingwell, D., 98

Leite, J.C.S.P., 106, 107

Lejk, M., 241, 272

Lewis, P.J., 599, 600

life cycle approach, **29–44**

Light, B., 70

Lincoln, T.J., 54

Linstone, H., 546, 548

Linux (computer software), 179

Litecky, C.R., 298

Lockheed Martin Management & Data Systems, 557

London Ambulance Service computer aided ambulance despatch system (LASCAD), 11, 106, 115, 189

Lotus Notes, 331–332

Lundeberg, M., 392, 516, 536, 604

Lycett, M., 161, 162

Lyytinen, K., 149, 614

M

MacCormack, A., 137, 138

Macredie, R.D., 101

Maddison, R.N., 567

Magal, S.R., 86

Malhotra, M., 64

Markus, M.L., 573, 574

Martin, James, 36, 128, 149, 340, 392, 435, 450, 451, 469, 470, 471, 485, 496

Maskell, B.H., 136, 137

Mason, R., 547

Mathiassen, L., 115, 117, 451

Mayer, R.E., 318, 319

McConnell, S., 98

McCracken, D.D., 121

McGrew, J.F., 552, 556, 557

McGuire, E.G., 557

McKeen, J.D., 152

McKnight, L.W., 19

McLoughlin, Jane, 213

McManus, J., 309

McMaster, T., 547

Melao, N., 64, 78

Mendelow, A.L., 67

Merise methodology, 36, 74, 110, 217, **426–434**, 559, 563, 573 *see also* blended methodologies

methodologies, 389–564
blended, 419–449
frameworks, 537–564
object-oriented, 451–468
organizational-oriented, 507–535
people-oriented, 487–506
process-oriented, 396–417
rapid development, 469–486
road map, 391

methodology comparisons, **591–613** *see also* methodology issues
comparison issues, 591–597
academic and/or practical reasons for comparison, 591
checklist for values, 593
contingency and different approaches, 595
ideal-type criteria can include

analysis and design separation, 592
communication, 592
design validity, 591
documentation, 591
enhanced productivity and quality, 592
environmental relevance, 592
goals and objectives, 592
participation, 592
performance evaluation, 592
planning and control, 592
product visibility within a boundary, 592
rules, 591
technical and non-technical systems integration, 593
importance of the philosophy of a methodology, 597
normative information model-based systems analysis and design (NIMSAD), 593
questions relating to the intended problem solver, 593, 594
problem situation, 593–594
problem-solving process, 593, 594
three-stage evaluation, 594–595
practical problems in comparing methodologies, 595–596
set of comparison criteria, 596–597
comparisons, **604–613**
model, 607–608
data flow diagram, 607, 608
process orientation, 608
outputs, 611–612
philosophy, **604–607**
domain, 606–607
distinction between strategy and specific problems, 606
objectives, 605–606
PI and ETHICS have wide objectives, 605
paradigm, 604–605
ISAC, IE and others belong to the science paradigm, 604–605
SSM and ETHICS adopt the system paradigm, 604
target, 607
general purpose or specialized, 607
size of organization, 607
practice, 612–613
analysis of participants, 612–613
skill levels, 613
user base, 612

product, 613
scope, 609–611
results of analysis of ten phases, 609, 610f
analysis, 610
feasibility, 6010
implementation, 611
logical design, 611
maintenance, 611
programming, 611
strategy, 610
techniques and tools, 608–609
as part of the methodology, 608
specific tools, 609
use of new techniques, 608–609
framework for comparing methodologies, 597–603
model, **602–603**
verbal, analytic, iconic and simulation, 602
outputs, 603
philosophy, **598–602**
domain, 601–602
objectives, 601
need for analysis and re-design, 601
science paradigm, 598–601
comprises reductionism, repeatability and refutation, 598
framework for analysing philosophies, 599, 600f
framing effect, 601
objectivist and subjectivist, 599–600
ontology and epistemology, 599
realism and nominalism, 599
target, 602
practice, 603
product, 603
scope, 603
seven elements of the framework, 597, 598f
technique and tools, 603
methodology issues, **567–589** *see also* methodologies; methodology comparisons
adopting a methodology, 572–576
differences between methodologies, 572–573
involvement of business managers in IT, 574
methods in action, 575
reasons for differences in performance, 576
repeatability of results, 575
definition of a methodology, 567–570
British Computer Society, 567–568

Checkland, 569
extended definition, 568
within the intellectual framework, 569–570
philosophy, 568
scope, 569
era of reassessment, 583–589
criticisms of methodologies, 583–586
adoption difficult, 585
difficult tools, 584
goals displaced, 584–585
impractical assumptions, 584
no improvements, 586
not contingent with project, 584
one-dimensional and inflexible, 584
over complexity, 584
problems with methods, 585
productivity disappointing, 583–584
skills requirement, 584
social issues, 585
future directions, 586–589
ad hoc development, 586–587
amethodical systems, 586
web-based applications, 586
contingency approach, 587–588
flexible but demanding, 587–588
continuing development of specific
methodologies, 587
rapid application development,
CommonKADS or customer
relationship management, 587
external development, 588
enterprise resource packages (ERP),
588
packages and outsourcing, 588
reasons for using methodology, 583
evolution and development, 576–589
early-methodology era, 577–578
information systems development
life cycle (SDLC), 577–578
methodology era, 578–583
commercial development, 583
consultancies, 579
development by academics, 582
expanding the scope, 580–581
strategy and planning, 580
filling the gaps in methodologies,
579, 581
gaining competitive advantage,
581–583
methodologies arising from practice
or developed from theory, 578

techniques of entity modelling or
data flow diagramming,
578–579
pre-methodology era, 576–577
system analysts role, 576–577
rationale for a methodology, 570–572
better development process, 571–572
differences with software products, 572
better end product, 570–571
components of quality, 570–571
standardized process, 572
improved specifications and easier
maintenance, 572
metodológica informática (MEIN), 559
Michell, V.A., 186
Microsoft:
Access, 348–358, 386
Office, 370
Project, 74, 342–348, 386
Visio, 339–343, 386
Milburn, M., 319
Miller, G.G., 139
Mills, S., 461
Mingers, J., 213, 549
Ministry of Defence, 473
Mirani, R., 84
Mitroff, I., 546, 547, 548
Moad, J., 64
modelling themes, **109–116**
data modelling, **111–113**
data analysis, 112
entity modelling, 112**f**
strengths, 112–113
object modelling, **113–116**
basis, 113–114
benefits, 114–115
integrates the system development,
115
reuse of software code, 115
unifies the development, 114
object orientation (OO), **113**
process modelling, **109–111**
data flow diagramming, 111
functional decomposition, 110**f**
independence of data, 110
structured English, 111
modified prototype method (MPM) for web
development, 128
Moran, N., 190
Morgan, G., 601
Motorola, 557

Moynihan, T., 306
Multiview methodology, 36, 53, 73, 231, 243, 481,
482, **537–549,** 582, 588 see also framework
methodologies
Mumford, Edith, 82, 108, 392, 487, 489, 490, 491,
492, 493, 494, 495, 505, 537
Munk-Madsen, A., 117
Murray, F., 548
Myers, G.J., 395, 547
N
National Audit Office, 9, 551
National Computing Centre (NCC), 35
net present value (NPV), 54
Nielsen, P.A., 117
Nilsson, A., 536
Nolan, R., 68, 69
non-functional requirements (NFR), 105–107
Norris, G., 193
Nunamaker, J., 310, 332
O
object modelling group (OMG), 163, 279
object orientation (OO), 113 see also object-
oriented techniques
object-oriented analysis (OOA), 15, 116, 391,
451–460
object-oriented methodologies, **451–468**
object-oriented analysis (OOA), 15, 116, 391,
451–460
defining attributes, 458**f,** 459
defining services, 459–460
service chart, 460**f**
finding class-&-objects, 452–454
criteria, 453–454
end result, 454
object is abstraction, 452
symbol for objects in a class, 452**f**
identifying structures, 454–457
gen-spec structure, 454, 455**f**
drawings of hierarchies, 455, 456**f**
testing, 455
generalization and specialization (gen-
spec), 454
whole-part structures, 456–457
notation, 457**f**
types of structure, 456
identifying subjects, 457–458
subject and structure layers, 457**f,** 458
Rational Unified Process (RUP), 460–467
architecture centric, 462
IBM involvement, 461–462
iterative and incremental, 462

process structure, 464**f**
use-case driven, 462
workflows, 465–467
analysis and design, 465
business modelling, 464–465
configuration and change management, 466
deployment, 466
environment, 467
implementation, 465
project management, 466
requirements, 465
test, 465–466
object-oriented techniques, **273–289**
object orientation, **273–279**
benefits, 279
classes of objects, 274
encapsulation, 277
analogy of an egg, 277**f**
example of customer update, 274, 275
hierarchy of classes, 274**f**
inheritance, 275, 276, 278
object-oriented programming
definition, **273**
network of objects, 278**f**
reuse of code, 278
Unified Modelling Language (UML)
defined, 279
goals for development, 279
multiple inheritance, 282
object modelling group (OMG), 279
Rational Unified Process (RUP), 283, 287
types of diagrams, 280–287
activity diagram, 286**f**, 287
class diagram, **280f, 281f, 282**
interaction diagrams, 283
sequence diagram, 283, 284**f**
statechart diagram, 284, 285**f**, 286
use case diagram, **282f, 283**
Odell, J.J., 435, 451
Office of Government Commerce, 530
Oliver, P.R.M., 359
Olle, T.W., 614
OMG *see* object modelling group
One.Tel (Australia), 14, 15
Online Learning Centre, xv
OO *see* object orientation
OOA *see* object-oriented analysis
Open Source Initiative, 179
open source software (OSS), 178–182 *see also*
external development

Open University Press, xix
Oracle software, 152, 368–372, 386 *see also*
toolsets
organizational-oriented methodologies, **507–535**
see also organizational techniques;
organizational themes
information systems work and analysis of
change (ISAC), **516–526**
activity studies, 519–521
cost/benefit analysis, 519–521
analysis of contributions, 519–521
co-ordination of subsystems, 519–521
information subsystems, 520–521
change analysis, 517–519
A-graph, 517, 518**f**, 520**f**
analysis of need for change, 518
description of objectives, 518
method steps, 517
recommend appropriate change, 518
data system design, 524–525
D-graph, 525
to meet the requirements specification, 525
program delimitation, 525
equipment adaptation, 525–526
E-graph, 526
emphasis on analysis and design, 526
five major phases, 517
importance of people, 516, **516**
information analysis, 522–524
C-graph, 523, 524**f**
component analysis, 522–523
I-graph, 522, 523**f**
precedence analysis, 522
process logic analysis, 523–524
property analysis, 524
Process Innovation (PI), **526–530**
design and build a prototype, 529–530
assess risk, 529
brainstorm design alternates, 529
change to the new system, 529
role of IT, 529–530
develop the business vision and process
objectives, 527
need for creative thinking, 527
identify the IT levers, 528–529
IT capabilities, 528–529
identify the processes to be re-engineered, 527–528
classification of processes, 527
support for legacy systems, 528

understand and measure the existing
process, 528
projects in controlled environments (PRINCE),
530–533
PRINCE2 for project control and
management, 530, 531**f**
eight processes, 531, 532**f**
reports, 532–533
Renaissance, **533–535**
evolution and re-engineering of legacy
systems, 533
process model suggests activities and
evolution strategies, 534–535
strategic, operational and service roles,
533–534
technical, economic and managerial
viewpoints, 533
Soft-Systems Methodology (SSM), **507–516**
actions recommended, 514
CATWOE mnemonic for criteria, 510, 512
Checkland's *Weltanschauung*, 509
conceptual models, 512, 513**f**
compared with reality, 513
feasible changes, 513–514
Mode 2 SSM
four activities, 515
mode 2 SSM, 514, 515**f**, 516
problem situation, 510–511
expressed, 510–511
rich picture diagram, 510, 511**f**
unstructured, 510
root definitions, 511–512
seven stages of SSM, 509**f**
systems thinking, 508
organizational techniques, **295–306** *see also*
organizational-oriented methodologies;
organizational themes
brainstorming, 295–296
case-based reasoning (CBR), **303–304**
derives from cognitive psychology, 303
and expert systems, 304
learning from cases, 303
critical success factors (CSF), **296–299**
analysis of goals and objectives, 298
example of use in hospital, 298–299
factors limited in number and measurable,
297
future analysis, **300–301**
flexibility analysis, 301
identifying effect of changes, 300–301
predicting lifespan, 301

lateral thinking, **295–296**
  and risk analysis, 296, 297**f**
risk analysis, **304–305**
  Synergistic Contingency Evaluation and
    Review Technique (SCERT), 304–305
    potential problems, 305
    stages are scope, response, parameters
      and interpretation, 305
scenario planning, **299–300**
  defined, 299
  Delphi approach and use of experts, 299
  morphological approach, 299–300
  in National Health Service, 300
Strengths, Weaknesses, Opportunities and
    Threats analysis (SWOT), **301–303**
  criticisms, 302–303
  defined, 301–302
  part of knowledge management, 302
  used in marketing strategy worksheet
    (MSW), 302
organizational themes, **51–77** *see also*
  organizational-oriented methodologies;
  organizational techniques
  business process re-engineering (BPR), **62–65**
    business process change, 65
    defined, 62
    failure of initiatives, 64
    framework of BPR, 65**f**
    influence in information systems, 64
    outcome of re-engineering, 63
    process characteristics, 63
    reasons for re-engineering, 63
    and total quality management, 65
  flexibility, **71–74**
    combined with robustness, 71–72
    in dealing with future developments, 73–74
    manoeuvres providing strategic flexibility,
      72
    needed to meet ill-defined requirements,
      72
    useful attribute, 71
  information systems planning, **65–68**
    information engineering, 67–68
    long, medium and short term planning, 66
    planning guidelines, 67–68
    tasks for an information strategy group, 67
  information systems strategy, **53–62**
    benefit realization, **55**, 56**f**
    business improvement by competitive
      advantage, **55**
    examples, 57

competitor-driven model, 60
  copying competitors causes problems,
    60
cost/benefit analysis, 54
Earl's multiple methodology, 60, 61**f**
  bottom-up analysis of systems, 60–61
  identification of IT opportunities, 62
  system audit grid, 61**f**, 62
  top-down analysis of the business, 60,
    61**f**
efficiency and effectiveness, 55–57
efficiency projects, defined, 53
framework of competitive strategy, 57, 58**f**
strategic role of IT, 57, 58**f**
technology-driven model, 59–60
  IT as a commodity, 59–60
  little evidence of commercial benefits,
    59
project management, **74–77**
  aims and objectives, 74–75
  benefits realization programme, **76**
  management of people, 77
  meeting quality standards, **76**
  monitoring progress, 76
  role of project leader, 75
  techniques to control projects, 74
stages of growth (SoG), **68–71**
  four elements to be tracked, 68–69
  Gibson and Nolan model, 68, 69**f**
  Hirschheim three stages, 69–70
systems theory, **51–53**
  boundary, defined, **51**
  decomposition, 51
  environment, **51**
  hard and soft systems, 53
  human activity purpose, 51, 52
  organizations as open systems, **52**
  purpose of an information system, 52
  system, defined, 51
Orman, L., 120, 121, 149
OSS *see* open source software
P
Page-Jones, M., 283
Palmer, S.R., 144
Parsons, T., 487
Patane, J.R., 190
Paulk, M.C., 14, 552, 554, 555, 557, 564
payback period, 54
people-oriented methodologies, **487–506** *see
  also* people techniques; people themes
  CommonKADS, **501–504**

context, concept and artefact, 501, 502**f**
  design model, 502
    system architecture, 503
  knowledge and communication models,
    502
    construction, 503
    knowledge transfer, 503, 504**f**
  knowledge management, 501
  organization, task and agent models, 502
effective technical and human
    implementation of computer-based
    systems (ETHICS), **487–496**
  applications, 495
  critical success factors, 496
  job satisfaction defined, 487
  measurements, 487–488
    efficiency, 488
    ethics, 488
    knowledge, 487
    psychology, 488
    task structure, 488
  participation, 488–489
  philosophy of the socio-techical approach,
    487
  QUICKETHICS (quality information from
    specific knowledge), 495
  steering committee and design group,
    490–494
    detailed work design, 494
    efficiency needs, 492
    existing system, 491
    future analysis, 492
    implementation and evaluation, 494
    job satisfaction needs, 492
    need for change, 491
    objectives and tasks, 491
    organizational design, 493–494
    specifying efficiency and job
      satisfaction objectives, 492
    system boundaries, 491
    technical options, 494
KADS, **496–501**
  application model, 498
  conceptual model, 500
  design model, 501
  model of co-operation, 498
  model of expertise, 498–499, 500**f**
    domain knowledge, 499
    inference knowledge, 499
    strategic knowledge, 500
    task knowledge, 499–500

models and processes, 496, 497**f**

organizational model, 496, 498

spiral life cycle model, 496, 497**f**

task model, 496

people techniques, **307–312** *see also* people-
oriented methodologies; people themes

joint application development (JAD), **310–312**

associated with rapid application
development (RAD), 310

examples, 312

meeting or workshop with facilitator, 310,
311

requires speed, right people and
commitment, 311

typical characteristics, 310–311

stakeholder analysis, **307–310**

privacy aspects, 309

problems in use, 309–310

stakeholder map, 308

stakeholders defined, 307–308

types of user, 307

people themes, **79–107** *see also* people-oriented
methodologies; people techniques

customer orientation, **96–97**

customer relationship management (CRM),
96

philosophy of CRM, 96

end-user computing (EUC), **84–87**

continuing role, 87

defined, 84

factors in successful use, 86

four types, 85

command-level, 85

end-user programmers, 85

functional support, 85

non-programming, 85

information centres (IC) to support EUC, **86**

maintenance problems, 85–86

expert systems, **87–92**

components, 88–90

blackboard and user interface, 90

explanation facility, 90

inference engine, 89

knowledge base, 88

language and rules, 89

shell, 89

defined, 87

descriptive and procedural knowledge, 91

factors important for success, 92

lack of business success, 90

problems in applications, 91–92

structure, 88**f**

knowledge management, **93–96**

Chief Knowledge Officer (CKO), 95

key drivers, 93–95

economic, 95

knowledge-centric, 93

organizational structure, 94

personnel, 94

process-focused, 94

technology, 94

knowledge defined, **93**

participation, 79–84

comparison of views on information
systems, 84**f**

dealing with reactions against a new
system, 79–80

encouraging involvement in the project,
80–81

ETHICS methodology, 82–83

joint application design (JAD), 83

joint requirements planning (JRP), 83

qualities to improve the human-computer
interface, 81

three levels of participation, 82

concensus, 82

consultative, 82

representative, 82

requirements, **97–107**

changing and evolving requirements, 104

communication culture gap, 101

cost of errors in definition, 98, 99**f**

non-functional requirements (NFR), 105–107

are extra constraints or qualities of the
system, 105

categories of NFRs, 106

difficult to find and define, 106–107

problem of definition, 98

real-world problems, 101–104

development, 103–104

production of the specification, 103

requirements capture, 101–103

the sign-off, 103

testing, 104

traditional requirements process, 99, 100**f**,
101

unknowable requirements, 105

PeopleSoft, 369

Peppard, J., 64

Petri nets, 431

Pfahl, D., 98

Phan, D.D., 557

Pidd, M., 64, 78

Plato, J.J., 127

Porter, M.E., 57, 58, 60, 299, 306

Pouloudi, A., 309

Prahalad, C.K., 10

Preece, I., 64

Pressman, R.S., 121, 174

Preston, L.E., 308

Pries-Heje, J., 137, 146

Primis Content Centre, xviii

PRINCE project management *see* projects in
controlled environments

Process Innovation (PI), **526–530,** 605 *see also*
organizational-oriented methodologies

process-oriented methodologies, **395–417** *see
also* process techniques

Jackson System Development (JSD), **407–417**

entity action step, 409

action criteria, 409

definition of an entity, 409

entity structure step, 410–411

structure diagram, 410**f**–412**f**

several accounts separately, 411**f**

several accounts simultaneously, 412**f**

single account, 410**f**

function step, 414–415

addition of functions to structure text,
414**f**

amended system speification diagram,
415**f**

implementation step, 416–417

sharing of processors, 416

system implementation diagram, 417**f**

initial model step, 411–414

pseudo code, 411–412

state vector connection, 413, 414**f**

structure text, 412, 413**f**

system specification diagram (SSD), 413**f**

time grain markers (TGM), 414

main phases are modelling, network and
implementation, 408–417

relationship with Jackson Structured
Programming (JSP), 407

system timing step, 416

structured analysis, design, and
implementation of information systems
(STRADIS), **395–402**

alternative solutions, 399–400

objectives, 399

report contents, 399–400

detailed study, 396–399

drafting of data flow diagrams (DFD), 397

system boundary, 397, 398**f**

types of user, 396–397

initial study, 396

physical design, 400–402

cost esimate, 401

tasks for completion, 402

transaction centred system, 400, 401**f**

transform centred system, 400, 401**f**

program modules and interfaces, 395

Yourdon Systems Method (YSM), **402–407**

essential modelling, 404–407

behavioural model, 405

context diagram, 404

data flow diagram, 405, 406**f**

entity relationship diagram, 406

environmental model, 404

levelling process, 405, 406**f**

feasibility study, 403–404

implementation modelling, 407

three phases, 402, 403**f**

process techniques, **243–271** *see also* process-oriented methodologies

action diagrams, **265–271**

examples, 265–267

admit student, 265**f**

concurrency, 266**f**

database operations, 266, 268**f**

inputs and outputs, 267**f**

process action block, 266**f**

repeated actions, 266**f**, 267**f**

selection process, 265**f**

used in Information Engineering, 265

data flow diagramming, **243–251**

context diagram, 248**f**

data flow, 244

data flow diagram (DFD) defined, **243**

data store, 245**f**, 246**f**

level 1 diagram, 248, 249**f**

lower level diagram, 250**f**, 251

process box, 244**f**, 245

rules, 247

source or sink, 246, 247**f**

defined, 246–247

decision tables, **253–255**

comparison with decision trees, 255

consolidation, 254, 255**f**

example, 254**f**

decision trees, **251–253**

general format and example, 252**f**

entity life cycle, 267–271

dependencies between events, 271**f**

identifying states of an entity, 270

Jackson Systems Development (JSD), 270

life cycle for student, 269**f**

symbols, 268**f**

technique of process analysis, 267

matrices, **263–265**

CRUD matrix, 264

document/department matrix, 264–265**f**

entity/function matrix/CRUD matrix, 263, 264**f**

function/event matrix, 263, 264**f**

structure diagrams, **258–260**

basic diagram, 258**f**

coupling and cohesion, 258

example, 259**f**

Jackson Systems Development (JSD), 259

Warnier–Orr diagram, 259, 260**f**

structured English, **255–258**

basic structure, 257**f**

defined, 256

example, 256**f**

functional decomposition, 256

general contruct, 256

THEN construct, 256**f**

structured walkthrough, **260–263**

benefits, 260–261

code reading, 262–263

defined, 260

dry running, 263

held at end of project stages, 261

organization, 261–262

Project Evaluation and Review Technique (PERT) charts, **290–292** *see also* project management techniques

project management, **74–77** *see also* organizational themes

project management techniques, **289–294**

estimation techniques, **289–290**

CoCoMo, **289**

function point analysis (FPA), **289–290**

work breakdown structure, **290**

Gantt charts, **292–293**

display of progress, 292**f**

Project Evaluation and Review Technique (PERT) charts, **290–292**

Microsoft Project tool, 291

network and critical path, 290**f**

reports and monitoring, 291–292

resource schedule, 291**f**

projects in controlled environments (PRINCE), 74, 75, 76, 392, **530–533** *see also* organizational-oriented methodologies

Prusak, L., 93, 108

Q

Quang, P.T., 450

R

RAD *see* rapid application development

Rakitin, S., 145

Randall, D., 547

rapid and evolutionary development theme, xiv, **119–148**

agile development, **134–145**

Agile Manifesto philosophy, 134–136

Agile Movement applications, 136–137

approaches to new product development, 137, 138**f**

characteristics of agile approaches, 139–145

collaboration and communication, 144

iteration, **139–140**

minimalism, 140

people orientation, 143–144

risk acceptance, 141–143

short life cycles, 140

emergence, 140–141

agile and CAS compared, 142**t**

complex adaptive systems (CAS), 140

implementation of agile approach, 141

limitations and criticisms, 145

principles reflect values, 136

evolutionary development, **119–123**

Boehm's spiral model of development, 121, 122**f**, 123

concept of iterations, 119, 120**f**

differs from SLDC, 121

rapid changes possible with evolution, 120–121

prototyping, **123–128**

advantages, 123–124

alternate forms, 125

design, 125

functional, 125

organizational, 125

performance, 125

process, 125

as basis for methodology, 126

criticisms, 126–127

expendable or evolutionary, 124

improved user acceptance, 124–125

modified prototype method (MPM) for web development, 128

objectives, 125
WISDM methodology, 128
rapid application development (RAD), 123,
128–134
advantages, 132–133
and cultural change, 134
incremental development, 129
joint application development workshops,
131–132
MoSCoW rules, **131**
Pareto principle, 131
prototyping, 132
sponsor and champion, 132
timebox development and traditional
development, 130**f**
timeboxing, 129**f**
toolsets, 132
web-based development, **145–148**
differences and similarities compared with
traditional projects, 147–148
early problems, 146
relevant concepts, 146–147
traditional and web-based applications,
148**t**
rapid application development (RAD), 36, 123,
128–134, 392, 587 *see also* rapid and
evolutionary development
rapid development methodologies, **469–486**
Dynamic Systems Development Method
(DSDM), **472–478**
consortium, 472–473
involves general concepts of RAD, 474
nine principles, 474
process, 474–478
business study, 476
feasibility study, 475–476
implementation, 477
iteration, 476
phases, 475**f**
project management, 478
roles for users, 478
timeboxes, 477**f**
Extreme Programming (XP), **479–481**
architectural spike and paired
programming, 479
designing, 480–481
developing the code, 481
planning, 480
productionizing, 481
user stories define requirements, 479
James Martin's RAD (JMRAD), **469–472**

characteristics, 469
construction, 471–472
skilled with advanced tools (SWAT), 471
cutover, 472
four phases, 470**f**
and information engineering, 469, 470**f**
requirements planning, 470–471
user design, 471
Web Information Systems Development
Methodology (WISDM), **481–484**
Dreamweaver, 481, 484
human-computer interface (HCI), 484
information analysis, 482
modification of Multiview, 481, 482**f**
organizational analysis, 482
technical design, 484
WebQual questions, 483**f**
work design, 483
Rational Software Corporation, 461
Rational Unified Process (RUP) methodology, 74,
116, 238, 287, **460–467** *see also* object-
oriented methodologies
Ravichandran, T., 162
Raymond, E.S., 180, 193
re-engineering existing systems, 380–381
Renaissance, 42, **533–535** *see also* organizational-
oriented methodologies
report generating package (RPGII), 37
return on investment (ROI), 54
reverse engineering *see* re-engineering existing
systems
Ringland, G., 299, 300, 306
Robertson, J., 98
Robertson, S., 98
Rockart, J.F., 84, 296
Rogers, E.M., 70
Rosenhead, J., 72, 213, 549
Ross, J.W., 11
Rothenberger, M., 162
Rumbaugh, J., 279, 451, 461, 468
Rumpe, B., 145
Russell, M., 105
Russo, N.L., 106, 132, 590, 612, 614

S
Sachs, P., 547
Sage, J., 117
Saiedian, H., 101
Sarson, T., 110, 231, 244, 339, 342, 390, 396, 402, 418,
607
Sawyer, P., 108
Scacchi, W., 180, 182

Scholes, J., 52, 78, 216, 319, 392, 509, 514
Schön, D., 548
Schreiber, G., 501, 503
Schwaber, K., 144
Schware, R., 191
Scott, K., 288
SCRUM methodology, 144
SDLC *see* information systems development life
cycle
SDM *see* system development methodology
Searle, J.R., 601
select software tools, 371, 386
Shadbolt, N., 505
Shah, H.U., 44, 117, 272
Shils, E., 487
Shorrock, D., 54
Short, J., 62, 63, 526, 528
Silk, D., 53
Silver, D., 310, 313
Simula (programming language), 113
Sinha, J., 191
Sligo, J., 557
Smalltalk (programming language), 113
Smith, A., 84
Smith, H.A., 143, 152
Smith, H.J., 308, 309
Smith, L.D., 126
Smith, R., 279
SODA *see* Strategic Options Development and
Analysis
Soft Systems Methodology (SSM), 37, 52, 199,
507–516, 582, 604 *see also* organizational-
oriented methodologies
Software Engineering Institute (SEI) Carnegie
Mellon, University, 552, 581
SoG *see* stages of growth
Sol, H.G., 614
Sommerville, I., 108
Sorensen, K., 19
Sparling, M., 162
SQL (computer language) *see* structured query
language
SSADM *see* Structured Systems Analysis and
Design Method
SSM *see* Soft Systems Methodology
stages of growth (SoG), **68–71** *see also*
organizational themes
Stahl, L., 585
Stapleton, J., 472, 473, 485
steering committee for information system
project, 14

Sterling Software, 364
Sterner, Th.A., 505
Stevens, W.P., 395
Stewart, J., 559
Stewart, K., 180
Stewart, T.A., 60
Stolterman, E., 590, 614
Stone, J., 387
Storey, V.C., 162
Stowe, M.W., 387
STRADIS *see* structured analysis, design and implementation of information systems
Strassman, P., 59
Strategic Options Development and Analysis (SODA), 213, **549–551** *see also* framework methodologies
Strengths, Weaknesses, Opportunities and Threats analysis (SWOT), 60, 67, **301–303**
structured analysis, design and implementation of information systems (STRADIS), 110, 158, 231, 342, 390, **395–402** *see also* process-oriented methodologies
Structured Query Language (SQL), 166, 368, 369
Structured Systems Analysis and Design Method (SSADM), 35, 36, 74, 110, 217, 270, 341, 342**f**, 395, **419–426,** 559, 563, 581 *see also* blended methodologies
Structured Systems Analysis (Gane and Sarson), 395
Subramanian, A., 193
Subramanian, G.H., 133
Sugumaran, V., 162
Sun Microsystems, 95
Sutherland, A.R., 70, 78
Swedish Royal Institute of Technology, 516
SWOT *see* strengths, weaknesses, opportunities and threats analysis
System Development Methodology (SDM), 559
systems development team, 14
systems theory, **51–53** *see also* organizational themes
Systems Thinking, Systems Practice! (Checkland), 508
Szalvay, V., 139

T

Takeishi, A., 192
Tardieu, Hubert, 426
Taylor, A.V., 595
Taylor, F.W., 493
Taylor-Cummings, A., 101
Tayntor, C.B., 86, 87

techniques, 199–326
  data, 217–240
  holistic, 199–215
  object-oriented, 274–289
  organizational, 295–306
  people, 307–312
  process, 243–271
  project management, 289–294
  road map, 197
  use in context, 315–326
techniques in context *see* use of techniques in context
Teorey, T.J., 117
Texas Instruments, 363, 435
themes in information systems development, 45–192
  engineering, 151–173
  external development, 175–192
  modelling, 109–116
  organizational, 51–77
  people, 79–107
  rapid and evolutionary development, 119–148
  'road map' of the themes, 47**f**
Tiwana, A., 93, 108, 504
tools, **331–358**
  database management, **348–358**
    Microsoft Access, 348–358
      basic options, 350**f**
      creating a query, 353**f**
      design of Access table, 350**f**
      documentation, 356, 357**f**
      field properties, 350, 351**f**
      form wizard, 355, 356**f**
      pie chart, 356, 358**f**
      query results and statements, 355**f**
      referential integrity rules for a query, 354**f**
      table analyzer wizard, 352**f,** 353**f**
      warning message, 351**f**
  drawing, **339–342**
    advantages of using tools, 339
    flowcharts, 341**f**
    methodology choices, 341**f**
    Microsoft Visio tool, 339–343
      data flow diagram, 343**f**
      look and feel, 340**f**
      SSADM symbols, 342**f**
      STRADIS symbols, 342**f**
    types of database drawing, 340**f**
    website design, 341**f**
  group ware, **331–334**

Lotus Notes and Groupsystems, 331–332
software applied to group working, 331
  aggregating opinion, 334**f**
  agreeing action plan, 335**f**
  brainstorming, 332**f**
  expressing opinions, 333**f**
  ideas generation, 332**f**
  videoconferencing, 331
project management, **342–348**
  Microsoft Project, 342–348
    activity reports, 347**f**
    calendar display, 345**f**
    completed tasks, 348**f**
    development in time, 346**f**
    display of data, 347**f**
    human resources, 346**f**, 348**f**
    interdependence of tasks, 344**f**
    PERT chart, 345**f**
    project summary, 349**f**
    start and end dates, 344**f**
    work breakdown structure, 343**f**
website development, **334–338**
  Dreamweaver software, **334–335**
    associated software, 337
    document window, 336**f**
    home page, 335
    HyperText Mark-up Language (HTML), 334–335, 337**f**
    site map, 338**f**
tools and toolsets, 327–388
  road map, 329
  tools, 331–358
  toolsets, 361–386
toolsets, **361–386**
  benefits, 375–381
    consistency checking, 377–378
    contribution to business strategy, 381
    designs closer to specifications, 377
    focus on analysis, 378
    maintenance reductions, 379–380
      accurate testing, 380
    management and control improvements, 376
    portability, 381
    productivity improvements, 379
      speed and economy, 379
    quality improvements, 376–377
    re-engineering existing systems, 380–381
    standards enforced, 378–379
  costs and requirements, 381–385
    education, training and consultancy, 382

integration into existing system, 382–384

management aspects, 384–385

overcoming incompatibility, 384

setting standards, 382

data dictionary systems (DDS), 362

definition, 361

framing influences, 385–386

restrictions caused by tools and
techniques, 386

horizontal and vertical integration, 361–362

information engineering facility (IEF),
363–368

analysis, 365–366

action diagrams, 366

entity relationship diagram, 365

construction and implementation, 367

design, 366–367

encyclopaedia, 367–368

planning, 364–365

create, read, update or delete
(CRUD), 365

matrix processor, 365

as rapid application development tool,
364

layout of integrated toolset, 362**f**

Oracle, **368–372**

2000 suites, 369–371

database management system, 369

Designer/2000, 370

Developer/2000, 370–371

Developer Suite 10g

comprises many tools, 371–372

Developer suite 10g, **371–372**

repository, 362–363

with enterprise information, 364**f**

Select Enterprise, **372–375**

class diagram, 375, 376**f**

features, 372

object animator, 375, 378**f**

related products, 373

repository administrator, 373, 374, 375**f**

sequence diagram, 375, 377**f**

tools menu, 374**f**

user interface, 373**f**, 374

Torvalds, Linus, 179

Total Acquisition Process (TAP), 561

Total Quality Management (TQM), 65

Towers, J.T., 359

Truex, D., 145, 158, 160

Truex, T., 585

Tudhope, D., 127

Turk, D., 145

Turner, P., 559, 564

Tversky, A., 319

U

Unified Modelling Language (UML), **279–287,** 460

see also object-oriented techniques

University of Stockholm, 516

US Department of Defense (DoD), 552

use of techniques in context, 315–325

blocks to problem cognition, 317–318

perceptual, emotional, cultural, 317–318

classification of techniques, 315, 316–317

framing effect a possible problem, 315

influences on problem cognition, 318–325

Gestalt psychology, 318–319

goal aspects, 320

language, 320

negative or positive bias, 319

paradigm/process influences, 321–322

closed and open, 322

representational influences, 321

hierarchical and structural, 321

structure influences, 319–320

support theory, 319

visual/language and paradigm/process
influences, 322–324

grouping of techniques, 323**f**, 324**t**

potential benefits, 316–317

advantages and use, 316

selection of suitable techniques, 317

V

Valentin, E.K., 302

Van de Velde, W., 505

van Heck, E., 308

Van Slooten, K., 160

Verrijn-Stuart, A.A., 614

Vidgen, R., 147, 148, 149, 313, 481, 484, 485, 546, 547

VisualBasic computer program, 83

Vitharana, P., 162

vorgehensmodell, 559

W

Walkley, N., 450

Walsham, G., 601

Ward, J., 73, 296

Warnier–Orr diagram, 259, 260**f**

Warren, I., 149, 151, 533, 536

Wastell, D., 548, 585

waterfall model, 31, 577 see also information
systems development life cycle (SDLC)

Watson, I.D., 304, 306

Weaver, P.L., 419, 450

web information systems development (WISDM),
481–484 see also rapid development
methodologies

Weber, C.V., 552, 564

Weinberg, V., 395

Weir, M., 487

Welke, R.J., 149

Weltanschauung as basis for system
requirements, 509, 541

Welti, Norbert, 392, 446, 448, 449, 450

Welti ERP development, **447–449** see also
blended methodologies

Westbrook, R., 303

Wetherbe, J.C., 70

Whitley, E.A., 585, 590, 613

Wiegers, K.E., 108, 145

Wielemaker, M.W., 10

Wielinga, B.J., 496, 500, 505

Wilkins, D.C., 505

Willcocks, L., 186, 187

Willis, R.R., 557

Wilson, B., 14, 208, 508, 516, 607

Wirth, N., 154

WISDM methodology, 128, 148, 390 see also rapid
development methodologies

Wood, J., 310, 313

Wood, J.R.G., 547

Wood, R., 149, 485

Wood-Harper, A.T., 149, 199, 319, 485, 537, 546, 547, 564

Woodcock, S., 141

World Wide Web (www), 5, 8

Wright, I., 193

Wynekoop, J.L., 132, 612

Wysocki, R.K., 64

X

Ximbiot (consultants), 182

XP see Extreme Programming

Y

Yalaho, A., 191, 192

Yeates, D., 78, 294

Yoon, Y., 92

Yourdon, E., 110, 113, 115, 123, 133, 273, 339, 341, 390,
392, 395, 403, 406, 418, 451, 452, 454, 455,
456, 457, 458, 459, 460, 468, 503

Yourdon Systems Methodology (YSM), 36, 110,
390, **402–407** see also process-oriented
methodologies

Z

Zach, M.H., 301, 302

Zmud, R.W., 70

Zuboff, S., 71